The Design and Performance of Road Pavements

The Design and Performance of Road Pavements

Paul Croney, MICE

David Croney, OBE, FICE, FIHT

Geotechnical and Highway Engineering Consultants

Third Edition

McGraw-Hill

New York San Francisco Washington, D.C. Auckland Bogotá
Caracas Lisbon London Madrid Mexico City Milan
Montreal New Delhi San Juan Singapore
Sydney Tokyo Toronto

Library of Congress Cataloging-in-Publication Data

Croney, David (date).
The design and performance of road pavements / David Croney and Paul Croney. — 3rd ed.
p. cm. Includes bibiliographical references and index.
ISBN 0-07-014451-6
1. Pavements—Great Britain—Design and construction.
2. Pavements—Great Britain—Maintenance and repair. 3. Pavements—Great Britain—Testing. I. Croney, Paul (date). II. Title.
TE251.C76 1997
625.8—dc21 97-25155
CIP

McGraw-Hill

A Division of The ***McGraw·Hill*** *Companies*

1 2 3 4 5 6 7 8 9 0 DOC/DOC 9 0 2 1 0 9 8 7

ISBN 0-07-014451-6

The sponsoring editor for this book was Larry Hager, the editing supervisor was Peggy Lamb, and the production supervisor was Pamela A. Pelton. It was set in Century Schoolbook by Priscilla Beer of McGraw-Hill's Professional Book Group composition unit.

Printed and bound by R. R. Donnelley & Sons Company.

This book is printed on recycled, acid-free paper containing a minimum of 50% recycled, de-inked fiber.

Contents

Part 3 Pavement Materials—Specification and Properties

Preface

During World War II there was a need for the rapid construction of forward airfields as U.S. and British forces advanced across Europe. Training courses for military engineers were organized in the 1940s at the the then Road Research Laboratory west of London. The lectures were given by the Laboratory staff augmented by a number of U.S. military engineers. To accompany the lectures a manual entitled "Soils, Concrete and Bituminous Materials" was written and issued to participants.

After the war it was decided to continue the lectures adapted to cover all aspects of road and airfield construction for civil purposes. Over the next 10 years four separate books were prepared by the Laboratory staff entitled

1. Soil Mechanics for Road Engineers
2. Bituminous Materials in Road Construction
3. Concrete Roads
4. Road Traffic

All these books were published by the U.K. Government publishing organization, Her Majesty's Stationery Office.

One of the present authors, who contributed to the above books while on the staff of the Road Research Laboratory, decided in 1977 to update most of the material in the above publications in a new book entitled *The Design and Performance of Road Pavements,* also published by the Stationery Office.

In 1991 a second edition of the book was needed to cover more recent developments and this was published by the McGraw-Hill Publishing Company in the U.K. in its International Series in Civil Engineering. This was under the joint authorship of David and Paul Croney.

In 1994 we were approached by McGraw-Hill in New York to write this abridged version based on U.S. Specifications. The opportunity has been taken to update the text.

David Croney
Paul Croney

Part

1

Introduction

Chapter 1

Overview

1.1. The 20th century, now drawing to its close, will be remembered for many events, both destructive and constructive. Among the more constructive it will be regarded as the period when mechanical road transportation was born and flourished. Heavy road haulage prior to World War I was almost the exclusive province of the steam traction engine. The development of large-bore internal combustion engines, accelerated by war needs, changed this situation, and the road engineer was soon faced with an annual growth of truck traffic of some 20 percent. Traditional methods of road construction and maintenance rapidly became obsolete.

1.2. In Europe, where a complex road network was already in place, the emphasis was on upgrading rather than on new construction. Initially there was resistance to the growth of motorized transport, largely on what would now be called environmental grounds. This led in Britain to a famous legal case in which a small local council attempted to ban motor vehicles from its area. The action eventually reached the High Court, which ruled that "Roads must be adapted to the traffic and not the traffic to the roads."

1.3. In the early part of the century most European countries set up research organizations to study road materials and the ways in which they might be improved. The need for an interchange of ideas across country boundaries was recognized as early as 1909, when the first meeting of the Permanent International Association of Road Congresses (PIARC) was held in Paris. Since that time regular meetings, at approximately 5-year intervals, have been held in various countries. In recent years these meetings have tended to become social gatherings rather than a forum for scientific debate.

1.4. The position in the United States has been very different from that in Europe. With the individual states often larger than European countries the need for new interstate highways created construction projects of a different order from those in Europe. The state highway departments have since the beginning of the motor transport era shouldered great responsibility. In an attempt to coordinate their work, the Association of State Highway Officials (AASHO) was created in 1914. Over the years this organization, together with the American Society for Testing and Materials (ASTM) founded in 1898, has assumed the responsibility for the preparation of specifications for road materials and testing procedures. In the late 1970s increasing interest in transportation research in the United States was recognized by the inclusion of "and transportation" in the AASHO title which subsequently became AASHTO. The major highway design/traffic experiment, known throughout the world as the AASHO Road Test was carried out during the period 1958–62, prior to the change of title. In this book this important experiment is described as the "AASHO Road Test," (*a*) because the description is factually correct and (*b*) to facilitate retrieval of the papers published throughout the world on the significance of the work. The specifications for the testing of materials for which AASHTO is responsible are constantly under review and throughout this book they are referred to by their AASHTO numbers. Where there are also equivalent ASTM specifications these are also quoted.

1.5. The scale of the road building program early focused attention on the role of the soil foundation in road performance. It is now more than 50 years since that fondly remembered character O. J. Porter, then chief engineer to the California State Highways Department, developed the California bearing ratio (CBR) test. It was originally intended to be used as a method for selecting granular base materials. Porter extended its use to the evaluation of soil foundations. On the basis of road performance investigations throughout the state, Porter developed performance curves A and B relating pavement performance to soil CBR values for medium and heavy traffic conditions. These were undoubtedly the first pavement design curves based on actual tests made on the foundation soil or subgrade.

1.6. During World War II, the U.S. Army Corps of Engineers adopted the CBR test to evaluate the strength of subgrades in connection with their field trials relating wheel loads with pavement thickness requirements. This work provided the design curves which were used for the construction of forward airfields during the war period. Since 1946, the CBR test has been adopted worldwide, often by engineers who affect to despise it on the grounds of empiricism. This brings to mind a statement made by Karl Terzaghi at one of his last lectures. He said that he could often learn more by seeing an iron stake driven into the ground than he could glean from a dossier of more sophisticated soil tests.

1.7. In Britain the approach to pavement design has for more than 60 years been made through full-scale experimental roads carrying heavy commercial traffic. In the late 1920s and early 1930s several new arterial roads out of London and a number of bypass roads around small towns in the south of England were in the planning stage. At that time little experience had been gained in the United Kingdom on concrete as a form of construction in major roads, and it was decided to build several of these new projects with concrete pavements which would include some experimental variations, such as slab length and load transfer devices at joints. These roads were kept under observation in relation to crack development and faulting at joints for more than 40 years, and two are still being monitored after 60 years. They have been given a thin replaceable asphalt surfacing to maintain adequate skid resistance.

1.8. Since the war a large number of flexible and concrete experimental roads have been built in Britain, some with recording weighbridges set into the road surface, recording the frequency and magnitude of axle loading. These experiments have provided the basis for the current pavement design standards in Britain. This method of gathering performance information has been continually criticized as being too slow both in the United Kingdom and in Europe generally, and there has been pressure to adopt accelerated loading procedures, involving, for example, circular test tracks. In the United States, the WASHO and AASHO experiments, on which current U.S. design standards are based, used linear test tracks with much accelerated trafficking on a 24-hour basis. In using accelerated testing for pavements and for constituent materials, it must be remembered that the properties of both bituminous mixtures and concrete are very time-dependent.

1.9. Asphaltic materials stiffen markedly as the bitumen hardens. In a properly designed flexible pavement this hardening continues while the pavement is carrying traffic, extending the life until the material can no longer accept the increasing tensile stress and cracking is initiated. Much the same thing happens in concrete, where increasing tensile strength with age much increases the fatigue strength. These effects cannot be fully included in accelerated testing programs.

1.10. The credit for introducing structural theory into pavement design must be given to H. M. Westergaard for his analysis of the stresses in concrete pavements, published in the late 1920s. The more complex problem of the analysis of stresses within multilayer asphaltic pavements could not be addressed until the finite-element concept provided computer analyses. We now know in a qualitative sense exactly what happens when a loaded axle passes over an asphalt pavement. The International Conferences on the Structural Design of Asphalt Pavements initiated by the University of Michigan in 1962 and held at 5-year intervals since have generated many hundreds of papers on this subject. The structural properties of all road materials, however, are very complex and variable, and it is unlikely that pavement designs based solely on theory will ever be feasible. A close marriage between theory and practical road experience appears to be the way forward.

1.11. However, it is becoming clear that the impetus behind research into road pavements is now waning throughout Europe, and one senses that the same applies in the United States. It seems therefore to be an opportune time to bring together American and British experience. This is the main purpose of this book.

Chapter

2

Historical Introduction to Road Construction

2.1. Whatever their motives may have been, the Romans must be accorded pride of place as pioneers in the art of road construction. Although it would be naive to compare the roads which they constructed more than 1800 years ago with modern highways, the sheer scale of their operations throughout the whole of Europe seems incredible even by today's standards. In Britain alone in the space of 150 years they drove some 3000 miles of principal roadways across the country, extending deep into Wales and north as far as Hadrian's Wall. As they advanced through the wet clay lands of western Europe thicknesses were modified to take into account the strength of the foundation. The layout seldom varied; two trenches were dug 5 m apart to act as drainage ditches, and the soil between was excavated down to a firm foundation on which a multilayer granular base was laid using materials locally available. Where feasible, the pavements were surfaced with flat quarried stone to give the appearance familiar to visitors to Pompeii. The engineers responsible for setting out the roads and supervising their construction would have known the elements of soil mechanics and were probably trained in what would now be called a school of military engineering.

2.2. The Roman roads in Europe were purely military and had no economic function in the lives of the indigenous population. Life in Europe during much of the first millennium A.D. was lived on a largely parochial basis, consisting of self-supporting enclaves between which there was little peaceful intercourse. Improved agricultural methods slowly changed this situation. The need to barter surplus crops led to the establishment by the 10th century of small market towns surrounded by satellite village communities. This meant that roads were again needed. Within the towns these were financed by levies on the householders, but road users were reluctant to maintain the rural roads, which for centuries remained close to impassable in winter.

2.3. A road system of this type presented an almost impossible barrier to long-distance coach travel, which the growing wealth of the 17th and 18th centuries encouraged, particularly between the European capitals. Various expedients to finance improvements, such as the levying of tolls and the setting up of turnpike trusts, were adopted with mixed success. Thomas Telford and John Louden Macadam were worthy products of the turnpike era in Britain, but in general the appointed engineers had little experience and training and the payment of tolls by no means guaranteed ease of travel.

2.4. As a road builder Telford preferred to use a bed of comparatively large pitched stone blinded by dry fines, while Macadam favored smaller angular material watered to assist granular interlock under the action of traffic. In the absence of methods of defining particle sizes, neither of these engineers was able to quantify his ideas in exact terms. This was left to the U.S. Army Corps of Engineers during World War II, when specifications for wet-mix graded stone and for the dry stone and fines process were drawn up to facilitate the construction of overseas military airfields.

2.5. Compaction of granular pavements in the early part of the 19th century was left to the traffic, and this resulted in an uneven surface liable to become waterlogged. Rolling as part of the construction process was first used in France in the 1830s and by Sir John Burgoyne in Britain a few years later. The rollers used at that period were of course horse-drawn. A roller weighing 5 tons needed a team of six horses, and the work was therefore difficult to control. Steam rollers were available by the 1860s, and the state of compaction achieved immediately improved. Figure 2.1 shows a remarkable steam roller constructed in 1871 by an American engineer named Ross. It had twin rolls with an impact compactor having five rams at the rear operated by a system of cams. There is no record of its performance, but in the 1950s the principle was reinvented in Britain to provide the compaction for a single-pass soil-cement train then being developed. It was not particularly effective.

2.6. Stone sett pavements. Throughout Europe from medieval times stone setts were the most commonly used form of pavement construction in towns and cities. The setts were of various sizes, but in

Figure 2.1 The Ross Ramming Machine.

London 3 by 8 by 9 inches deep was favored. The upper surface was crowned to give a better foothold for horses. The setts were originally laid on a granular foundation, but differential settlement became a problem under heavy traffic and a lime-concrete or cement-concrete foundation up to 12 in thick was used. A great deal of research was carried out in France, Germany, and Britain to locate sources of local stone which gave a satisfactory balance between wear under the action of steel tires and polishing under the action of horses' hooves. After several hundred years of service, stone setts are not entirely extinct in London, particularly in the vicinity of the Tower of London.

2.7. Brick pavements. Stone sett pavements never gained the popularity in the United States experienced in Europe. American engineers preferred the much greater degree of standardization which could be achieved in hard-burnt bricks. Pavements of such bricks were, for example, widely used in New York, Chicago, and the cities of the Mississippi valley where brick-making clays were more available than stone aggregate. In 1899 an American commission was appointed to investigate the future of brick paving. An article published in the *New York Municipal Journal* in 1914 stated that more than 6 million square yards of brick paving had been laid.

2.8. In Europe the use of brick pavements was largely restricted to western Holland, where they were widely used over natural sand foundations until the 1960s. The pavements were releveled at regular intervals to counteract traffic deformation, simply by lifting and adding more sand. Specialist gangs were used, moving from site to site.

2.9. Wood block pavements. Wood block pavements were introduced in many European cities after 1850 as a less noisy alternative to stone setts. They were also widely used in Australia. The blocks were similar in size to stone setts, and again a great deal of research was carried out to determine the most suitable timbers and their pretreatments. Wear was greater than with stone setts, but the much lower cost meant that this form of paving was economical. The blocks were generally laid on a lime or cement mortar bed. The use of wood blocks in London continued until the 1950s. In later years their life was extended by tar spraying and chipping. This waterproofed the pavement and ensured an adequate skid resistance. In London during severe rainstorms wood block pavements sometimes floated if proper attention had not been given to drainage beneath the blocks.

2.10. Wood block paving does not appear to have been used extensively in the United States. However, an experimental length of 100 yards was laid in New York's Broadway in 1835, and at much the same time a similar length was laid in Philadelphia. For this work hexagonal blocks 6 in across and 8 in deep were used, and they were laid on a lime-concrete bed. At both sites the life was only about 2 years. The process reappeared some 30 years later as the Nicolson improved wood pavement. The blocks were then rectangular of the size used in London, but they were laid on what was described as an elastic foundation. This consisted of two layers of soft timber each 1 inch in thickness impregnated with hot tar. The blocks were laid in transverse rows separated by narrow wooden fillets. A hot tar/sand mixture was used to fill the joints. An area of 6000 square yards was laid under heavy traffic in King William Street, London. After 2 years severe mud pumping occurred, leading to vertical deflections under traffic in excess of 1 in. The pavement was removed after 3 years.

2.11. Asphalt pavement. Asphalt was first used as a surfacing for roads in Paris in 1858, when the Rue Begere was surfaced with Val de Travers asphalt. Two years later three sides of the Palais Royal were similarly treated. (The English novelist Charles Dickens was a frequent visitor to Paris at that time, and in an essay published in the *Uncommercial Traveller* of 1860 he refers to the cleanliness of asphalt surfacings. It may be that he was an interested observer of these early experiments.)

2.12. In 1870, the first asphalt pavement was laid in London's Threadneedle Street. This work was rapidly extended, and by 1875 most of the major roads in the commercial "Square Mile" of the city had surfacings of Val de Travers asphalt. The asphalt used in both Paris and London was derived by crushing natural limestone impregnated with bitumen. The powder derived contained about 6 to 8 percent bitumen. It was heated on site to about 250°F before being spread to the required thickness, generally between 1.5 and 2.5 in. The compaction was carried out using heavy rammers heated in a roadside brazier. This method of laying continued in use well into the 1930s.

2.13. In the United States asphalt surfacings were very widely used after 1870. From Washington, for example, it was reported that an asphalt pavement laid in Pennsylvania Avenue in 1873 was still in very good condition in 1890. By 1888, 74 miles of asphalt and coal-tar distillate pavements had been laid in that city alone. A Captain Symons in charge of the work in Washington concluded that a two-layer construction with a coal-tar binder in the lower course topped by asphalt gave the most durable form of pavement. This may well have been the origin of a base course or binder course beneath an asphalt wearing course.

2.14. Throughout the development of asphalt pavements in the 19th century it was generally assumed that the asphalt surfacing made no significant contribution to the strength of the pavement. This was the function of the base. In most cases the base was a concrete layer 6 to 9 in thick. A form of dry lean concrete was used which was compacted by the roller used to compact the foundation.

2.15. Asphalt pavements of the above type had a very low skid resistance. In London an army of street cleaners was employed to keep the roads free of horse droppings and to distribute grit as necessary.

Concrete Paving

2.16. Experimental lengths of concrete paving designed to carry traffic without an asphalt surfacing were laid in Scotland in 1865 and 1866. Over a period of 3 years no wear or other deterioration was observed, but an inspection after 10 years indicated considerable surface damage. This was regarded as evidence that unsurfaced concrete roads became brittle with age and should not be used except for light traffic. It is in fact very likely that the surface of both roads suffered from frost spall during two abnormally cold winters in the early 1870s.

2.17. This experience did not encourage the development of concrete roads in Britain. There was also considerable, and to some extent justifiable, prejudice against the use of concrete in built-up areas where access to underground services was a constantly recurring problem. However, in 1913 the Second International Road Congress was held in London and a number of American engineers read papers on concrete road construction. In particular A. N. Johnson, the state engineer to the Illinois Highway Commission, stressed the point that river gravel was an ideal aggregate for concrete pavements but that its use in bituminous materials was often suspect. At the February 1914 National Conference on Concrete Road Building held in Chicago, the Port of London Authority was represented. As a consequence, they were able to prepare a detailed specification for port roads using concrete construction. World War I generated the need for a rapid expansion of the London docks, and some miles of concrete road were constructed. Despite the rather poor ground, the roads carried the heavy dock traffic without distress for many years.

Effect of Motor Transport on Rural Road Construction

2.18. By 1910 it had become clear that motor transport had come to stay and that water-bound macadam rural roads, in their normally poorly maintained condition, would need to be upgraded as a matter of urgency. This was too costly an operation to be wholly financed by the local authorities concerned, and the greater part of the cost was provided from central government funds. In Britain parliament set up the Road Board in 1909 to advise local authorities on the best ways to pave their roads and to provide the necessary funds. The board was also required to carry out any research necessary to fulfill its advisory role. In the decade 1909 to 1919 despite World War I some 150,000 miles of country roads were surfaced at a total cost of little more than 1 million pounds per year. The treatment generally consisted of scarifying the existing water-bound macadam, adding stone where necessary, and then spraying with hot coal tar to form a type of penetration macadam. This proved very effective and provided an excellent base course for future strengthening.

2.19. The problem in the United States over the same period was very different from that in Europe. Although the existing road network was less dense, the distances involved were very much greater. To draw attention to the foundering state of the interstate highways a convoy of heavy military vehicles attempted in 1919 to travel from Washington, D.C., to San Francisco. A young army officer, Dwight Eisenhower, was a member of the convoy. An enormous program of road building was carried out in the 1920s and 1930s, including both asphalt and concrete construction. This called for mechanization on a very large scale, and for this reason developments in earthmoving and compaction plant, in bituminous and concrete pavers have mainly originated in the United States. Europe has tended to concentrate on smaller types of plant suited to both construction and maintenance.

2.20. Research in road design and construction has been worldwide. Most countries in Europe have established research organizations and laboratories. In the United States most of the states carry out research. The work is to some extent coordinated by the American Association of State Highway and Transportation Officials (AASHTO), which is also responsible for drawing up standards and specifications. Research is also sponsored by bitumen and cement producers.

2.21. Since 1909, the Permanent International Association of Road Congresses has held regular meetings, generally in Paris, to encourage road engineers to present papers and interchange their ideas. The International Society of Soil Mechanics and Foundation Engineering meets in different countries every 5 years and generally devotes one session to road foundations and subgrades. More recently the International Conferences on the Structural Design of Asphalt Pavements organized every 5 years, initially by the University of Michigan and more recently by the International Society for Asphalt Pavements, have provided a valuable forum for the discussion of developments in flexible road design.

Chapter

3

Modern Pavements and the Principles of Pavement Design

The Pavement

3.1. The pavement is the structure which separates the tires of vehicles from the underlying foundation material. The latter is generally the soil, but it may be structural concrete or a steel bridge deck. Pavements over soil are normally of multilayer construction with relatively weak materials below and progressively stronger ones above. Such an arrangement leads to the economic use of available materials.

3.2. In Europe pavements have traditionally been classified as flexible or rigid; the former consist of unbound compacted stone under a bituminous surfacing, and the latter of a concrete slab laid on a shallow granular bed. In the United States the more descriptive terms asphaltic concrete and Portland cement concrete pavements are often used. In this book pavements in which a normal-strength concrete slab provides the major component of the strength will be referred to as concrete (even when they have a bituminous surfacing) and all other pavements will be termed bituminous or flexible.

3.3. The implied difference of flexibility between the two forms of pavement is misleading, in the sense that the same theoretical or structural method of design can be applied to both types, although some factors involved in such analyses may be more important in one type than in the other.

3.4. Where only a poor-quality granular material is readily available for use in flexible pavements it is sometimes mixed with a small quantity of cement to provide a foundation for the bituminous material. Experience with cement-bound materials of this type in flexible pavements is generally disappointing. This matter is discussed further in Chaps. 12 and 18.

Pavement Layers

3.5. Flexible pavements consist of three main layers, the bituminous surfacing, the base (or road base), and the subbase. The surfacing is generally divided into the wearing course and the binder course laid separately. The base and subbase may also be laid in composite form using different materials designated the upper and lower base or subbase. Where the soil is considered to be very weak, a capping layer may also be introduced between the subbase and the soil foundation. This may be of an inferior type of subbase material, or it may be the upper part of the soil improved by some form of stabilization (e.g., with lime or cement). The soil immediately below the subbase (or capping layer) is generally referred to as the subgrade, and the surface of the subgrade is termed the formation level.

3.6. Concrete pavements normally consist of two layers only, the concrete slab and the subbase. The slab may be laid in composite form using different aggregates in the upper and lower layers. Upper and lower subbase layers and a capping layer may also be used, but this is unusual with concrete pavements.

3.7. Concrete pavements may be reinforced with steel mesh or they may be unreinforced (often referred to as plain). Unreinforced concrete pavements have frequent transverse joints (approximately 5 m apart) to prevent thermal cracking. Reinforced concrete pavements have less frequent joints (15 to 35 m apart). The function of the reinforcement is to prevent any cracks which form from opening. Continuously reinforced concrete pavements have much heavier reinforcement, and joints are used only when necessary for construction purposes. The heavy reinforcement is intended to distribute cracks uniformly along the length of the pavement, the intention being to prevent isolated wide cracks. The length of concrete between joints is generally referred to as a slab or a bay.

3.8. The elements of flexible and concrete pavements as defined above are shown in Fig. 3.1.

FLEXIBLE PAVEMENT LAYERS

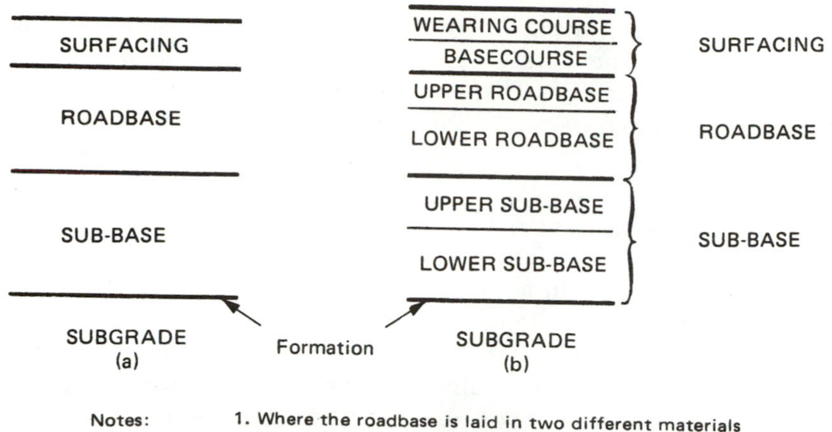

Notes:

1. Where the roadbase is laid in two different materials as in (b) the pavement is often referred to as COMPOSITE
2. Where bituminous or unbound stone or gravel materials are used exclusively in the pavement it is often referred to as FULLY FLEXIBLE

CONCRETE PAVEMENT LAYERS

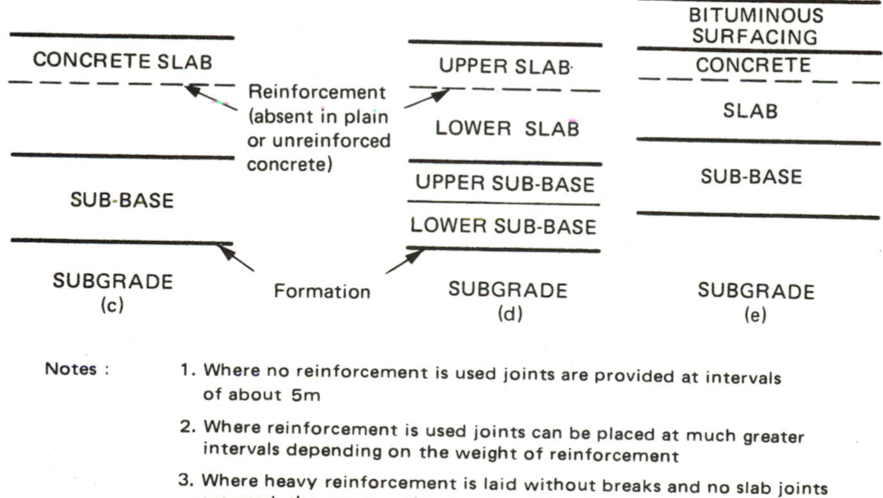

Notes :

1. Where no reinforcement is used joints are provided at intervals of about 5m
2. Where reinforcement is used joints can be placed at much greater intervals depending on the weight of reinforcement
3. Where heavy reinforcement is laid without breaks and no slab joints are used the construction is referred to as continuously reinforced concrete
4. Where two aggregates are used in the slab as in (d) above, two-layer construction is essential and the break is made at the reinforcement level if present

Figure 3.1 Components of flexible and concrete pavements.

The Principles of Pavement Design

3.9. Before designing a pavement it is essential to understand the processes which influence the behavior of any pavement under the action of time and traffic. These processes are rather different for flexible and concrete pavements and are considered separately below.

3.10. Flexible pavements. As soon as traffic begins to flow over a flexible pavement, permanent deformation will start to develop in the area of the wheel tracks followed by the commercial vehicles. This permanent deformation in a well-designed pavement is fairly evenly distributed between the asphaltic materials, the unbound base and subbase, and the subgrade. In bituminous materials it may arise from additional compaction under traffic and from sideways displacement. In unbound materials and the subgrade it will be due to traffic compaction, and only in very underdesigned pavements will shear in the subgrade be involved.

3.11. Wheel loads subject all pavement layers to vertical compressive stress. The wearing course, the binder course, and any bituminous base material will also be subject to tensile stress as the wheel load passes. The magnitude of this tensile stress in each layer will be determined by the effective modulus of elasticity of the layer and will be greatest at the bottom of the layer. Lower bituminous layers will be subject to smaller tensile stress. Unbound granular materials used in bases and subbases cannot accept significant tensile stress, and the structure of such layers will relax under load, so reducing the effective elastic modulus of the materials.

3.12. The transient deflection of the pavement under the passage of a wheel load has also been shown to be shared by the pavement layers and the subgrade.

3.13. The viscosity of the bitumens used in flexible pavements increases markedly with time over a period in excess of 10 years. This means that the stiffness or effective elastic modulus of the wearing course, the binder course, and any other bituminous element will increase with time, causing a marked decrease in the transient deflection under traffic. In a well-designed pavement the deflection may be halved compared with the early life value over a period of about 5 years.

3.14. As the bituminous elements of a flexible pavement harden, the materials will attract increased tensile stresses from traffic loading.

Structural failure will generally be initiated by fatigue cracking in the wearing course, followed by similar cracking in the binder course and any other bituminous layers.

3.15. It follows that any form of accelerated testing of pavements or bituminous pavement materials is likely to indicate pavement lives very much shorter than would occur in practice. In applying structural theory to pavement design it also follows that tests for structural properties, i.e., elastic modulus and fatigue life carried out on freshly made samples, will be largely irrelevant. This brings into question the value of any form of accelerated testing for pavement structures carried out, for example, on circular test tracks or on road pavements subjected to accelerated traffic loading, as used in the WASHO and AASHO road tests.

3.16. Concrete pavements. Concrete pavements are not susceptible to surface deformation under traffic, and the normal process of structural deterioration is indicated by cracking. Wheel loads give rise to tensile stress in the underside of the slab, which in an underdesigned pavement give rise to fatigue cracking. Both the compressive and flexural strength of concrete increase considerably with age, and this process continues significantly up to a life of 10 years. The increase in flexural strength will of course increase the tensile stress at the underside of the slab generated by a wheel load. Tests show, however, that the fatigue life increases much more rapidly with time than is the case with the tensile stress. It therefore follows that with concrete as well as flexible pavements accelerated testing is very likely to underestimate lives achieved under normal traffic.

The Two Approaches to Pavement Design

3.17. An engineer proposing to design a road pavement must have available design standards relating traffic, design life, and pavement thickness for various forms of pavement. Such design standards can be developed empirically by observing the long-term performance of existing roads or can be approached more fundamentally (but generally less reliably) using structural theory. These approaches are developed fully in Chaps. 17–19 and 23–25 and are reviewed very briefly below.

3.18. The empirical approach. This approach involves the laying of a large number of experimental pavement sections on heavily trafficked normal highways, representing the full range of road materials currently in use. This approach has been used in the United Kingdom since the 1940s. At most of the sites chosen recording weighbridges have been used to give the axle load spectra and the total number of axle loads carried. The experiments are generally continued for periods up to 30 years.

3.19. Similar studies have been made in some of the states of the United States, but they have not been coordinated very fully. Preference has been given to accelerated loading on closed loops on which a wide variety of pavements have been laid. This approach was used in the WASHO Road Test carried out in Idaho in 1953–1954 and in the AASHO Road Test in 1958–1960 carried out in Illinois. These experiments are described in detail in Chap. 17, but it may be pointed out here that the concentration of traffic in both these tests into a comparatively short period meant that aging of the pavement materials was not taken into account. This undoubtedly led to much shorter lives for the various bituminous and concrete pavements than would have been found under less concentrated traffic operating over a much longer period.

3.20. The theoretical approach. The theoretical approach to pavement design has been around for a long time—at least 50 years. All the International Conferences on the Structural Design of Asphalt Pavements organized by the University of Michigan since 1962 and more recently by the International Society for Asphalt Pavements have devoted at least 70 percent of their deliberations to the application of theory to pavement design. Most of the contributions have been from universities throughout the world and not from the people who design, build, and maintain real roads carrying real traffic. One of the main objectives of this book is to bring together theory and practice and to show how far theory can be used to forecast pavement life. There is no doubt that theory will explain why asphalt and concrete pavements behave as they do. The question is whether theory is also of use in a quantitative sense.

3.21. Modern finite-element programs can be used to evaluate the stress patterns induced by wheel loads within any pavement structure. The procedure involves the appropriate structural properties of the road materials and of the soil foundation. The problem is that the elastic properties and the fatigue properties are changing throughout the life of the pavement and these changes must be incorporated in the theoretical approach before theory and practice give the same answers. This matter is discussed fully in Chaps. 23 and 24.

Responsibilities of the Design Engineer

3.22. Road construction is, and should be regarded as, a partnership between the engineer, the client, and the contractor. In major international contracts it is not at all unusual for the engineer to come from one country, the client to be in another country, and the contractor to be a consortium from two other countries. This is not an easy situation and it demands firmness, tact, and above all knowledge and expertise on the part of the engineer. On the engineer falls the responsibility to formulate the design, to prepare the drawings, and to provide a detailed specification of all the materials to be used. The engineer must also at every stage of the design have a clear understanding of exactly how to carry out the work involved. No specification should be written, for example, which entails operations which it may be impossible to carry out at the site because of the soil or environmental conditions; recent litigation and arbitration proceedings indicate that it is no longer possible for the engineer to take the attitude that it is the contractor's responsibility to carry out an operation required by the contract which the contractor has signed, even if events show it to be impossible. The engineer would be well advised to draw attention, in a preamble to the contract documents, to any difficulties which may arise, particularly in relation to soil conditions and climate, and to give any advice felt to be relevant to the type of plant likely to be inoperable on the site.

3.23. The engineer should, without fail, place on file all the detailed reasoning relating to each stage of the design procedure. In the event of litigation this may be required as evidence, perhaps several years after the work has been completed.

Basic Information Necessary to the Design of Pavements

3.24. The basic pieces of information which the engineer must have before starting to design a road or industrial pavement are as follows:

1. Climatic environment (rainfall and temperature) at the site
2. A detailed knowledge of the soil conditions
3. Constitution and volume of the traffic to be carried

The first of these requirements is dealt with in Chap. 6, the second in Chap. 9, and the third in Chap. 8.

Chapter

4

Design Life—Performance and Failure Criteria

Design Life

4.1. Roads seldom become redundant. Even when a route is duplicated by a bypass, freeway, or motorway, the old road continues to carry traffic, although the volume may be temporarily reduced. The concept of design life has to be introduced to ensure that a new road will carry the volume of traffic associated with that life without deteriorating to the point where reconstruction or major structural repair is necessary.

4.2. In the fifties and sixties economists were advocating the use of short initial design lives for roads on the assumption that the interest on the money saved on the first cost would finance subsequent maintenance costs. For flexible pavements lives as short as 15 years were proposed. The calculations on which such conclusions were based included the traffic delay costs involved in repair work, and these, when discounted, were shown to be negligibly small. Such arguments are now known to be invalid. The cost of constructing either a flexible or a concrete road is little affected by increasing the design life by a factor of 2. This is because the increase in thickness of the pavement is comparatively small, and because the cost of a major highway includes many factors such as earthworks, bridges, and drainage which are independent of the thickness of the pavement.

4.3. In recent years, public protest has made it clear that road users are no longer prepared to accept the frequent lane closures on major highways which inevitably occur when short design lives are adopted.

Furthermore, the lane-switching and contraflow conditions which result from lane closures on freeways are increasingly becoming a cause of multivehicle accidents.

4.4. For roads in Britain the currently recommended design lives are 20 years for flexible pavements and 40 years for concrete. Many engineers feel that for all major roads the design life for both forms of construction should be 40 years. It is true that under heavy traffic an adequate skid resistance is difficult to maintain for more than 15 years on flexible pavements and about 20 years on concrete surfaces. This necessitates surface dressing, or the provision of a thin overlay. This work can be, and frequently is, undertaken at night, with little interruption to daytime traffic. Renting individual lanes or carriageways to the contractor for this type of work has been found to be an effective way of reducing closure times, although close supervision by the client's engineer is essential.

Performance and Failure Criteria

4.5. Both flexible and concrete roads should be designed and constructed to provide, during the design life, a riding quality acceptable for both private and commercial vehicles. There should be no significant ponding of water, and the skid resistance should be maintained at a level appropriate to the type of road. Acceptable levels of riding quality and skid resistance are defined in Chaps. 26 and 27.

4.6. The assumption is often made that road pavements begin to deteriorate as soon as they are opened to traffic. This is true for underdesigned pavements, but where the design life is of the order of 20 years or more, there should be no visible deterioration for the first 5 years. If there is, then serious problems must be expected in the later life. This is in part due to the increase in strength which both bituminous materials and concrete experience during their early life. If the traffic stresses are excessive because of faulty design, the pavement cannot take advantage of this increase in strength.

Failure Criteria for Flexible Pavements

4.7. Deterioration of flexible pavements arises from deformation under traffic loading, generally associated, in the later stages, with cracking. Such deformation is associated with heavy commercial vehicles; the contribution of private cars and light commercial vehicles is negligible. Figure 4.1*a* shows the characteristic behavior of flexible pavements. The surface deformation is shown for one carriageway of

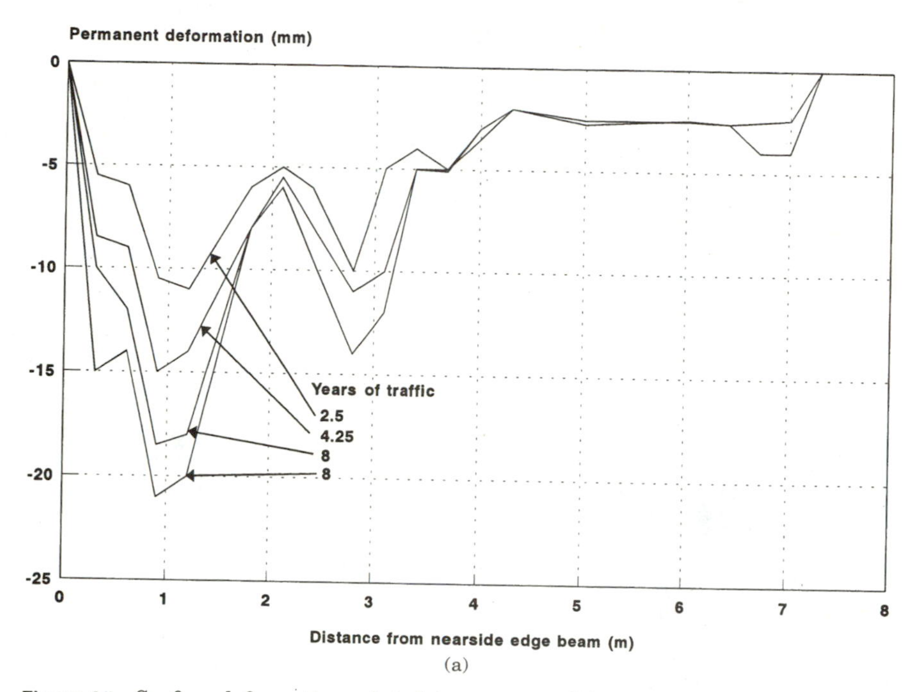

(a)

Figure 4.1 Surface deformation related to pavement life. (*a*) Variation with distance from nearside edge.

a dual two-lane highway. The pavement consisted of 100 mm of asphalt laid on a 150-mm crushed stone base. The measurements which were made after various intervals of time show the deformation measured from the original surface level. The road carried about 2000 heavy commercial vehicles per day, and the design life was expected to be about 10 years. The deformation is largely restricted to the nearside wheel track of the slow lane, which carried more than 90 percent of the heavy commercial vehicles. This is mainly because the wheel loads are more concentrated in this area. Contributory factors may arise from the crossfall and the ingress of moisture from the verge.

4.8. Figure 4.1*b* shows the development of maximum deformation in the nearside wheel tracks of a road carrying heavier traffic. In this case the performances of two adjacent pavements with different road bases are compared. The age of the road is expressed both in years and in millions of standard axles (msa) carried (see Chap. 8). The pavement with a 200-mm wet-mix stone base had a life of about 18 years, while an adjacent pavement with a 150-mm base of bitumen macadam should have an estimated life in excess of 30 years.

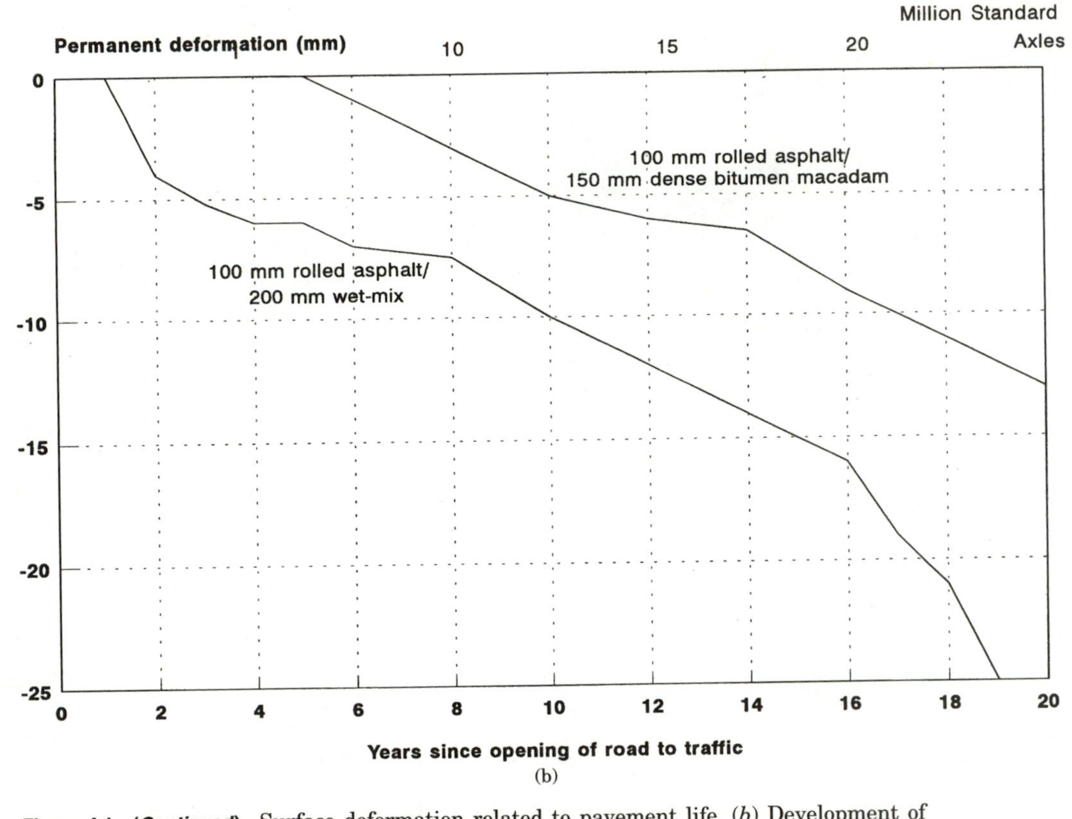

(b)

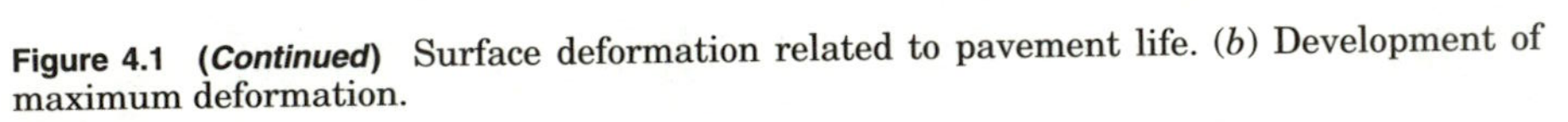

Figure 4.1 ***(Continued)*** Surface deformation related to pavement life. (*b*) Development of maximum deformation.

4.9. Experience shows that once the permanent deformation exceeds 15 mm there is an increasing probability of cracking in the wheel tracks. Water entering the cracks is then likely to accelerate failure. In Britain, for flexible pavements, a maximum deformation of 25 mm in the wheel tracks has been defined as the failure condition, and a maximum deformation of 15 to 20 mm is regarded as the optimum condition for remedial work, such as the provision of an overlay or replacement of the surfacing. These figures relate to measurements made from the original level of the surface. In practice, measurements of rutting are more likely to be made with a 2-m straightedge. For such measurements, the failure condition will be represented by a 20-mm gap under the straightedge, and the optimum condition for remedial work, by a 12- to 18-mm gap.

4.10. The measurement of deformation from the original level of the pavement involves transverse leveling against a deep benchmark prior and subsequent to trafficking. In experimental work, metal studs are set into the wearing course of the surfacing at transverse intervals of about 300 mm. Routine leveling involves the closure of at least one traffic lane. Even where a straightedge is used to assess rut depth, measurements are hazardous unless the traffic is diverted.

4.11. Figure 4.2*a* shows an example of a flexible road which has 20 mm of deformation in the nearside wheel track. This pavement is in a critical condition close to failure. In Fig. 4.2*b* the pavement has already failed with a deformation of approximately 30 mm under the straightedge.

4.12. Flexible pavements which are called upon to carry much heavier traffic loads than their design would permit often crack as a result of the large elastic deflections which occur. This condition can cause breakup of the surface and give rise to potholing, before appreciable permanent deformation has occurred.

4.13. As part of the American Association of State Highway Officials (AASHO) Road Test carried out in the United States early in the sixties, a rating system, known as the present serviceability index (PSI), was developed to classify the condition of pavements. This is discussed in detail in Chap. 17. However, the failure condition for flexible roads, defined above, corresponds to a PSI value of between 2 and 2.5.

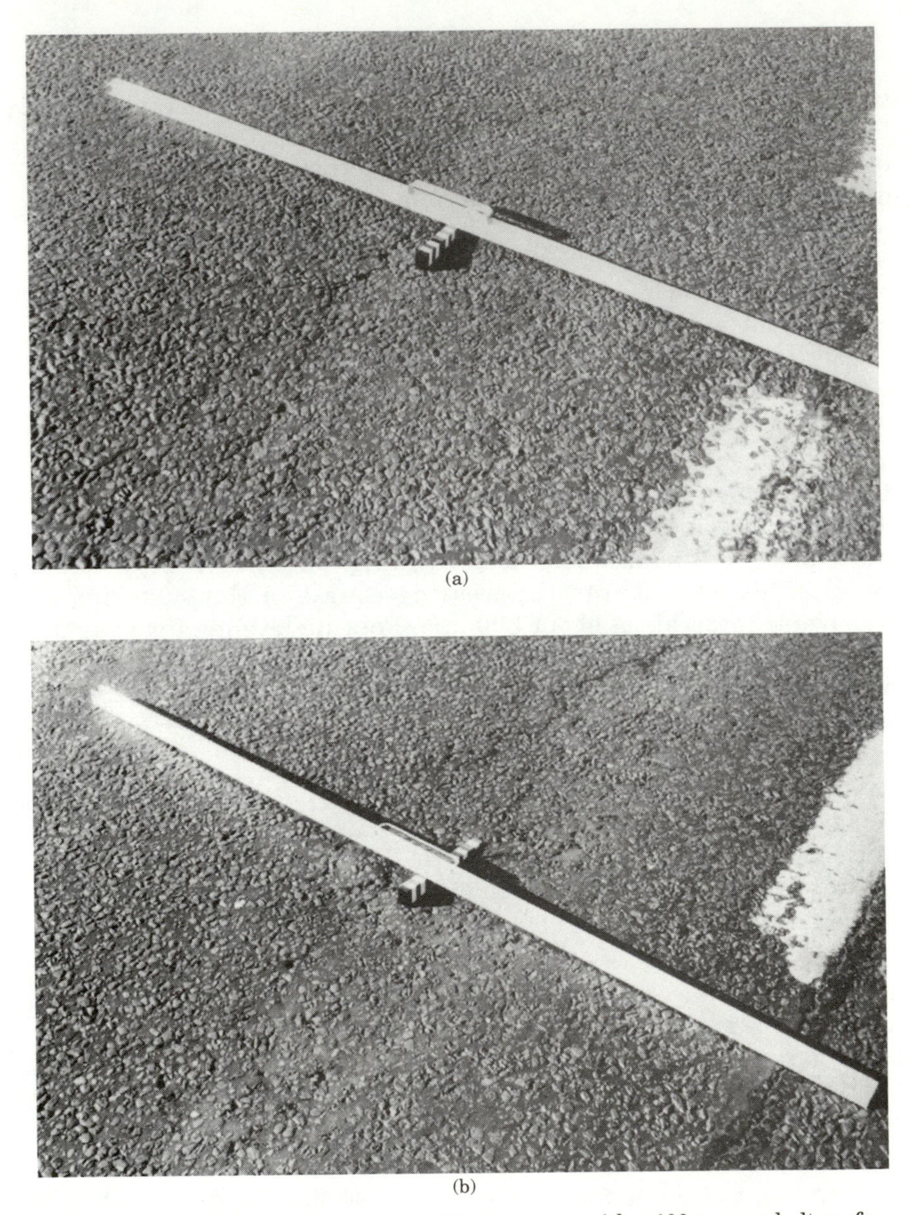

(a)

(b)

Figure 4.2 (*a*) Critical condition in a flexible pavement with a 100-mm asphalt surfacing and a 75-mm dense bitumen macadam base. (*b*) Failure condition in a flexible pavement with a 100-mm asphalt surfacing and a 75-mm dense tarmacadam base.

Failure Criteria for Concrete Pavements

4.14. Performance criteria for concrete pavements in relation to riding quality and skid resistance are the same as those for flexible pavements as indicated in Par. 4.5. However, much greater care is generally needed in the construction of concrete roads to ensure that these requirements are met. If the strength of the concrete and the thickness of the slabs are sufficiently great, it is relatively easy to lay a concrete road which will last virtually forever. However, it is unlikely to meet with the client's approval if it has a poor riding quality and a low skid resistance. Failure criteria for reinforced and unreinforced concrete pavements are different and are dealt with separately below.

4.15. Failure criterion for reinforced concrete pavements. Reinforced concrete pavements are expected to show some cracking during their design life, which should not be less than 40 years. The function of the reinforcement is to keep the cracks closed so that wheel loads are transferred across the cracks without overstressing the reinforcement. If the thickness or strength of the concrete is underdesigned for the traffic to be carried, then abrasion at the cracks and subsequent failure of the reinforcement due to rusting or "necking" will permit wide cracks to develop, leading eventually to failure. If the dowel bars across expansion and contraction joints are badly placed, then relative movement between adjacent slabs will not occur and this will contribute to cracking and failure.

4.16. For experimental concrete pavements in Britain a failure condition corresponding to a total length of cracking of 250 m per 100 m of lane width has been adopted. This includes all the following types of cracking:

- Hair cracks, which often become apparent only when the concrete is drying and which are normal features of reinforced concrete
- Fine cracks, which are less than 0.5 mm wide at the surface of the concrete
- Narrow cracks, which are between 0.5 and 1.2 mm wide at the surface
- Wide cracks, of width exceeding 1.2 mm at the surface

Figure 4.3 shows results from a concrete road carrying heavy commercial traffic, which included lengths of concrete pavements of thicknesses 175 and 200 mm. For each of these thicknesses two weights of reinforcement were used, as indicated on the figure. The 175-mm slabs with the lighter reinforcement failed, using the above definition

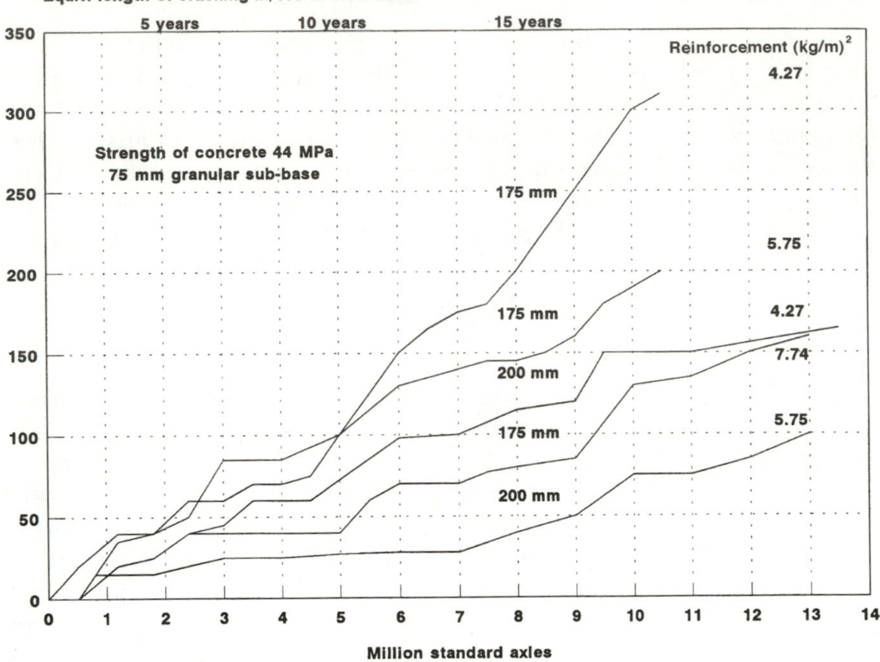

Figure 4.3 Development of cracking in reinforced concrete pavements.

of failure, after 16 years. The same thickness with the heavier weight of reinforcement is likely to give a 20-year life. The 200-mm slabs with the lighter reinforcement will have a life approaching 40 years. In this example the performance is also expressed in terms of standard axles, as in Fig. 4.1*b*.

4.17. Figures 4.4 and 4.5 show examples of areas of reinforced concrete pavement having average cracking of 500 and 200 m of total cracking per 100 m of lane width. Most of the cracking is in the wide category.

4.18. Failure criterion for unreinforced concrete roads. A failure criterion for unreinforced pavements is more difficult to quantify. The slabs are made short between contraction joints (5 m or less) to reduce the probability of thermal cracking. They are then designed to be thick enough to resist traffic cracking. If a crack occurs, it tends to widen rapidly and granular interlock is lost. Detritus entering the crack tends to cause spalling and water entering the crack results in loss of strength in the subbase and sometimes in pumping of fines. The

Figure 4.4 Appearance of local area of 125-mm-thick reinforced concrete slab with average cracking of 500 m per 100 m of the lane width.

Figure 4.5 Appearance of local area of 150-mm-thick reinforced concrete slabs with average cracking of 200 m per 100 m of the lane width.

Figure 4.6 Cracks in 200-mm-thick unreinforced concrete slab.

commonest cause of such cracking is low-strength concrete. It is inevitable that in a large contract a few cracks will occur during the early life, owing perhaps to sticking dowel bars. This can be overcome by taking out the affected bays and reconstructing them. If, however, the cracking in the early life affects more than one in four of the bays, consideration must be given either to reconstruction or to the provision of a thick overlay of bituminous material. Figure 4.6 shows the type of cracking typical of unreinforced concrete pavements. Initial transverse cracks are rapidly joined by longitudinal cracks, accompanied by progressive spalling.

4.19. Failure criterion for continuously reinforced concrete roads. Continuously reinforced concrete pavements rely on heavy reinforcement, without gaps for joints, to distribute uniformly a large number of fine cracks, which the reinforcement holds closed. This requires much heavier reinforcement than is normally used in jointed reinforced concrete pavements. In terms of the transverse cross-sectional area of the slab the area of the reinforcement is generally between 0.6 and 1 percent. For a 200-mm slab this corresponds to 9.6 to 16 kg/m^2 reinforcement. Figure 4.7 shows the type of cracking expected on a well-designed and constructed pavement using continuous reinforce-

Figure 4.7 Typical cracking in continuously reinforced concrete pavement.

ment. Even a single wide crack in a pavement of this type represents failure and requires urgent attention. The normal cause is fracture of the welds in the reinforcement or the use of a low-strength batch of concrete. The whole area around the crack must be broken out and the broken bars rewelded prior to the relaying of the concrete.

Chapter

5

Concrete versus Flexible Construction

Introduction

5.1. Both in the United States and to a lesser extent in Europe a situation of fairly friendly rivalry has existed for many years between those who favor bituminous roads and those who favor concrete. The publications of the Asphalt Institute and the Portland Cement Association to some extent reflect this rivalry in the United States.

5.2. Apart from use on some housing estate developments and some port installations concrete was not widely used for roads until the mid-1920s in Britain. At that time a program of new arterial roads in the southeast of England was devised by the recently created Ministry of Transport. Most of these roads were constructed between 1928 and 1937, largely to U.S. specifications. A number of these roads are still carrying traffic beneath a bituminous surfacing. Several have remained under detailed observation until comparatively recently.

5.3. In the 1950s and 1960s a great deal of research was carried out to develop methods for the accurate placement of reinforcement to withstand the movement of various concreting trains. Similar work related to the rigid fixing of dowel bar cages to resist movement under the passage of the train. These methods formed the basis of a method specification. Under pressure from contractors this type of specification was soon abandoned in favor of an end product specification, with generally unsatisfactory results.

5.4. In Continental Europe, particularly in Belgium, France, and Germany, concrete construction has been more successful than in the

United Kingdom, possibly because of a greater expertise in the use and control of concreting trains.

Costs

5.5. The cost of building a major concrete road to the very high standard necessary for present-day traffic is generally rather greater than the corresponding cost of an equivalent flexible road. However, the availability of materials more suited to one or other of the two forms of construction will influence this situation.

Life Expectancy

5.6. A well-designed and properly constructed concrete road has the potential for a very long structural life with low maintenance costs. Experience shows that such a road designed for a 40-year life is in fact likely to have a much longer structural life, although a renewable bituminous overlay may be necessary to maintain adequate skid resistance. Although it is possible to design flexible pavements with a life expectancy of 40 years, some structural maintenance must be expected during that period. As with concrete pavements, surface treatment will be necessary to retain skid resistance.

It can be argued that the higher cost of a concrete pavement is offset by the longer maintenance-free life. To some extent this is true, provided the concrete pavement is built to the required standard. However, in Britain at the present time this is not always the case and the engineer responsible for the construction of a major machine-laid concrete road must be prepared to put in a great deal of effort at every stage of the construction phase.

Riding Quality

5.7. Before the introduction of bituminous pavers in the thirties the riding quality of concrete roads, laid either by hand or by rail-guided paving trains, was superior to that being obtained on hand-laid bituminous pavements. With bituminous pavers now being used to lay bases and subbases as well as surfacings, there are at least four passes of the machine, each contributing to an improved riding quality. During the same period, slip form pavers have largely replaced concreting trains, often with some deterioration of riding quality. The result is that today it is easier to guarantee a good riding quality with flexible construction than is the case with concrete, particularly if adequate care is not taken in forming the joints.

Road Noise

5.8. To maintain high-speed skid resistance there has been a tendency in the United Kingdom to require a coarse brushed or grooved finish for the surface of concrete roads. This has resulted in enhanced road noise, both within and outside road vehicles, which has been found to annoy drivers and residents close to the road. This has increased prejudice against concrete roads in Britain.

Construction Experience

5.9. Any large contracting organization will be experienced in flexible road construction, and provided the work is supervised by an energetic and knowledgeable resident engineer, a satisfactory end product should result. Outside the United States the number of contractors experienced in the construction of modern concrete roads is more limited. For a major concrete road project it would be unwise to employ a contractor without experience of modern concreting machinery. For this reason, the contractor should be drawn from a nominated list of contractors with this expertise. It should also be the design engineer's responsibility to ensure that the resident engineer and his or her senior staff are thoroughly conversant with the paver to be used, well before concreting commences. The contract should also require the construction of trial bays well in advance of the commencement of paving. It should be the duty of the resident engineer to ensure that this requirement is strictly enforced.

Competition between Concrete and Flexible Road Construction

5.10. A spirit of competition between concrete and flexible road construction originated in the United States at least 50 years ago. It spread to Europe in the forties. The main protagonists have been the trade organizations concerned with the production and marketing of bitumen and cement. To an extent this competition has been healthy insofar as it has encouraged research into both forms of pavement construction. However, the marginal advantage which flexible pavements have over concrete in relation to first cost has led to a demand from the concrete side for relaxation of the rather tight specification advocated for concrete roads. This has led to the almost exclusive use of unreinforced concrete and to reductions in both slab and base thickness. These in turn have contributed to the construction of some unsatisfactory concrete roads and have affected the image of that form of construction with road users.

Part

2

Basic Design Data

Chapter

6

Climatic Data

Introduction

6.1. Before commencing the design of any form of pavement it is the responsibility of the design engineer to become familiar with all aspects of the climatic conditions under which the pavement will operate. This is particularly important if the construction is to be in an unfamiliar area of the world. Although a preliminary idea of the climate can generally be obtained from a good world atlas, information from local meteorological stations should be obtained. Failure to do this has led to many cases of expensive litigation.

6.2. The influence of rainfall, evaporation, and temperature on pavement specifications is discussed below.

Rainfall and Evaporation

6.3. Rainfall is of importance in the design process for roads in three main respects:

1. The construction of earthworks
2. Determination of subgrade strength
3. Surface water drainage

6.4. Construction of earthworks. The earthworks specification for a contract will normally define, either directly or in terms of a standard compaction test, the moisture content range between which the earthworks are to be compacted. It is the responsibility of the design engineer to ensure that the contractor meets the requirements of the specification during at least a sufficient part of the contract period. This means that if the excavated soil is required to be dried prior to being incorporated in the earthworks, the rate of evaporation must be significantly greater than the rainfall during the period of the earthwork construction.

6.5. Rainfall data will normally be available for almost any site worldwide, but information on evaporation is likely to be minimal, and empirical methods based on field studies made generally in connection with agriculture have to be used. The simplest such method is that proposed by Thornthwaite.[1] He developed the equation

$$E_p = 1 \cdot 6\left(\frac{10t}{I}\right)^a \tag{6.1}$$

where E_p = estimated evapotranspiration, in of water
t = mean monthly air temperature, °C
I = annual total of $(t/5)^{1 \cdot 514}$ calculated for each month
$a = 0.000000675(I)^3 - 0.0000771(I)^2 + 0.01792(I) + 0.49239$

Using this equation, the moisture balance can be calculated on a monthly basis. Periods which have an excess of rainfall over evaporation are likely to be difficult for earthworks, and drying by scarification will normally be possible only in months where evaporation predominates.

6.6. Subgrade strength. The natural moisture content of the soil will determine the subgrade strength to be used in the design of the pavement. At the design stage after receiving the site investigation report, it is the responsibility of the engineer to estimate the moisture content and corresponding strength of the subgrade and also to ensure that this moisture content will not be exceeded at the time the subgrade is finally prepared and covered by the subbase. To ensure that these various requirements are met, the engineer may need to specify the periods of the year when earthworks and pavement construction may be carried out.

6.7. Surface water drainage. The maximum intensity of rainfall is required for the design of the surface water drainage system of the road. The latter will normally be designed for a 10- or 20-year storm. The period chosen should be agreed on with the client and a record placed on file. Surface water drainage should be given special attention in areas of low rainfall. In parts of the Middle East, for example, the annual rainfall may be less than 100 mm, but half of this may fall in a single afternoon. In such circumstances if the client decides against a surface water drainage scheme the danger of washouts affecting verges and embankment slopes should be clearly stated by the engineer and recorded on file.

Temperature

6.8. Temperature is important in the design of flexible pavements because of its influence on the stiffness of bituminous materials. In concrete roads it creates thermal stresses and in this way affects the thickness requirements for slabs and the spacing of joints.

6.9. Influence on the stiffness of bituminous materials. All bituminous materials decrease in stiffness as the temperature increases over the working range of below 0°C to above 40°C. The stiffness is also influenced by the constitution of the mix and the hardness of the binder used. At low temperatures, to minimize thermal contraction cracking a relatively soft binder and a high binder content would be used, whereas for hot conditions the hardness of the binder would be increased and the binder content reduced to minimize plastic flow in the materials.

6.10. Influence of temperature on the design of concrete roads. The warping stresses in concrete roads are generated by the diurnal temperature changes, and the longitudinal foundation restraint stresses are determined principally by the annual temperature fluctuation. The effect of these stresses is reduced by shortening the slab length. Since these stresses are, at some periods of the day and of the year, additive to the traffic stresses, they also influence the slab thickness requirements. This is discussed in detail in Chap. 24.

Depth of Frost Penetration

6.11. Freezing in certain types of silty soils tends to cause an upward movement of water as the zero isotherm passes downward. This may result in frost heave and in a much increased moisture content in the upper subgrade. While the pavement is frozen, the stiffness is much increased, but the thaw will be accompanied by what may be a potentially large loss of strength. The design of the pavement should as far as possible take into account the heave which may accompany freezing and the loss of strength following the thaw (see Chap. 9).

The Climate of the United States

6.12. In a relatively small country such as the United Kingdom, climatic differences are not large, although for pavement design purposes it is usual to divide the area into north, south, and central zones, largely on the basis of temperature. The situation in the United States is very different, because of the size of the individual states and the influence of the Atlantic and Pacific Oceans.

6.13. For the convenience of American engineers, Table 6.1 has been compiled summarizing the climate of the states of the United States. This is intended to be a guide relating to temperature, rainfall, and depth of frost penetration. Because of the extent of many of the states, a range of climatic factors is given. The surface temperatures are for unsurfaced soil, and temperatures in bituminous surfacings may be above or below those quoted. Each state has its own meteorological department, and those records should be consulted with regard to specific problems. Based on Table 6.1, the following conclusions can be drawn:

1. The following states are likely to have annual frost penetration of 300 mm or more: Alaska, Colorado, Connecticut, Delaware, District of Columbia, Idaho, Kansas, Maryland, Massachusetts, Michigan, Minnesota, Missouri, Montana, Nevada, New Jersey, New Mexico, North Carolina, Ohio, Oklahoma, Pennsylvania, Rhode Island, South Carolina, Tennessee, Virginia, West Virginia, Wisconsin, and Wyoming.
2. The following states are likely to have summer surface temperatures of pavements in excess of 25°C: California, Colorado, District of Columbia, Florida, Georgia, Idaho, Illinois, Indiana, Kansas, Kentucky, Louisiana, Maryland, Michigan, Minnesota, Mississippi, Missouri, Montana, Nebraska, Nevada, New York, North Carolina, Oklahoma, Oregon, Rhode Island, South Carolina, South Dakota, Tennessee, Texas, Utah, Virginia, West Virginia, and Wyoming.
3. States with an annual rainfall in excess of 1000 mm are: Alabama, Connecticut, Delaware, District of Columbia, Florida, Kentucky, Louisiana, Michigan, Mississippi, Missouri, New Jersey, North Carolina, Oklahoma, Pennsylvania, South Carolina, Tennessee, Washington, and West Virginia.

TABLE 6.1 Summary of Climatic Data for Individual States of the United States

State	Surface temperature,* C°		Depth of frost* penetration, mm	Annual rainfall,* mm
	January	July		
Alabama	10 to 15	30	460	1524
Alaska	−30 to −10	10	permafrost	—
Arizona	5 to 15	30 to 35	0	130 to 500
Arkansas	5 to 10	25 to 30	0	1270
California	10 to 25	30	0	120 to 760
Colorado	−5 to 5	25 to 30	300 to 400	250 to 500
Connecticut	−5	20	600 to 900	1020
Delaware	−3	20	300 to 600	1270
District of Columbia	−5	25	600 to 900	1270
Florida	15 to 20	30	0	1270 to 1520
Georgia	10 to 15	30	0	890
Idaho	−10 to 0	25	460 to 600	500
Illinois	−5 to 0	25	300 to 750	1000
Indiana	−5 to 5	25	300 to 600	760 to 1000
Iowa	−5 to 0	25	680 to 900	1000
Kansas	−5 to 5	25 to 30	300 to 750	500 to 1000
Kentucky	5	30	0	1000 to 1250
Louisiana	10	30	0	1250
Maine	−10	15	600 to 900	960
Maryland	0 to 5	25	0	1020
Massachusetts	−5	20	600 to 900	1250
Michigan	−5	25	600 to 900	1000
Minnesota	−10 to −20	15 to 25	1200 to 1370	635
Mississippi	10 to 15	30	0	1520
Missouri	−5 to 5	25 to 30	600	1020
Montana	−10 to −5	20 to 25	900 to 1200	360 to 500
Nebraska	−5	25	900	250 to 630
Nevada	5	30	0	<200
New Hampshire	−10	15	900 to 1050	960
New Jersey	−5	20	900	1200
New Mexico	5 to 10	35	0	380 to 500
New York	−5	25	900	900
North Carolina	5 to 10	25	0	1200
North Dakota	−20 to −15	20	1200 to 1400	500
Ohio	−5	25	400	760 to 1000
Oklahoma	5	25 to 30	0	1020
Oregon	−5 to 5	20	300 to 500	500 to 1500
Pennsylvania	−5	25	300 to 450	900
Rhode Island	−5	25	300 to 450	900
South Carolina	10	25	0	1270
South Dakota	−10	25	900 to 1050	500
Tennessee	5	30	0	1250
Texas	5 to 10	30 to 35	0	250 to 1000
Utah	0 to 10	25 to 30	0	120 to 500
Vermont	−10	15	900 to 1200	760
Virginia	0 to 10	25	0	1020
Washington	−5 to 5	20	600	500 to 1600 (on coast)
West Virginia	0 to 5	25	0	1020
Wisconsin	−20 to −5	20	900 to 1200	960
Wyoming	−5 to 0	25	900 to 1100	500

*A range is shown where large differences occur across the state.

Reference

1. Thornthwaite, C. W.: An Approach towards a Rational Classification of Climate, *Geog. Rev.,* vol. 38, pp. 55–94, 1948.

Chapter

7

Geological Data—Site Investigation

Introduction

7.1. Once the need for a road between two points has been established, the exact line which it will follow must be decided. In developed urban situations minimum interference with existing buildings and services will be the main consideration, while in developed rural areas environmental considerations will be a major factor. In developing countries the starting point may well be an aerial survey followed by a ground contour survey. The line will then be selected to avoid natural barriers as far as possible. The levels chosen for such rural roads will depend on the acceptable gradients and the amount of cut and fill which these gradients will dictate. The final levels chosen will also depend to some extent on the nature of the soils found in the subsequent site investigation.

Preparation of the Site Investigation Report

7.2. The engineer is responsible for writing, or approving, the site investigation report which will form part of the contract documents on which the contractors will base their tenders.

7.3. If the site is in a country or a location with which the engineer is not familiar, the starting point of the site investigation should be the geological drift maps of the area. A surprising amount of geological information worldwide is now available on data bank. The Geological Survey and Museum in Britain can generally give worldwide information on what is available for particular sites and where it can be obtained or consulted.

7.4. The temperate soils of North America and Europe have been extensively studied and their properties are well documented. Less is known about tropical volcanic and laterite soils. For such areas it is advisable to make a literature survey. The soils journal *Geotechnique*, published by the Institution of Civil Engineers in London, contains many papers on volcanic and lateritic soils.

Scope of the Site Investigation Report

7.5. Having established the soil conditions of the area, it is the engineer's responsibility to obtain reliable information concerning the soil types which will be found

1. At foundation level in cut areas
2. Beneath the embankments in fill areas
3. In cut areas to evaluate the suitability of the excavated soil for use in embankments

7.6. The boring and sampling work will normally be carried out by a specialist organization appointed and supervised by the engineer. The borings must be sufficiently closely spaced to locate any significant quantities of unsuitable or contractually difficult materials present along the route. Deep borings are expensive and the temptation is to reduce the frequency to keep costs down. Engineers generally attempt to safeguard their position by stating in the preamble to the site investigation document that it is provided only for the guidance of the contractor and that it is the latter's responsibility to verify the information given by further borings if considered desirable. In the event of the contractor's entering into litigation on the grounds of inadequate or incorrect information in the site investigation report, this escape clause now has little influence on judges and arbitrators. They recognize that a contractor with perhaps only 3 to 4 weeks to complete a tender cannot possibly arrange for an independent soil survey, particularly if the contract relates to a distant country.

7.7. Borings should be made at a maximum spacing of 5 per kilometer and these should be supplemented by intermediate bores if obvious changes of soil strata are found between the initial boreholes. Borings should be continued to a depth of at least 3 m below the proposed formation level in areas of cut and to a depth in excess of 3 m below existing ground level in areas of fill. Separate boreholes may be required to evaluate the soil conditions beneath bridge and other structural foundations.

7.8. Samples weighing at least 2 kg should be taken at depth intervals of 500 mm in fine-grained soils, for classification tests. In granular soils this weight should be increased to 10 kg. These samples should be placed in airtight containers immediately after excavation, and at the same time a representative sample of not less than 0.5 kg should be taken at the same depths for moisture content determination. These samples should be placed in preweighed sample containers and the wet weight determined without any loss of moisture by evaporation. The oven-dry weight can be determined at a convenient later time.

7.9. All the boreholes should be temporarily capped and checked for groundwater level at intervals until the equilibrium condition is reached.

Tests to Be Carried Out on Samples

7.10. Tests will normally be carried out by the site investigation organization, as directed by the engineers. Too often, insufficient attention is given by engineers when deciding exactly what tests should be carried out. Moisture content determinations should be made on all the samples taken, as indicated in Par. 7.8. The appropriate classification tests should also be made on all the samples. Very frequently, moisture content and classification tests are made on different samples, so that the contractor has no indication of how the natural moisture content is related to the classification tests. The classification tests for cohesive soils are the liquid and plastic limits, and for granular soils the wet-sieve particle size determination is carried out (see Chap. 10).

7.11. To establish the degree of compaction to be specified in earthworks and in the subgrade, laboratory compaction tests relating density and moisture content should be carried out on at least half of the samples taken. Both standard and heavy compaction should be specified (see Chap. 9).

7.12. In the design of flexible pavements engineers will need to know the California bearing ratio (CBR) of the subgrade after the earthworks have been completed. For this they will need to know the relationship between moisture content, dry density, and CBR value. This involves preparing CBR test samples at a range of moisture contents and dry densities. The work involved is considerable and the testing program will normally be confined to the main soil types found at the site. A limited number of triaxial tests will be needed to check the stability of the embankment and cutting slopes, and consolidation tests may also be necessary if highly compressible soils are present.

7.13. Detailed laboratory testing is best planned after the engineer has considered the moisture content and classification test results. In the past this has been difficult if the design is being prepared at a great distance from the project site. However, the advent of the fax transmission system means that the results of moisture content and classification tests can be transmitted and considered before the decision on further testing is made. This could lead to much-improved site investigation reports.

7.14. The amount of money allocated to site investigation will obviously be related to the size of the contract. Experience shows that a large proportion of the disputes which arise between engineers and contractors originate in what the contractor regards, often with good reason, as an inadequate site investigation. Such disputes, if they end in litigation, can be very costly to the client, in relation to the cost of well-planned and well-executed site investigations.

Chapter

8

Road Traffic and Axle Loading

Introduction

8.1. Traffic, and particularly commercial traffic, provides the main input to any pavement design problem. However, comparatively little effort is devoted, within research organizations and universities, to establishing the volume of traffic on different types of road and the associated axle load spectra. Despite its importance, the collection of this type of information is regarded as unglamorous compared, for example, with the application of structural theory to the computation of stresses within pavement structures. The proceedings of International Conferences on the Structural Design of Asphalt Pavements (1962–1987) contain several hundred papers, only two of which deal specifically with the constitution and distribution of traffic on real roads.

8.2. In the decade 1970 to 1980 a great deal of effort was devoted by the Road Research Laboratory in the United Kingdom to collecting traffic data necessary to the design of highway pavements. The information given in this chapter relating to the distribution of commercial traffic between the carriageway lanes, and the influence of private car traffic on that distribution, is based on work carried out during that period. Although since that time the volume of commercial traffic has increased by 1 or 2 percent per annum, the distribution of the traffic across the carriageway lanes has not changed materially.

8.3. Such evidence as is available suggests that the distribution of traffic across carriageway lanes in continental Europe is similar to that in the United Kingdom, and it seems probable that this is also the case in the United States.

8.4. The subject of traffic is dealt with in this chapter under the following headings:

1. Types of commercial vehicles in use in relation to the prevailing limits of maximum axle load and gross vehicle weight
2. Collection of traffic data
3. Distribution of commercial vehicles and private cars between the traffic lanes
4. Types of commercial vehicles and their axle loading
5. Traffic loading expressed in terms of standard axles

Types of Commercial Vehicles in Use in Relation to the Prevailing Limits of Maximum Axle Load and Gross Vehicle Weight

8.5. The comparatively small axle loads imposed by private cars and light panel trucks cause virtually no structural damage to modern road pavements. Damage is primarily caused by the heavier axle loads associated with large commercial vehicles. For this reason every country legislates for a maximum axle load limit and a maximum gross vehicle weight, neither of which should in theory be exceeded. In the United States, because of the wide climatic variations together with industrial considerations, each state fixes its own limits for axle and vehicle loading. Table 8.1 summarizes the maximum axle load and maximum gross vehicle weight applying to each of the states. The gross weights for some of the states include notifiable loads and loads restricted to designated routes.

8.6. Table 8.2 gives similar information for European countries together with current proposals for changes prepared by EEC committees. The mean maximum axle load for the states in the United States given in Table 8.1 is 9.3 tonnes, with a range of 8.2 to 10.9 tonnes. This is a little less than the current European value. The mean gross vehicle weight for the United States is close to 39 tonnes. This is the same as the current U.K. limit but less than the EEC proposals for five- and six-axle vehicles. It is reasonable therefore to assume that the behavior of commercial vehicles in the United States

TABLE 8.1 Details of Axle and Vehicle Loading Permitted by the Various States of the United States

State	Maximum length, m, tractor and semitrailer	Maximum weight, tonnes		
		Single axle	Tandem axle	State maximum gross weight*
Alabama	18.3	9.1	18.1	41.9
Alaska	19.8	9.1	15.4	49.5
Arizona	19.8	9.1	15.4	36.3
Arkansas	18.3	8.2	14.5	33.2
California	18.3	9.1	15.4	36.3
Colorado	21.3	9.1	16.3	38.4
Connecticut	16.8	10.2	16.3	36.3
Delaware	18.3	9.1	18.1	36.3
District of Columbia	16.8	10.0	16.8	33.2
Florida	16.8	9.1	18.1	36.3
Georgia	18.3	8.2	16.3	36.3
Hawaii	17.7	10.9	15.4	40.3
Idaho	19.8	9.1	15.4	47.8
Illinois	18.3	8.2	14.5	83.2
Indiana	16.8	8.2	14.5	33.2
Iowa	18.3	9.1	15.4	36.3
Kansas	19.8	9.1	15.4	38.6
Kentucky	16.8	9.1	15.4	37.2
Louisiana	19.8	9.1	15.4	36.3
Maine	18.3	10.0	15.4	36.3
Maryland	16.8	10.2	18.1	36.3
Massachusetts	18.3	10.2	16.3	47.1
Michigan	18.3	9.1	15.4	74.4
Minnesota	18.3	9.1	15.4	36.3
Mississippi	18.3	8.2	15.5	36.3
Missouri	16.8	8.2	15.5	33.3
Montana	18.3	8.2	15.5	47.8
Nebraska	19.8	9.1	15.4	43.1
Nevada	21.3	9.1	15.4	58.5
New Hampshire	18.3	10.2	16.3	36.3
New Jersey	16.8	10.2	15.4	36.3
New Mexico	19.8	9.8	15.6	39.2
New York	18.3	10.2	16.3	36.3
North Carolina	16.8	8.6	16.3	36.3
North Dakota	19.8	9.1	15.4	47.8
Ohio	18.3	9.1	15.4	36.3
Oklahoma	19.8	9.1	15.4	40.8
Oregon	18.3	9.1	15.4	47.8
Pennsylvania	18.3	10.2	15.4	36.4
Rhode Island	16.8	10.2	16.3	44.9
South Carolina	18.2	9.1	16.3	36.6
South Dakota	21.3	9.1	15.4	43.1
Tennessee	16.8	8.2	15.5	32.7
Texas	19.8	9.1	15.4	36.3

*Includes, for some states, notifiable loads and loads restricted to designated routes.

TABLE 8.1 Details of Axle and Vehicle Loading Permitted by the Various States of the United States (Continued)

State	Maximum length, m, tractor and semitrailer	Maximum weight, tonnes: Single axle	Tandem axle	State maximum gross weight*
Utah	19.8	9.1	15.4	55.4
Vermont	18.3	10.2	16.3	36.3
Virginia	16.8	9.1	15.4	34.4
Washington	16.8	9.1	15.4	47.8
West Virginia	16.8	9.1	15.4	40.6
Wisconsin	18.3	9.1	15.2	36.3
Wyoming	25.9	9.1	16.1	45.8

*Includes, for some states, notifiable loads and loads restricted to designated routes.

in relation to pavement performance will be similar to that in Europe and the United Kingdom.

8.7. It has been shown that increasing the permitted gross weight of commercial vehicles can result in a saving up to 5 percent in haulage costs. Provided that the increase in gross vehicle weight is accompanied by an increase in the number of axles, the damage to the road pavement will not be increased. However, additional axles may involve an increase in vehicle length, and this has given rise to considerable opposition by environmental groups in Europe.

Collection of Traffic Data

8.8. The collection of traffic data is necessary

1. To assess the traffic-carrying capacity of different types of road
2. To examine the distribution of traffic between the available traffic lanes
3. In the preparation of maintenance schedules for in-service roads
4. In the forecasting of expected traffic on a proposed new road from traffic studies on the surrounding road system

8.9. The simplest form of traffic census involves separate counts of both private and commercial vehicles on each of the traffic lanes

TABLE 8.2 Maximum Axle Loads and Gross Vehicle Weights (GVW) for European Countries. With Typical Axle Load Configurations

Country	Configuration (axle weights, tonnes)	GVW (tonnes)	Maximum axle load (tonnes)
		4 axle	
France	6 12 20	38.0	12
Germany	6 10 20 (> 2m)	36.0	10
Italy	6 12 22 (> 2m)	40.0	12
U.K.	6 10.2 16.3	32.5	10.2
U.K.	5 7.1 20.4 (> 2m)	32.5	10.2
Proposed for EEC	6 11 18	35.0	11
		5 axles, 3-axle trailer	
France	6 12 20	38.0	12
Germany	6 10 22	38.0	10
Italy	6 12 26	44.0	12
U.K.	6 10 22	38.0	10
Proposed for EEC	6 11 23	40.0	11
		5 axles, 2-axle trailer	
Proposed for EEC (s.d. 3-axle tractor)	6 11/7 18	42.0	11
Proposed for EEC (3-axle, double-drive tractor)	6 18 18	42.0	11
		6 axles	
Netherlands	6 18 26	50.0	9
Proposed for EEC (single drive)	6 11/7 20	44.0	11
Proposed for EEC (double drive)	6 18 20	44.0	11

available. The results should be recorded on an hourly basis. Light commercial vehicles of unladen weight less than 1.5 tonnes are normally recorded as private cars. This will include all panel-type vans. Because the traffic flow generally varies with the day of the week, a census should be carried out at least over a 7-day period.

8.10. The result of a 24-hour count carried out on one carriageway of a six-lane industrial freeway in the United Kingdom is shown in Table 8.3. A simple count of this type, differentiating only between private cars and commercial vehicles, would require at least one observer for each of the three traffic lanes, each equipped with a hand-held counter. To cover the 24-hour period, a total of nine observers would be required.

8.11. A more comprehensive type of traffic survey entails classifying the commercial vehicles in terms of the number of axles and whether the chassis is rigid or articulated (semitrailer). The full classification system used in the United Kingdom is shown in Fig. 8.1. Whether the axles are fitted with single wheels or with dual wheel assemblies is noted only when the survey is made in connection with the application of structural theory to a particular pavement.

8.12. At a number of sites in Britain permanent recording weighbridges have been installed across the traffic lanes of a representative sample of industrial and urban freeways and on several heavily trafficked truck roads.

8.13. Traffic engineers often assess the capacity of roads in terms of "passenger car units" or pcu's. This unit is intended to take into account the relative lane occupancy of commercial vehicles and private cars. For urban roads each commercial vehicle is rated as 3 pcu's and each private car or light van as 1 pcu. For a four-lane dual freeway a maximum of 1500 pcu/h on each carriageway is regarded as appropriate, and for a six-lane dual freeway this figure is increased to 2250 pcu/h. For all-purpose roads these figures are reduced to 550 and 950 pcu/h. Applying the figure of 2250 pcu/h to the traffic census shown in Table 8.3, it can be deduced that during the period 8 A.M. to 6 P.M. the road will be overcrowded, reaching 3620 pcu/h between 10 and 11 A.M. This overcrowding will result in congestion and a marked reduction of vehicle speeds. The traffic census shown in Table 8.3 was carried out several years ago, and the situation has deteriorated to the point where additional lanes must be added.

TABLE 8.3 Distribution of Vehicles between Carriageway Lanes on One Carriageway of a Six-Lane Industrial Motorway

Time period	Commercial vehicles only					Private cars only						
	Number of vehicles			Percentage of vehicles		Number of vehicles				Percentage of vehicles		
	SL*	CL*	Total	SL	CL	SL	CL	FL*	Total	SL	CL	FL
0–1	50	10	60	83	17	234	303	108	645	36	47	17
1–2	45	6	51	88	12	168	164	69	401	42	41	17
2–3	51	1	52	98	2	99	99	19	217	46	46	8
3–4	71	2	73	97	3	75	52	9	136	55	38	7
4–5	100	7	107	93	7	80	51	8	139	58	37	5
5–6	162	11	173	94	6	84	69	11	164	51	42	7
6–7	316	35	351	90	10	111	184	47	352	32	55	13
7–8	329	10	339	97	3	207	741	166	1114	18	67	15
8–9	373	148	521	72	28	258	748	90	1096	24	68	8
9–10	394	271	668	59	41	242	558	1143	1943	12	29	59
10–11	373	252	625	60	40	248	489	1010	1747	14	28	58
11–12	341	235	576	59	41	279	211	819	1309	21	16	63
12–13	297	194	491	60	40	282	453	637	1372	21	33	46
13–14	283	120	403	70	30	281	496	487	1264	22	39	39
14–15	264	124	388	68	32	303	510	480	1293	23	39	38
15–16	247	117	364	68	32	303	572	477	1352	22	43	35
16–17	277	122	399	69	31	277	647	530	1454	20	44	36
17–18	240	119	359	67	33	460	711	679	1850	25	38	37
18–19	169	102	271	62	38	442	428	806	1686	26	26	48
19–20	148	96	244	61	39	353	517	571	1441	24	36	40
20–21	120	53	173	69	31	353	464	429	1246	28	38	34
21–22	116	39	155	75	25	314	357	361	1032	30	35	35
22–23	88	16	104	85	15	308	297	291	896	34	33	32
23–24	84	11	95	88	12	303	149	67	519	58	29	13

*SL = slow lane, CL = center lane, FL = fast lane (in the case of three-lane carriageways no heavy commercial vehicles are permitted to use the fast lane).

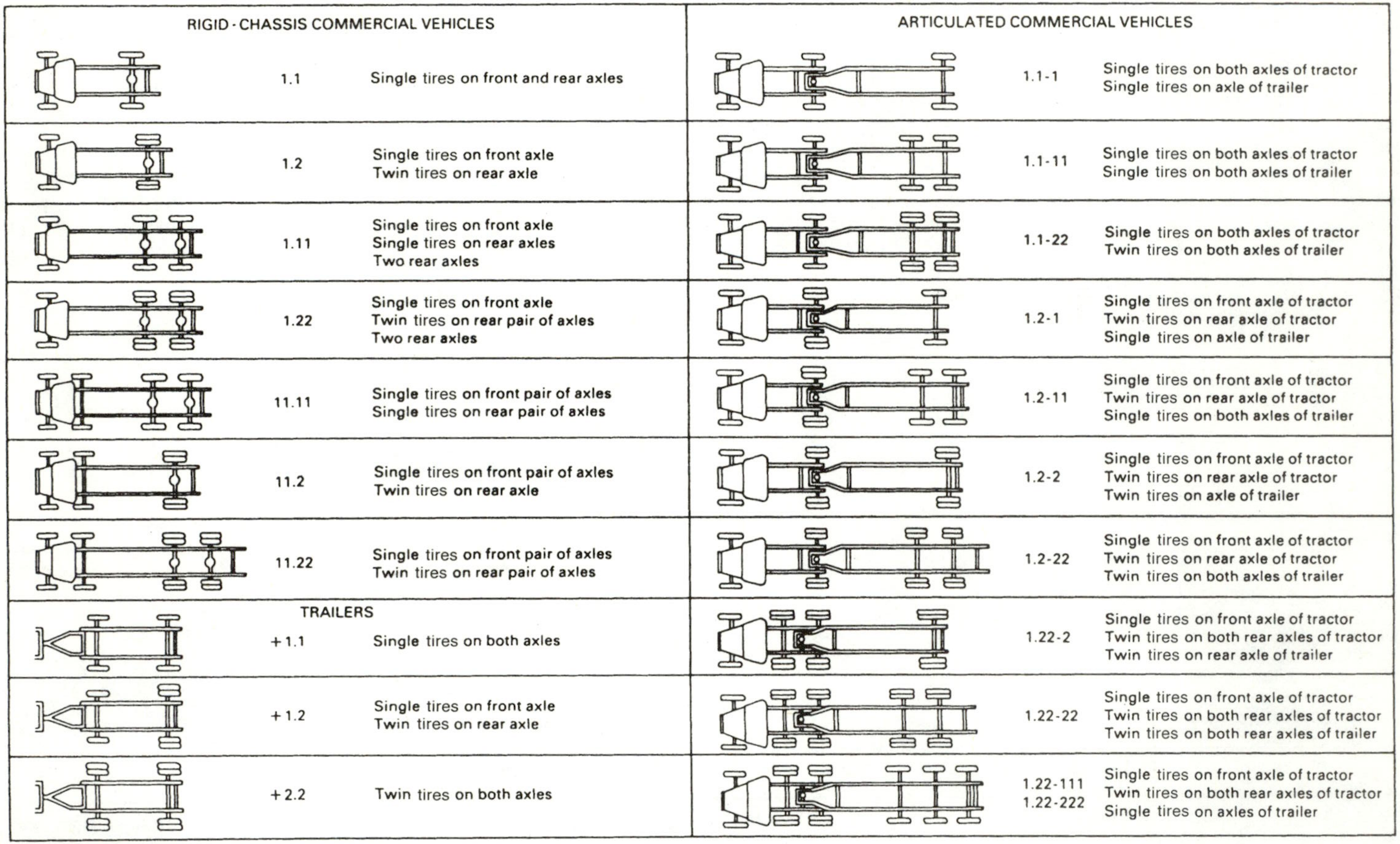

RIGID-CHASSIS COMMERCIAL VEHICLES		ARTICULATED COMMERCIAL VEHICLES	
1.1	Single tires on front and rear axles	1.1-1	Single tires on both axles of tractor Single tires on axle of trailer
1.2	Single tires on front axle Twin tires on rear axle	1.1-11	Single tires on both axles of tractor Single tires on both axles of trailer
1.11	Single tires on front axle Single tires on rear axles Two rear axles	1.1-22	Single tires on both axles of tractor Twin tires on both axles of trailer
1.22	Single tires on front axle Twin tires on rear pair of axles Two rear axles	1.2-1	Single tires on front axle of tractor Twin tires on rear axle of tractor Single tires on axle of trailer
11.11	Single tires on front pair of axles Single tires on rear pair of axles	1.2-11	Single tires on front axle of tractor Twin tires on rear axle of tractor Single tires on both axles of trailer
11.2	Single tires on front pair of axles Twin tires on rear axle	1.2-2	Single tires on front axle of tractor Twin tires on rear axle of tractor Twin tires on axle of trailer
11.22	Single tires on front pair of axles Twin tires on rear pair of axles	1.2-22	Single tires on front axle of tractor Twin tires on rear axle of tractor Twin tires on both axles of trailer
TRAILERS +1.1	Single tires on both axles	1.22-2	Single tires on front axle of tractor Twin tires on both rear axles of tractor Twin tires on rear axle of trailer
+1.2	Single tires on front axle Twin tires on rear axle	1.22-22	Single tires on front axle of tractor Twin tires on both rear axles of tractor Twin tires on both rear axles of trailer
+2.2	Twin tires on both axles	1.22-111 1.22-222	Single tires on front axle of tractor Twin tires on both rear axles of tractor Single tires on axles of trailer

Figure 8.1 Method of classifying axle types.

Distribution of Commercial Vehicles and Private Cars between the Traffic Lanes

8.14. Figure 8.2 shows the distribution of the traffic between the traffic lanes for the census data given in Table 8.3. There is no clear evidence that the presence of the commercial traffic is having much influence on the distribution of private cars. During periods of heavy commuter traffic there is, however, competition for use of the fast lane.

8.15. From a number of 7-day surveys made on the same industrial freeway referred to in Table 8.3, the relationship between the total traffic per hour on the carriageway and the percentage of the total traffic on the three lanes has been studied. Results are shown in Fig. 8.3. For total traffic of about 1600 vehicles per hour each of the three lanes is carrying about one-third of the traffic. When the total traffic exceeds 2000 vehicles per hour, the fast lane is carrying the most traffic.

8.16. If commercial vehicles only are considered, Fig. 8.4 shows the relationship between the commercial traffic on the slow and center

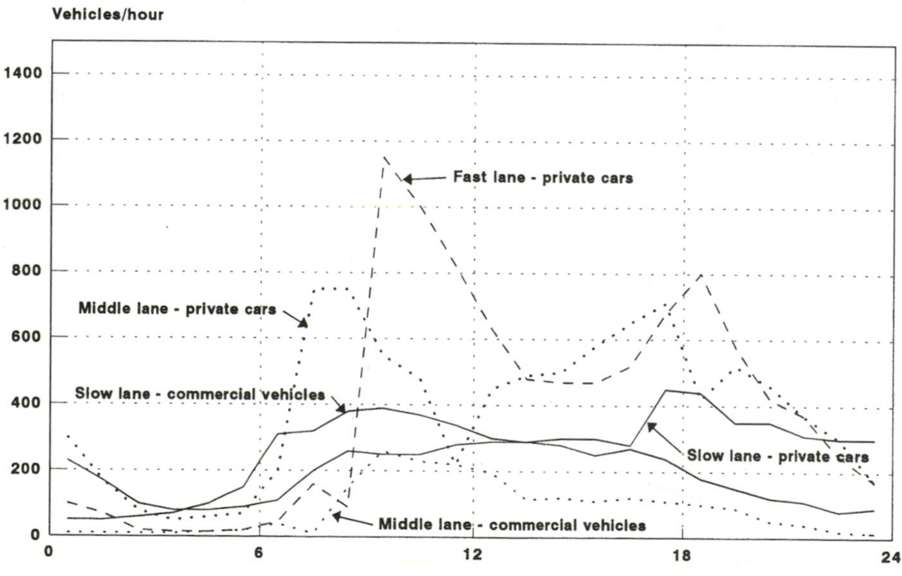

Figure 8.2 Distribution of vehicles between carriageway lanes on one carriageway of a six-lane industrial freeway.

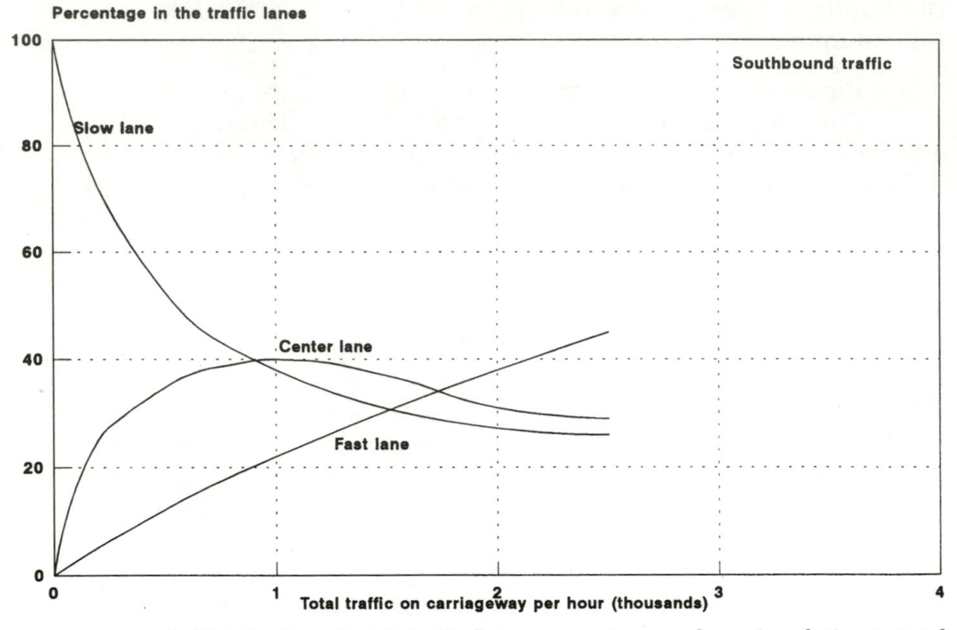

Figure 8.3 Distribution of total traffic between carriageway lanes in relation to total hourly flow on industrial freeway.

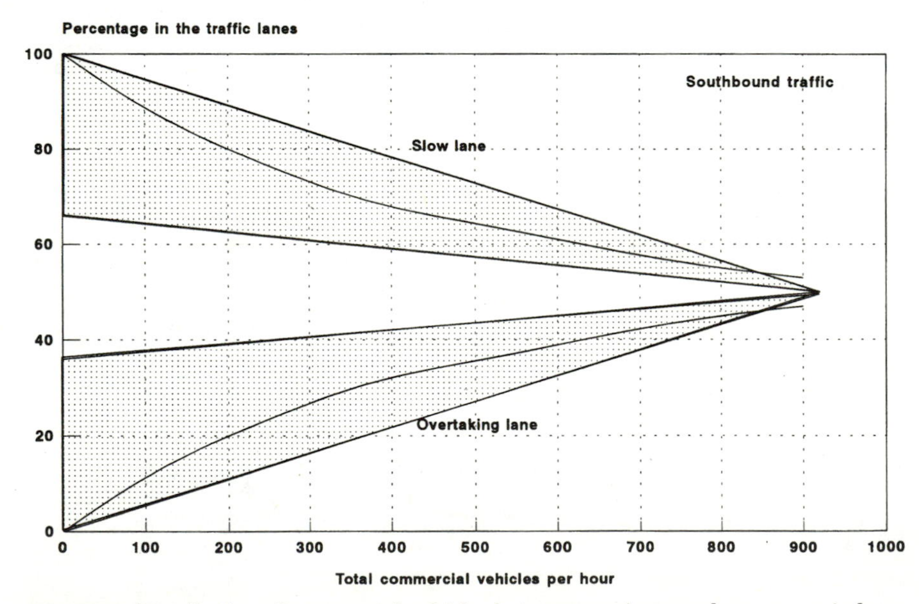

Figure 8.4 Distribution of commercial vehicles between carriageway lanes on an industrial freeway.

lanes. There is considerable scatter in this relationship between observations at different six-lane freeway sites. This is indicated by the shaded areas in Fig. 8.4. There is not a great deal of evidence relating to total commercial traffic flows in excess of 800 vehicles per hour, but it seems doubtful if equality in the commercial traffic flow between the slow and center lanes of a three-lane dual freeway is ever reached. The probable trend is shown superimposed on the shaded areas on Fig. 8.4.

8.17. Figures 8.5 and 8.6 give similar information for a six-lane urban freeway carrying more commuter traffic. The form of the relation is very similar; however, the approximate 30 percent split of traffic between the three lanes occurs at a much higher total traffic intensity than is the case in Fig. 8.3. The split of commercial traffic between the slow and overtaking lanes is very similar to that in Figs. 8.4 and 8.6.

8.18. For four-lane freeways and limited-access dual carriageway roads, commercial vehicles use both slow and overtaking lanes. Figure 8.7 shows that the lanes carry approximately the same mixed traffic when the total traffic on the carriageway reaches about 1000 vehicles per hour. For higher intensities of traffic the overtaking lane

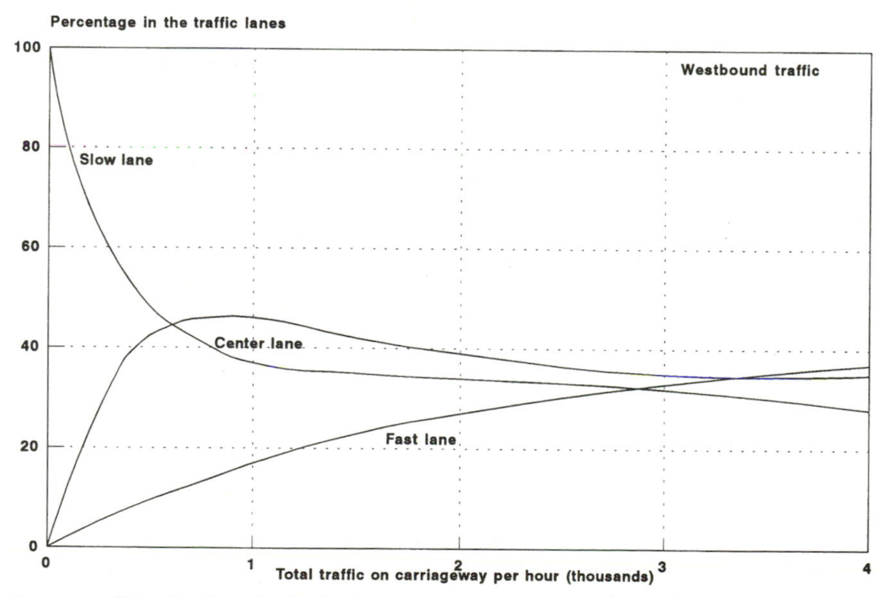

Figure 8.5 Distribution of vehicles between carriageway lanes in relation to total hourly flow on an urban freeway with much commuter traffic.

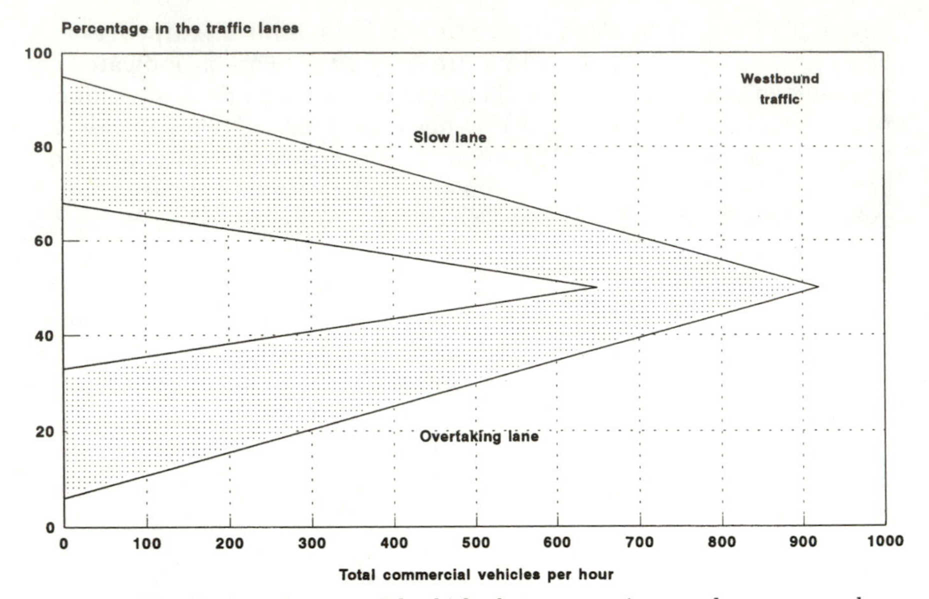

Figure 8.6 Distribution of commercial vehicles between carriageway lanes on an urban freeway.

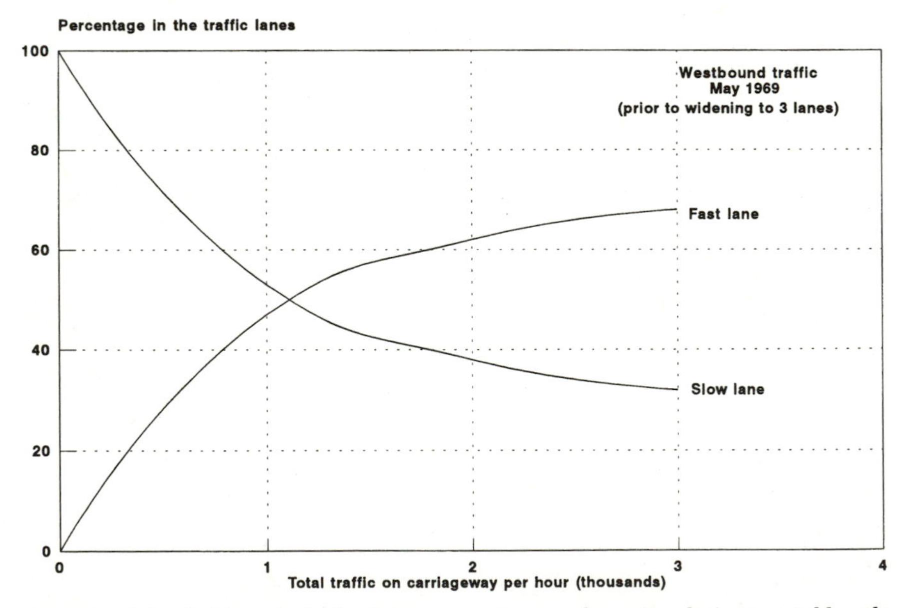

Figure 8.7 Distribution of vehicles between carriageway lanes in relation to total hourly flow on a four-lane urban freeway.

carries an increasing proportion of the total traffic. Such roads seldom carry more than 600 to 800 commercial vehicles per hour. The distribution of such vehicles between the lanes is shown in Fig. 8.8.

8.19. The pavement engineer normally designs the pavement to cater to the amount of commercial traffic likely to be carried by the slow lanes during the design life of the road. Figure 8.9 shows the percentage of the daily total of commercial traffic carried by the slow and overtaking lanes for all forms of dual freeway road. For single freeways carrying traffic in both directions it can be assumed that both lanes will carry one-half of the cumulative total of commercial vehicles to be carried during the design life.

8.20. In the United Kingdom the amount of commercial traffic carried by roads varies with the day of the week in the manner shown in Fig. 8.10. A close approximation to the average daily commercial traffic is obtained by adding four times the Monday traffic to the combined Friday, Saturday, and Sunday traffic and dividing this total by 7. In the United States, the much greater distances covered by commercial vehicles may result in a much smaller difference between weekend traffic and daily traffic.

Types of Commercial Vehicles and Their Axle Loading

8.21. In the last 40 years, commercial vehicles have increased in size and in the number of axles per vehicle. There has also been a marked trend away from rigid chassis vehicles to the more versatile articulated form.

8.22. For U.K. conditions these changes are shown in Fig. 8.11. The trend toward multiaxled vehicles means that the average number of axles per commercial vehicle has changed in the manner shown in Fig. 8.12.

8.23. The most satisfactory method of measuring axle loads is by a weighbridge set into the surface of the pavement. To avoid unacceptable dynamic effects, the pavement on both sides of the weighbridge must be maintained in a smooth condition. Figure 8.13 shows a weighbridge set into the slow lane of a flexible pavement. In this case it is set in a small concrete transition length.

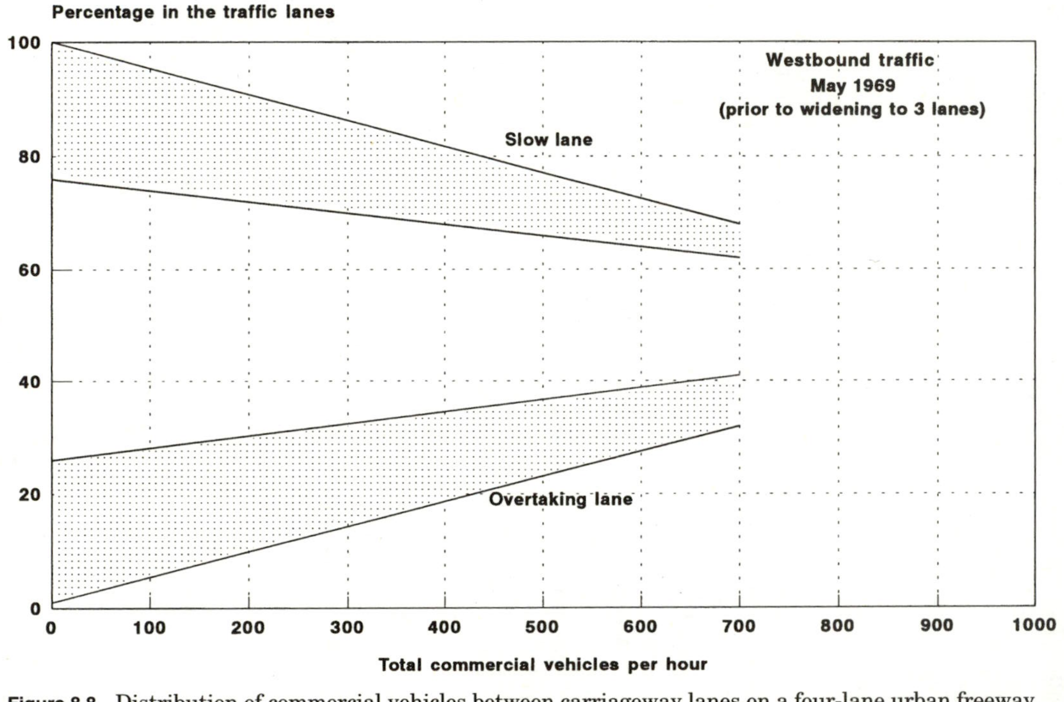

Figure 8.8 Distribution of commercial vehicles between carriageway lanes on a four-lane urban freeway.

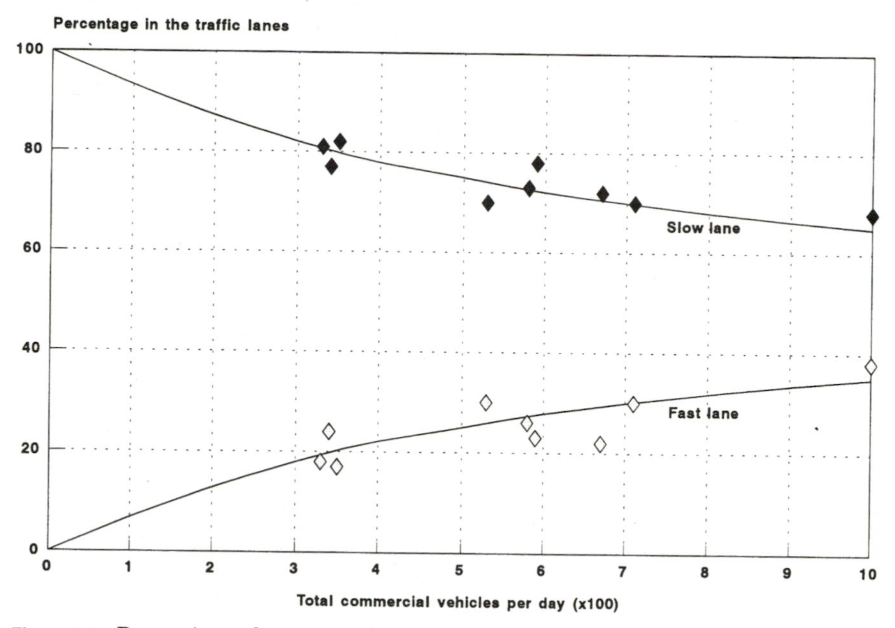

Figure 8.9 Proportions of commercial vehicles in each traffic lane on a daily basis.

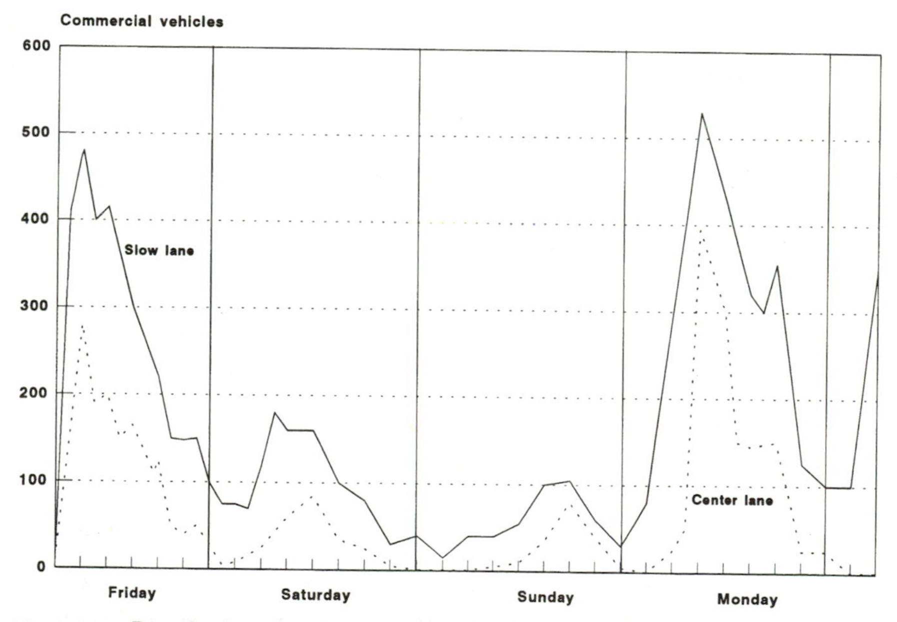

Figure 8.10 Distribution of commercial vehicles between traffic lanes of a six-lane dual industrial freeway in the United Kingdom (southbound traffic only).

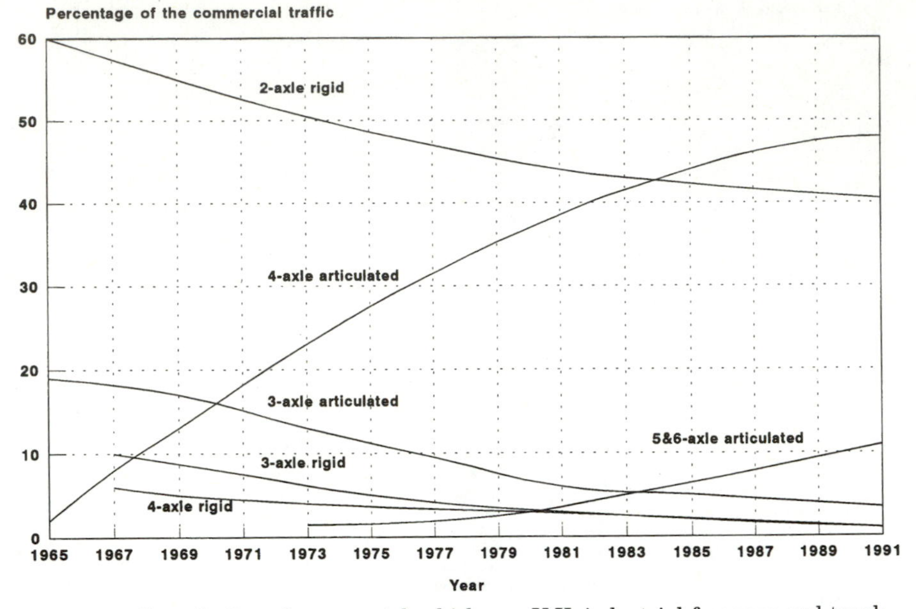

Figure 8.11 Constitution of commercial vehicles on U.K. industrial freeways and trunk routes.

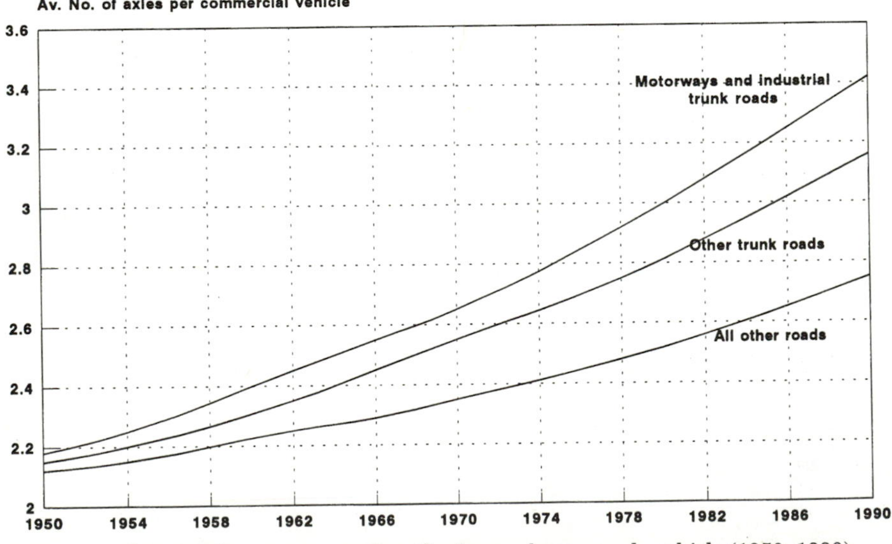

Figure 8.12 Increase in average number of axles per heavy goods vehicle (1950–1990).

Figure 8.13 Weighbridge platform modules in position.

8.24. The weighbridge requires a shallow pit in the traffic lane over which the measurements are to be made. The pit accommodates an appropriate number of platform modules, each 0.6 m square, to cover the recording width required. Generally more than half the lane width is covered to give half-axle loads, but the full width can be covered if necessary. Each platform module is supported on four load cells, and the output is connected to a preamplifier and then to a digitizer and classifier with an hourly printout.

8.25. Weighbridges of this type have been installed in a number of industrial freeways and trunk roads. Normally the measurements are restricted to the slow traffic lane, but at some sites the work has been extended to the overtaking lane.

Traffic Loading Expressed in Terms of Standard Axles

8.26. Figure 8.14*a* shows the axle load spectrum measured by a weighbridge in the slow lane of a three-lane dual freeway in Britain

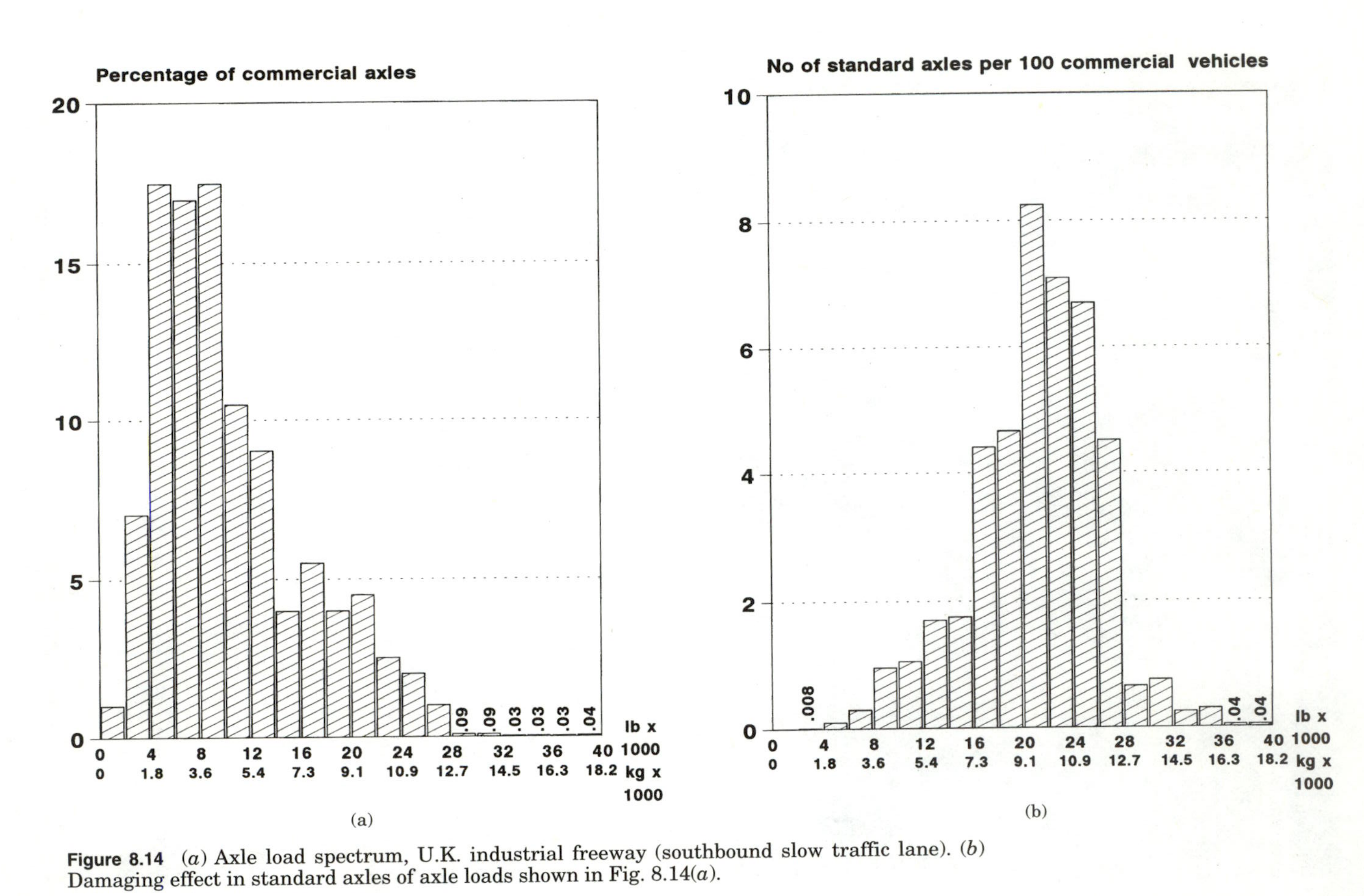

Figure 8.14 (*a*) Axle load spectrum, U.K. industrial freeway (southbound slow traffic lane). (*b*) Damaging effect in standard axles of axle loads shown in Fig. 8.14(*a*).

TABLE 8.4 Equivalence Factors and Damaging Power of Different Axle Loads

Axle load		
kg	lb	Equivalence factor
910	2,000	0.0002
1,810	4,000	0.0025
2,720	6,000	0.01
3,630	8,000	0.03
4,540	10,000	0.09
5,440	12,000	0.19
6,350	14,000	0.35
7,260	16,000	0.61
8,160	18,000	1.0
9,070	20,000	1.5
9,980	22,000	2.3
10,890	24,000	3.2
11,790	26,000	4.4
12,700	28,000	5.8
13,610	30,000	7.6
14,520	32,000	9.7
15,420	34,000	12.1
16,320	36,000	15.0
17,230	38,000	18.6
18,140	40,000	22.8

in 1973. In Figure 8.14*b* each of the axle-load bands has been expressed in equivalent 8.16-t (18,000-lb) axles. The equivalence factors used are shown in Table 8.4. These equivalence factors were derived from the 1962 AASHO Road Test. That test is discussed in detail in Chap. 17.

8.27. Figure 8.15 shows, for the road referred to in Fig. 8.14, the increase in the number of standard axles per commercial vehicle since 1960. This increase is due in part to the increase in the number of axles per commercial vehicle and in part to the more intensive use of commercial vehicles.

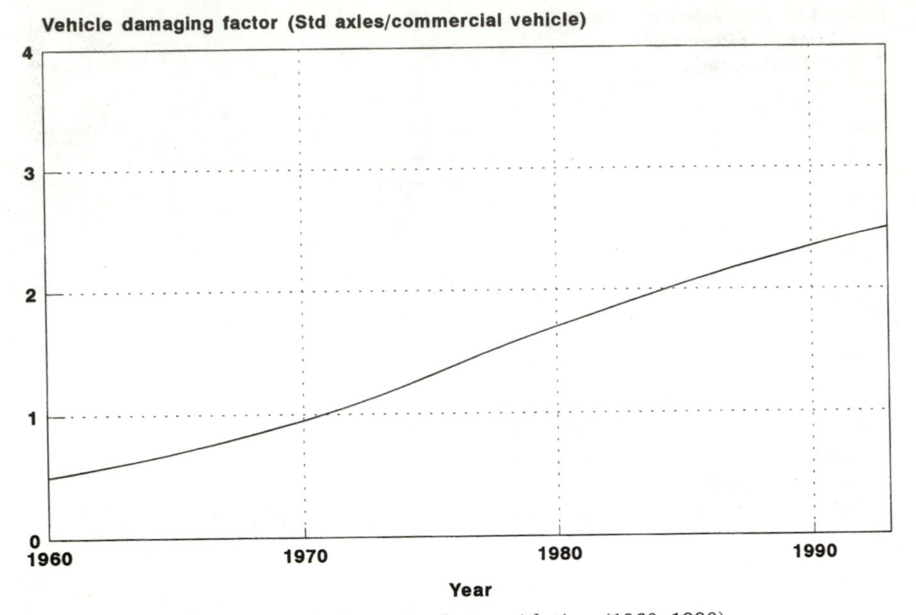

Figure 8.15 Increase in vehicle damaging factor with time (1960–1990).

8.28. The procedure for estimating future traffic for purposes of pavement design is dealt with in Chap. 21.

Part

3

Pavement Materials—Specification and Properties

Chapter

9

The Soil Foundation

Introduction

9.1. When a loaded wheel moves over the surface of a multilayer pavement, the magnitude of the generated stresses decreases from layer to layer, and at the same time the duration of the stress pulse increases with depth. The stress reduction effected by any layer depends on the stiffness and thickness of that layer. A major function of the pavement is to reduce the stresses transmitted to the subgrade to a level which the soil will accept without significant deformation.

9.2. This means that the structural design of the pavement should be equated to the axle loads to be carried and the effective modulus of elasticity of the subgrade. This provides the basis for structural design procedures discussed in detail in Chaps. 23 and 24. The shear strength of soil is mainly of interest to the road engineer in connection with the design of the slope angles of the embankments and cuttings necessary to the construction of modern high-speed roads.

9.3. Soil is the most variable material with which the civil engineer has to deal; large changes can occur in the type and condition of the soil over distances of only a few meters. The comparatively low unit cost of road pavements does not permit the detailed soil testing and evaluation usual with other, more compact civil engineering projects, such as high-rise buildings and dams. Pavement engineers must generally be content with an average assessment of the soil conditions over a long length of carriageway. They must accept that this may lead eventually to some areas of the finished road requiring structural maintenance before the end of the design life, or before general strengthening by an overlay is necessary. On a small job, closely under the control of the engineer, it may be possible to vary the design to take into account soil changes found during the progress of the work, but on major projects the design must be based largely on the preliminary soil survey and the geology of the site as discussed in connection with the site investigation in Chap. 7.

The Constitution of Soil

9.4. Soil consists of a mass of weathered mineral particles of various shapes and sizes, between which there is water and generally some air. Near the surface most soils contain organic matter derived from the decomposition of vegetation, but in poorly drained areas, strata consisting almost entirely of vegetable matter may occur at various depths below the surface. This material, which is loosely described as peat, presents many problems to the road engineer.

9.5. In thinking of soil, the mind almost inevitably concentrates on the solid particles. However, it cannot be too strongly emphasized that the engineering properties of soils depend on all three constituents—solid matter, water, and air. Generally speaking, the type of soil can be defined in terms of size and shape of particles, but the state of the soil, which determines the engineering characteristics, can be defined only in terms of the solid, liquid, and air contents.

9.6. Since in all foundation problems a knowledge of the type and state of the soil is necessary, it is essential that these terms should be defined as concisely as possible.

Particle Size Distribution

9.7. The type of soil is generally specified in terms of the particle sizes present. The particles in soil range from stones several centimeters across down to colloidal material too fine to be susceptible to direct measurement. The normal method of representing the particles in a soil (sometimes called the grading) is in terms of the percentages of particles by weight finer than specified sizes, as shown in Fig. 9.1. Since the small particles play a particularly important part in determining the properties of soil, it is usual, as shown in the figure, to use a logarithmic scale for particle size. This has the effect of spreading the smaller sizes.

9.8. The normal procedure used to determine the particle size distribution of soil involves sieving for the particles larger than 0.06 mm and sedimentation for the smaller particles. These procedures are

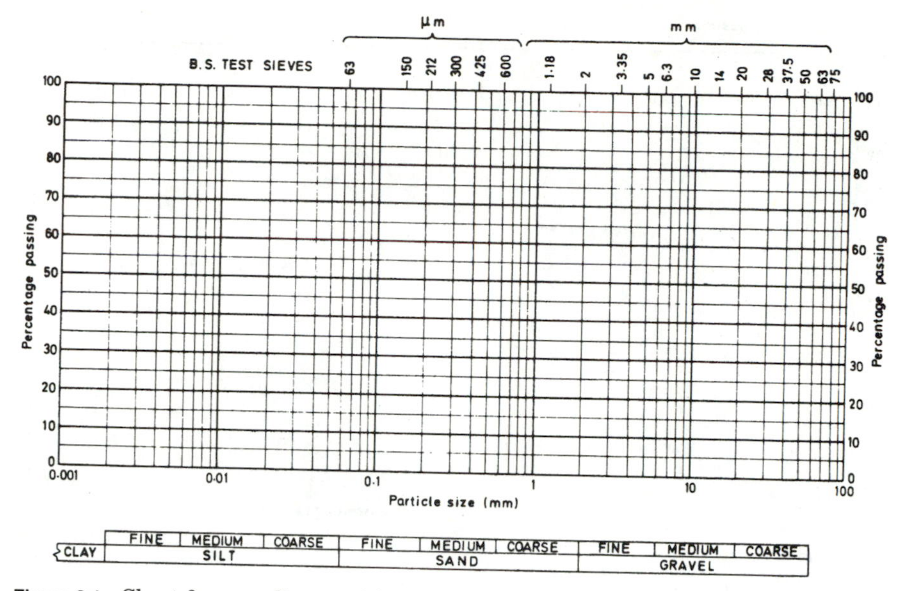

Figure 9.1 Chart for recording particle size distribution.

described in AASHTO Designation T88-86 and as Tests 7A–7D of BS 1377:1975. The sedimentation process is too complex for most site laboratories, and the properties of the silt and clay fractions are generally assessed by plasticity tests as described in Par. 9.40.

9.9. Certain broad terms are used to describe the shape of the particle distribution curve. A well-graded soil is characterized by a wide range of particle sizes in proportions which give a smooth curve for the particle size distribution. In poorly graded soils, although a wide range of sizes may be present, the soil is deficient in certain intermediate particle sizes, resulting in a "stepped" distribution curve. A uniformly graded soil (sometimes referred to as a single-size soil) contains only a very narrow band of particle sizes. Examples of well-graded, poorly graded, and uniformly graded soils are shown in Fig. 9.2.

9.10. The particle size spectrum is divided into four arbitrary groups: gravel, sand, silt, and clay. The maximum and minimum particle sizes included in each of these groups are as follows:

Gravel	Particles between 60 and 2 mm
Sand	Particles between 2 and 0.06 mm
Silt	Particles between 0.06 and 0.002 mm
Clay	All particles smaller than 0.002 mm

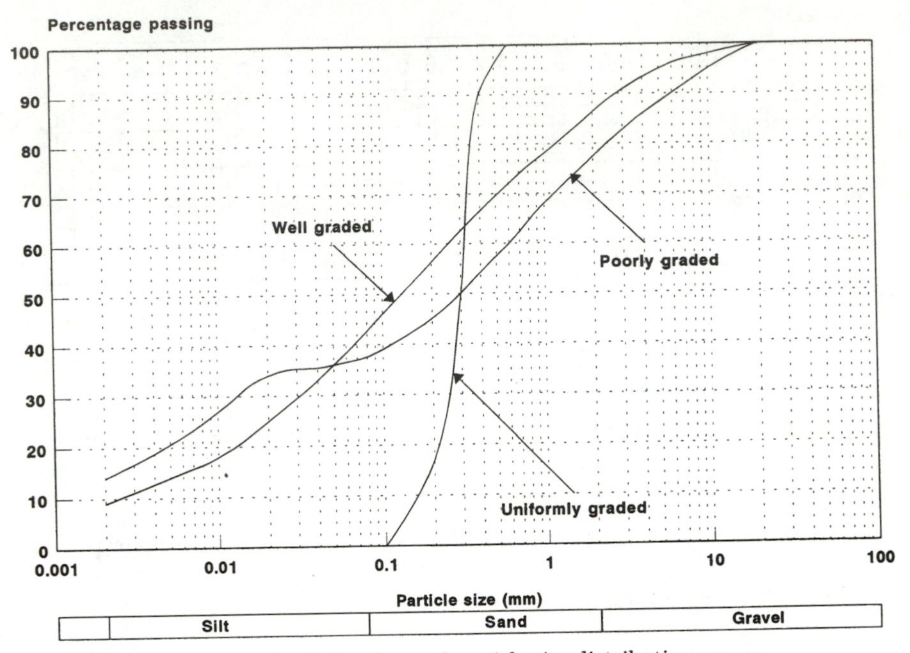

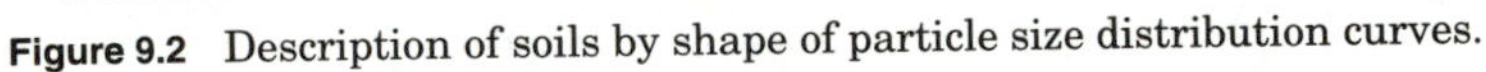

Figure 9.2 Description of soils by shape of particle size distribution curves.

The sand and silt groups are further subdivided into coarse, medium, and fine fractions as shown below:

Coarse sand Particles between 2 and 0.6 mm
Medium sand Particles between 0.6 and 0.2 mm
Fine sand Particles between 0.2 and 0.06 mm
Coarse silt Particles between 0.06 and 0.02 mm
Medium silt Particles between 0.02 and 0.006 mm
Fine silt Particles between 0.006 and 0.002 mm

These subdivisions are shown beneath Fig. 9.1.

9.11. Although these subdivisions of the particle size spectrum are in some measure based on observed differences in the behavior of the groups, it is best to regard them as purely arbitrary.

Description of Soils in Terms of Particle Groups

9.12. Most soils have particle size distributions which span several of the grading groups referred to in Par. 9.10. Thus a soil whose grading

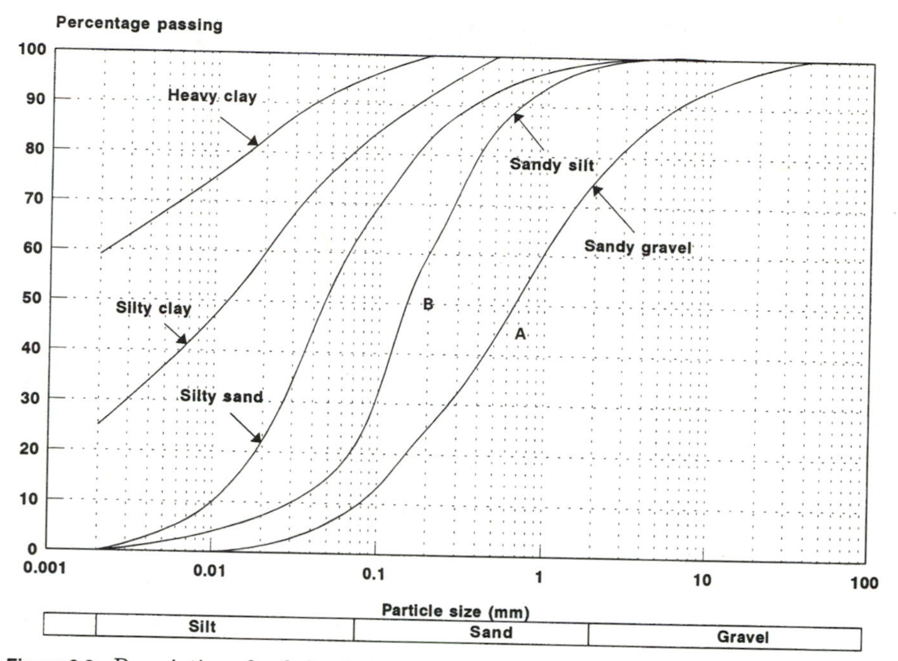

Figure 9.3 Description of soils by their particle size components.

included all particle sizes would be referred to as a gravel-sand-clay. Where the particles fall predominantly in two groups, ambiguities can arise in the descriptions used in different countries. In the United States the main grading group is given first followed by the secondary group. Thus, if we refer to Fig. 9.3, a soil with grading curve A would be classified as a sand-gravel, and the other grading curves included in the figure would be classified in the manner indicated on the diagram.

9.13. In the United Kingdom the terms sandy gravel and silty sand are often used. When this procedure is used, the adjective should be formed from the lesser size group, e.g., curve A in Fig. 9.3 would be described as a gravelly sand and curve B as a sandy silt. A more precise indication of the constitution of the soil can of course be given if the proportions by weight in each size group are tabulated.

Relative Size and Surface Area of Particles

9.14. The behavior of soils, particularly in relation to changing moisture conditions, depends to a great extent on the relative sizes of the particles present and their surface areas. A clearer picture of these factors than is given by a mere statement of equivalent particle diameters

TABLE 9.1 Size and Surface Area of Soil Particles

Particle group	Equivalent diameter, mm	Approximate number of particles in a 2-mm cube	Total surface area of particles in a 2-mm cube, mm^2
Gravel	2	1	12.5
Sand	1	8	25
Silt	0.03	340,000	950
Clay	0.002	8,000M	24,000

can be obtained by considering the number of particles of different sizes which can be contained in a given volume, and their surface areas, assuming them to be spherical. In Table 9.1 the number of such particles which can be contained in a 2-mm cube is given together with the corresponding total surface area of the particles. The particle sizes selected are the minimum of the gravel group (2 mm), the means of sand and silt groups (approximately 1 and 0.03 mm) and the maximum of the clay group (0.002 mm). The loosest degree of packing consistent with contact between the particles and the enclosing cube is assumed.

Since the clay particles tend to be flat and elongated rather than spherical, the number and surface area of particles in this group quoted in Table 9.1 will be an underestimate.

Nature of the Soil Particles

9.15. Inorganic material. The gravel, sand, and silt arise from the disintegration of the parent rocks by weathering. They are roughly cubical in shape and may be transported from their place of origin by water or wind. The clay particles differ from the other particle fractions both in chemical composition and in their physical properties. Chemically, they consist of hydrated alumino-silicates which are formed during the leaching processes to which the coarser particles of the rock minerals are subjected. Among the minerals which occur in clay particles are forms of kaolinite, montmorillonite, and mica.

9.16. Physically, the clay particles are elongated and lamellar, which together with their small size and large specific surface area (see Table 9.1) imparts to clay soils their properties of plasticity and compressibility referred to in Pars. 9.37 and 9.43.

9.17. The specific gravity of soil particles is generally between 2.6 and 2.8; lower values usually indicate the presence of organic matter and higher values the presence of metallic ores. The particle specific gravity of fine-grained soils is determined by the specific gravity bottle method, and the pycnometer method is used for coarse-grained soils. Both methods are described in AASHTO T100-86 or ASTM (American Society for Testing and Materials) D854-83 and in British Standard 1377.

9.18. Organic material. The organic material is derived mainly from the plant residue and is generally concentrated in the top 30 cm of the soil. However, leaching in sandy soils may cause soluble constituents to be extracted and deposited lower down. The distribution of organic deposits such as peat is conditioned by geological factors and may extend to much greater depths.

9.19. The composition of the organic matter is dependent on the plant cover and on the extent to which decomposition has progressed. Thus the organic constituents of forest soils are derived largely from leaf mold, whereas in pasture soils the organic matter is derived from grass leaves and roots. In some cases the plant materials can be recognized visually, while in other cases the decomposition may have proceeded so far that the original plant structure has largely disappeared, leaving a dark, amorphous material called humus. Freshly decomposed organic matter and humus have different characteristics in that the former consists of macroparticles and fibers which are relatively inert, whereas humus is acidic and colloidal in nature and absorbs water readily.

9.20. The organic content of soil immediately below road structures should not exceed 4 percent. A method of measuring the organic content of soil is described in AASHTO T267-86 and British Standard 1377:1975, Test 8.

The Solvent Action of Soil Water

9.21. In addition to its important influence on the properties of soil, which is discussed later in this chapter, the soil water is of concern to the engineer because of its action as a solvent. The principal materials in solution are (1) soluble salts, generally in the form of sulfates of sodium, magnesium, and calcium, and (2) finely divided organic matter.

Sulfates in the soil water affect the properties of the soil and of structures in contact with it, by

1. Attacking concrete and other materials containing cement
2. Disrupting the soil structure by crystallization
3. Corroding metals and particularly iron pipes buried in the soil

In general, only (1) and (3) are likely to be important. Special precautions such as the use of sulfate-resistant cements and the protection of iron pipes with an impervious coating have to be taken when the sulfur trioxide content of the groundwater exceeds 50 parts in 100,000. The determination of the sulfur trioxide content is described in British Standard 1377:1975, Test 9.

9.22. The organic matter in soil water is probably not in a true state of solution but rather consists of colloidal matter in suspension. In this way, organic matter can be leached out of the topsoil and deposited in the subsoil below, resulting in dark zones. It is possible that soluble organic matter also influences the redistribution of mineral elements of the soil, since iron is known to form soluble complexes with certain organic compounds. Thus iron may be removed from some parts of the soil mass and deposited lower down in the form of concretions around siliceous particles, giving rise to an iron pan.

The State of the Soil

9.23. The strength properties of soil which are of importance to the engineer can be only very loosely related to the type of soil as defined above in terms of the particle size distribution. They are much more related to the manner in which the particles are packed together (the density) and the amount of water between the particles (the moisture content). These factors define the state of the soil.

9.24. An engineer preparing a specification either for the construction of earthworks or to give the condition which is required in the subgrade must carry out sufficient testing, as part of the site investigation, to ensure that the specification not only can be achieved under the prevailing climatic conditions using the type of plant normally available to a civil engineering contractor, but will also give the strength and stability required by the design. The required state of the soil may be expressed by limits of (1) the moisture content, (2) the dry density, or (3) the maximum air content in the compacted material. It is important for the engineer to understand the interrelation of these factors. This is considered in the following paragraphs in conjunction with Fig. 9.4, which shows the weights and volumes of dry solids, water, and air in a volume V of wet soil of weight W.

Dry Density

9.25. The weight of dry material in unit volume of wet soil is defined as the dry density ρ_d and is usually expressed in Mg/m^3. If we refer to

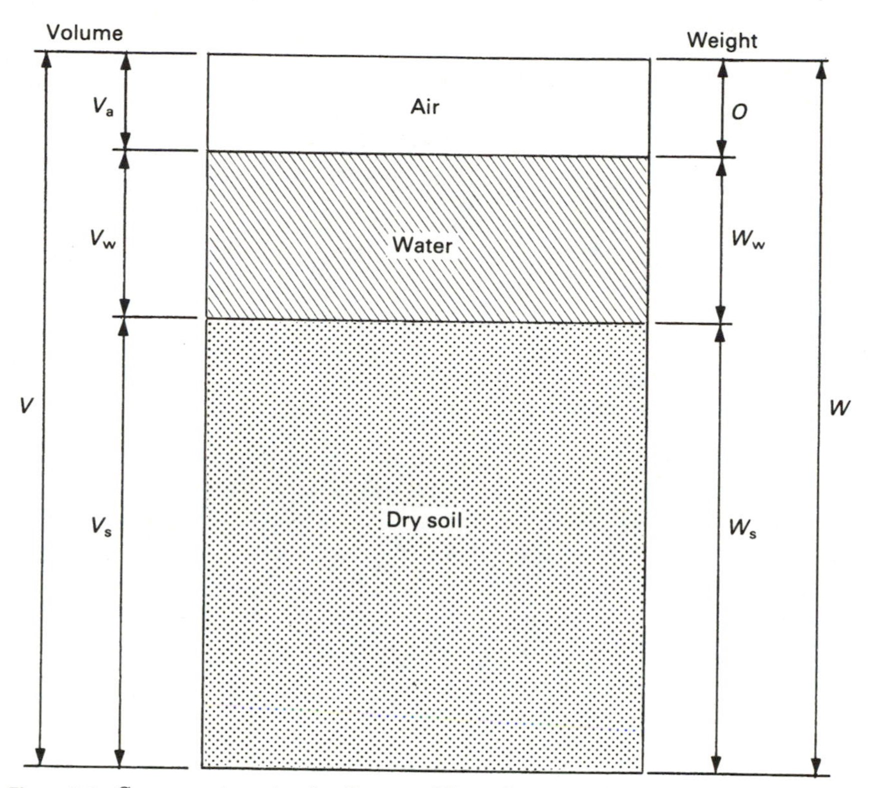

Figure 9.4 Component parts of soil mass. (The volumes and weights shown are defined in the text starting at Par. 9.25.)

Fig. 9.4, which shows the weights and volumes of dry solids, water, and air in a volume V of wet soil of weight W, then

$$\rho_d = \frac{W_s}{V} = \frac{W_s}{V_s + V_w + V_a} \tag{9.1}$$

Moisture or Water Content

9.26. The weight of water expressed as a percentage of the weight of dry solids is termed the moisture or water content w. In both these definitions the weight of dry solids is defined as the weight of the soil after drying for 24 h at a temperature of 105°C. Hence,

$$w = \frac{W_w}{W_s} \times 100 \tag{9.2}$$

Percentage Air Voids (Air Content)

9.27. The space in soil not occupied by solid material is termed the voids, and the space occupied by air is referred to as the air voids. The percentage ratio of the volume of air to the total volume of soil is defined as the percentage air voids or air content a, and it follows that

$$a = \frac{V_a}{V} \times 100 = \frac{V_a}{V_s + V_w + V_a} \times 100 \tag{9.3}$$

Relationship between Dry Density, Moisture Content, and Percentage Air Voids

9.28. The relation between these quantities can be deduced from Eqs. (9.1) to (9.3):

$$\rho_d = \frac{W_s}{V_s + V_w + V_a}$$

hence

$$\frac{1}{\rho_d} = \frac{V_s}{W_s} + \frac{V_w}{W_s} + \frac{V_a}{W_s}$$

$$= \frac{1}{\rho_s} + \frac{w}{100\rho_w} + \frac{V_a}{V\rho_d}$$

where ρ_w and ρ_s are the densities of water and of the soil particles, and

$$\frac{1}{\rho_d}\left[100 - \left(\frac{V_a}{V} + 100\right)\right] = \frac{100}{\rho_s} + \frac{w}{\rho_w}$$

hence

$$\frac{1}{\rho_d}(100-a) = \frac{100}{\rho_s} + \frac{w}{\rho_w}$$

or

$$\rho_d = \frac{\rho_w(1 - a/100)}{1/G_s + w/100} \tag{9.4}$$

where G_s is the particle specific gravity.

In a saturated soil the voids are completely filled with water and $a = 0$. For such a soil,

$$\rho_d = \frac{\rho_w}{1/G_s + w/100} \tag{9.5}$$

The above definition of a saturated soil should be noted carefully. The term "saturated" does not imply that the soil will not take up any more water, since soils containing clay particles may, even in the saturated condition, absorb more water by the process of swelling. The term "saturated" is therefore not sufficient to define the state of a soil.

9.29. The relationship between dry density, moisture content, and percentage air voids given in Eq. (9.4) can be represented graphically by a family of curves as in Fig. 9.5, which refers to a particle specific gravity of 2.70, i.e., $\rho_w = 1$ Mg/m^3, $\rho_s = 2.70$ Mg/m^3. Each curve represents a different air content. The limiting curve corresponding to zero air content is often referred to as the saturation line or the zero air voids line.

9.30. The state of any soil can be represented by a point on a diagram such as Fig. 9.5 drawn to the appropriate particle specific gravity. Every point on the diagram to the left of the saturation line will represent a different state of the soil. It will be seen that if any two of the factors—dry density, moisture content, and percentage air voids—are known, then the third is established. Further, a change in the

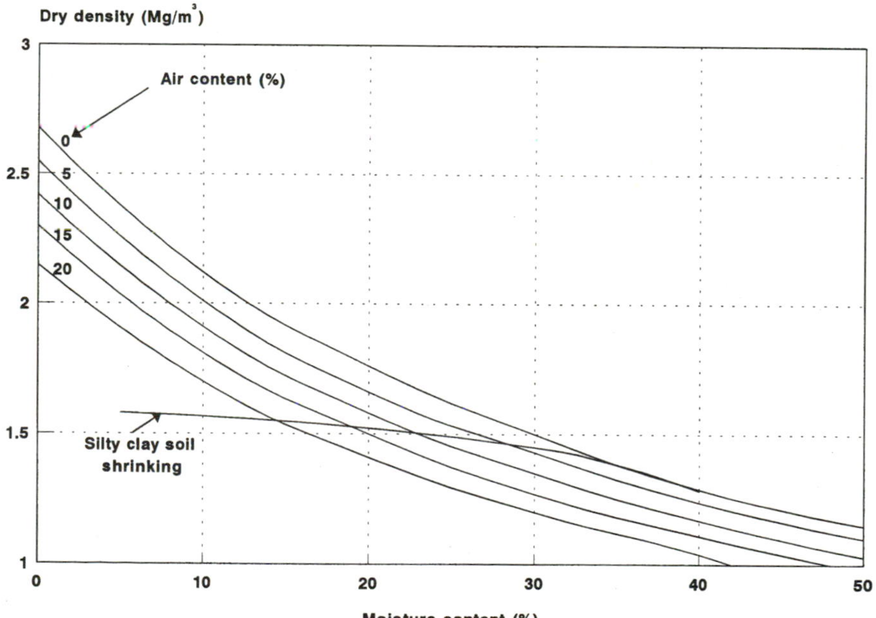

Figure 9.5 Relationships between dry density, moisture content, and percentage air content.

state of the soil can be represented by a line or curve on the diagram. Thus if the particles of a soil are made to move closer together without any change of moisture content (defined as the process of compaction), then the corresponding change of state is represented by a straight line parallel to the axis of dry density, which will cross the family of curves representing different air voids conditions, showing that as the dry density is increased the air content must in this case decrease. Similarly, if water is removed from a saturated clay soil by the application of a compressive load (defined as the process of consolidation), then the change of state is represented by the saturation line. The curve actually shown in Fig. 9.5 represents a silty clay soil shrinking because of the removal of water by evaporation. In this case both the dry density and the air content of the soil increase as the moisture content decreases. It is important to realize that in a diagram such as Fig. 9.5 the state of the soil cannot be represented by a point to the right of the saturation line, since this would imply compressibility either of the water or of the soil particles.

9.31. For the convenience of engineers, a set of charts relating dry density and moisture content for air voids of 0, 5, 10, 15, and 20 percent and particle specific gravities of 2.6, 2.65, 2.7, 2.75, and 2.8 is included as an appendix to this chapter.

Bulk or Wet Density

9.32. Both the moisture content and the dry density of soil can be defined on a wet weight basis, the water content being expressed as a percentage of the wet weight and the density as the weight of wet mineral in unit volume. In civil engineering practice it is most unusual to express moisture content in this way, but the bulk or wet density is often required, as, for example, in the calculation of overburden pressures. If we refer back to Fig. 9.4, the bulk density is given by the following:

$$\rho_w = \frac{W}{V} = \frac{W_s + W_w}{V_s + V_w + V_a} \tag{9.6}$$

Relationship between Bulk Density and Dry Density

9.33. The relation between these quantities follows from their definitions:

$$\frac{\rho_w}{\rho_d} = \frac{W}{V} \times \frac{V}{W_s} = \frac{W}{W_s} = \frac{W_s + W_w}{W_s} = 1 + \frac{w}{100}$$

Therefore

$$\rho_w = \rho_d\left(\frac{100 + w}{100}\right) \tag{9.7}$$

Voids Ratio

9.34. Another term sometimes used in expressing the state of soil is the voids ratio e, defined as the volume of voids per unit volume of matter. Hence, if we refer again to Fig. 9.4,

$$e = \frac{V_w + V_a}{V_s} \tag{9.8}$$

or, in a saturated soil,

$$e = \frac{V_w}{V_s} = \frac{\rho_s W_w}{\rho_w W_s} = \frac{G_s W_w}{W_s} = \frac{G_s w}{100} \tag{9.9}$$

Relationship between Voids Ratio and Dry Density

9.35. It follows from the definitions that dry density and voids ratio are simply related:

$$\frac{1}{\rho_d} = \frac{V_w + V_a}{W_s} + \frac{V_s}{W_s}$$

$$= \frac{e \cdot V_s}{W_s} + \frac{V_s}{W_s}$$

$$= \frac{1 + e}{\rho_s}$$

or

$$\rho_d = \frac{\rho_s}{1 + e} \tag{9.10}$$

Thus the state of the soil can be expressed in terms of voids ratio and moisture content as an alternative to dry density and moisture content.

Porosity

9.36. The porosity n of a soil is defined as the ratio of the volume of voids to the total volume of the soil. Thus:

$$n = \frac{V_w + V_a}{V} \tag{9.11}$$

Plasticity of Clay Soils

9.37. The particles in soil are normally covered by a layer of water the thickness of which will increase as the soil is wetted from the dry condition. In the case of clay particles the thickness of the water films is also influenced by the particular types of clay mineral involved. Further, some clay minerals have an expanding lattice structure into which water can enter to cause expansion of the clay mineral itself. Montmorillonite is the commonest form of such minerals likely to be found in practice.

Liquid and Plastic Limits

9.38. Because of the very large number of water films associated with the small, flat clay particles (Table 9.1), any soil with a clay content exceeding about 15 percent exhibits the properties of plasticity and cohesion. Over a certain moisture range the particles will slide over each other, the action being to shear rather than break the interconnecting water films. This gives rise to an appearance of plasticity when the soil is "worked." At higher moisture contents the clay will deform under its own weight and behave like a viscous liquid.

The moisture content at which the soil changes from the plastic to the liquid state is termed the liquid limit LL. As the moisture content is reduced below the plastic range the soil becomes friable owing to breaking of the moisture bonds as the soil is worked. The moisture content at which this change occurs is referred to as the plastic limit PL. The moisture content range between the liquid and plastic limits is termed the plasticity index PI. Hence,

$$\text{LL} - \text{PL} = \text{PI} \tag{9.12}$$

9.39. Two other relations are sometimes used to define the moisture content w in terms of the plasticity properties:

The liquidity index LI is defined as follows:

$$\text{LI} = \frac{w - \text{PL}}{\text{PI}} \tag{9.13}$$

The consistency index CI is defined as follows:

$$CI = \frac{LL - w}{PI} \tag{9.14}$$

Determination of the Liquid and Plastic Limits

9.40. The determination of the liquid and plastic limits (sometimes called the Atterburg limits or the index tests) is discussed in detail in AASHTO Standards T89-86 and T90-86 and BS 1377. The soil is first mixed to a high moisture content and a pat is placed in the cup of the liquid limit machine. The cup is raised by a cam and allowed to fall sharply onto the hard rubber baseplate as the handle of the machine is turned. The number of turns or "blows" required to close a groove, cut through the soil from front to back, is observed. The test is repeated at decreasing moisture contents, and the moisture content at which the groove is closed by 25 blows is arbitrarily defined as the liquid limit. After further drying the soil is rolled out on a glass plate using the palm of the hand. The lowest moisture content at which the soil can be rolled into a thread of 3 mm diameter without breaking is termed the plastic limit. Full details of the sample preparation and the test procedures to measure the index tests are given in the standards referred to above.

Cohesion

9.41. A granular soil has little strength in tension, but the water bonds between the clay particles which give a clay soil its plasticity also bond the particles together and give a significant strength in tension. This strength is termed the cohesion c.

Relation between the Plasticity Index and Clay Content of Soils

9.42. Table 9.2 shows values of the clay content of a wide range of soils, together with the plasticity properties and the particle specific gravity. The table includes both British and overseas soils. The data are shown graphically in Fig. 9.6. There is a great deal of scatter which arises mainly from the different clay minerals present in the soils. Soils with very active clay minerals such as montmorillonite tend to fall below the line, and those with less active minerals are above the line. The mean relation indicated is useful as an approximate guide to the clay content of soils of known plasticity index.

TABLE 9.2 Clay Content, Particle Specific Gravity, and Results of Index Tests on Various Soils

Soil description	Place of origin	Clay content, %	Results of index tests PL, %	LL, %	PI, %	Particle specific gravity
British soils:						
Lias clay	Shipton, Oxon.	39	24	60	36	2.71
Weald clay	Ditchling, Sussex	62	25	68	43	2.74
Oxford clay	Cumnor, Berks.	56	24	72	48	2.74
Kimmeridge clay	Sunningwell, Berks.	67	24	77	53	2.74
London clay	Heathrow, Middx.	60	26	78	52	2.73
Gault clay	Steyning, Sussex	56	25	70	45	2.68
Gault clay	Stoke Mandeville, Bucks.	57	29	81	52	2.68
Gault clay	Aylesbury, Bucks.	69	34	102	68	2.70
Gault clay	Shaves Wood, Sussex	60	32	121	89	2.65
Silty clay	Harmondsworth, Middx.	34	24	43	19	2.70
Silty clay	Harmondsworth, Middx.	33	10	43	24	2.72
Sandy clay	Harmondsworth, Middx.	24	19	27	8	2.70
Clay silt	Rippon, Yorks.	19	17	28	11	2.67
Clay silt	Wellington, Som.	12	19	33	14	2.75
Overseas soils:						
Decomposed granite	Pretoria, South Africa	20	42	65	23	2.75
Silty clay	Habbaniya, Iraq	45	28	58	34	2.78
Sandy, silty clay	Thornhill, Zimbabwe	40	25	57	32	2.71
Silty clay	Heany, Zimbabwe	20	20	41	21	2.76
Silty clay	Heany, Zimbabwe	32	22	48	26	2.76
Clay	Kumalo, Zimbabwe	58	27	60	33	2.77
Sandy, silty clay	Kumalo, Zimbabwe	34	24	51	27	2.60
Clay	Sudan	54	26	95	69	2.72
Silty clay	Trinidad	19	27	40	13	2.72
Silty clay	Trinidad	26	17	36	19	2.68
Clay	Trinidad	45	16	51	35	2.73
Clay	Trinidad	60	24	71	47	2.80
Clay	Kenya	63	34	81	47	2.65
Clay	Kenya	75	44	101	57	2.65
Sandy, silty clay	Grenada	24	23	63	40	2.78
Silty clay	Ghana	36	21	65	44	2.83
Clay	Kenya	79	70	104	34	2.79
Clay	Kenya	83	44	87	43	2.85
Clay	Uganda	59	21	57	36	2.76
Silty clay	Kenya	22	23	38	15	2.56

Swelling and Shrinkage of Clay Soils

9.43. Increase in thickness of the water films between the clay particles in soils results in an increase in volume, and there is therefore a relation between volume and moisture content for all soils with a significant clay content.

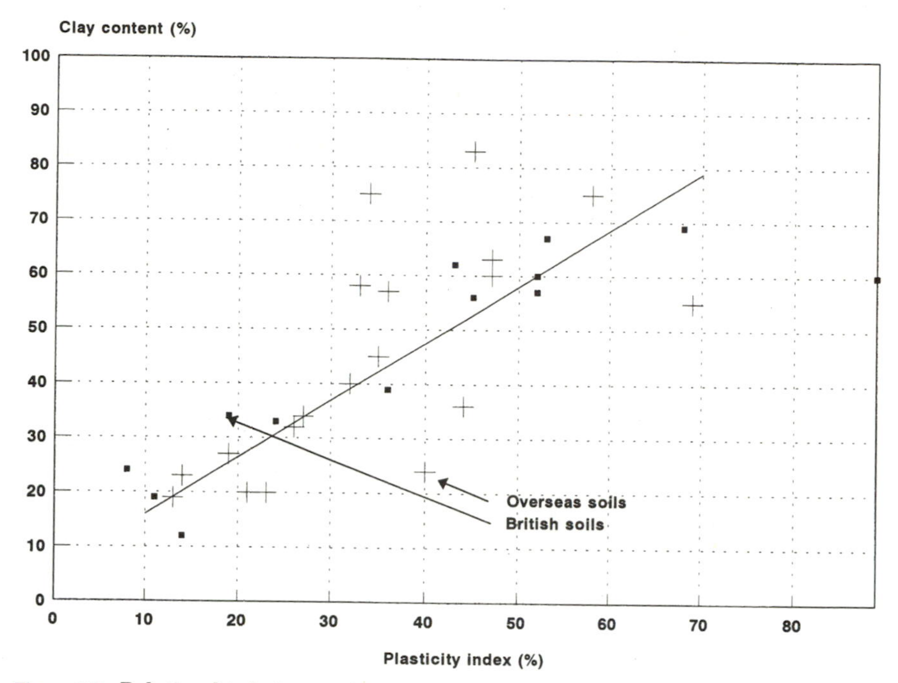

Figure 9.6 Relationship between clay content and plasticity index.

9.44. Shrinkage is expressed as the observed change of volume per 100 g of dry soil as indicated in Fig. 9.7. As the soil dries in the saturated condition, the relation follows the zero air voids or saturation line. As the larger particles come into contact, air enters the soil structure and a further change in moisture content results in a volume change smaller than the volume of water removed. Close to zero moisture content the rate of shrinkage tends to zero.

The Shrinkage Limit

9.45. The moisture content at which the saturation line meets the horizontal, corresponding to the volume at zero moisture content, is arbitrarily termed the shrinkage limit, although it does not represent a limiting condition with regard to shrinkage.

Measurement of Shrinkage

9.46. The shrinkage curve for soils is often measured by a mercury displacement method (AASHTO T92-86), but a preferable procedure,

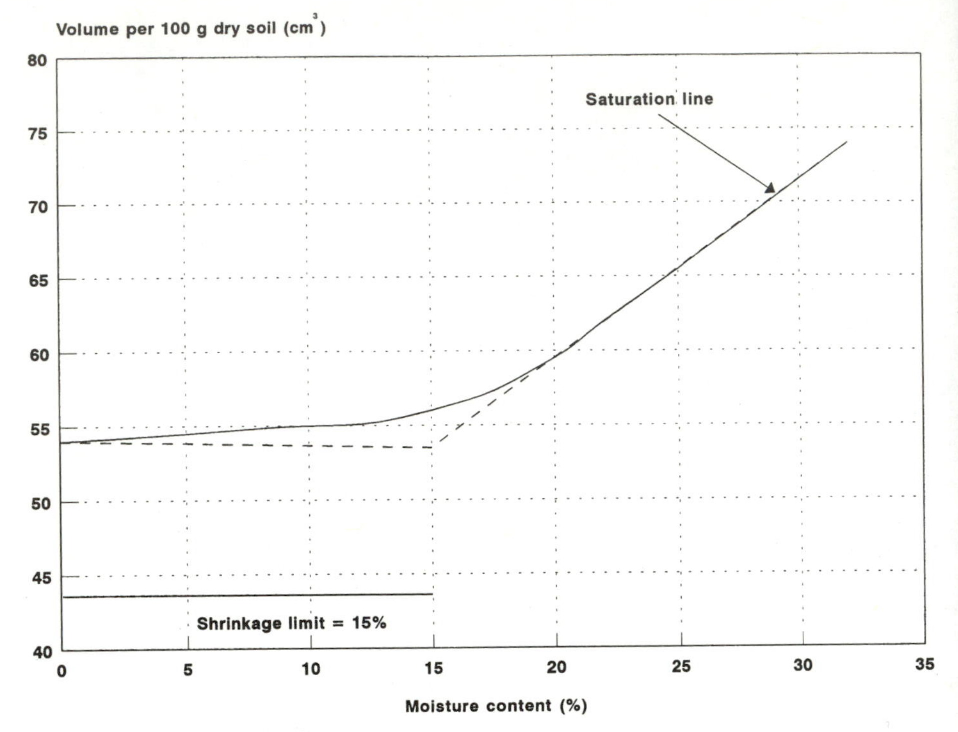

Figure 9.7 Shrinkage of soil.

because it imposes no external loading on the soil, is to use an optical profile projector operating at a magnification of 50 times.[1] The outline of a cylindrical sample of the soil about 15 mm in diameter and 15 mm high is projected on the screen, and the mean height and mean diameter are deduced by direct measurement. The soil is allowed to dry slowly on the rotating stage of the projector, and a series of measurements of weight and volume are obtained. The moisture contents are subsequently deduced from the dry weight of the sample.

9.47. Figure 9.8*a* shows the shrinkage curves for a gault clay, for a London clay, and for a silty clay determined in the above manner. In Fig. 9.8*b* the curves are plotted in an alternative form to show the change in dry density with moisture content. Provided the soil is wetted or dried slowly, there is no significant hysteresis between the wetting and drying conditions.

9.48. The shrinkage limits for the three soils considered in Fig. 9.8*a* and *b* are as follows: the gault clay 10.5 percent, the London clay 18.3

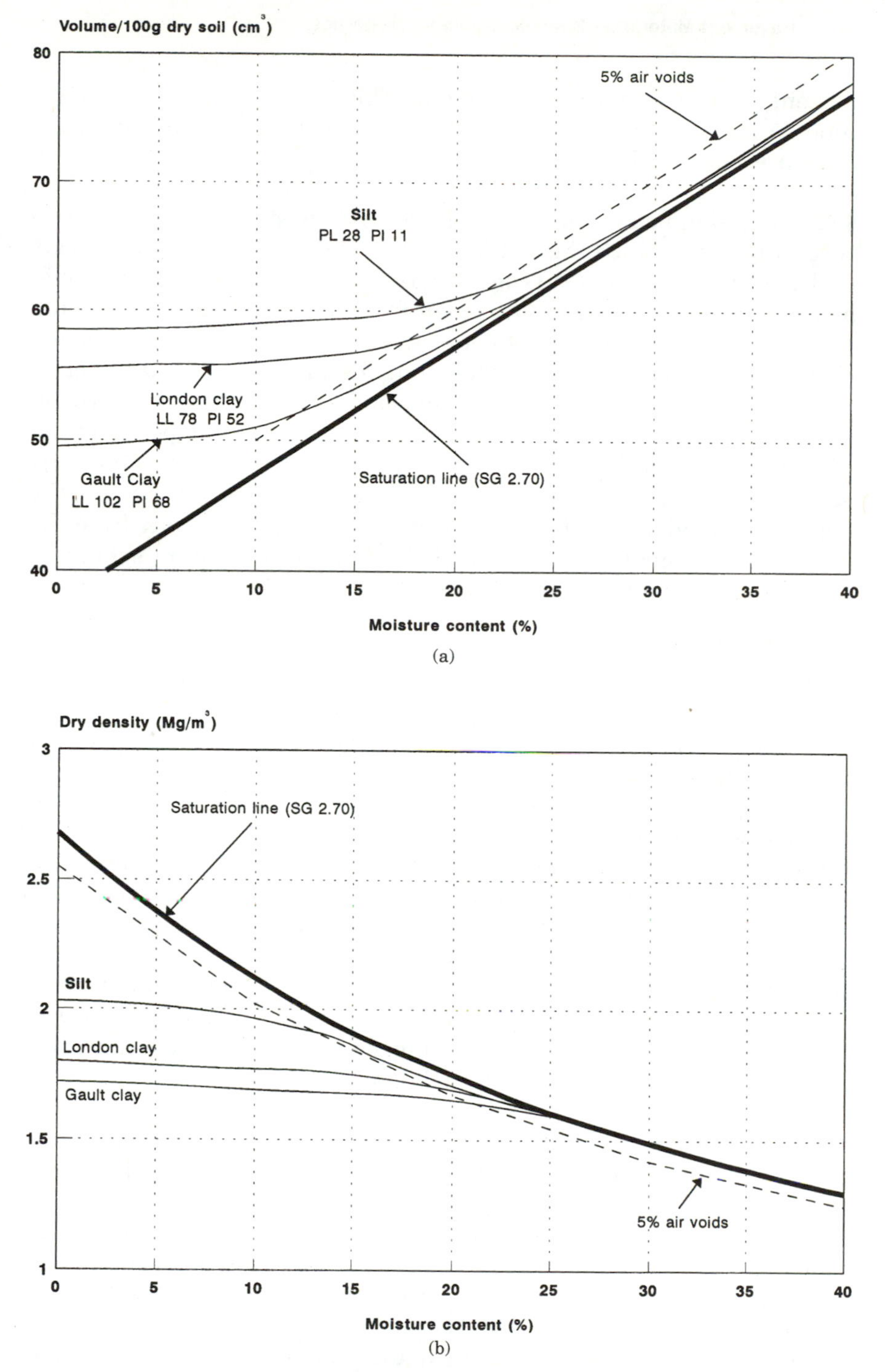

Figure 9.8 (*a*) and (*b*) Alternative ways of showing the shrinkage of clay soil.

percent, and the silty clay 21 percent. This shows that the shrinkage limit tends to decrease with the clay content, although the clay minerals present are also important.

9.49. The significance of shrinkage depends in practice on the moisture range over which the soil is likely to fluctuate. For the silty clay under a road pavement a typical moisture content fluctuation would be between 13 and 20 percent, which from the shrinkage curve would correspond to a volume change of less than 2 percent. For the London clay the range could be 24 to 30 percent, giving a volume change of 13 percent. For the gault clay the range could be 30 to 45 percent, corresponding to a volume change of 25 percent. Thus the engineer would anticipate no shrinkage problems with the silty clay, but potentially serious problems from the gault clay. To minimize such problems the engineer should specify that vegetation should be kept away from the verges of pavements laid on heavy clays, or consideration should be given to the use of impermeable shoulders.

Soil Classification

9.50. Various methods of soil classification have been developed to relate soil types to civil engineering properties. The oldest is the Casagrande system, which dates from the forties. Soils were classified into groups, based in the case of noncohesive soils on the particle size distribution and for cohesive soils on the values of liquid and plastic limits. At best, a classification system can give only the broadest indication of the engineering behavior, and attempts which have been made to introduce further subdivisions into the Casagrande system have led to further complications without improving its validity. The problems arise from the fact that the state of the soil is often much more important than its type. In the last 15 years the tendency in both the United States and Europe has been to revert to classification procedures not directly linked to soil behavior.

9.51. In the United States the use of the Casagrande system has largely been superseded by AASHTO Designation M145-82, Recommended Practice for the Classification of Soils and Soil Aggregate Mixtures for Highway Construction Purposes. This is simpler than the Casagrande method and involves seven main soil groupings (A-1 to A-7), as described in Table 9.3. These are divided into subgroups shown in Table 9.4. The various groupings are defined in terms of the soil gradings, and the liquid and plastic limits. The information given relating to the engineering properties of the soils is less comprehensive than is the case of the Casagrande system.

TABLE 9.3 Classification of Soils and Soil-Aggregate Mixtures

	General classification						
	Granular materials (35% or less passing 0.075 mm)			Silt-clay materials (more than 35% passing 0.075 mm)			
Group classification	A-1	A-3*	A-2	A-4	A-5	A-6	A-7
Sieve analysis. Percentage passing:							
2.00 mm (No. 10)							
0.425 mm (No. 40)	50 max.	51 min.					
0.075 mm (No. 200)	25 max.	10 max.	35 max.	36 min.	36 min.	36 min.	36 min.
Characteristics of fraction passing 0.425 mm (No. 40):							
Liquid limit				40 max.	41 min.	40 max.	41 min.
Plasticity index	6 max.	N.P.	†	10 max.	10 max.	11 min.	11 min.
General rating as subgrade	Excellent to good			Fair to poor			

*The placing of A-3 before A-2 is necessary in the "left-to-right elimination process" and does not indicate superiority of A-3 over A-2.
†See Table 9.4 for values.

TABLE 9.4 Classification of Soils and Soil-Aggregate Mixtures

	General classification										
	Granular materials (35% or less passing 0.075 mm)							Silt-clay materials (more than 35% passing 0.075 mm)			
	A-1			A-2							A-7
Group classification	A-1-a	A-1-b	A-3	A-2-4	A-2-5	A-2-6	A-2-7	A-4	A-5	A-6	A-7-5, A-7-6
Sieve analysis. Percentage passing:											
2.00 mm (No. 10)	50 max.										
0.425 mm (No. 40)	30 max.	50 max.	51 min.								
0.075 mm (No. 200)	15 max.	25 max.	10 max.	35 max.	35 max.	35 max.	35 max.	36 min.	36 min.	36 min.	36 min.
Characteristics of fraction passing 0.425 mm (No. 40):											
Liquid limit				40 max.	41 min.	40 max.	41 min.	40 max.	41 min.	40 max.	41 min.
Plasticity index		6 max.	N.P.	10 max.	10 max.	11 min.	11 min.	10 max.	10 max.	11 min.	11 min.*
Usual types of significant constituent materials	Stone fragments, gravel and sand		Fine sand	Silty or clayey gravel and sand				Silty soils		Clayey soils	
General rating as subgrade	Excellent to good							Fair to poor			

*Plasticity index of A-7-5 subgroup is equal to or less than LL minus 30. Plasticity index of A-7-6 subgroup is greater than LL minus 30.

9.52. British Standard 5930:1981, Code of Practice for Site Investigation, uses essentially the Casagrande classification system with some further subdivision of the soil types as indicated in Table 9.5 (Table 8 of the British Standard). Cohesive soils are classified in terms of the slightly modified plasticity chart shown here as Fig. 9.9 (Fig. 31 of the British Standard). BS 5930 contains a great deal of important information relating to soil sampling and the visual classification of core samples.

Compaction of Soil

9.53. Compaction is the process by which air is excluded from a soil mass to bring the particles closer together and thus increase the dry density. The effect is generally, but not necessarily, to increase the strength of the soil. Distinction must be drawn between the meaning of the terms compaction and consolidation. The latter, already referred to in more detail earlier, is the process by which water is excluded from a saturated clay soil by the action of a sustained load.

9.54. Under the application of a small stress, soil may behave elastically, any strain being fully recovered when the stress is removed. With greater stresses, compaction occurs which increases the strength of the soil to a level where further strain is resisted. At still higher stresses, the soil compacts to a state where no further strength can be mobilized, and this is followed by shearing at constant volume.

9.55. If a loose soil is compacted by the application of a fixed amount of energy then the dry density achieved is related to the moisture content of the soil. The curve relating the two factors shows a maximum dry density at a particular moisture content generally referred to as the "optimum moisture content" for that method of compaction or that amount of compacting energy. The shape of the curve relating compaction and moisture content can be deduced from the strength properties of the soil. Figure 9.10*a*, shows the relationship between unconfined compressive strength, determined on compacted cylinders of a cohesive soil of known dry density, and the moisture content. If it is assumed that energy is applied to initially loose samples of the same soil to produce maximum strengths of 100, 200, 300, and 400 kN/m^2, the dry density–moisture content curves corresponding to each compaction condition would be as indicated in Fig. 9.10*b*. In each case the curve reaches a peak at the moisture content corresponding to saturation. A qualitative explanation of the shape of dry density–moisture content relationship curves often put forward is that on the dry side of the optimum moisture content the soil water acts as a lubricant, while on the wet side it acts as a displacer.

TABLE 9.5 British Soil Classification System for Engineering Purposes*

Soil groups		GRAVEL and SAND may be qualified sandy GRAVEL and gravelly SAND, etc., where appropriate	Group symbol		Subgroups and laboratory identification: Subgroup symbol	Fines (% less than 0.06 mm)	Liquid limit %	Name
COARSE SOILS (less than 35% of the material is finer than 0.06 mm)	GRAVELS (more than 50% of coarse material is of gravel size (coarser than 2 mm))	Slightly silty or clayey GRAVEL	G	GW	GW	0 to 5		Well-graded GRAVEL
				GP	GPu GPg			Poorly graded/uniform/gap-graded GRAVEL
		Silty GRAVEL	G-F	G-M	GWM GPM	5 to 15		Well-graded/poorly graded silty GRAVEL
		Clayey GRAVEL		G-C	GWC GPC			Well-graded/poorly graded clayey GRAVEL
		Very silty GRAVEL	GF	GM	GML, etc.	15 to 35		Very silty GRAVEL; subdivide as for GC
		Very clayey GRAVEL		GC	GCL GCI GCH GCV GCE			Very clayey GRAVEL (clay of low, intermediate, high, very high, extremely high plasticity)
	SANDS (more than 50% of coarse material is of sand size (finer than 2 mm))	Slightly silty or clayey SAND	S	SW	SW	0 to 5		Well-graded SAND
				SP	SPu SPg			Poorly graded/uniform/gap-graded SAND
		Silty SAND	S-F	S-M	SWM SPM	5 to 15		Well-graded/poorly graded silty SAND
		Clayey SAND		S-C	SWC SPC			Well-graded/poorly graded clayey SAND
		Very silty SAND	SF	SM	SML, etc.	15 to 35		Very silty SAND; subdivided as for SC
		Very clayey SAND		SC	SCL SCI SCH SCV SCE			Very clayey SAND (clay of low, intermediate, high, very high, extremely high plasticity)
FINE SOILS (more than 35% of the material is finer than 0.06 mm)	Gravelly or sandy SILTS and CLAYS (35% to 65% fines)	Gravelly SILT	FG	MG	MLG, etc.			Gravelly SILT; subdivide as for CG
		Gravelly CLAY		CG	CLG CIG CHG CVG CEG		<35 35 to 50 50 to 70 70 to 90 >90	Gravelly CLAY of low plasticity of intermediate plasticity of high plasticity of very high plasticity of extremely high plasticity
		Sandy SILT	FS	MS	MLS, etc.			Sandy SILT; subdivide as for CG
		Sandy CLAY		CS	CLS, etc.			Sandy CLAY; subdivide as for CG
	SILTS and CLAYS (65% to 100% fines)	SILT (M-SOIL)	F	M	ML, etc.			SILT; subdivide as for C
		CLAY		C	CL CI CH CV CE		<35 35 to 50 50 to 70 70 to 90 >90	CLAY of low plasticity of intermediate plasticity of high plasticity of very high plasticity of extremely high plasticity
Organic soils		Descriptive letter O suffixed to any group or subgroup symbol			Organic matter suspected to be a significant constituent. Example MHO: Organic SILT of high plasticity			
Peat		Peat soils consist predominantly of plant remains which may be fibrous or amorphous						

*Extract from BS5930:1981 reproduced with permission of BSI.

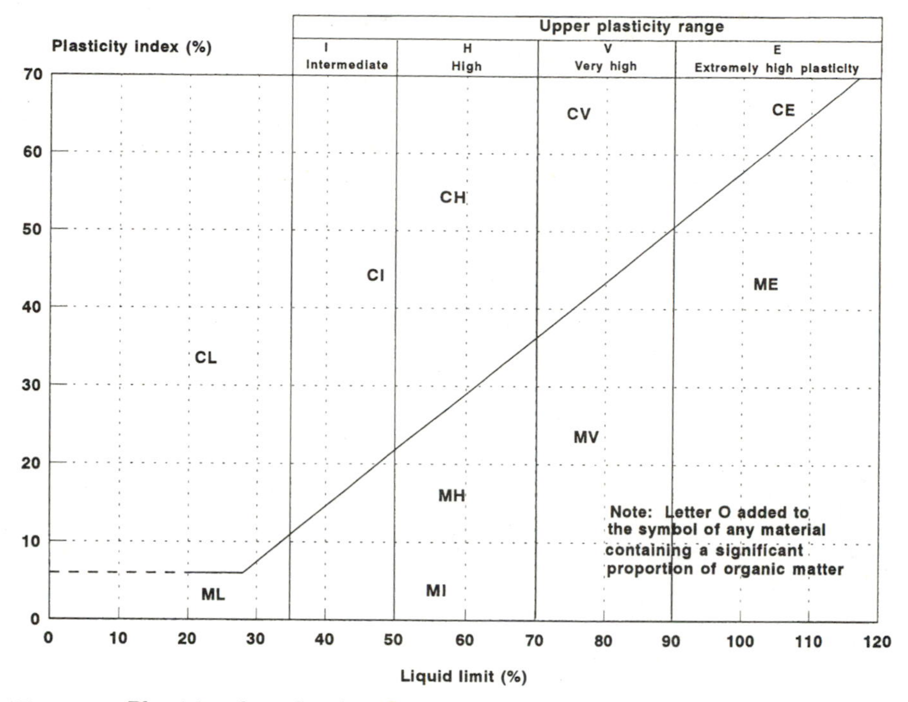

Figure 9.9 Plasticity chart for classification of fine soils and the finer part of coarse soils.

9.56. The form of the curves shown in Fig. 9.10*b* illustrates the basic properties of relationships of this type, viz., that the maximum dry density increases and the optimum moisture content decreases with increasing compactive effort.

Laboratory Compaction Tests

9.57. To provide an indication of the compactibility of soils in the field, standardized laboratory compaction tests were developed in the United States some 50 years ago. In the first of these tests the soil is compacted in three approximately equal layers in a cylindrical mold 101.6 mm (4 in) in diameter and 116 mm (4.58 in) in height. A rammer having a 51-mm (2-in) diameter end face and a weight of 2.5 kg (5.5 lb) is used to compact the soil. It is allowed to fall a distance of 305 mm (12 in) 25 times on each layer of the soil, a uniform distribution over the area of the mould being obtained as far as is practicable. The weight of the soil, after the top surface has been struck off level with the top of the mould, is obtained and the dry density is calculated from the volume of the mould and the measured moisture content of the soil. Tests are normally carried out at a range of moisture con-

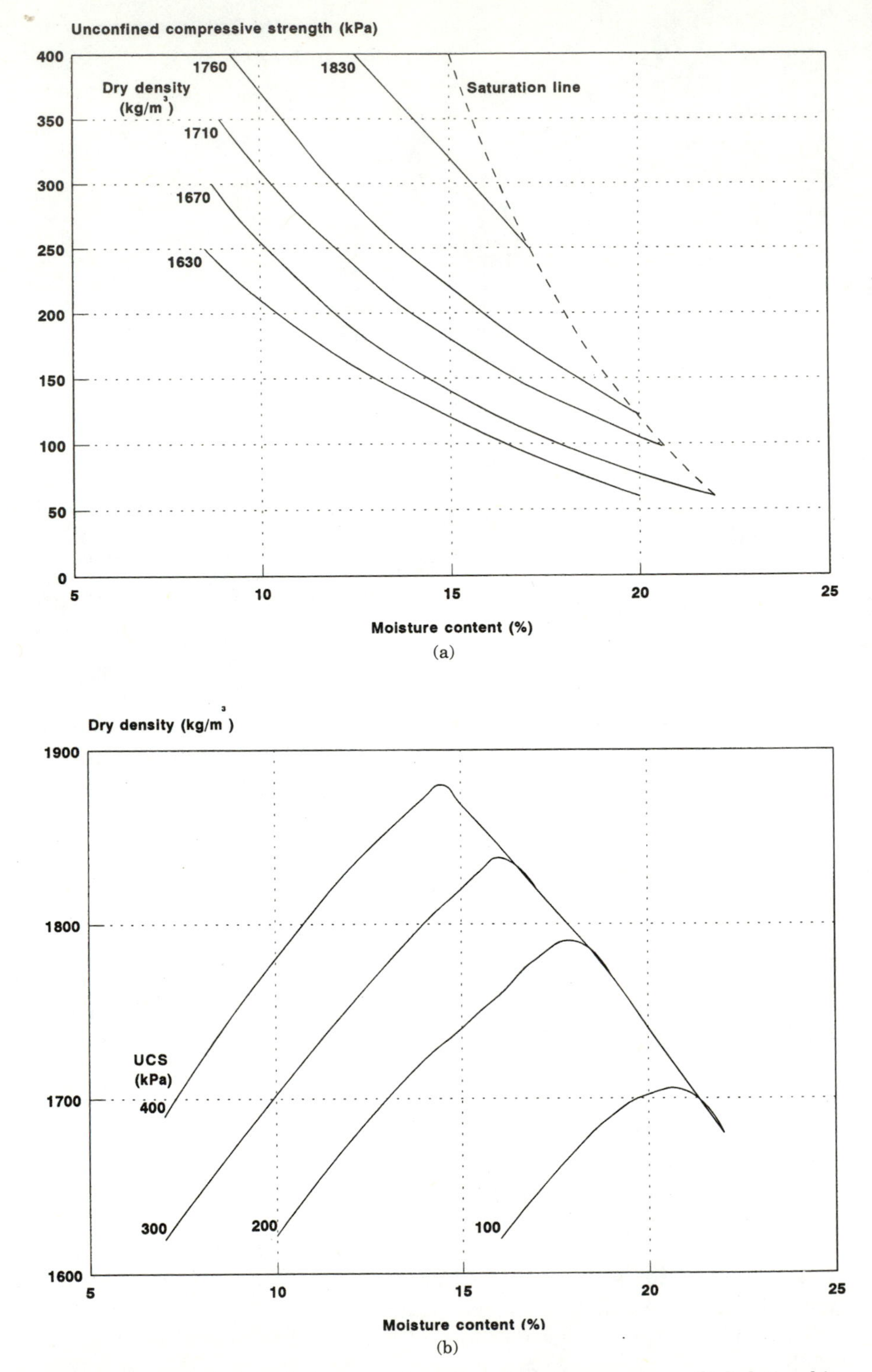

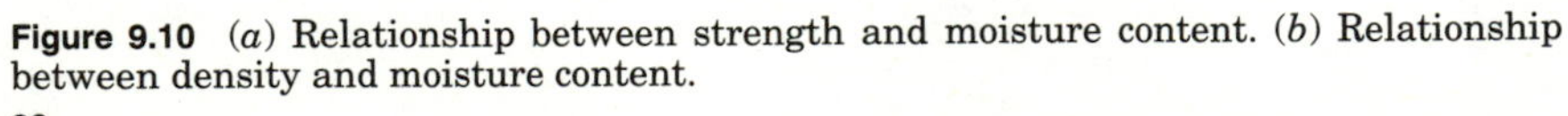
Figure 9.10 (*a*) Relationship between strength and moisture content. (*b*) Relationship between density and moisture content.

tents to give the density–moisture content relationship. This test was originally referred to as the Proctor test, after the originator, but it is now designated as AASHTO Test T99-86, as ASTM Test D698, or as British Standard 1377:1975, Test 12.

9.58. The advent of heavier compaction plant led to the need for a test which would produce greater densities. In this test the same size of mould is used but the soil is compacted in five layers using a rammer of weight 4.54 kg (10 lb), dropped 25 times on the surface of each layer from a height of 457 mm (18 in). This test is designated as AASHTO Test T180-86, as ASTM Test D1557, or as British Standard 1377:1975, Test 13.

9.59. The two compaction tests are generally distinguished by the adjectives "normal" and "heavy." As would be expected from Fig. 9.10*b*, the maximum dry density for the heavy compaction test is greater than for normal compaction and the optimum moisture content is lower. In the field the two compaction tests are carried out using hand rammers, but in the laboratory it is usual to use a mechanized form of the equipment.

9.60. The optimum moisture contents derived from these laboratory tests provide a useful indication of the range of moisture content suitable for field compaction, for most soils. However, a third compaction test has in recent years been introduced in the United Kingdom which relates laboratory and field compaction more closely in the case of granular soils with only a small silt and clay content. This is the vibrating hammer test in which the soil is compacted in three layers in a larger steel mould (152 mm in diameter and 127 mm high). The compaction equipment is a vibrating hammer operating at a frequency of 25 to 45 Hz. The hammer is applied to each layer of soil using a circular steel tamper of diameter 145 mm for a period of 60 s. Full details of the test procedure are given in BS 1377:1975, Test 14. This test is particularly useful in developing moisture content and density targets for granular road base and subbase materials.

The Performance of Compaction Plant

9.61. In the United Kingdom since the early fifties, new and existing compaction plant has been evaluated using four typical British soils having the gradings shown in Fig. 9.11. The relationship between dry density, moisture content, and the number of passes of the plant has been studied and comparison made with the results of the laboratory

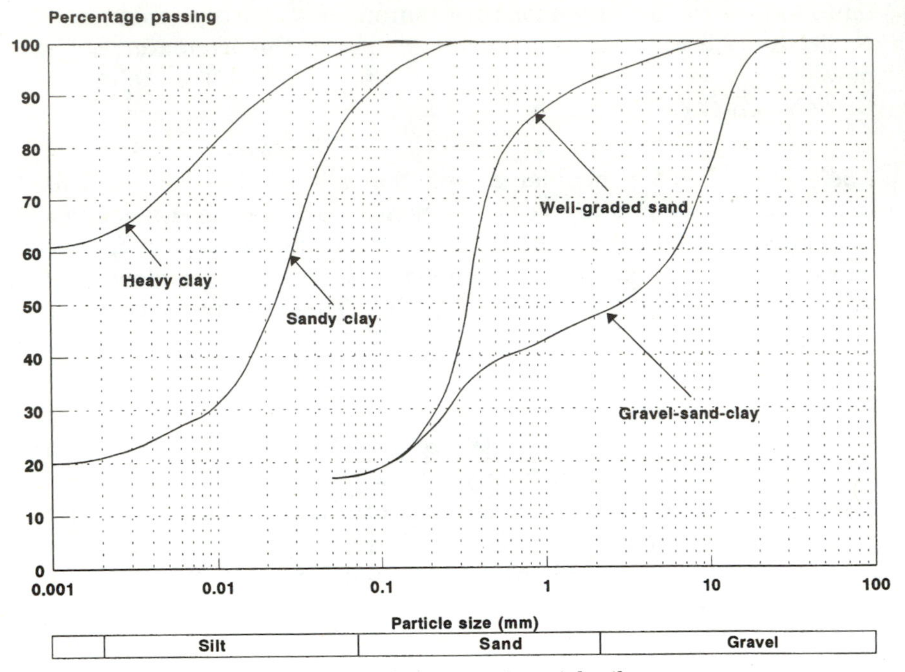

Figure 9.11 Particle size distributions for compaction trial soils.

compaction tests referred to in Pars. 9.57 to 9.60. Typical results are shown in Fig. 9.12*a* to *d*.

9.62. Derived from this work, Table 9.6 shows the number of passes and thickness of compacted layer necessary to give approximately 90 percent of BS "heavy" compaction (AASHTO "modified" compaction) at the field moisture content for a wide range of plant when compacting (1) cohesive soils, (2) well-graded granular and dry cohesive soils, and (3) uniformly graded soils and granular materials. The table refers to the natural moisture content of the soils under British conditions of climate.

9.63. Table 9.6 is reproduced from the U.K. Department of Transport Specification for Road and Bridge Works (1976 edition). In the 1986 edition the table is expanded to cover the compaction of granular subbase and road base materials. For the compaction of subgrades, as distinct from bulk earthworks, it is recommended that the number of passes of suitable plant should be doubled. With this type of specification, control on site is largely by observation of plant movements, but some density measurements in situ are desirable.

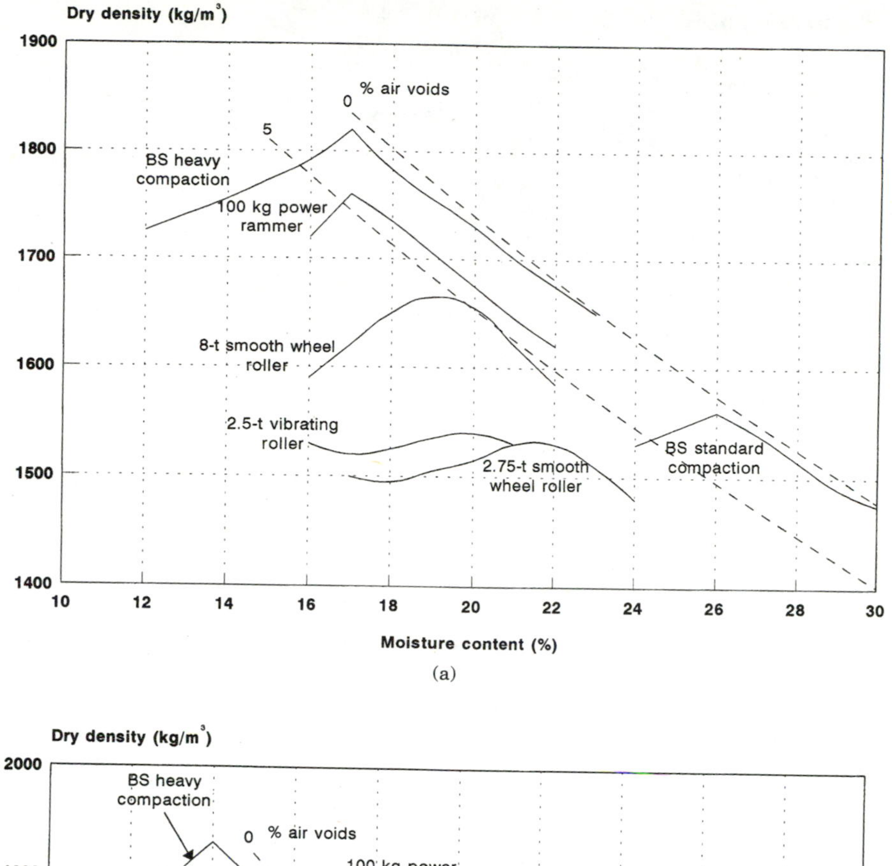

(a)

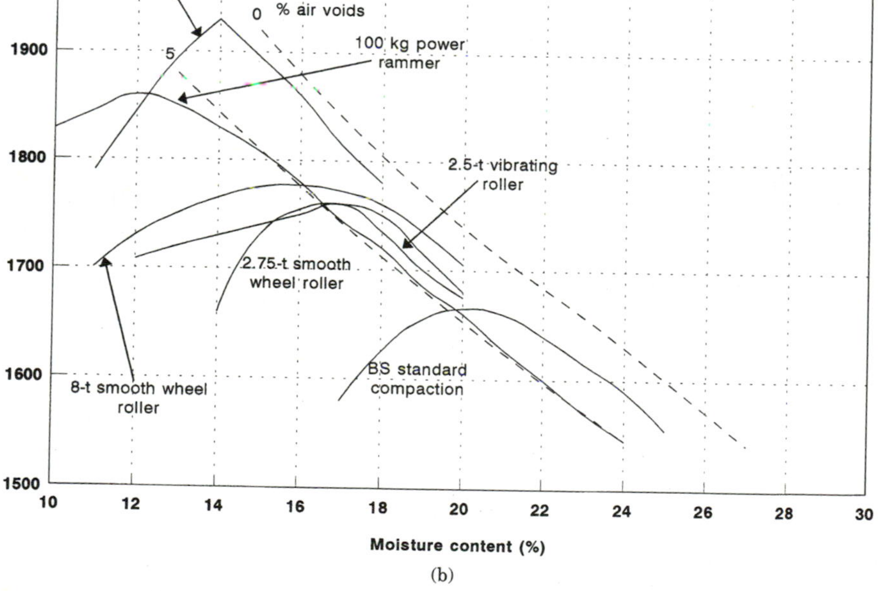

(b)

Figure 9.12 Compaction studies on (*a*) heavy clay, (*b*) silty clay, (*c*) sand, and (*d*) gravel-sand-clay.

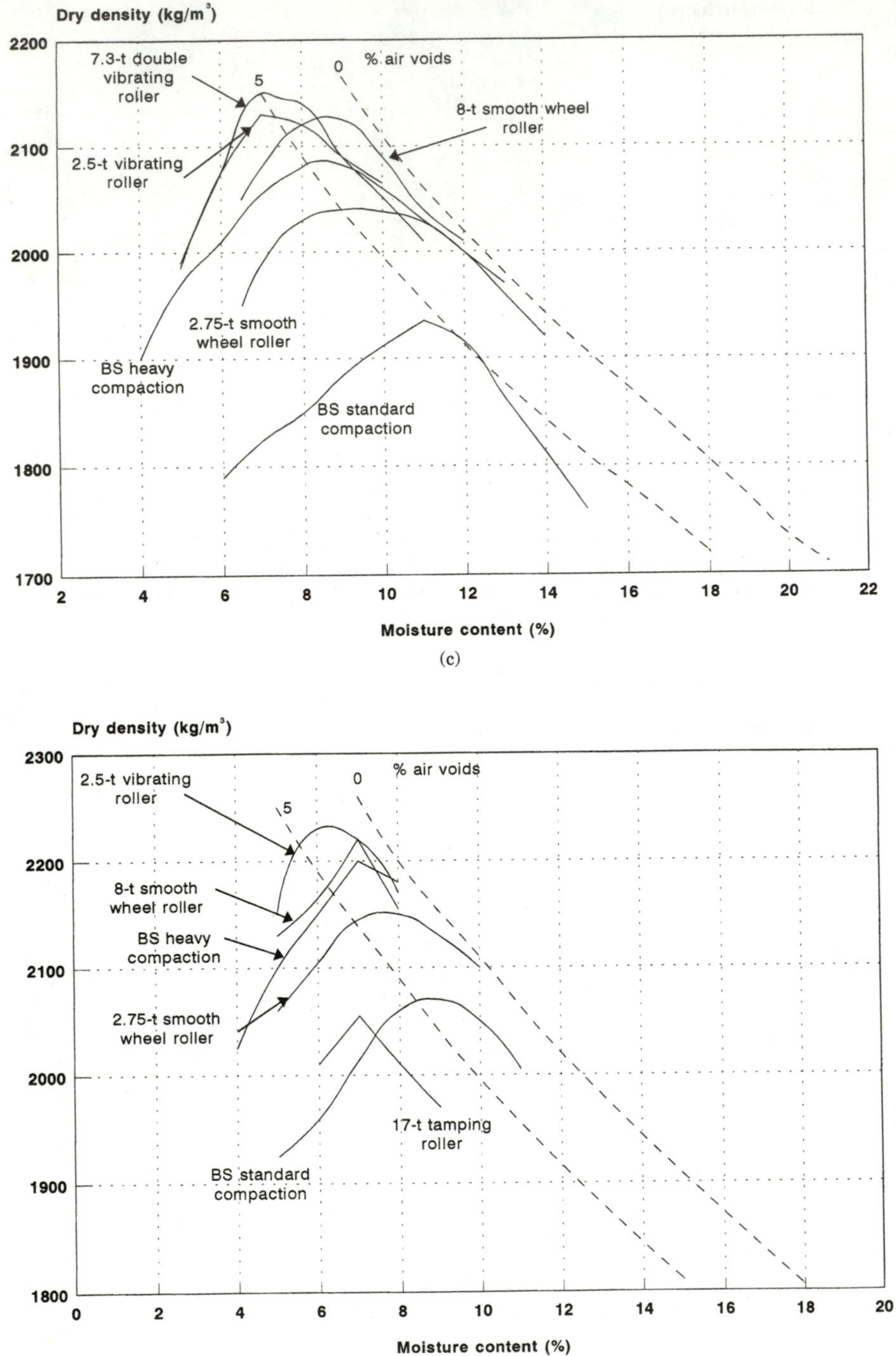

Figure 9.12 (*Continued*) Compaction studies on (*a*) heavy clay, (*b*) silty clay, (*c*) sand, and (*d*) gravel-sand-clay.

TABLE 9.6 Performance of Compaction Plant

Type of compaction plant	Category	Cohesive soils		Well-graded granular and dry cohesive soils		Uniformly graded material	
		D*	N*	D	N	D	N
Smooth-wheeled roller	Mass per meter width of roll:						
	Over 2100 up to 2700 kg	125	8	125	10	125	10†
	Over 2700 up to 5400 kg	125	6	125	8	125	8†
	Over 5400 kg	150	4	150	8	Unsuitable	
Grid roller	Over 2700 up to 5400 kg	150	10	Unsuitable		150	10
	Over 5400 up to 8000 kg	150	8	125	12	Unsuitable	
	Over 8000 kg	150	4	150	12	Unsuitable	
Tamping roller	Over 4000 kg	225	4	150	12	250	4
Pneumatic-tired roller	Mass per wheel:						
	Over 1000 up to 1500 kg	125	6	Unsuitable		150	10†
	Over 1500 up to 2000 kg	160	5	Unsuitable		Unsuitable	
	Over 2000 up to 2500 kg	175	4	125	12	Unsuitable	
	Over 2500 up to 4000 kg	225	4	125	10	Unsuitable	
	Over 4000 up to 6000 kg	300	4	125	10	Unsuitable	
	Over 6000 up to 8000 kg	350	4	150	8	Unsuitable	
	Over 8000 up to 12,000 kg	400	4	150	8	Unsuitable	
	Over 12,000 kg	450	4	175	6	Unsuitable	
Vibrating roller	Mass per meter width of a vibrating roll:						
	Over 270 up to 450 kg	Unsuitable		75	16	150	16
	Over 450 up to 700 kg	Unsuitable		75	12	150	12
	Over 700 up to 1300 kg	100	12	125	12	150	6
	Over 1300 up to 1800 kg	125	8	150	8	200	10†
	Over 1800 up to 2300 kg	150	4	150	4	225	12†
	Over 2300 up to 2900 kg	175	4	175	4	250	10†
	Over 2900 up to 3600 kg	200	4	200	4	275	8†
	Over 3600 up to 4300 kg	225	4	225	4	300	8†
	Over 4300 up to 5000 kg	250	4	250	4	300	6†
	Over 5000 kg	275	4	275	4	300	4†
Vibrating-plate compactor	Mass per unit area of base plate:						
	Over 880 up to 1100 kg	Unsuitable		Unsuitable		75	6
	Over 1100 up to 1200 kg	Unsuitable		75	10	100	6
	Over 1200 up to 1400 kg	Unsuitable		75	6	150	6
	Over 1400 up to 1800 kg	100	6	125	6	150	4
	Over 1800 up to 2100 kg	150	6	150	5	200	4
	Over 2100 kg	200	6	200	5	250	4
Vibro-tamper	Mass:						
	Over 50 up to 65 kg	100	3	100	3	150	3
	Over 65 up to 75 kg	125	3	125	3	200	3
	Over 75 kg	200	3	150	3	225	3
Power rammer	Mass:						
	100 up to 500 kg	160	4	150	6	Unsuitable	
	Over 500 kg	275	8	275	12	Unsuitable	
Dropping-weight compactor	Mass of rammer over 500 kg: Height of drop:						
	Over 1 up to 2 m	600	4	600	8	450	8
	Over 2 m	600	2	600	4	Unsuitable	

*D = maximum depth of compacted layer (mm); N = minimum number of passes.

†Indicates that the number of passes specified must be applied by a roller and towed by a track-laying tractor, not by self-propelled machine.

The Moisture Condition Value (MCV) of Soil

9.64. In Par. 9.54 it is stated in effect that if compactive energy is applied to a soil at a fixed moisture content then the strength of the soil will increase as a result of the increased density until the soil is incapable of mobilizing further strength at that moisture content. Any additional energy will then cause shear at constant volume and density. In field compaction, the effort expended in shearing the soil will be wasteful and in the case of clay soils it may destroy the soil structure and lead eventually to a long-term increase in moisture content and reduction in strength. It is therefore desirable that soils should not be overcompacted.

9.65. In the late seventies, A. W. Parsons, working in the United Kingdom, developed a simple field test which assesses the limit of compactibility of soils, which he termed the moisture condition value (MCV) test.[2,3] The test has proved most useful (1) in assessing the suitability of cohesive soils for use in earthworks, (2) in providing a rapid field method for determining moisture content during the progress of works without the need for oven drying, and (3) in assessing the trafficability of soil by civil engineering plant and vehicles. Since the test is comparatively new it is described in some detail here.

9.66. The equipment for the moisture condition test is shown in Fig. 9.13. The soil is compacted by a falling rammer as in the standard compaction tests discussed in Pars. 9.57 and 9.58. The difference is that in the MCV equipment the cylindrical rammer has a diameter of 97 mm and it operates in a heavy steel mould of diameter 100 mm and height 200 mm, with a plastic sealing disk 99 mm in diameter between the rammer and the soil. Significant shear of the soil after compaction is thus prevented by the confining conditions. The mass of the rammer is 7 kg and the height of the drop between the soil and the underside of the rammer is controlled at 250 mm using the Vernier scales fixed to the rammer and the side rails. The mass of soil used is normally 1.5 kg and it is broken down through a 20-mm sieve.

9.67. The soil to be tested is placed in the mould with the plastic disk lying on its surface. The mould is mounted in the machine and the rammer lowered to the surface of the soil; the penetration of the rammer into the mould is noted from the scale engraved on the rammer surface and the automatic release mechanism is adjusted to give a drop of 250 mm. The additional penetration of the rammer for various numbers of blows is recorded as indicated in the table above Fig. 9.14. The height of drop is corrected to 250 mm, as is necessary during the application of the blows.

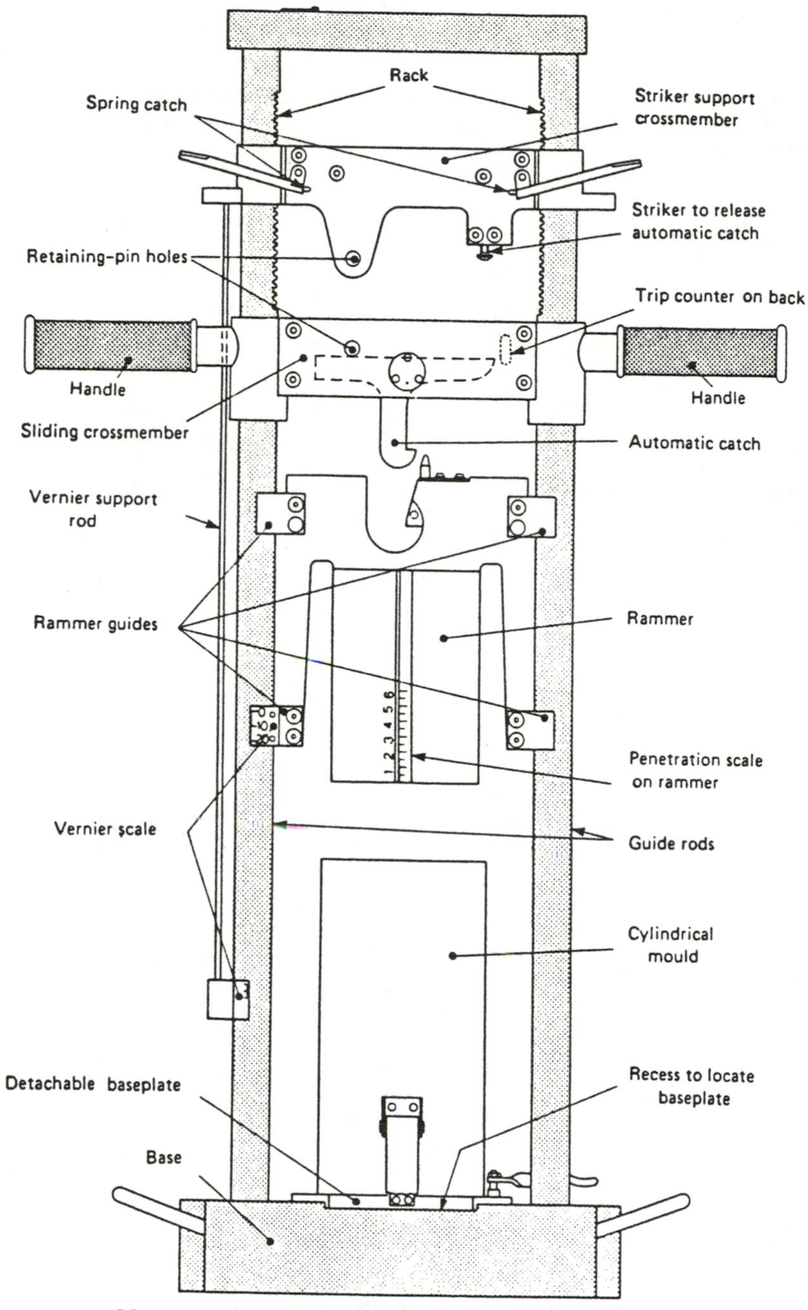

Figure 9.13 Moisture condition apparatus.

Soil: Heavy clay		Moisture content: 26.3 percent
Number of blows of rammer n	Penetration of rammer into mould, mm	Change in penetration with additional $3n$ blows of rammer, mm
1	41	33.5
2	57.5	33
3	67	33.5
4	74.5	26.5
6	84	17
8	90.5	10.5
12	100.5	0.5
16	101	
24	101	
32	101	
48	101	

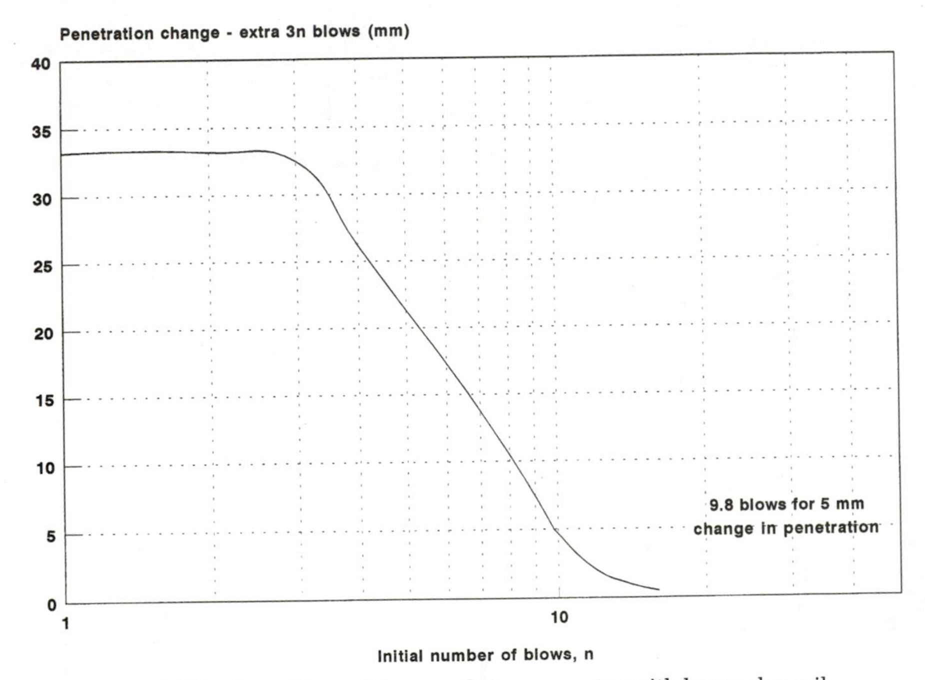

Figure 9.14 Calibration of the moisture condition apparatus with heavy clay soil.

9.68. In the interpretation of the test, the penetration of the rammer at any given number of blows is compared with the penetration for four times as many blows and the difference in penetration determined. This "change in penetration" is plotted against the lower number of blows in each case. This is shown in the table in Fig. 9.14 and in the graph, which shows the number of blows on a logarithmic scale. To avoid predicting the point in the plotted relation when the change in penetration reaches zero, a change in penetration of 5 mm has been arbitrarily accepted as indicating the point beyond which no significant change of density occurs. The MCV is defined as 10 times the logarithm (to the base 10) of the number of blows corresponding to a change of penetration of 5 mm on the plotted curve. In the example shown in Fig. 9.14 the number of blows for 5 mm change in penetration is 9.8; this would correspond to an MCV of $10 \times \log 9.8 = 9.9$.

9.69. Figures 9.15 and 9.16 show this procedure carried out for a heavy clay of liquid limit 70 percent and plastic limit 27 percent and for a sandy clay of liquid limit 40 percent and plastic limit 20 percent. In each case tests were made over a wide range of moisture contents.

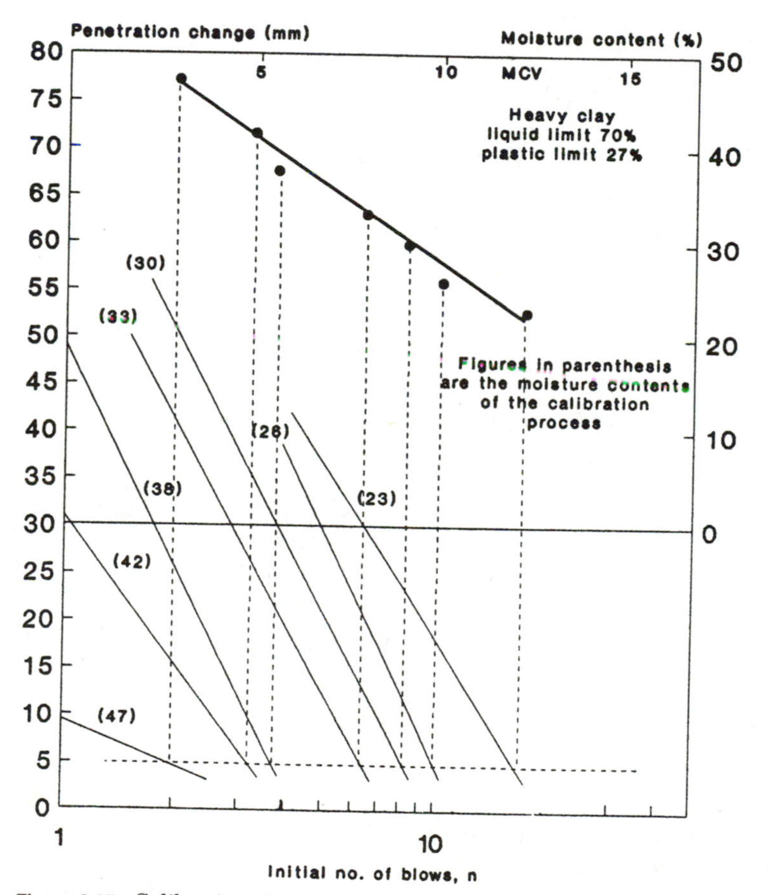

Figure 9.15 Calibration chart for the moisture condition apparatus with heavy clay.

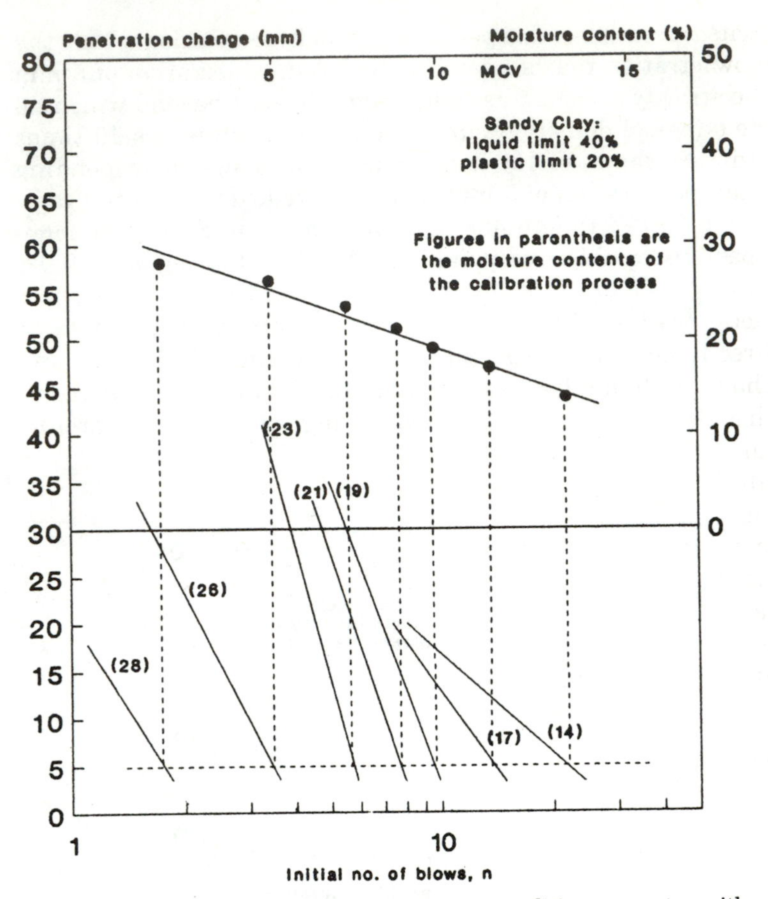

Figure 9.16 Calibration chart for the moisture condition apparatus with sandy clay.

The number of blows corresponding to a change of penetration of 5 mm gives MCVs which can be read off the scale shown at the top of the diagrams. Because the lower scale showing the initial number of blows is logarithmic, the MCV scale becomes linear. For both soils the relationship between the moisture content of the soil and the MCV is shown to be closely linear. This means that the MCV test can be used in the field as a very rapid method of checking the moisture content of the soil, since tests can be performed in a few minutes from the back of a site vehicle.

9.70. It is current British practice in the case of cohesive soils under U.K. climatic conditions to specify an upper limit of moisture content of 1.2 times the plastic limit at the time of compaction. For the two soils referred to in Figs. 9.15 and 9.16 this would correspond to moisture contents not greater than 32.5 and 24 percent, respectively. The figures show that for the heavy clay the minimum MCV would be 8 and for the sandy clay 6.2. For a site where a variety of clays were likely to be encountered, a mean value of 7 would be appropriate and this would be included in the contract documents to define the required state of compaction of the earthworks.

9.71. As part of the development of the MCV test procedure a study was made of the relationship between undrained shear strength c_u (determined by the vane test, see Par. 9.100) and MCV for the two soils referred to in Figs. 9.15 and 9.16.[2] The results shown in Fig. 9.17 confirm that there is a linear relationship between log c_u and MCV. Since both scales of the diagram are logarithmic, it follows that shear strength and number of blows required to give full compaction are linearly related. This confirms the statement implied in Par. 9.54 that compaction is effected by repetitions of a compactive effort until a shear strength sufficient to resist further compaction is created at the prevailing moisture content. This relationship between shear strength and MCV is important because it indicates that the MCV of cohesive soils is a reliable guide to their ability to support the passage of site traffic, i.e., to their trafficability.

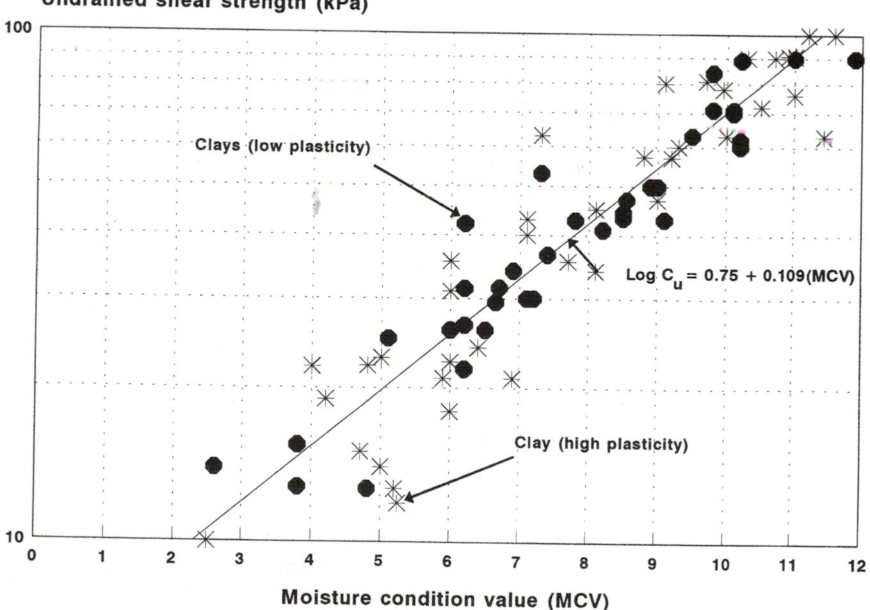

Figure 9.17 Relationship between undrained shear strength and MCV for clays of intermediate and high plasticity.

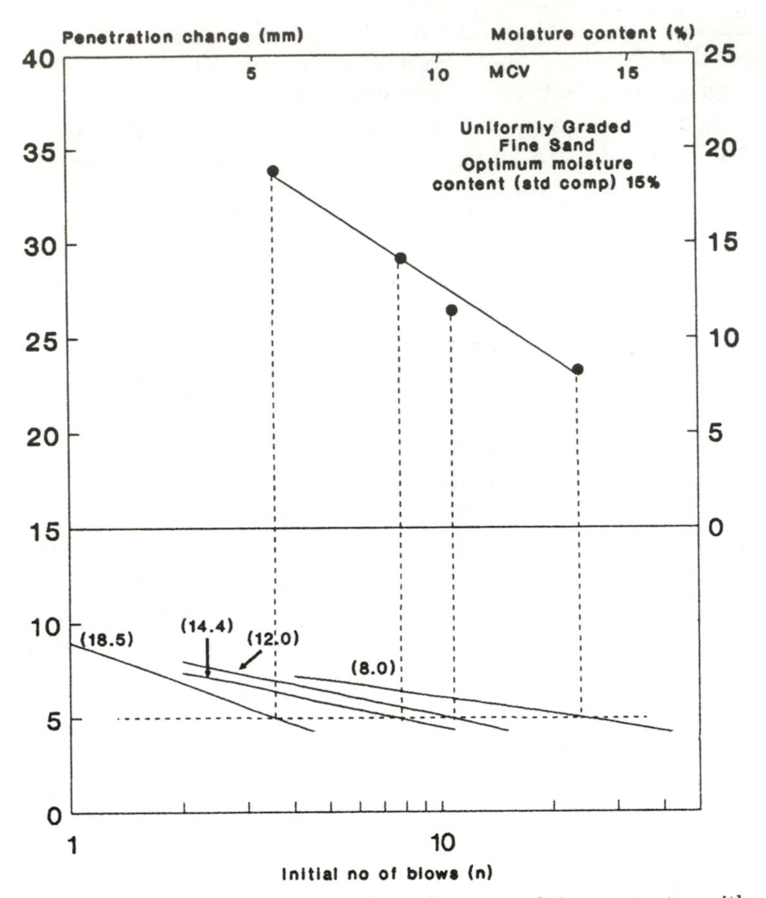

Figure 9.18 Calibration chart for the moisture condition apparatus with uniformly graded fine sand.

9.72. MCV tests have also been carried out on granular soils.[4] Figures 9.18 and 9.19 show results for a uniformly graded fine sand and for a well-graded sand. For such soils in the United Kingdom the upper limit of moisture content permitted in earthworks is 1.5 percent above the optimum moisture content for the standard compaction test (see Par. 9.57). For the two soils to which Figs. 9.18 and 9.19 refer, this would correspond to maximum moisture contents of 16.5 and 10.5, respectively. The diagrams show that the corresponding MCVs would be 7 and 8, i.e., very similar to those deduced for the cohesive soils. No data are given in the published papers for the relationship between shear strength and MCV for the granular soils, but it seems probable that the relationship would be very similar for all the soils.

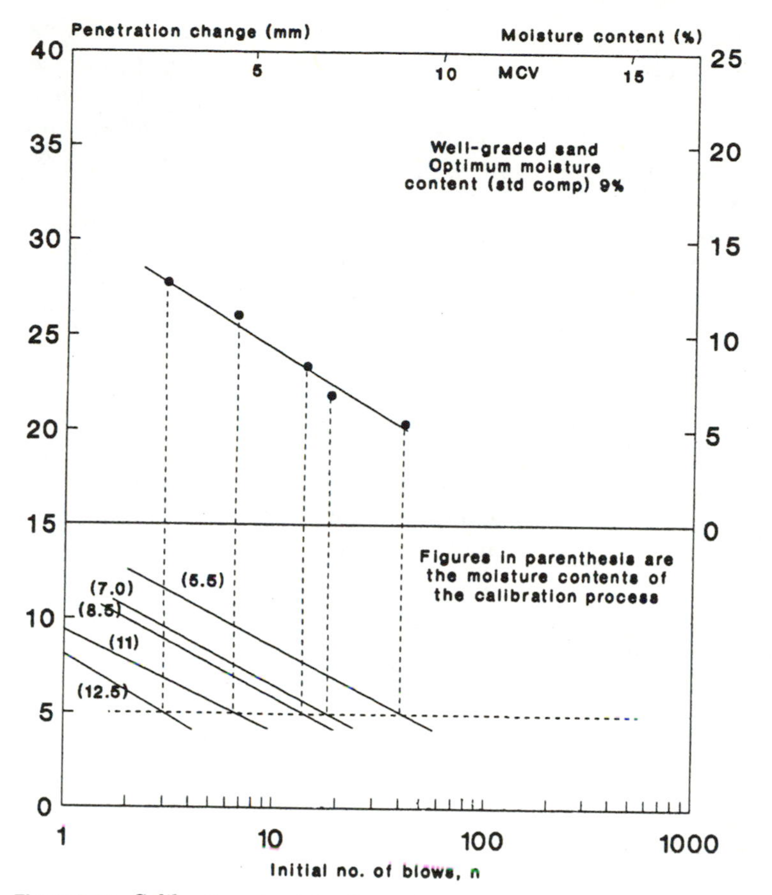

Figure 9.19 Calibration chart for the moisture condition apparatus with well-graded sand.

The Engineer's Responsibility for Defining the Suitability of Soils for Use in Earthworks

9.73. Certain materials such as peat, highly organic soils from swamps and bogs, materials susceptible to spontaneous combustion, and soils contaminated with hazardous chemicals should be excluded from earthworks. In wet climates it may also be prudent to exclude clays of very high liquid limit (e.g., >90 percent), which may prove impossible to handle under such circumstances. Normally only small quantities of soil unsuitable for these reasons will be involved in a major road contract.

9.74. In preparing the design and specification for a major road scheme the engineer must as far as is possible balance the quantities of cut and fill by the choice of the vertical alignment. At one time this was considered to be a primary requirement only in developed countries. It is now accepted that economic development in Third World countries depends much on the speed and capacity of trucks,[5] and permissible gradients are accordingly being reduced. The engineer has therefore for any contract to set limits of acceptability which will enable an experienced contractor to make use of the majority of the excavated material in the construction of the embankments. These limits will vary from contract to contract and they cannot therefore be defined in the form of a standard specification. In a wet climate, the best that a contractor can achieve is to move the soil from the embankment with no change of moisture content. In practice, the engineer will be wise to assume that there will be a small increase in moisture content of 1 to 2 percent involved in this operation. In a dry climate, loss of moisture may be a problem and the engineer may need to set a limit of moisture content which will involve artificial wetting of the soil during the compaction process.

9.75. The upper limit of moisture content which the engineer selects for fill will influence the permissible slopes of the embankments and the strength of the soil used in designing the pavement. These decisions have to be made at the design stage and emphasize the need for a thorough site investigation supported by an adequate test program on the materials involved.

9.76. A local-authority engineer working in a comparatively limited area soon develops expertise in specifying moisture content and density requirements for local soils. However, civil engineering is now international and engineers are increasingly putting in bids for design and supervision contracts in locations of which they have little or no experience. Under these circumstances it is the responsibility of the engineer to become thoroughly familiar with all aspects of the climate and geology of the site and to organize a site investigation which will give all the information necessary to the production of a workable earthworks specification. Failure to recognize this has led to some very expensive cases of litigation in recent years.

The Distribution and Movement of Water in Soil

9.77. The suction of soil. Water is held in soil by surface tension and adsorption forces, which impart to it a negative pressure or suction with respect to atmospheric pressure.[1] The magnitude of this suction increases rapidly as the soil dries out to give a relationship between

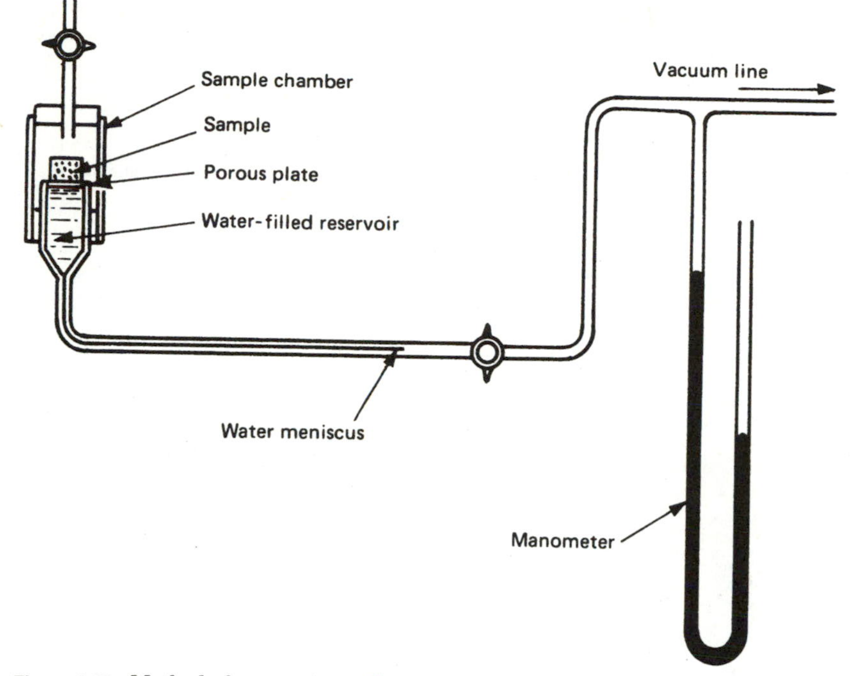

Figure 9.20 Method of measuring soil suction.

suction and moisture content which can be investigated at low suctions by the apparatus shown in Fig. 9.20. A small sample of the soil is placed on a ceramic plate sealed into a water-filled reservoir, connected to a flow tube to which a vacuum can be applied. The ceramic plate has a porosity which allows water to pass, but not air, at the range of suctions used. In this way the suction of the soil moisture is balanced by the applied suction, and, by testing soils over a range of moisture contents, the relation between the two factors is explored. Typical suction–moisture content relations for granular and cohesive soils are shown in Fig. 9.21. The soils were tested in the drying condition.

9.78. Pore water pressure and suction. If the surface of soil in the field is covered to prevent the effects of rainfall and evaporation an equilibrium moisture distribution is reached with respect to the position of the water table. The pore water pressure which can be measured directly by piezometers is zero (i.e., atmospheric) at the level of the water table, and increases linearly with depth below the water table. Above the water table there is a similar linear decrease of pore water pressure, which means that it is negative with respect to atmospheric pressure. The soil water at a particular depth is also

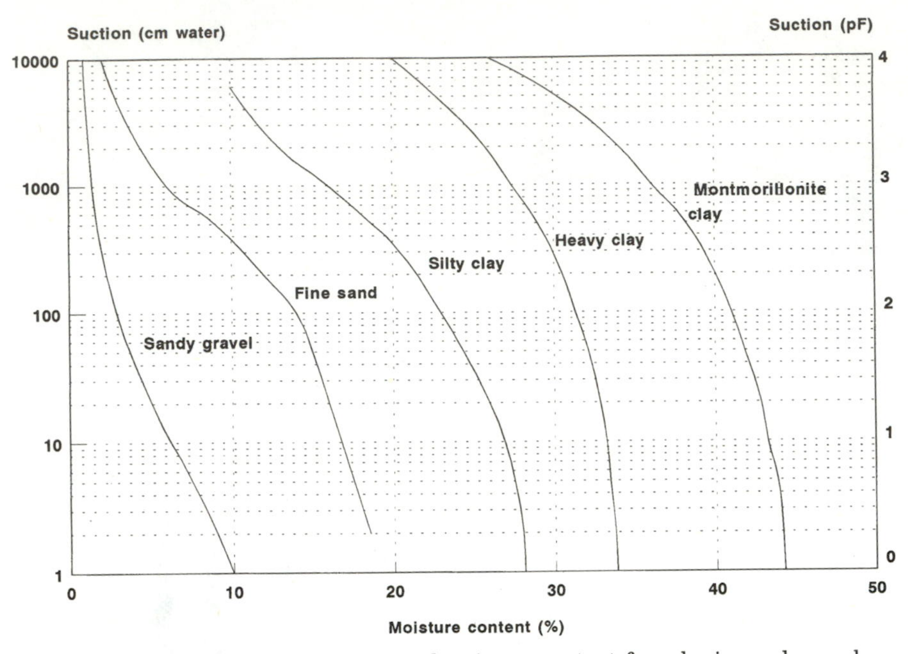

Figure 9.21 Relation between suction and moisture content for cohesive and noncohesive soils (drying condition).

subject to overburden pressure from the wet soil above. In heavy clay soils the whole of the overburden pressure acts on the soil water, but in granular soils intergranular contacts support the overburden and no part is carried by the soil water. For intermediate soil types part only of the overburden pressure is imparted to the soil water. The suction of the soil, as defined above, acts to retain the water in the soil and the overburden pressure acts to exclude it. Thus if u is the pore water pressure, s is the suction, P is the overburden pressure, and α is the proportion of the overburden pressure acting on the soil water (the compressibility factor), then

$$u = s + \alpha P \tag{9.15}$$

The suction s will be negative and the overburden pressure P will be positive. Above the water table, s will numerically exceed αP and the pore water pressure will be negative. Below the water table the reverse will be the case. In heavy clays all the overburden pressure is carried by the soil water and $\alpha = 1$. In purely granular soils all the overburden pressure is taken by intergranular contacts and $\alpha = 0$. For other soils its value is between 0 and 1, depending on the plasticity index as indicated in Table 9.7.

TABLE 9.7 Relationship between Compressibility Factor and the Plasticity Index of Cohesive Soils

Plasticity index, %	Compressibility factor α
10	0.15
15	0.27
20	0.40
25	0.55
30	0.70
35	0.80
>35	1.00

9.79. The density of wet soil is close to twice the density of water. Thus if u and s in Eq. (9.15) are also expressed in centimeters of water, the equation can be used in conjunction with the appropriate soil suction curve to deduce the equilibrium moisture content above the water table. Such calculations have been made in Fig. 9.22 for all the five soils referred to in Fig. 9.21 assuming the water table to be at a depth of 2 m below the surface. For the sands α was assumed to be 0 and for the two heavy clays a value of 1 was taken. For the silty clay 0.5 was assumed.

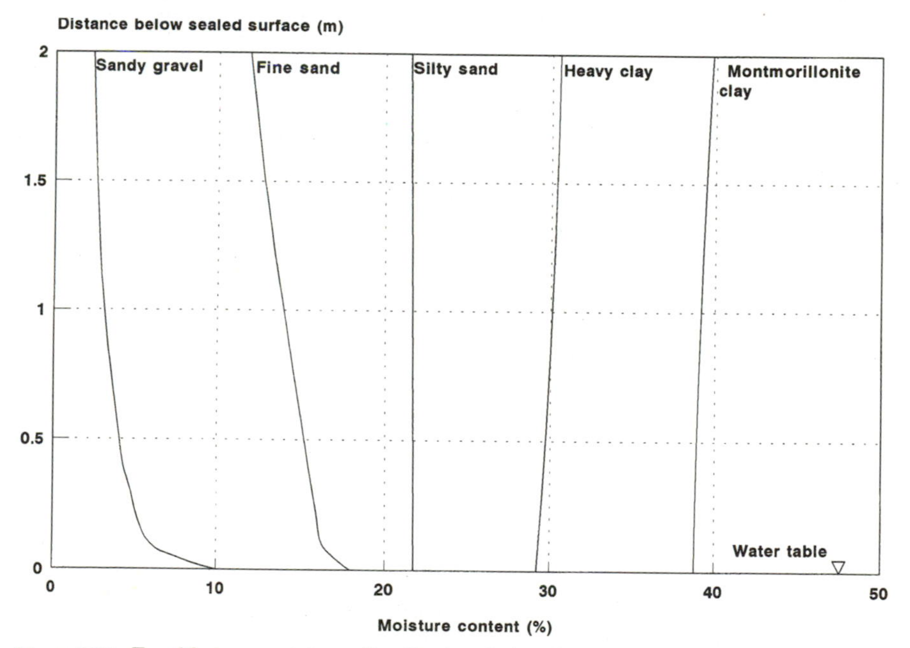

Figure 9.22 Equilibrium moisture distribution deduced from suction curves in Fig. 9.21 (water table depth 2 m).

9.80. For the sands the moisture content increases with depth as the level of the water table is approached. With the heavy clays the moisture content decreases with depth owing to compression of the soil structure, while for the silty clay the moisture content is substantially constant with depth. These moisture distribution curves show that soils with very different moisture contents would be able to coexist in close proximity to one another.

Subsoil Drainage

9.81. It is important to realize that subsoil drains placed above the level of the water table cannot effect a reduction in moisture content. Such a reduction can be achieved only by lowering the water table. The suction–moisture content relationship for a soil in conjunction with Eq. (9.15) can be used to calculate the effect of lowering the water table on the moisture distribution. Figure 9.23 shows the effect of lowering the water table in the heavy clay referred to in Figs. 9.21 and 9.22 from the surface to depths of 1 and 2 m. The first increment produces a substantial reduction of moisture content close to the surface but for the second increment the effect is much smaller. In gener-

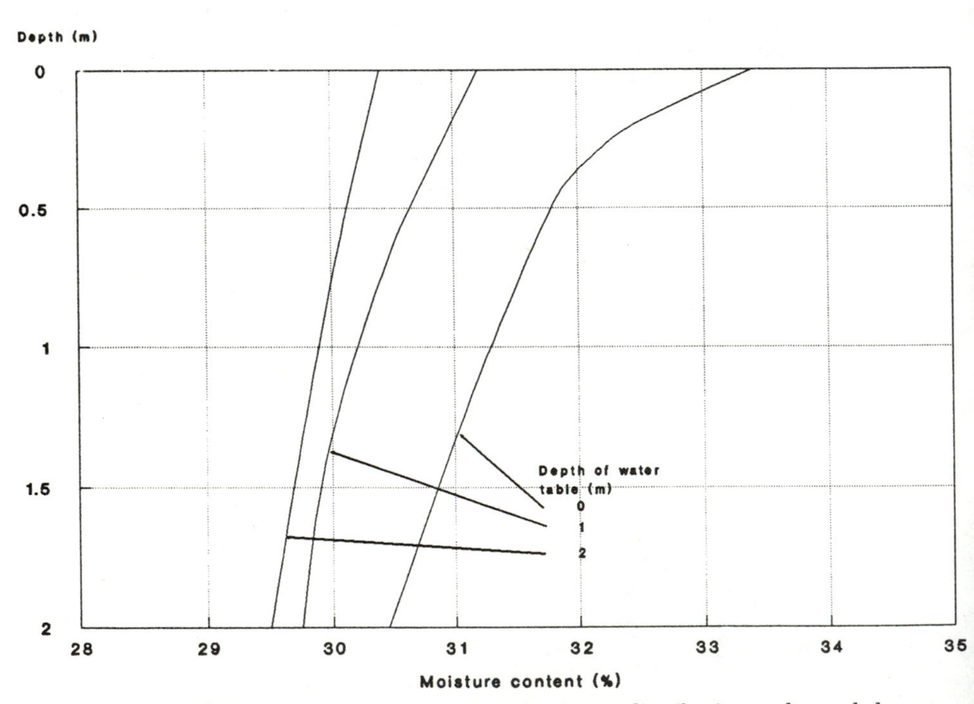

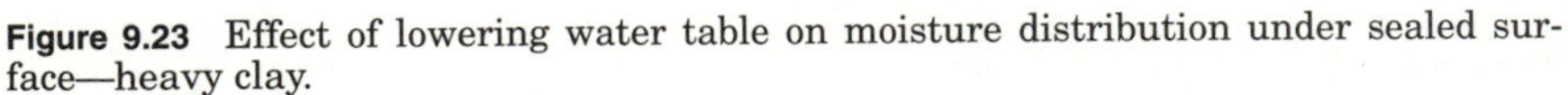
Figure 9.23 Effect of lowering water table on moisture distribution under sealed surface—heavy clay.

al, subsoil drainage needs to be considered if there is any danger of the water table rising closer to the surface than 1 to 1.5 m.

9.82. Various methods of calculating the draw-down effect of subsoil drains laid in road verges are available.[5] They depend on a knowledge of the permeability of the soil which in practice is likely to be variable because of fissuring. Figure 9.24 shows actual measurements of draw-down from side drains placed close to the edge of a concrete pavement built in a very high water table situation, on a heavy clay soil. In this case side drains 1.5 m deep would have been required to ensure that the water table was maintained at least 1 m below the underside of the pavement. To prevent silting up of the subsoil drainage system by fines washed in from the surrounding soil, the pipes should be laid in a selected filter material. This should be continuously graded and the grading curve should satisfy the following requirements:

$$\frac{\text{15 percent size of filter material}}{\text{85 percent size of the subgrade soil}} < 5$$

$$\frac{\text{15 percent size of filter material}}{\text{15 percent size of the subgrade soil}} > 5$$

(By the 15 percent size is meant that size of particle corresponding to the 15 percent ordinate of the particle size distribution chart; see Fig. 9.1.)

Consolidation of Clay Soils

9.83. It follows from Par. 9.78 that if a heavy clay soil in moisture equilibrium with a water table is loaded at the surface then the over-burden pressure at any point within the soil will be increased by the loading pressure applied and as a consequence the suction will be increased in accordance with Eq. (9.15), and the moisture content will

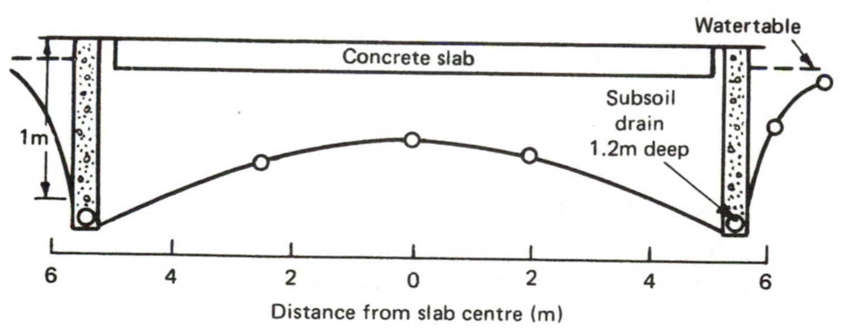

Figure 9.24 Measured draw-down curve in London clay.

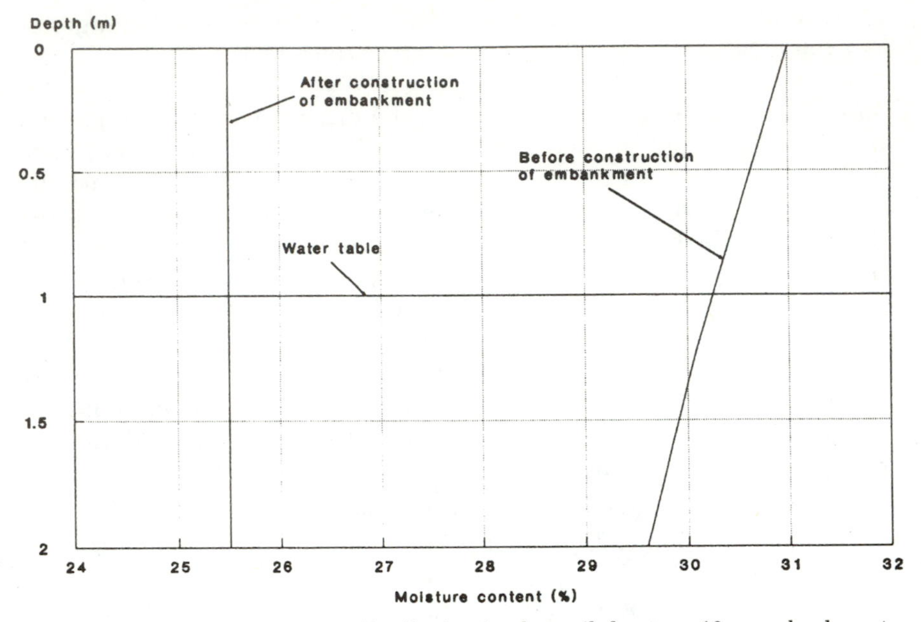

Figure 9.25 Change in moisture distribution in clay soil due to a 10-m embankment construction.

be decreased by an amount which can be deduced from the moisture content–suction curve. This process is termed consolidation. Taking again the example of the heavy clay referred to in Fig. 9.23, if an embankment 10 m high were constructed on the surface of the soil, with the water table maintained at a depth of 1 m, then the average suction in the soil would increase to 2200 cm of water, and Fig. 9.25 indicates a fall in moisture content from about 31.0 to 25.5 percent. The corresponding reduction of 5 percent would result in a settlement of about 200 mm in the 2 m of soil.

9.84. Consolidation can be estimated more simply from the laboratory consolidation test in which a disk of the soil mounted between ceramic porous plates is progressively loaded and the change of thickness measured. Full details of the method are given in AASHTO T216-83, ASTM D2430-80, and BS 1377:1975, Test 17. An indication of the rate of consolidation can also be obtained from this test by studying the consolidation–time relationship. However, such one-dimensional tests generally overestimate the settlement time when compared with field observations. In the case of embankments placed on compressible soils, most of the settlement occurs during the construction phase and settlement after paving is small. However, it may become apparent where embankments meet piled structures such as bridge abutments.

Surcharging is sometimes used to accelerate settlement at such points. The construction of cuttings in heavy clay soils results in a release of overburden, and some upward movement due to swelling of the soil must be expected. This can affect measurements during setting out.

The Effect of Climate on the Moisture Distribution of Soil

9.85. Influence of rainfall and evaporation. The two factors evaporation and rainfall largely determine the moisture condition of a soil in the field. Figure 9.26, which relates *a* and *b* to an area of grassland in southern England, close to London Airport (Heathrow), illustrates this point. For a number of years in the fifties the moisture condition of the silty clay, both exposed and beneath concrete pavements, was studied in relation to climatic factors and the level of the water table.[1] The results shown refer to two consecutive years, 1954 and 1955. The first was one of unusually high summer rainfall and the second was normal in this respect. Throughout 1954, rainfall exceeded evaporation except for a short period between mid-June and mid-July, there was very little seasonal drying, and the water table remained at a depth of 3 m. During 1955, evaporation exceeded rainfall during the period June to September and the moisture content of the top 0 to 600 mm of soil was reduced by about 12 percent and at a depth of 900 to 1200 mm the reduction was 3 percent. This reduction of moisture content was accompanied by a fall of the water table of about 300 mm. This illustrates that even in temperate climates the annual moisture balance between rainfall and evaporation can have a major effect on earthwork construction and subgrade preparation. Figure 9.27 shows the distribution of moisture content with depth in February 1955 and September 1955 under grass cover and for comparison is shown the distribution in November 1955 under an adjacent concrete slab laid several years earlier. A similar distribution was reached under another concrete slab laid when the soil was very dry, as in September 1955. However, in this case the change in moisture content to the equilibrium condition was very slow and was completed only after 5 to 6 years. This was attributed to the very low permeability of the dry soil.

9.86. In an attempt to generalize the conclusions from this work to a wide range of climatic conditions a concurrent program of tests was started at 10 airfields widely distributed in tropical and semitropical situations. Monthly measurements of moisture content were made at different depths both in uncovered soil and under runways or taxiways for a period of about 2 years.[6] Details were obtained of rainfall, evaporation, and temperature at all the sites. It was found that where the

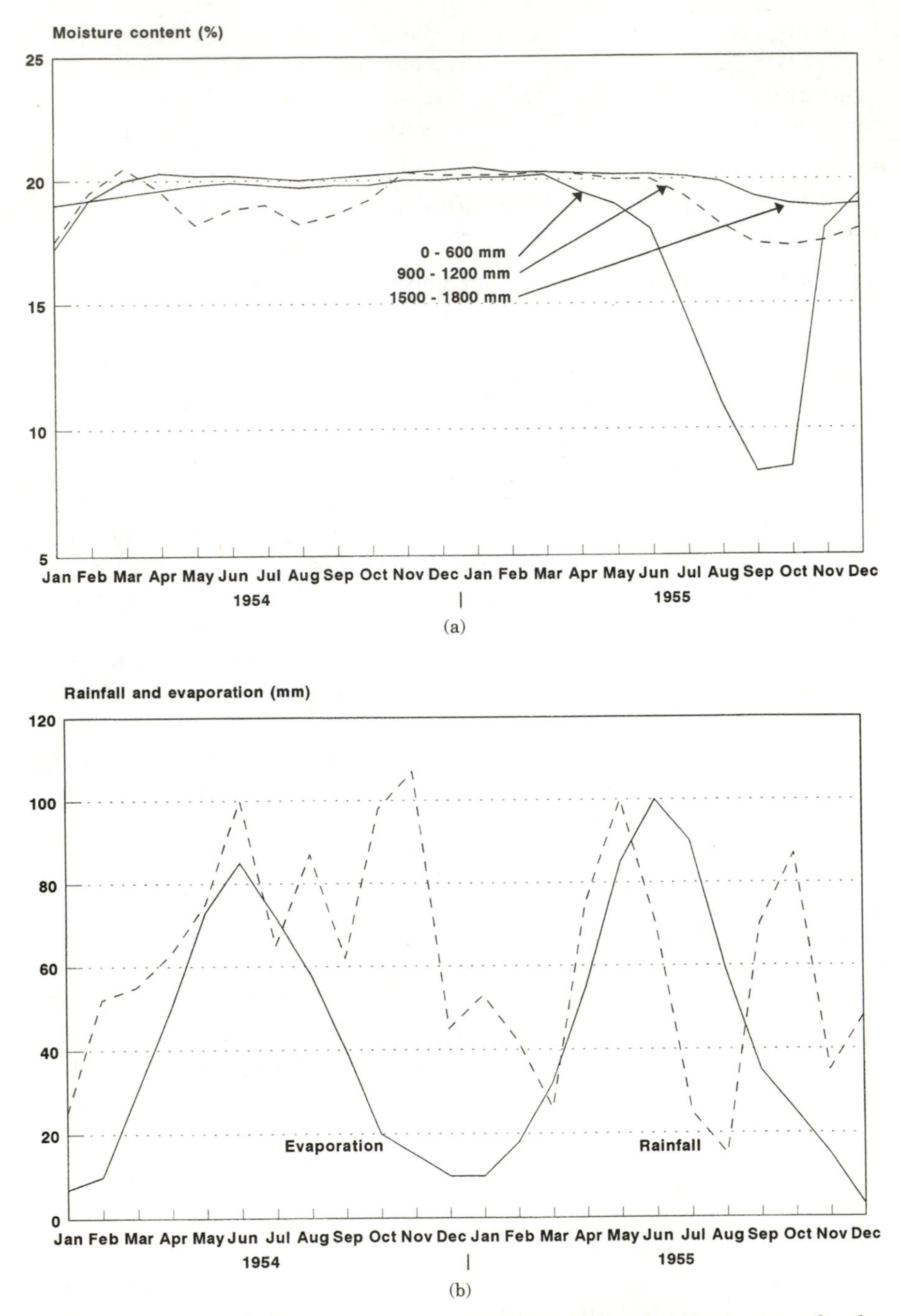

Figure 9.26 (*a*) Influence of moisture balance on the moisture distribution in a silty clay soil. (*b*) Influence of evaporation and rainfall on the moisture balance at the same site.

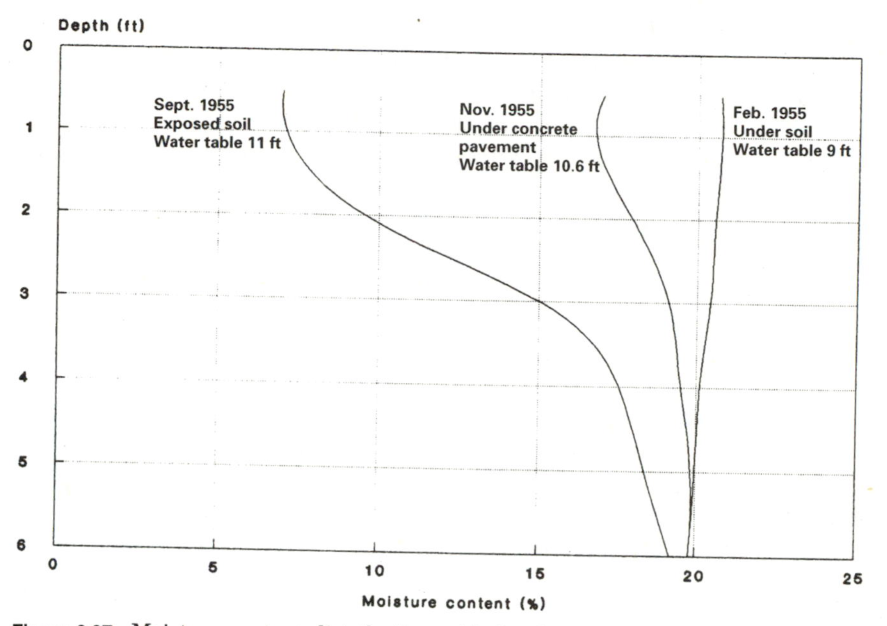

Figure 9.27 Moisture content distribution with depth in silty clay soil at a site near London Airport (Heathrow).

water table was within 10 m of the surface the theoretical methods of calculating the equilibrium moisture distribution under the pavement could be applied, and the moisture distribution in uncovered soil was determined by the moisture balance between rainfall and evaporation. Where the water table was deeper, generally in very arid areas, the moisture distribution with depth was similar under the pavement and in the uncovered soil, and both appeared to be controlled by the atmospheric humidity. Two cases of a high water table (1 to 2 m) were found where the rainfall was negligible. In these cases the high water table was found to arise from the proximity of the sea or of a river. No conclusive evidence emerged to indicate a buildup of moisture beneath pavements as a result of water vapor movements caused by temperature differences. The depth of the water table affected the magnitude of "edge effects" resulting from the migration of water from the verges of paved areas and confirmed the importance of impermeable shoulders in maintaining equilibrium moisture conditions under pavements.

Frost Heave in Soils

9.87. Some soils and certain granular materials are susceptible to frost heave as the zero isotherm passes through the road structure into the subgrade. This heave, particularly if it is not uniform over the area of the pavement, is likely to cause cracking. However, more important is the fact that after the thaw the pavement foundation may be left in a very weak state and be subject to rapid breakup under traffic. The thermodynamics of the freezing process in moist porous materials are now fairly clearly understood.[7] At temperatures above freezing point, the water in such materials has a negative pressure or suction which results from the surface tension and adsorption forces by which the water is retained (see Pars. 9.77 to 9.78). This suction increases rapidly with decreasing moisture content, and for this reason it is often expressed in terms of the logarithmic pF scale on which the common logarithm of the suction expressed in centimeters of water is equivalent to the pF value. If the temperature of the material is reduced a little below the freezing point of free water, water within the pores freezes until the suction of the water left unfrozen rises to a value which inhibits further freezing at that temperature. There is thus a relation between the temperature depression below 0°C and the suction of the unfrozen water. This relation, in terms of the pF scale, is

$$\text{pF} = 4.095 + \log t \tag{9.16}$$

where t is the temperature depression. This relationship is shown in Fig. 9.28.

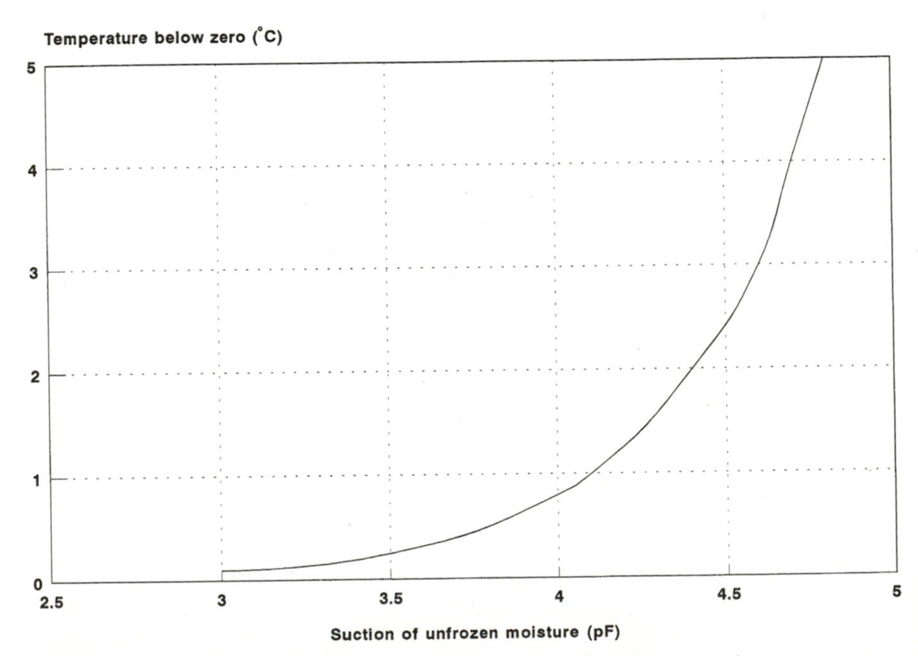

Figure 9.28 Relationship between freezing point and soil moisture suction.

9.88. It follows that if soil or any other porous material in hydrostatic equilibrium with its surroundings is affected by local freezing then the equilibrium will be disturbed and there will be a tendency for water to move toward the freezing zone, defined as the zone in which temperatures are below 0°C. The significance of the moisture movement can best be discussed in terms of the physical properties of typical road foundations.

9.89. Figure 9.29*a* shows the distribution of temperature with depth in late January 1963, beneath a concrete pavement in southern England. The water table at this site was at a depth of about 1.8 m, and prior to freezing the negative pore-water pressure of the soil below the pavement was in approximate equilibrium with this level of water table. The equilibrium suction expressed on the pF scale is shown in Fig. 9.29*b*. The relationship shown in Fig. 9.28 can be used to estimate the effect on the pore-water pressure distribution of the temperature gradient of Fig. 9.29*a* as shown in Fig. 9.29*b*. Above the level of the zero isotherm there is a rapid increase in suction to a value between 10 and 100 times that below the freezing zone. This suction gradient will tend to draw water from the unfrozen soil into the freezing zone to form ice lenses and give rise to frost heave. The distribution of suction below the freezing zone shown in Fig. 9.29*b*

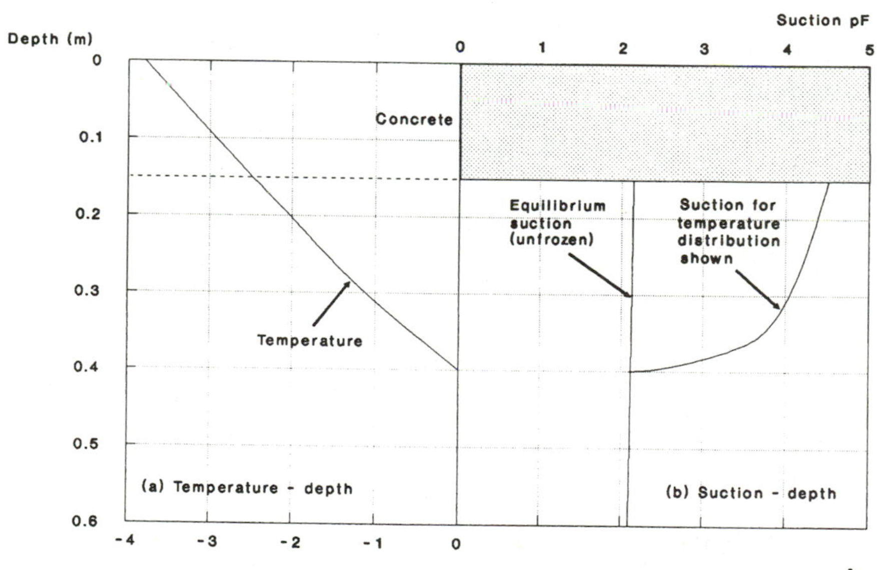

Figure 9.29 Temperature distribution and suction beneath a concrete road, January 1963.

would of course be modified temporarily by any flow of moisture arising from the suction gradients.

9.90. In the frost heave process the suction developed in the freezing zone provides the necessary pressure gradient to cause moisture flow. The rate at which the associated moisture migration occurs depends on the resistance to flow and hence on the prevailing permeability. Two permeabilities are involved: that of the unfrozen material through which the water must pass and that of the freezing front, defined arbitrarily as the thin boundary layer of the frozen zone. The exact mechanism involved in the formation of ice lenses is still a matter of conjecture. However, it is generally agreed that ice crystals form in the soil pores close to the boundary of the frozen zone and that these crystals grow from water drawn through what has been defined here as the freezing front. If the permeability of the unfrozen material is too low to permit any significant movement of water under the pressure gradient prevailing, heave cannot occur. Within the freezing front the permeability is controlled by the amount of water unfrozen at the prevailing temperature. Figure 9.21 shows relationships between suction and moisture content for two granular and three clay soils. For the suction between pF 3 and pF 4 prevailing in the proximity of the zero isotherm the heavy clay would contain about 25 percent of unfrozen water, the silty clay 12 percent, and the sand about 2 percent. The very small amount of unfrozen water in clean granular soils renders those materials virtually impermeable to moisture flow through the freezing front, so that frost heave in such materials is inhibited. In heavy clays, despite the relatively large proportion of unfrozen water, the natural permeability is in general too low to allow significant upward migration of water through the material during the relatively short periods of freezing associated with British winters. It is important to realize that frost heave can only continue while the zero isotherm is moving downward. Once downward movement has ceased additional heave would lift the zero isotherm back into the frozen zone. In Britain the maximum depth of frost penetration is likely to be reached in a few days. In the northern United States it is shown in Table 6.1 that frost penetration to a depth of 1 m is not uncommon. This depth of penetration will occur over several months during which progressive heave will occur where the soil is frost-susceptible.

9.91. It would be expected that if clay fines were added to a clean granular material then the liability of the composite material to frost heave would increase up to an optimum fines content. Beyond this point the addition of more clay would decrease the permeability in the

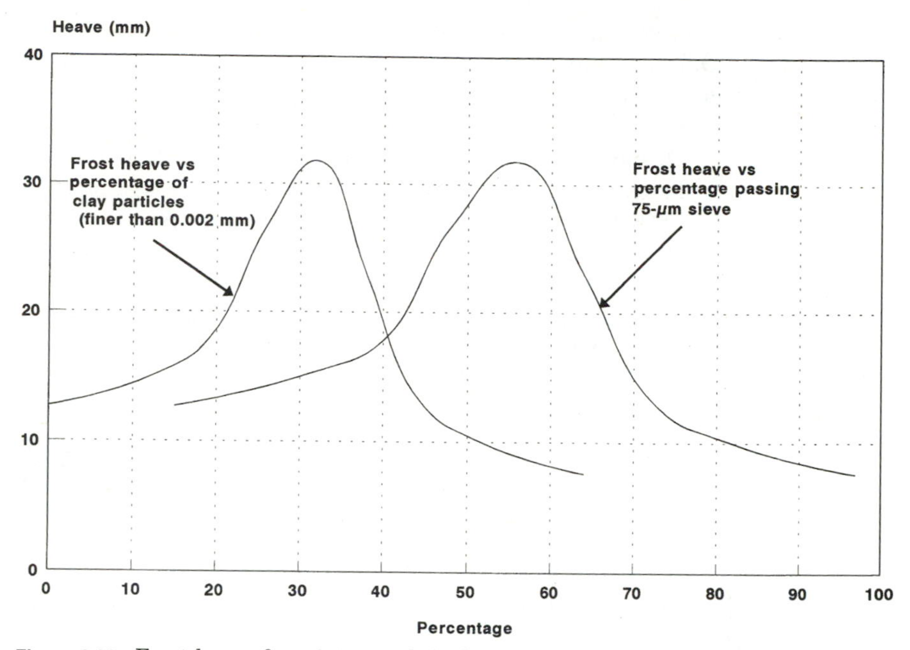

Figure 9.30 Frost heave for mixtures of single-size sand and heavy clay in various proportions.

soil below the zero isotherm and cause a progressive reduction of heave. The results of experiments to verify this are shown in Fig. 9.30. A clay soil, with a clay content of 60 percent and liquid and plastic limit values of 78 and 26 percent, respectively, was dried and ground and mixed with a fine, single-size sand to give the required proportions. Moisture was added during the mixing and the samples were compacted to the maximum dry density and optimum moisture content for standard compaction (2.5-kg rammer). The samples were tested for frost heave using the procedure described below. The heave shows a marked peak when plotted either against the percentage in the mix passing the 75-μm sieve or the percentage of clay particles (finer than 0.002 mm).

The British Test for Frost Susceptibility

9.92. The test used in Britain to assess frost susceptibility[8] is based on one originally developed in the United States by Taber.[9] Compacted cylindrical samples, 102 mm in diameter and 152 mm long, of the material under test are frozen from one end while the other is in contact with water maintained at a constant temperature of + 4°C. The samples are usually compacted at the optimum moisture content and maximum dry density of the standard compaction test (see Par. 9.57), although for some clays the natural (as-dug) moisture content is used with compaction to 5 percent air voids.

9.93. Figure 9.31 shows the test cabinet, which is designed to accommodate nine specimens. After compaction and extrusion the curved surface is covered with waxed paper, end faces being left uncovered. The specimens are then placed in metal carriers provided with a porous ceramic base, which is in contact with water in a base tank when the samples are lowered into the test cabinet. The water in this tank is maintained at 4°C during the test. The samples are surrounded by dry sand which extends to the upper face of the samples. The top face of each sample is covered by a waxed disk, supporting a light brass pushrod operating through a metal bar fixed to the top of the cabinet. The heave is recorded from the movement of the top of the pushrod.

9.94. After equilibrating at room temperature for 24 hours, the cabinet is transferred to a cold room operating at a constant temperature of −17°C. Heave is recorded daily for a total period of not less than 250 hours, and the heave–time curve is plotted for each sample. It is important to realize that this test is not intended to simulate practice. The high water table (approximately 70 mm below the level of the zero isotherm) represents a very severe condition. Furthermore, the specimens are allowed during freezing to rise above the level of the surrounding sand fill and for this reason the zero isotherm is falling throughout the test with respect to the top of the samples. (However, heaving is stopped immediately if the sand level is brought up to the level of the samples at any time during the test period.) Therefore, the test, as normally carried out, is assessing the combined permeability through the freezing front and in the unfrozen part of the specimen. These are the factors which control whether or not heave will occur in practice in relation to the length of the cold spell. Typical test results are shown in Fig. 9.32. The silty clay soil (brickearth) has sufficient fines to depress the freezing point in the freezing front and allow water to pass through to form ice lenses, and the soil beneath the freezing zone is sufficiently permeable to permit an upward movement of water from the water table. The result is considerable heave after 250 hours. The London clay, on the other hand, will permit water to pass through the freezing front, but the low permeability in the bulk of the material severely restricts the upward movement of water. The importance of the bulk permeability is illustrated in this example by the inclusion of a comparatively thin layer of

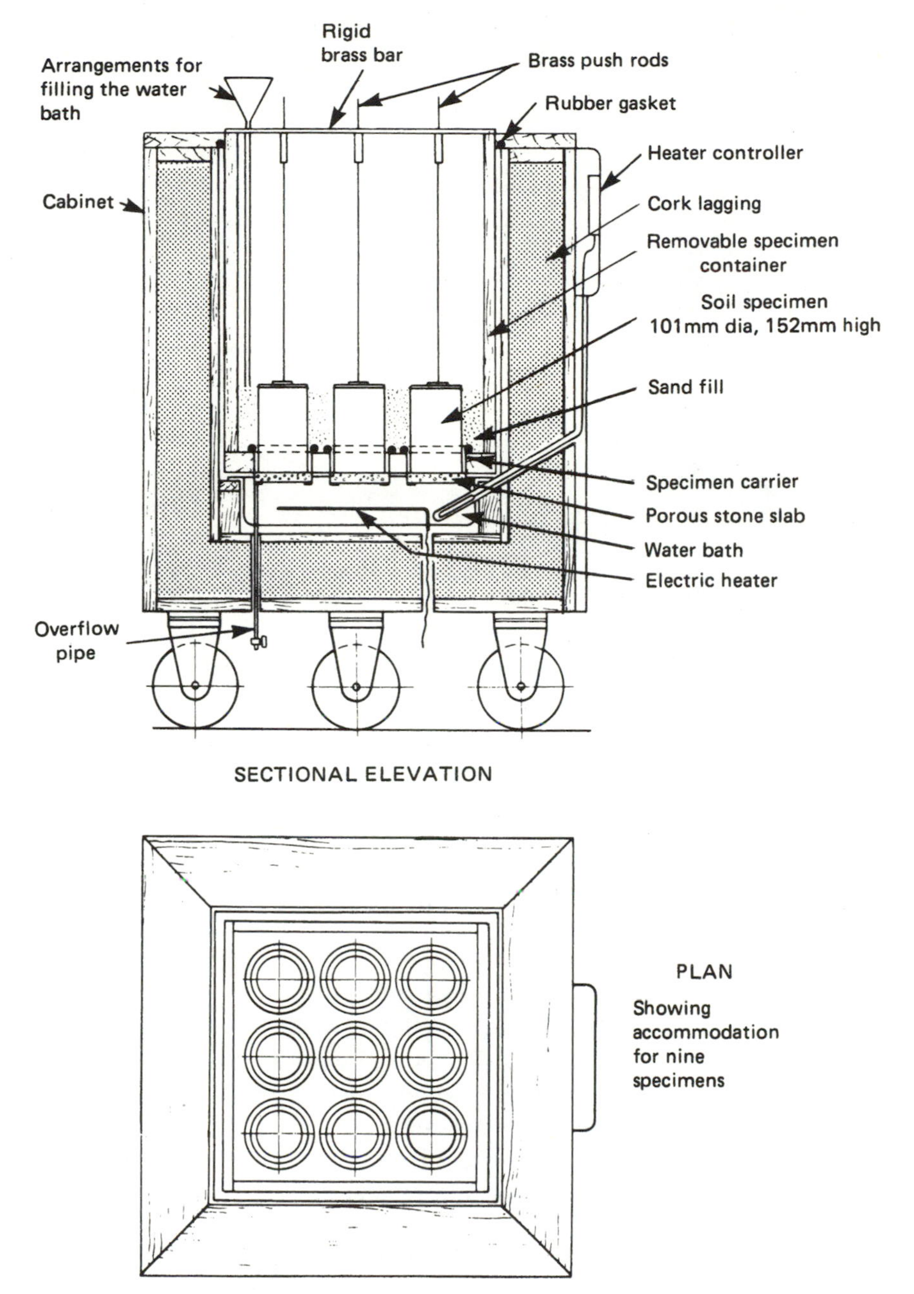

Figure 9.31 Details of frost heave cabinet.

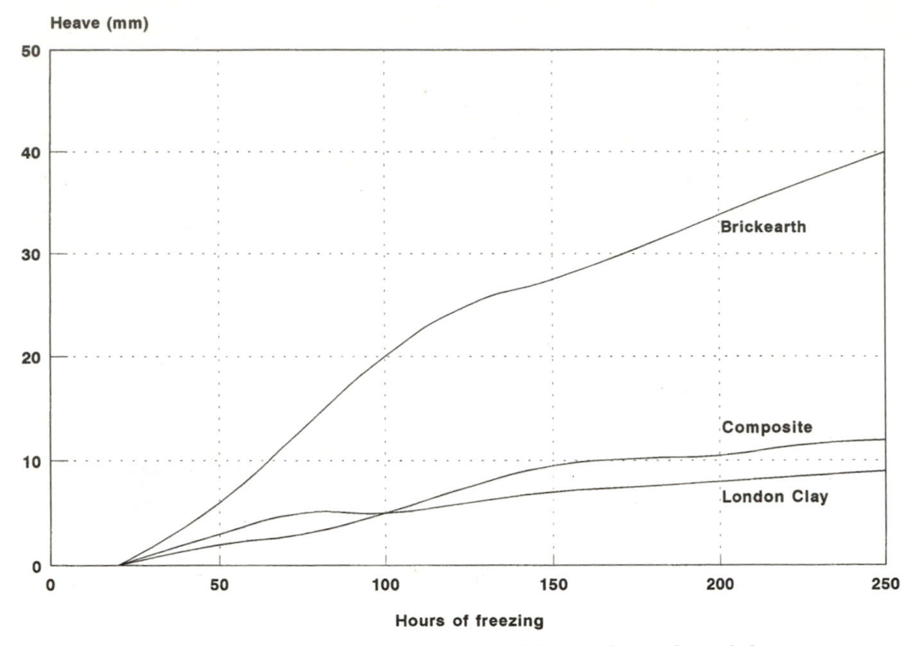

Figure 9.32 Comparison of heaves in silty clay and heavy clay soils with heave in composite sample.

London clay between the brickearth and the water table. This considerably reduces the heave in the composite sample.

9.95. During very severe British winters, which occur on average at 10-year intervals, the test has been applied to soils and granular materials which have been involved in cases of severe frost damage to roads, and to others where there has been little or no heave or damage. From this work it was concluded that materials which heaved 13 mm or less during a 250-hour test period were satisfactory, that materials which heaved between 13 and 18 mm were marginally frost-susceptible, and that those which heaved more than 18 mm should be classified as frost-susceptible. Since a main function of a subbase is to replace frost-susceptible soil, the same criteria were subsequently applied to subbase and road base materials.

9.96. Chalks and soft oolitic limestones are liable to be frost-susceptible, owing to the high permeability of the parent rock, and their use at depths likely to be affected by frost must be avoided. Various additives including cement have been found to reduce frost heave. Chemicals such as calcium lignosulfonate and sodium tripolyphosphate will inhibit frost heave with concentrations of 0.5 percent by weight. However, the use in practice of such chemicals is unlikely to be economic.

9.97. In the United Kingdom, frost penetration very rarely exceeds 450 mm and for some years it has been normal practice to replace any frost-susceptible soil within that depth range with non-frost-susceptible subbase material. In the northern states of the United States, where Table 6.1 shows that there may be frost penetration of 1.5 m it would be good practice to use a capping layer, over a frost-susceptible subgrade, using a granular material stabilized with the addition of cement or bitumen. Normally only a very small additive content is required to prevent frost heaving.

9.98. Since the fifties the concept of frost index has been used to express the severity of very cold periods. The frost index is defined as the product of the number of freezing days and the average daily air temperature below 0°C. Thus two consecutive days each with an average daily air temperature of −2°C would represent a frost index of 4°C days. A summation of this type based on daily 24-hour air temperatures during a continuous cold spell will give the frost index for that period. A paper by Johnson, Beck, et al. of the U.S. Army Corps of Engineers quoted by Sherwood and Roe[10] has related the frost penetration to the frost index for a range of different types of soil (see Fig. 9.33). This gives a useful broad guide to possible frost penetrations.

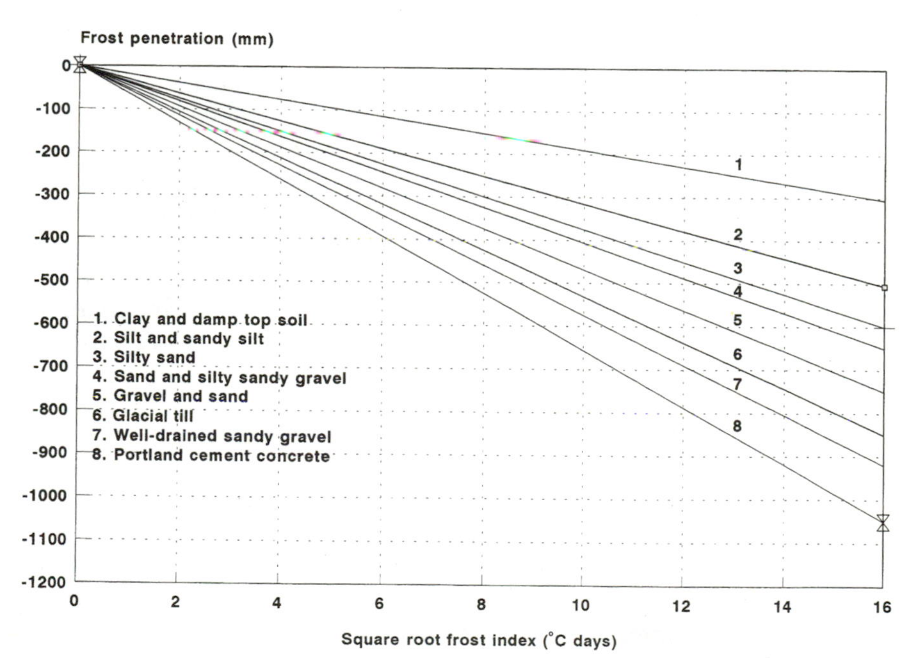

Figure 9.33 Relationship between frost index and frost penetration into snow-free homogeneous materials.

The Strength of Soil

9.99. General. The shear strength of soil is generally expressed in the following form:

$$s = c + \sigma_n \tan \phi$$

where s = shear strength
c = cohesion
ϕ = angle of shearing resistance
σ_n = stress normal to the shear plane

For cohesive soils the shear strength is determined indirectly by triaxial compression tests carried out under the appropriate conditions of drainage. Such tests are described in AASHTO Test T234-85 (ASTM Designation D2850-70), and in less detail in BS 1377:1975, Test 21.

9.100. In clay soils free of stones, direct measurements of shear strength can be made using the rotating field vane, in which a small cruciform vane is pushed into the soil and the shear stress calculated from the torque required to rotate it. This test is described in AASHTO Designation T233-76 (1981) or BS 1377:1975, Test 18. For coarse cohesive soils large shear boxes are generally used.

9.101. Stability of slopes of cuttings and embankments. In designing slope angles for cuttings and embankments, it is normally assumed that failure occurs on a circular arc, and that it develops when the moment tending to cause rotation of the soil mass cannot be resisted by the shear strength of the soil acting round the slip circle. In considering the stability of earth dams, where the risks involved in failure are so great, a great deal of analytical work relating to safe slopes is essential. The number of cuttings and embankments in modern road projects precludes a similar detailed analysis, particularly where soil conditions are far from homogeneous. Slopes are therefore based on experience and the generalized findings of research.

9.102. In 1937, Taylor produced a simple design procedure for slope angles in terms of c and ϕ. This was based on total stress and the role of the water table was not considered. In 1960, Bishop and Morgenstern[11] produced a similar analysis based on effective stress, in a form particularly suited to computer analysis. Symons[12] has since published a relation between life and critical slope for clay cuttings of various heights. The results,[13] which provide a useful basis for the design of clay slopes, are shown in Fig. 9.34. A recent survey of slope behavior on the freeway system of the United Kingdom[13] has

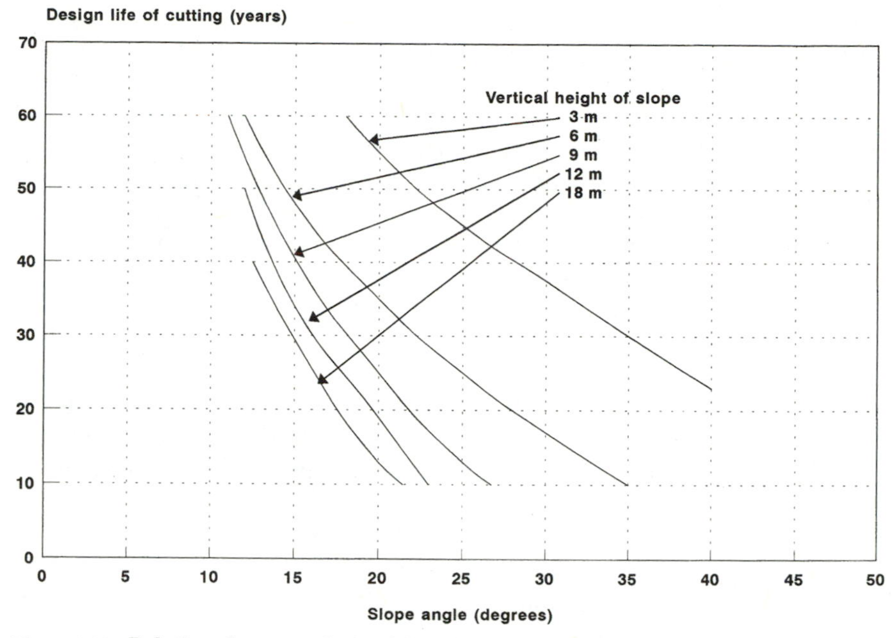

Figure 9.34 Relations between design life and slope angle for overconsolidated clay cuttings.

enabled a number of useful conclusions to be reached relating long-term slope stability and soil type. The conclusions relating to slope angles and soil types are summarized in Table 9.8.

9.103. Cuttings and embankments in granular soils present no serious problems and slopes of 1:1.5 will normally be adequate for all

TABLE 9.8 Geologies with a High Percentage of Failure

Geology	Percentage of failure	Predominant slope angle
Embankments:		
Gault clay	8.2	1:2.5
Reading beds	7.6	1:2
Kimmeridge clay	6.1	1:2
Oxford clay	5.7	1:2
Lower Keuper sandstone	4.9	1:1.5
London clay	4.4	1:2
Cuttings:		
Gault clay	9.6	1:2.5
Enville beds	5.8	1:2.5
Oxford clay	3.2	1:2
Reading beds	2.9	1:3
Bunter pebble beds	2.3	1:2
Lower Old Red sandstone—St. Maughan's group	1.7	1:2

heights. For sandy and silty clay soils the slope angle widely used is between 1:2 and 1:1.5, depending on the plasticity of the soil and the height.

9.104. Although cuttings in chalk and soft limestones give the appearance of stability when nearly vertical, such slopes are subject to frost erosion. Slopes in such materials should not be steeper than 50°, to allow vegetation to establish itself. Even with such a slope, provision should be made by fencing to catch any debris loosened by frost. Most slips which occur in highway embankments are surface slides which arise mainly from water entering poorly compacted soil on the face of the slopes. In the construction of embankments particular care should be taken in the edge compaction as the layers of the embankment are constructed. Vegetation helps to stabilize all slopes, but on clay embankments the vegetation should not be deep-rooted.

9.105. Where an area has a history of landslides, or where roads are to be cut in sidelong ground, the engineer is advised to have a thorough geological examination of the site made before deciding on the line and the level of the road.

The Strength of Subgrades

9.106. A well-designed flexible road pavement would be expected to show a permanent deformation of little more than 20 to 30 mm after a life of 20 years. The deformation in the subgrade would then be less than 10 mm, corresponding perhaps to the passage of 100 million standard axles. It follows that under the imposed stress regime the soil needs to behave elastically, and that shear strength is not directly the factor which defines a satisfactory subgrade. In designing a pavement structurally the elastic modulus of the subgrade and its Poisson ratio are the foundation properties needed. However, these are complex properties to measure, and, bearing in mind the variability of soils over comparatively small distances, it would not be economically feasible to use them directly to assess subgrade suitability. It follows therefore that if pavements are designed empirically on past experience or fundamentally using elastic theory, a relatively simple test procedure is needed, the results of which can be related by experiment to the structural properties.

The California Bearing Ratio Test

9.107. Various penetration type tests were developed in the United States for evaluating soils, the most enduring of which has been the

California Bearing Ratio (CBR) test, which dates from the twenties and was adapted by the U.S. Army Corps of Engineers for airfield design in the early forties. The CBR test seems to have been first used in the California State Highways Department by O. J. Porter, engineer to the department. Like many tests developed and modified over a long period, the method of sample preparation and testing varied considerably in the early years and Mr. Porter's papers are not precise, particularly relating to when and whether samples were soaked prior to testing and at what stage and under what conditions measurements in situ were permitted to be used.

9.108. During World War II the method was introduced into Britain in connection with airfield construction and immediately after the war it began to be used in road design. From Britain its use has spread to a number of European countries. Because of ambiguities in the test procedure, the decision was taken in the United Kingdom to test samples at the dry density and moisture content likely to be achieved in the field, without soaking. For clay soils, where the mould confinement had little effect on the measured value, an in situ rig was developed to monitor in situ the subgrade during construction. Where time allowed the relationships between CBR, dry density, and moisture content were studied for the principal soils of a contract.

9.109. The equipment and test procedure are described in detail in BS 1377:1975, Test 16, and under AASHTO Designation T193-81 (1986). The equipment is of the form and dimensions shown in Fig. 9.35. The soil under test is compacted into the mould at the moisture content and dry density which it is estimated will be achieved in the prepared subgrade, and the penetration test is carried out at the standard rate. A loading frame is used to give the necessary reaction and a proving ring measures the load.

9.110. Typical load–penetration curves are shown in Fig. 9.36. The loads required to cause penetrations of 2.5 and 5 mm are recorded and expressed as ratios of the loads to cause the same penetrations in a "standard" crushed rock material, the load–penetration curve for which is also included in Fig. 9.36. For various reasons the initial part of the penetration curve may be concave, as shown for Test 2 in the diagram. In such a case the load–penetration curve is projected back to the horizontal axis, as shown in the figure, to give the intercept A. This intercept is added to the standard penetrations of 2.5 and 5 mm when evaluating the loads equivalent to those penetrations for the materials under test. No such addition is of course made in obtaining the corresponding loads for the standard crushed stone

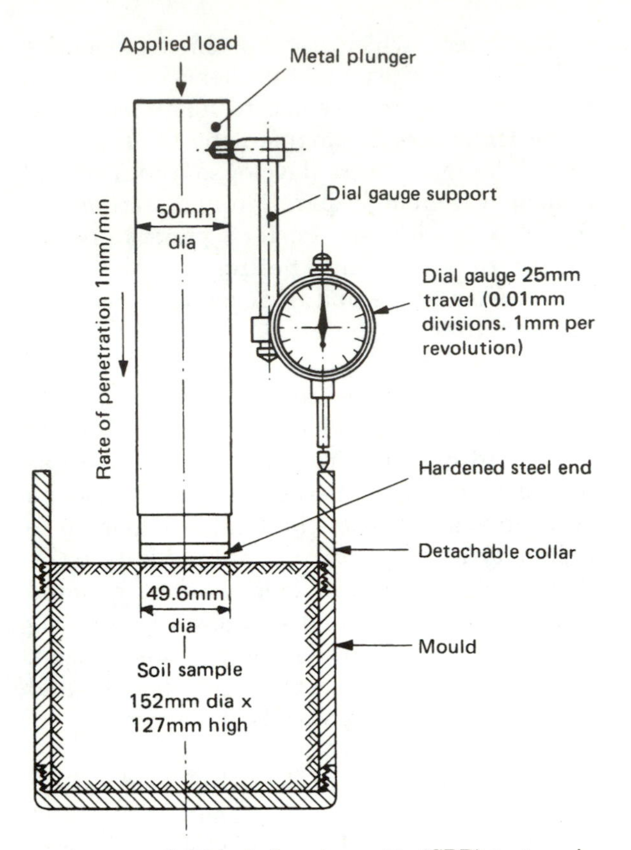

Figure 9.35 California bearing ratio (CBR) test equipment.

material. The larger of the ratios corresponding to the 2.5- and 5-mm penetrations is normally taken as the CBR value of the material for the test conditions used. The test is usually carried out on materials with a maximum particle size of 20 mm. If the material to be tested contains 10 percent or less by weight coarser than 20 mm then this fraction can be removed without seriously underestimating the strength of the soil. The test is not really suitable for soils or other materials containing more than 10 percent coarser than 20 mm. However, under such circumstances tests on the fraction passing 20 mm carried out at a range of moisture contents will give an indication of whether the material as a whole will lose strength markedly if the moisture content is raised.

9.111. The use of a standard curve for crushed rock arises from the fact that the test was originally designed to assess the quality of fine

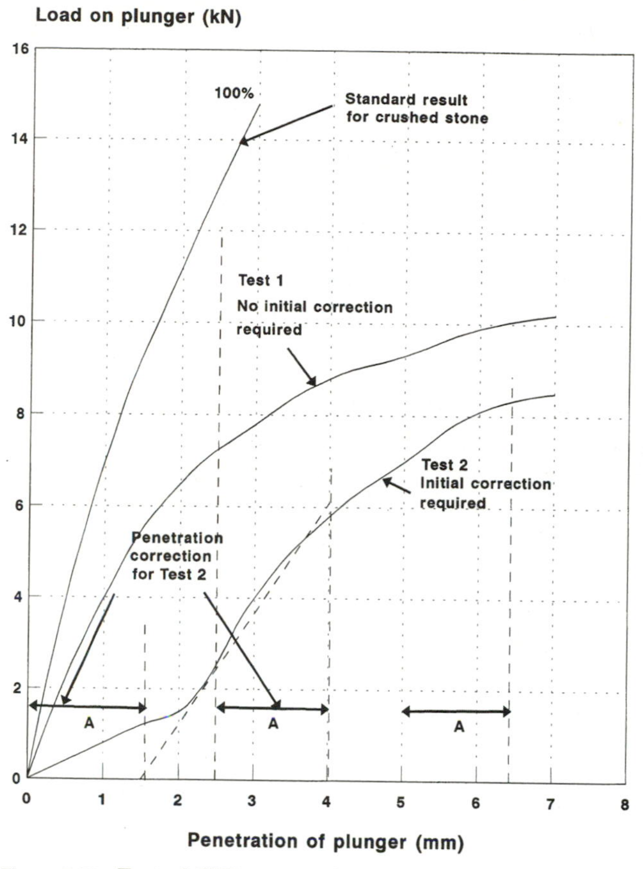

Figure 9.36 Typical CBR test results.

crushed-rock base materials in the state of California. At a later stage, when the test was used on subgrades, this method of expressing the results was continued, so that the majority of clay soils in their natural conditions have a CBR value less than 10 percent.

9.112. Just before World War II tentative design thickness curves for pavements with crushed stone bases were in use in the United States in conjunction with the CBR test. These were based on tests made on a variety of existing pavements judged to have reached a critical structural condition. The two existing curves represented "heavy" and "light" traffic conditions, expressed in terms of the maximum wheel loads likely to use the two categories of road. A few years later this enabled the Boussinesq equations for vertical stress to be used to extrapolate these design curves to cover the much higher wheel loads associated with military aircraft. These curves were modified after full-scale experimental checks.

9.113. After the war it was decided to adopt the CBR test as the basis for the design of flexible pavements in Britain. Because of uncertainties relating to the exact test procedure and to the method used to ascribe wheel loads to the original curves, the decision was taken to prepare new design curves linked to CBR tests carried out at the equilibrium moisture content and dry density conditions expected under the road pavement. A program of testing on existing roads together with a measure of extrapolation led to the curves shown in Fig. 9.37 based on the traffic intensity expected to use the road. These curves were used to give the total thickness of pavement required, the constitution in terms of subbase, road base, and surfacing being established from a comprehensive program of full-scale road experiments. This procedure remains the basis of the current design for flexible pavements used in the United Kingdom.

9.114. A study has been made by E. H. Davis of the influence of dry density and moisture content on the CBR of a wide range of soils from a heavy clay to a sandy gravel.[14] The gradings of the soils are shown

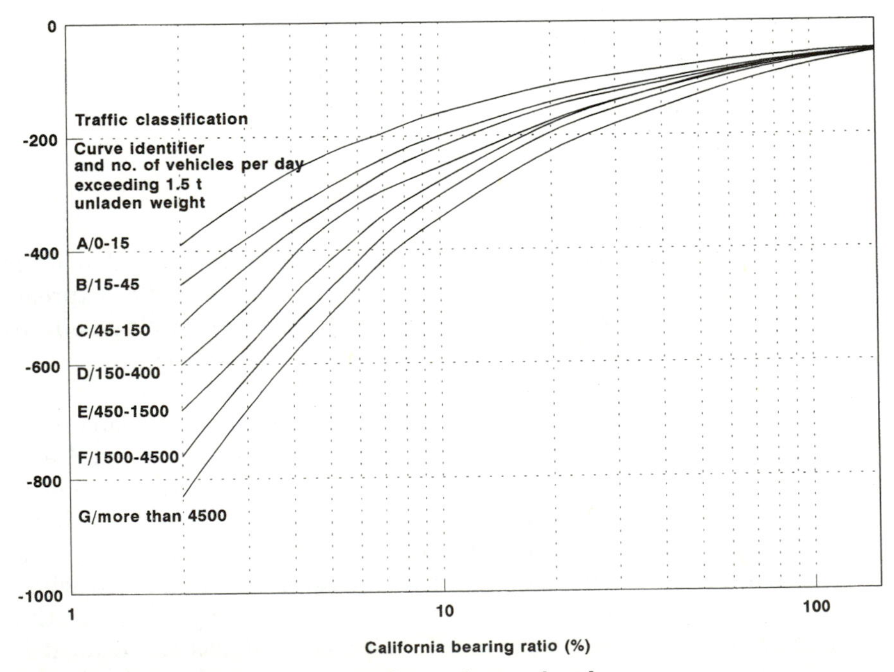

Figure 9.37 CBR design curves for different classes of road.

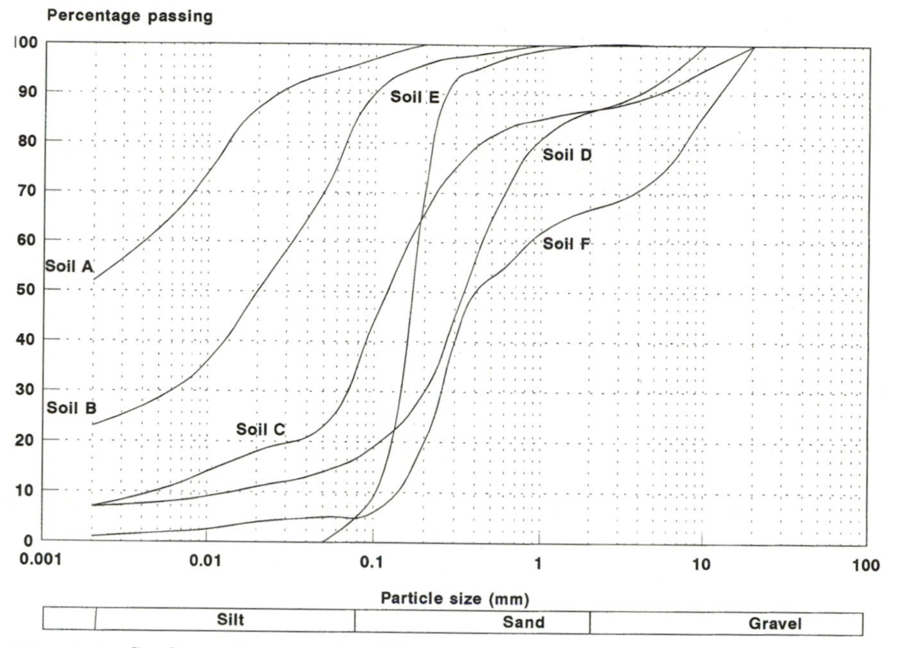

Figure 9.38 Gradings of soils used in CBR test program.

in Fig. 9.38 and the results of classification tests in Table 9.9. The relations between CBR, moisture content, and dry density for soils A, B, C, and D are given in Fig. 9.39.

9.115. Figure 9.39 shows that the relationship between log CBR and moisture content is linear over the likely field moisture content range. Exactly what shapes the curves are, as they approach the sat-

TABLE 9.9 Details of Four Soils Used in CBR Investigation

					BS compaction test	
Soil	Location	Liquid limit, %	Plasticity index, %	Specific gravity of soil	Maximum dry density, kg/m³	Optimum moisture content, %
Heavy clay	Staines, Middx.	75	42	2.76	1554	26
Sandy clay	Harmondsworth, Middx.	31	11	2.76	1794	16
Silty clay	Rippon, Yorks.	24	3	2.62	1874	13
Well-graded sand	Heritingforbury, Herts.		Nonplastic		2002	10

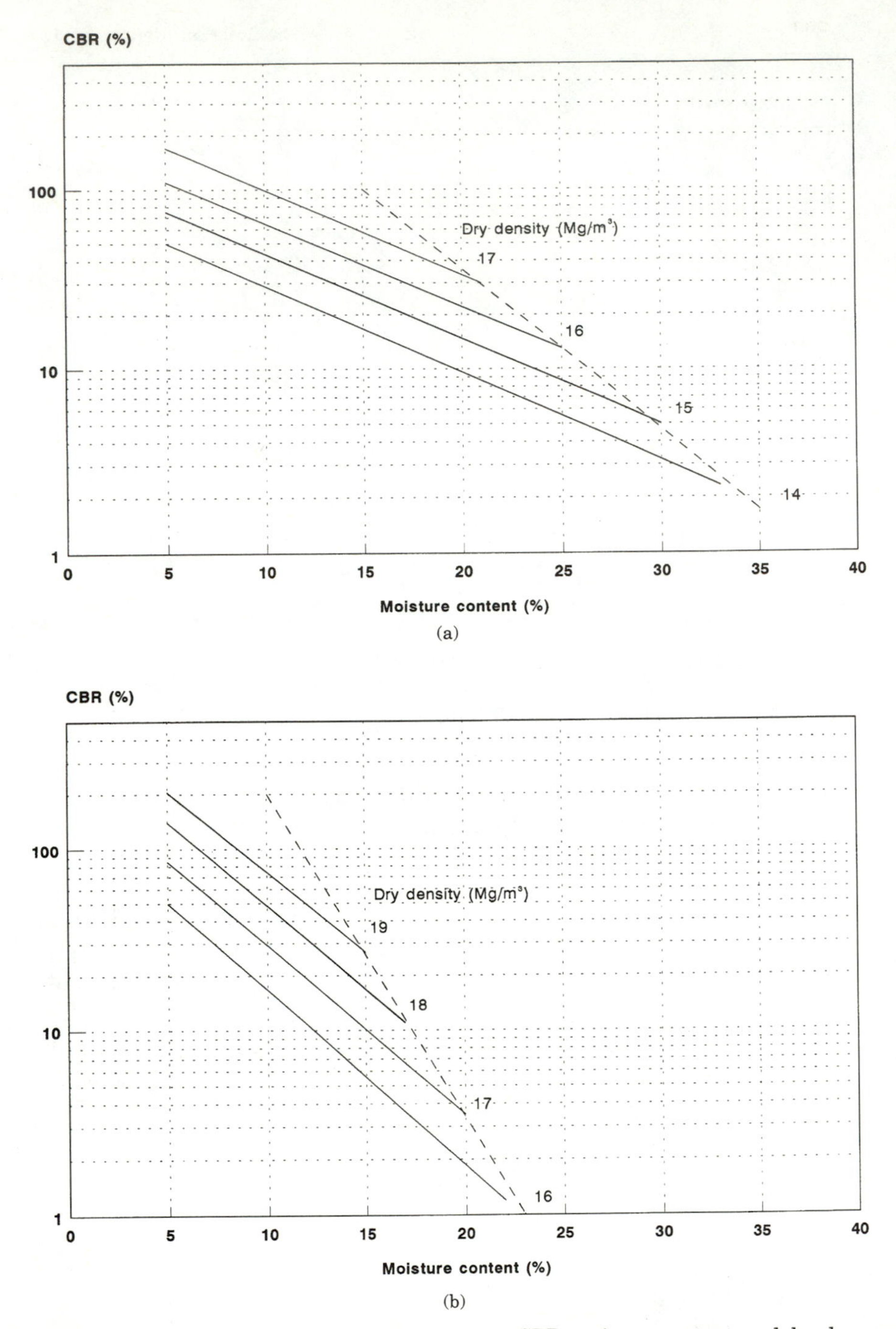

Figure 9.39 (*a*) Laboratory measurements relating CBR, moisture content, and dry density for a heavy clay (soil A). (*b*) Laboratory measurements relating CBR, moisture content, and dry density for a silty clay (soil B).

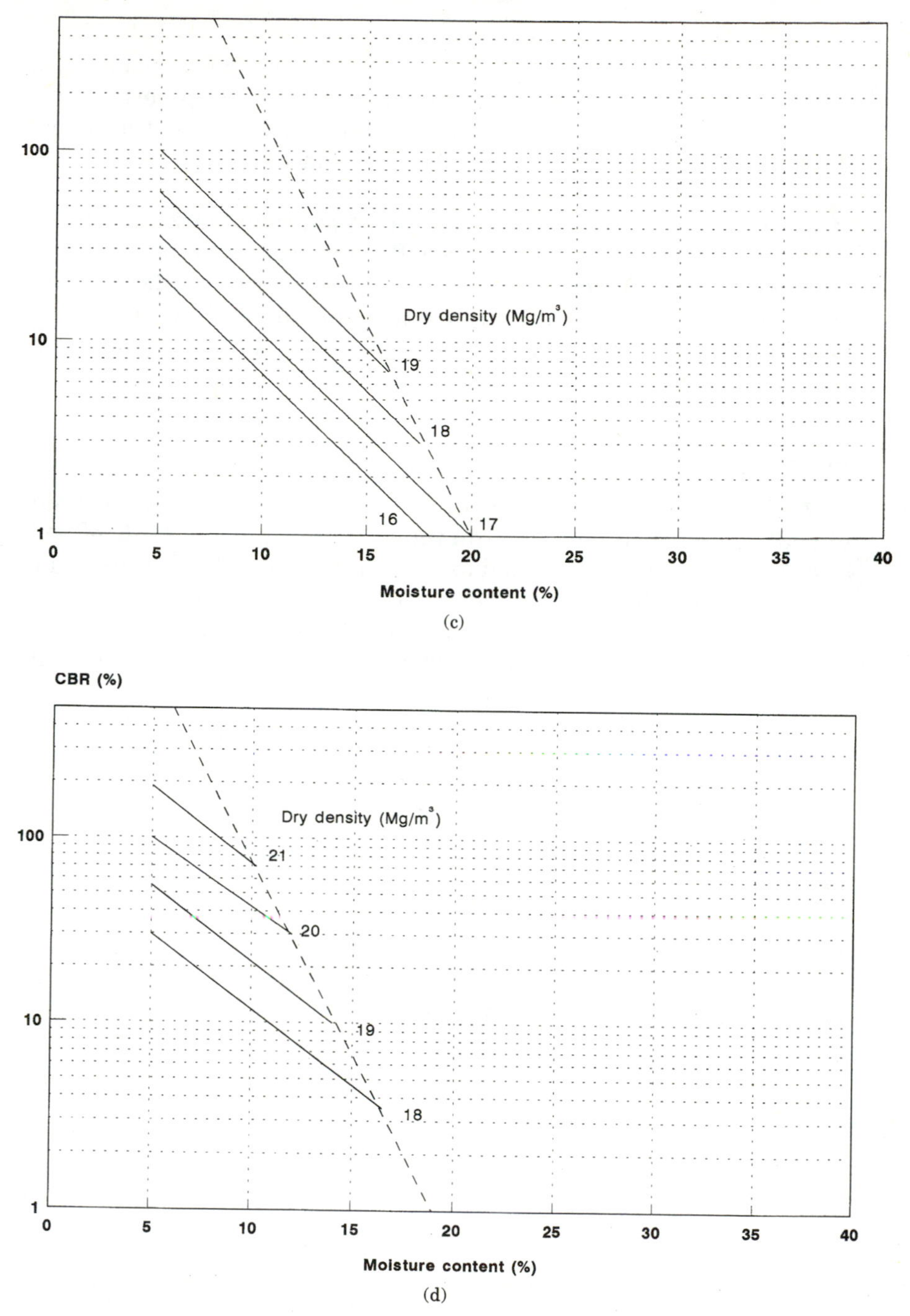

Figure 9.39 (*Continued*) (*c*) Laboratory measurements relating CBR, moisture content, and dry density for a silty sand (soil C). (*d*) Laboratory measurements relating CBR, moisture content, and dry density for a well-graded sand (soil D).

uration line, is difficult to explore experimentally. They appear to join the saturation line, but it is more likely that they drop vertically as the pore water pressure becomes positive with respect to atmospheric pressure. This is not a matter of particular concern.

9.116. An engineer preparing designs for pavement thickness should, at the site investigation stage, have CBR–moisture content–dry density curves such as those shown in Fig. 9.39 prepared for the principal soils on the site. Then, on the basis of the moisture content and density requirements specified by the engineer for the subgrade, it will be possible to deduce the CBR value and design the pavement accordingly.

Estimation of the CBR from Soil Properties

9.117. Black, regarding the CBR test as a bearing capacity test using a small plate,[15,16] produced a method for relating the CBR of soils with their bearing capacity and for cohesive soils with the consistency index and the suction. From the suction curves for a wide range of British soils he related the consistency index (see Par. 9.39) of remoulded soils with the suction in the manner shown in Fig. 9.40.

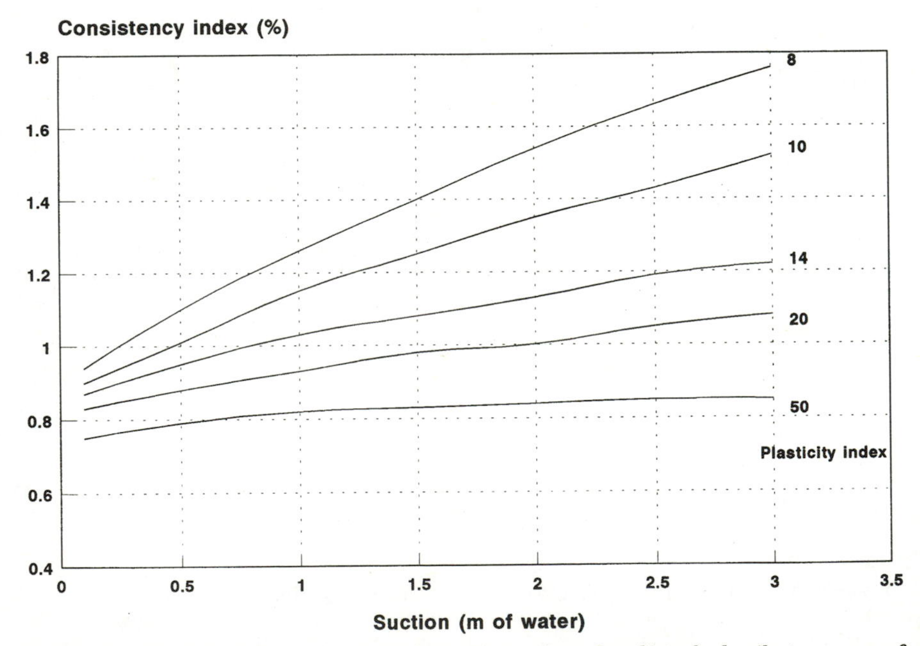

Figure 9.40 Variation of consistency index with suction of undisturbed soil at a range of plasticity indexes.

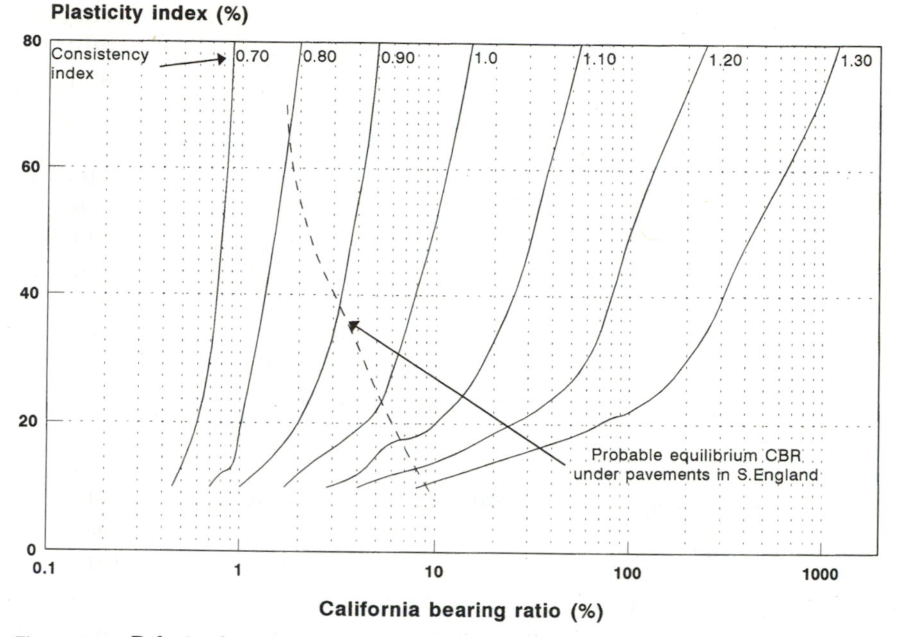

Figure 9.41 Relation between CBR and plasticity index at various plasticity indexes.

Using the bearing capacity approach, he then related the plasticity index and the CBR as in Fig. 9.41.

9.118. To illustrate the use of these graphs the case will be considered of a pavement 550 mm thick on a soil of plasticity index 30 percent with a water table 600 mm below formation level. The average density of the pavement material is assumed to be twice that of water. If we refer to Table 9.7 for a soil of plasticity index 30 percent, then the value of the compressibility factor α will be 0.7. From Eq. (9.15),

$$u = s + \alpha P$$

at formation level, with the water table 60 cm below, $u = -60$ cm of water,

$$\alpha P = 2 \times 55 \times 0.7 \text{ cm of water}$$

thus

$$s = 77 + 60 = 137 \text{ cm of water}$$

From Fig. 9.40 for a suction of 137 cm of water the consistency index is 0.92, and from Fig. 9.41 the CBR is approximately 3 percent.

9.119. In situ CBR tests. The CBR test can be carried out in situ and rigs to fit at the rear of suitably loaded site vehicles are available. The degree of confinement of the soil in laboratory tests and in those conducted in situ is clearly different. This influences the stress distribution under the plunger, and the load–penetration curves. For heavy clays and for other cohesive soils having an air content of 5 percent or more, the difference between the results of laboratory tests and those of tests in situ is small. For other cohesive soils and most granular materials the difference is much larger and tests in situ should not be used to verify the quality of workmanship in relation to specification requirements.

9.120. Relationship between CBR and moisture condition value MCV. Parsons has published the relationships shown in Fig. 9.42 between CBR and MCV (see Pars. 9.64 to 9.72).[17] They relate only to cohesive soils. Where MCV tests are included in a site investigation report these relationships give a useful indication of the CBR values likely to be achieved in subgrades.

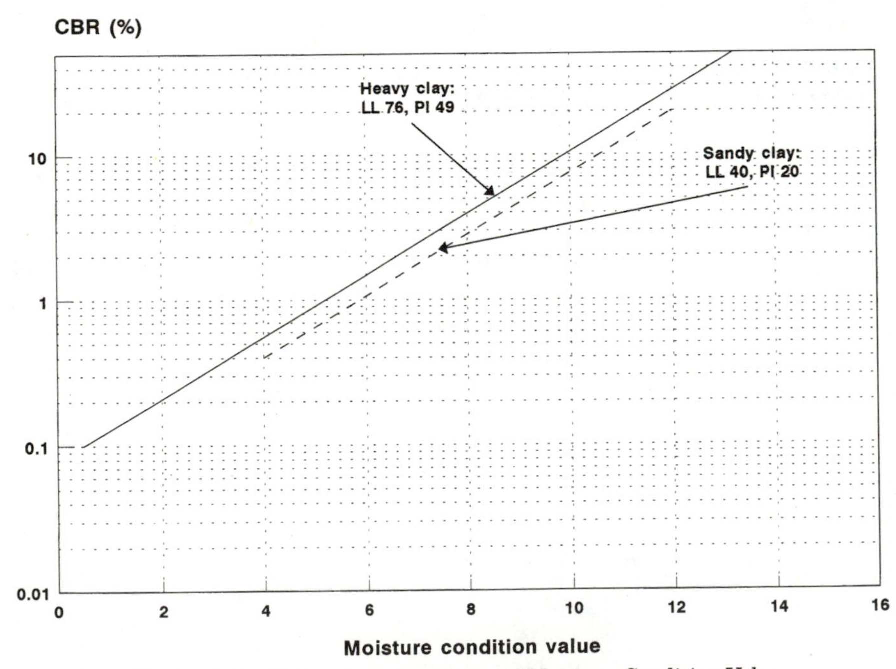

Figure 9.42 Theoretical relations between CBR and Moisture Condition Value.

9.121. Plate bearing test. Plate loading tests are sometimes specified, particularly in connection with the evaluation of subgrades for concrete pavements. The load–deformation relationship using a 760-mm-diameter plate was used by Westergaard[18] to define what he called the modulus of subgrade reaction k.

9.122. The standard 760-mm-diameter plate is normally about 16 mm thick. To increase the stiffness, plates of 660 and 560 mm diameter are often used on top of the standard plate. A movable trailer, loaded up to 30 tonnes, provides a suitable reaction against which the plate is loaded hydraulically. Care is necessary to seat the plate accurately and hand leveling using a straightedge is necessary on clay soils. On granular soils which are difficult to level, a thin layer of well-graded sand is used as a bedding. Alternatively, a quick-setting plaster bed a few millimeters thick can be used. Gantries, located as far as possible outside the zone of influence of the plate and the trailer wheels, support dial gauges set to record the plate settlement at four equally spaced locations around the perimeter of the plate. A proving ring, or alternatively the pressure in the hydraulic system, is used to measure the load. Figure 9.43 shows a typical test result. The modulus of subgrade reaction k is normally calculated for a mean

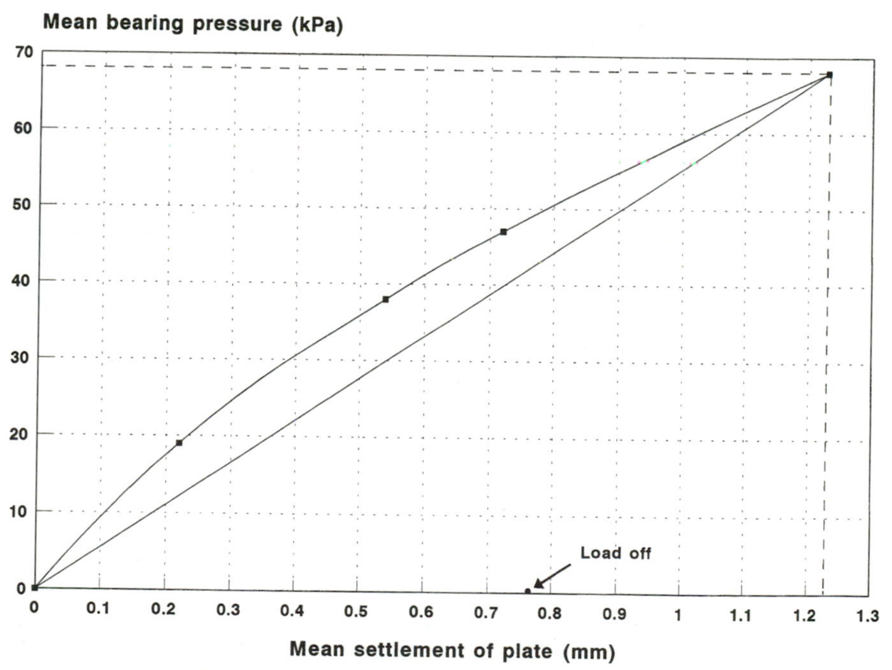

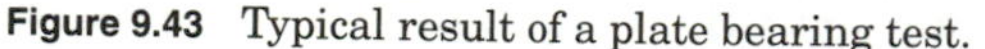
Figure 9.43 Typical result of a plate bearing test.

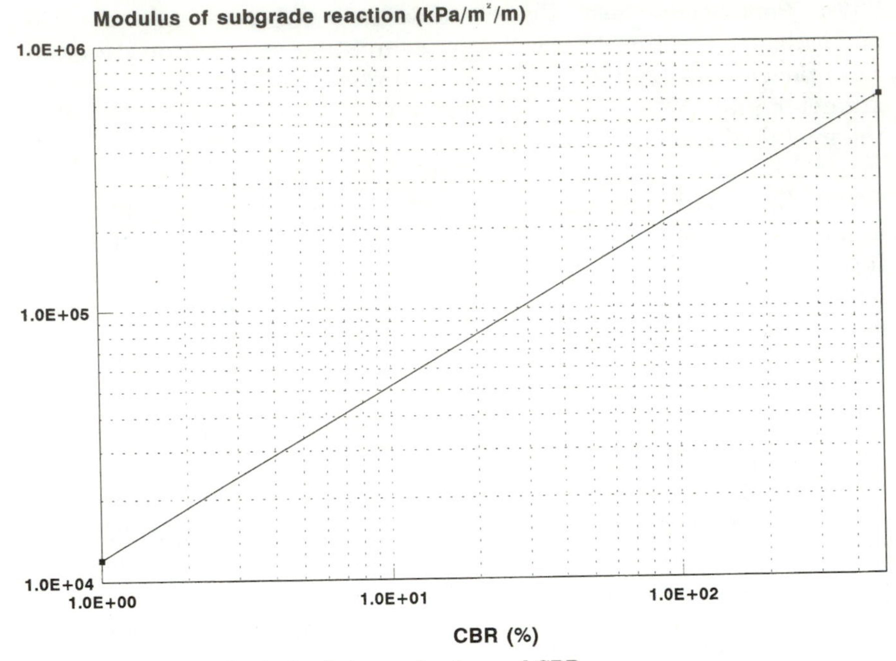

Figure 9.44 Empirical relation between *k* value and CBR.

plate deflection of 1.25 mm. The load–deformation curve is not in general linear and on removal of the load at the completion of a test the greater part of the deflection is found to be nonrecoverable, as is indicated on Fig. 9.43. Therefore, the test, as normally carried out, is not an elastic one and it cannot be closely related to the elastic modulus of the soil. There is an approximate relationship between the *k* value and CBR when the soil is uniform in depth. This empirical relation is shown in Fig. 9.44.

9.123. The value of the modulus of subgrade reaction depends critically on the size of plate used. Clearly, plates smaller than the standard diameter of 760 mm are easier to use and require a smaller loading rig. Figure 9.45 shows a curve based on American experimental evidence relating the measured *k* value with plates of various sizes.[19]

The Elastic Properties of Soil

9.124. The thickness requirements which form part of a road construction contract must be specified in terms of subgrade strength tests which can be carried out by both the engineer and the contractor, using equipment which will be normally available in a site labora-

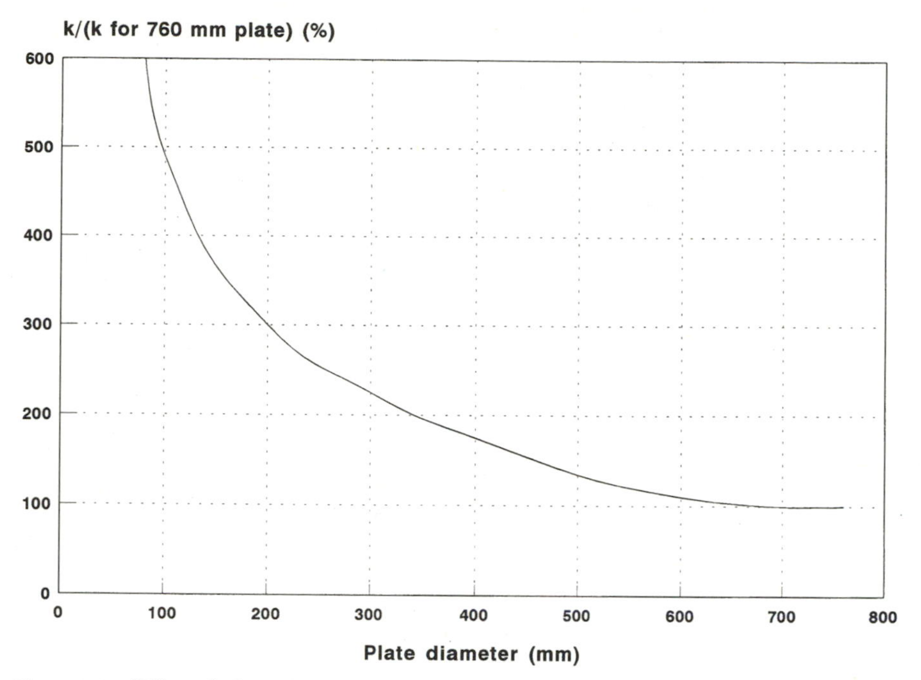

Figure 9.45 Effect of plate size on apparent modulus of subgrade reaction.

tory and which can be operated by trained site staff. For this reason pavement design procedures throughout the world have been developed around such tests as the CBR. Such procedures need to be validated by long-term experience before they are adopted. In the United Kingdom since World War II numerous pavement design experiments have been constructed on heavily trafficked in-service highways and some of these experiments are now more than 30 years old (see Chap. 18). Such a method of developing pavement design procedures is highly reliable, but it does not lend itself to detailed studies of the influence of changes of subgrade support or of pavement material specifications, beyond those incorporated in the original experiments. For this reason efforts have been made since the twenties to generalize available experience using elastic theory.

9.125. The structural analysis of pavements based on elastic theory requires a knowledge of the elastic modulus and Poisson's ratio of soil foundations and of subbases, bases, and surfacing materials. The remainder of this chapter summarizes present knowledge regarding the elastic properties of soils. Similar discussions at the close of Chaps. 11 to 15 deal with the elastic properties of the other constituents of pavements.

9.126. The wave velocity technique provides the most reliable method of studying the elastic properties of subgrades in situ. For laboratory testing repeated loading triaxial tests are normally used. These methods are discussed below.

Wave Velocity Techniques

9.127. A vibratory impulse acting at a point on the surface of a material infinite and homogeneous with respect to area and depth gives rise to a complex system of vibrations throughout the material. These include Rayleigh waves which are propagated close to the surface. The velocity of the Rayleigh waves in such a material is related theoretically to the shear modulus of the material and this in turn to the Young's modulus E and Poisson's ratio v of the material. These relationships are

$$V_R^2 = p^2 \cdot g \frac{G}{\rho} \qquad (9.17)$$

where V_R = velocity of Rayleigh wave
ρ = density of material
p = a constant depending on Poisson's ratio, but close to 0.95
g = acceleration due to gravity

The modulus of elasticity E is obtained from the shear modulus using the equation

$$E = 2(1 + v)G \qquad (9.18)$$

9.128. Jones in the early fifties pioneered the use of this method of testing subgrades and subsequently road bases and subbases.[20,21] Vibrations are produced by an electromagnetic vibrator driven by a variable-frequency oscillator and a power amplifier of suitable gain. The vibrator is placed on a prepared area of the surface soil, so that close contact is obtained, and a straight line is established extending from the vibrator for a distance of about 10 m. With the vibrator running at a frequency above 40 Hz, a seismic geophone pickup is moved away from the vibrator and successive positions are established along the test line for which the vibrations from the pickup are in phase with those from the vibrator. This condition is indicated on a cathode ray display. The distance between successive points located in this manner is equal to one wavelength and the wave velocity is established from the equation

$$V = nL \qquad (9.19)$$

where n is the frequency of vibrations being used in the test. At frequencies below about 150 Hz the wavelength and phase velocity measured in this way may be distorted by vibration waves other than surface Rayleigh waves. However, as the frequency of the vibrator is increased the phase velocity assumes a constant value corresponding to that of the Rayleigh wave.

9.129. Figure 9.46 shows results obtained between June 1955 and May 1956 on an area of sandy clay which was exposed to the weather but free of vegetation. The E values calculated from four sets of measurements are shown in the figure. The suction at the time of the May 1956 measurements was equal to approximately 180 cm of water (pF 2.3). For the earlier three sets of measurements, the suction varied between 30 and 60 cm of water (pF 1.5 to 1.8).

9.130. More comprehensive tests of this type were subsequently carried out by Jones using the facility, and the four soils, used at the TRRL in the United Kingdom for the testing of compaction plant (see Par. 9.61). In this work the elastic modulus deduced using the test procedure described above was related to CBR tests carried out in situ on the surface of the soils. The results are shown in Fig. 9.47.

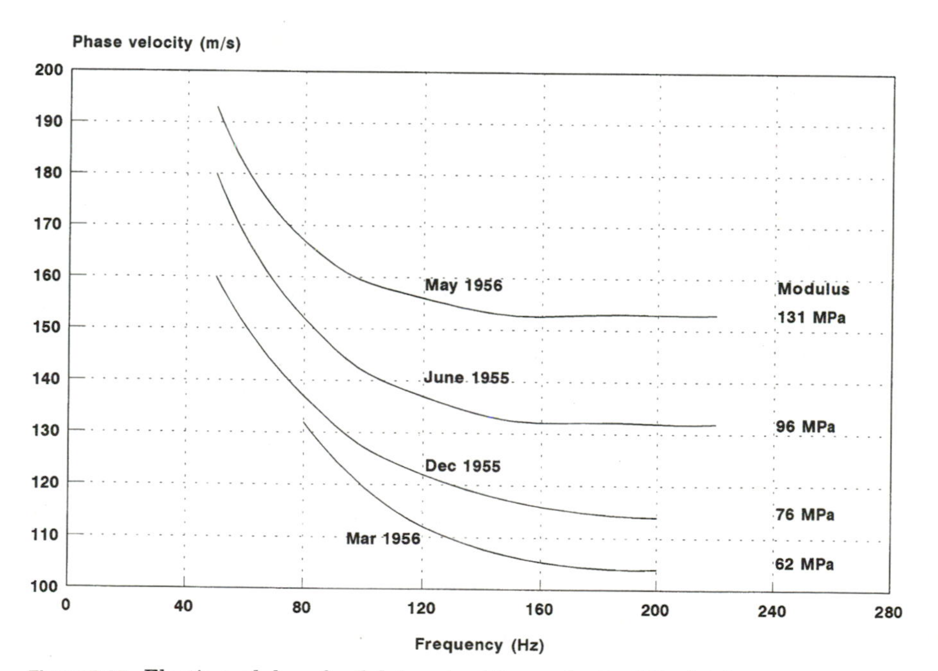

Figure 9.46 Elastic modulus of soil determined from velocity of Rayleigh waves.

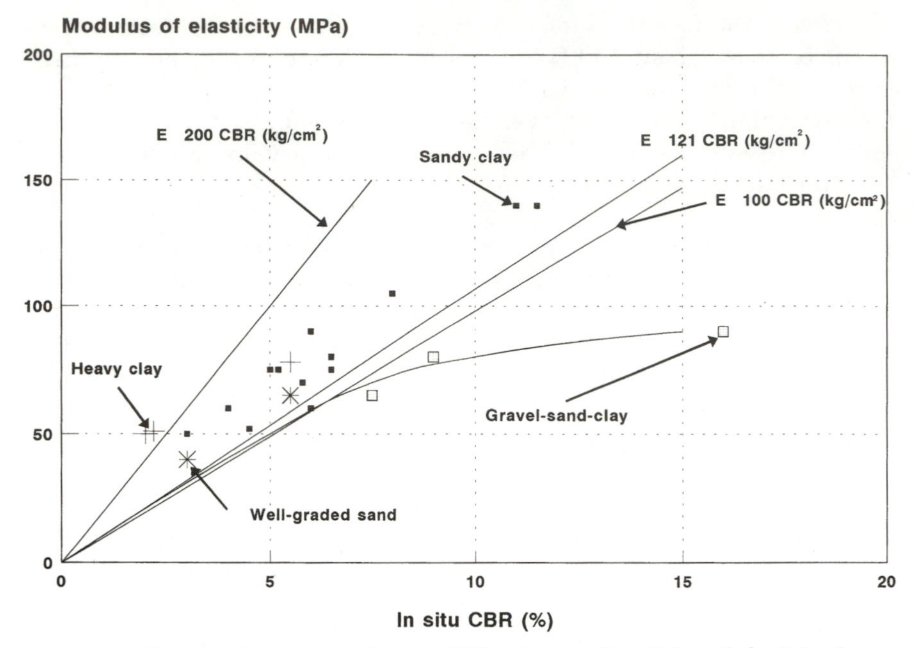

Figure 9.47 Relationship between in situ CBR values and modulus of elasticity (wave velocity measurements).

The procedure was subsequently used at a number of sites where full-scale pavement design experiments were being constructed. The intention was to provide information on the elastic properties of the soils for use in subsequent structural analyses. Fig. 9.48 shows results from one such experiment. The soil at this site was predominantly a heavy clay which included pockets of boulder clay of the same type but with a variable content of stone. The liquid limit of the clay was 62 percent and the plasticity index 41 percent.

9.131. Figures 9.47 and 9.48 show that the relationship between CBR in situ and modulus of elasticity determined by the wave velocity method falls between straight lines represented by E = 100 CBR and E = 200 CBR, when E is expressed in kg/cm^2 or approximately E = 10 CBR and E = 20 CBR when E is expressed in MN/m^2. However, the relationship for more granular soils, again determined in this manner, is not linear and falls below the E = 100 CBR line. This applies to the gravel-sand-clay of Fig. 9.47 and to the more stony boulder clay in Fig. 9.48.

Repeated Loading Triaxial Tests

9.132. In the early sixties a relatively simple repeated loading triaxial machine was constructed at the Transport and Road Research

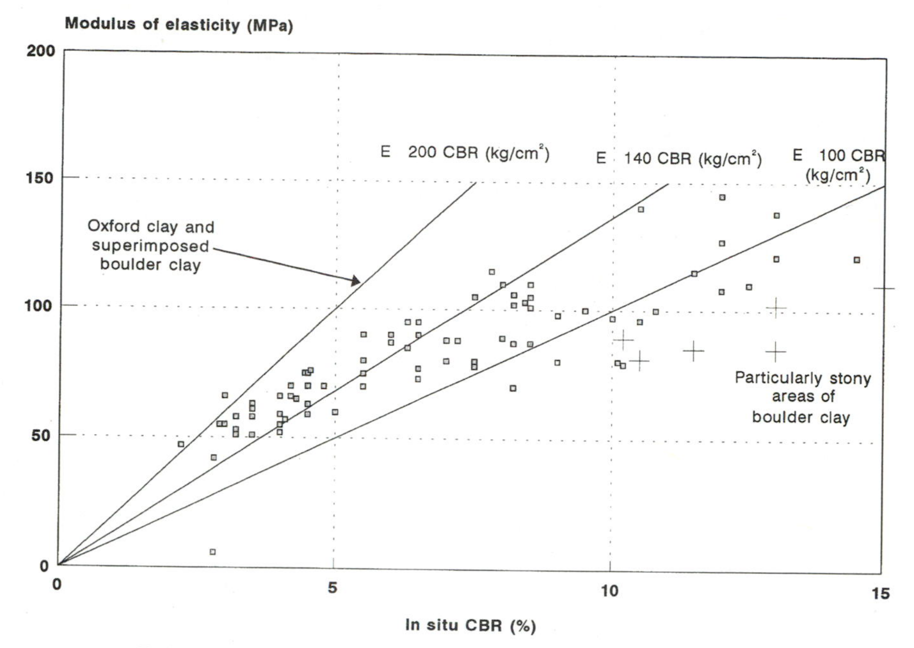

Figure 9.48 Relationship between in situ CBR values and modulus of elasticity (wave velocity), Alconbury Hill.

Laboratory.[22] Both the deviator stress and the cell pressure were pulsed in phase, to simulate the stress conditions in road subgrades. The machine was purely mechanical and was based on the loaded lever and lifting-cam principle.

9.133. Figure 9.49 shows the results of a test made with this machine on remoulded London clay of liquid limit 76 percent and plastic limit 27 percent. The clay was mixed at a moisture content of 40.4 percent and then recompacted to zero air voids in a standard triaxial mould. As would be expected, the material was very weak when set up in the machine. The test conditions were as shown in the figure. Although the test was undrained, there was a more than threefold increase in the measured modulus between 100 and 400,000 stress applications. The corresponding axial permanent deformation is shown in Fig. 9.50.

9.134. This test was carried out to illustrate some of the difficulties which can arise in interpreting repeated loading triaxial tests on soils unless very careful attention is given to the sample preparation and the stress regime used in the tests. The comparatively large permanent deformation after 400,000 applications of stress shown in Fig. 9.50 was not due to compaction, since the clay was fully saturated. The sample deformed laterally owing to the low cell pressure used.

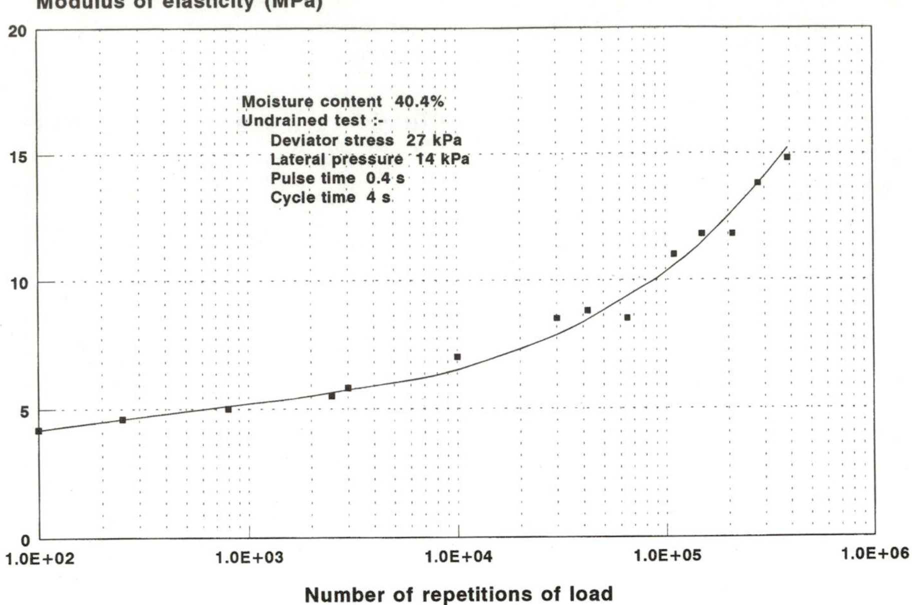

Figure 9.49 Change in elastic modulus of remoulded London clay with repeated loading.

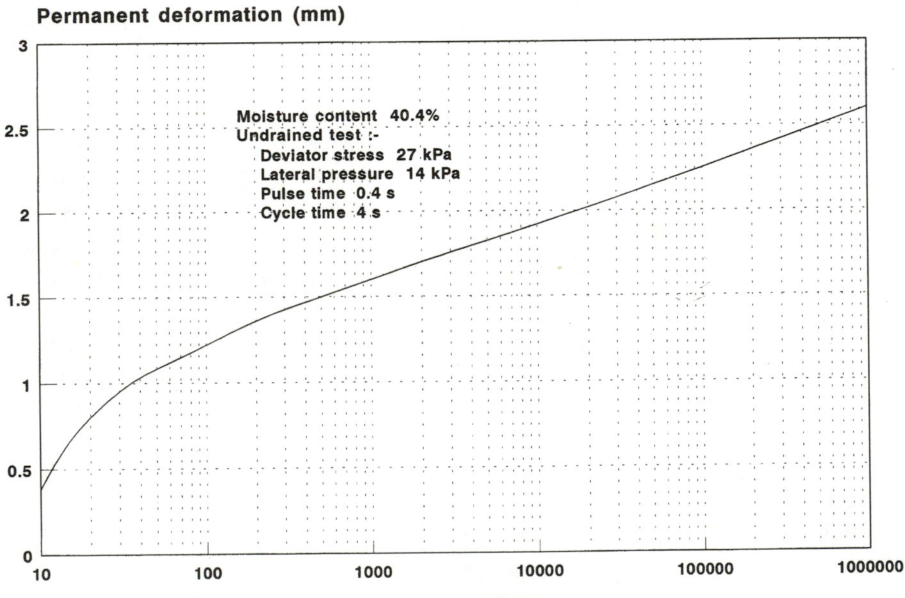

Figure 9.50 Permanent axial deformation of remoulded London clay during repeated loading test.

Had such a weak soil been used as a subgrade then the total thickness of the pavement would have been very considerable and a higher cell pressure would have been appropriate.

9.135. The increase in the elastic modulus of the soil with repetitions of stress arises from the rather complex changes which would have occurred in the suction of the soil during the test. The suction of soil and its relationship to moisture content and soil type has been discussed in Pars. 9.77 to 9.80. For the London clay used in the repeated loading test, Fig. 9.51 shows relevant relationships between suction and moisture content. Because of the rapid increase in suction which arises when a soil dries to a low moisture content, it is usual to express suction levels on a logarithmic scale. The scale which has been used by agricultural scientists is the pF scale (already referred to in Par. 9.87). On this scale the pF value is the common logarithm of the suction expressed in cm of water, i.e., pF 1 equals 10 cm of water and pF 2 equals 100 cm of water. Curve A in Fig. 9.51 refers to undisturbed London clay drying from its natural moisture content. Curve C refers to the same soil slurried with water to the sedimentary condition, and then allowed to dry naturally. Curve B relates to the remoulded soil used in the repeated-loading test. This was a field sample remoulded with additional water to a moisture content of 40 to 42 percent before being extruded for the triaxial test. It will be noted that all three curves come together at a suction between pF 4 and 5, which relates to the preconsolidation pressure to which the soil has been subjected in its past history.

9.136. Figure 9.51 also includes curve D. This is not a suction–moisture content relationship, but defines the suction which a sample of the soil, at a given moisture content, will assume on shearing. Research at the TRRL[23] has shown that when a sample of soil is sheared at constant moisture content its suction will be represented by curve D irrespective of its suction prior to shearing. There is a "unique" line of this type for every cohesive soil. Thus, if the soil prepared for the repeated loading triaxial test had a suction of pF 1 at its initial moisture content of 41 percent then its condition would be represented by point b_1 on curve B. On shearing at constant moisture content the condition would be represented by point b_2 on curve D. If the same soil were drying from a slurried condition, the suction at a moisture content of 41 percent would be represented by point b_3 on curve C. On shearing the suction in this case would also fall to point b_2 on curve D. Since both the liquid and plastic limit tests involve shearing the soil, the corresponding suctions lie on curve D as shown in the figure.

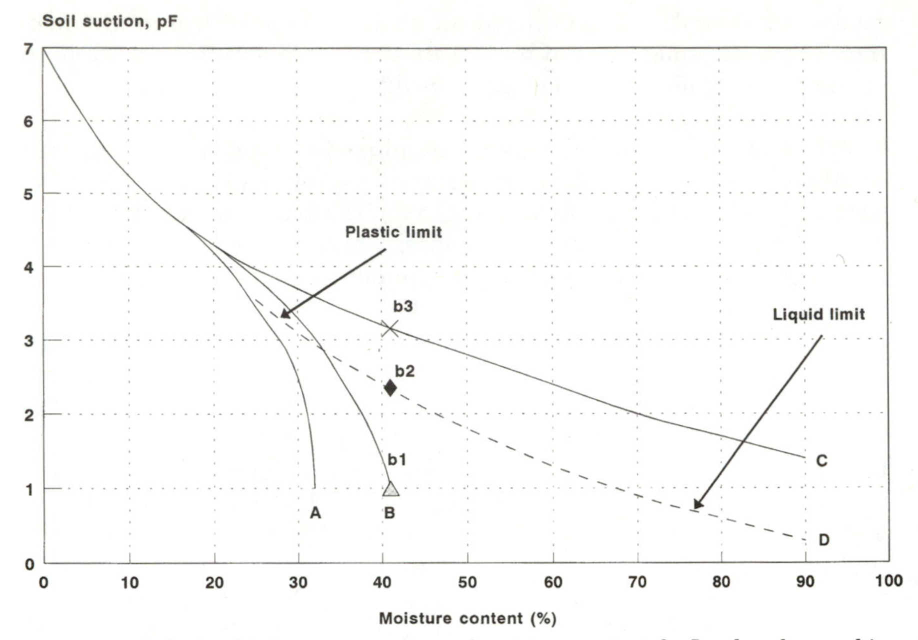

Figure 9.51 Relationship between suction and moisture content for London clay used in repeated loading tests.

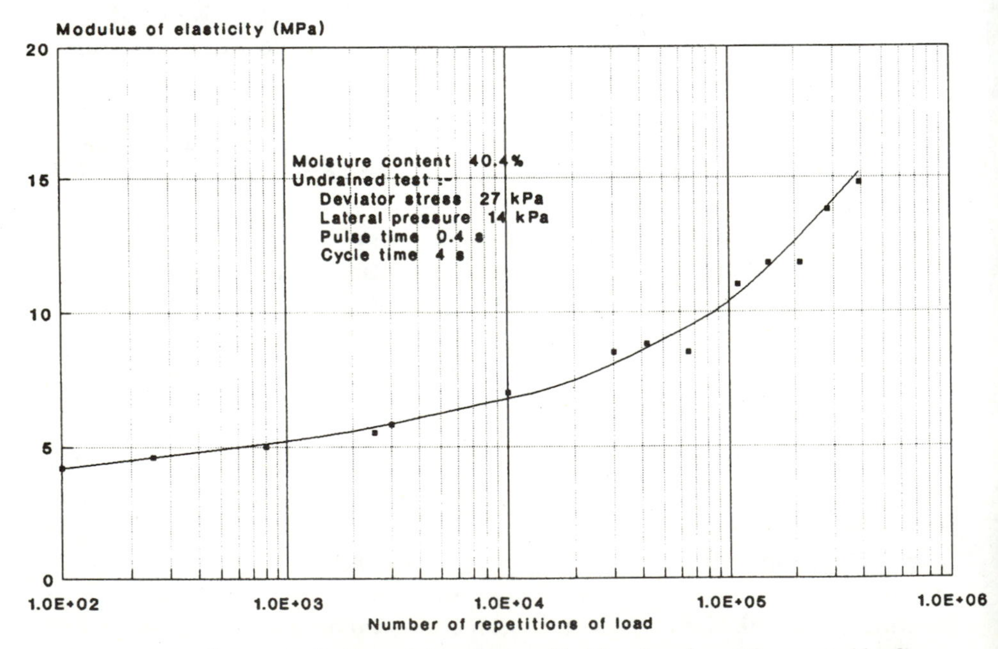

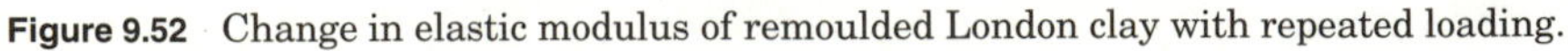

Figure 9.52 Change in elastic modulus of remoulded London clay with repeated loading.

9.137. In the repeated loading test on London clay, Fig. 9.50 shows that the significant deformation of the sample occurred as a result of repeated loading and although the sample did not exhibit shear failure the suction progressively increased toward the value represented by curve D, with a resulting increase in shear strength and modulus of elasticity as indicated in Fig. 9.52. (Further tests on cubical samples of wet heavy clays subject to shear between opposite faces showed that a progressive increase in suction occurred as the shear angle increased. A shear angle of about 30° was necessary to increase the suction to curve D.)

9.138. In practice, the determination of the elastic modulus of a soil would be carried out on samples remoulded to the moisture content and density conditions likely to be present in a prepared subgrade, and the interpretation of the results would be less difficult than in the case illustrated above. However, it is normal to make determinations at several moisture contents and densities to explore the sensitivity of the results to these factors.

9.139. An excellent example of how the elastic modulus of soils should be researched in connection with the structural analysis of

TABLE 9.10 Summary of Test Data Relating to the Determination of Elastic Modulus of Soils Using Repeated Triaxial Loading

Soil	LL, %	PI, %	Moisture content, %	CBR, %	Cell pressure, kN/m^2	Deviator stress, kN/m^2	M_r, MN/m^2	Source reference
Oxford clay	56	37	28.2	2*	13.5	15	22	24
			13.5	30*	13.5	15	195	
London clay	75	42	40	2†	14	27	12	22
Silty clay (San Diego test)	38	17	11.6		27	82	219	25
			13.9		27	82	240	
			15.5		27	81	166	
			17.6		27	40	103	
Silty clay (Marl)	32	14	19.6§	14‡	38	15	70	26
			19.6§	20‡	38	15	140	
Fine sand	—	—	10.5	10*	15	30	86	24

*Based on site measurements.
†Calculated using Black's method; see Pars. 9.117 and 9.118.
‡Based on suction measurements.
§Drained tests.

road pavements is contained in the Asphalt Institute Report RR-71 published in connection with the San Diego experimental base project in 1977. The influence of moisture content, confining pressure, and deviator stress was examined in detail. It was shown that confining pressure and deviator stress had a relatively small effect compared with moisture content. Table 9.10, which gives typical modulus values for a range of soils, includes some values from the San Diego experiment.

Appendix to Chapter 9

The following charts relate dry density and moisture content for soils of particle specific gravity 2.60 to 2.80.

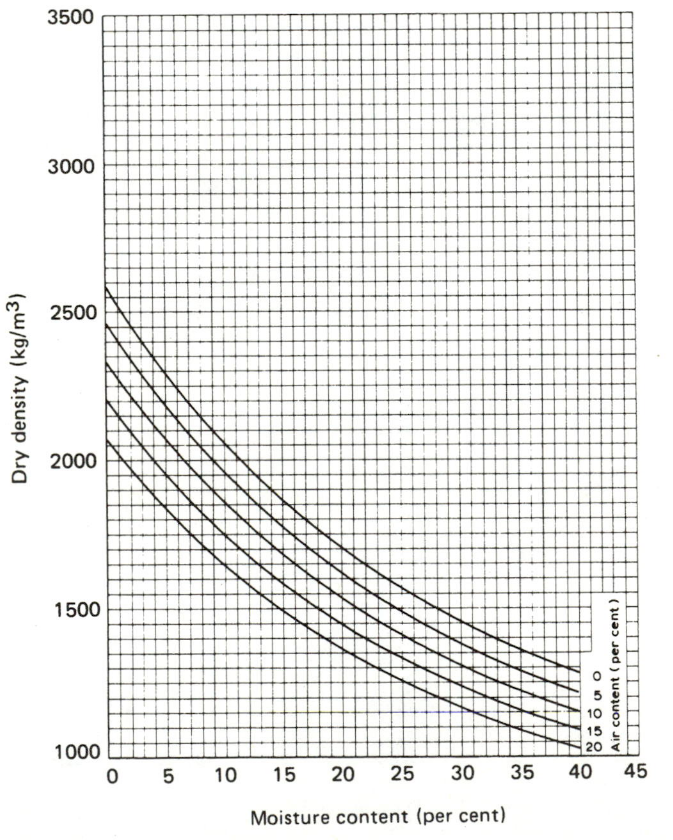

Chart 9.1 Specific gravity 2.60.

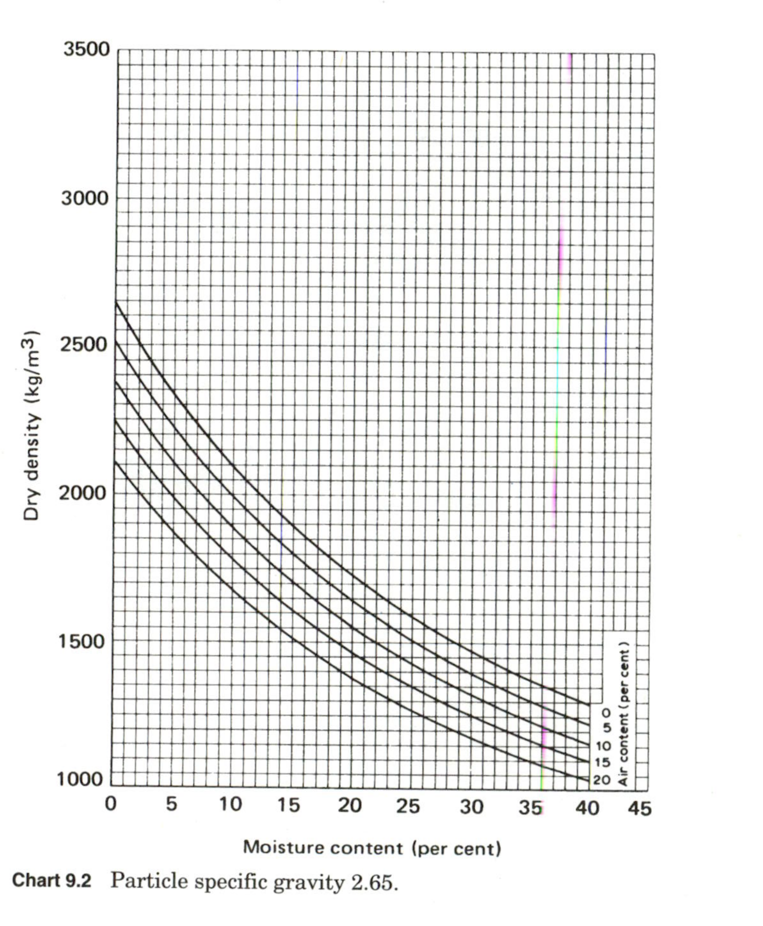

Chart 9.2 Particle specific gravity 2.65.

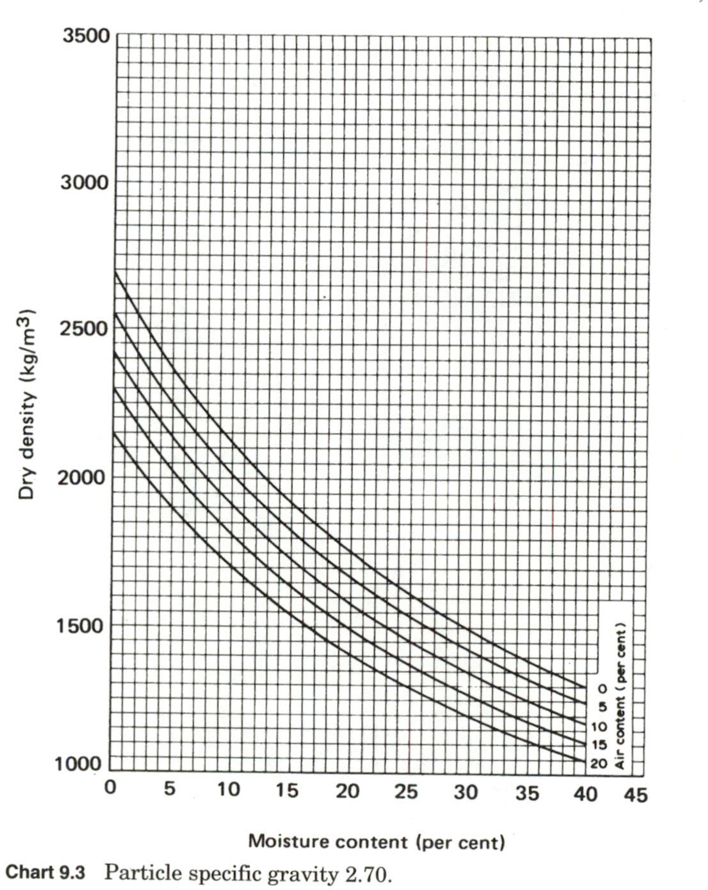

Chart 9.3 Particle specific gravity 2.70.

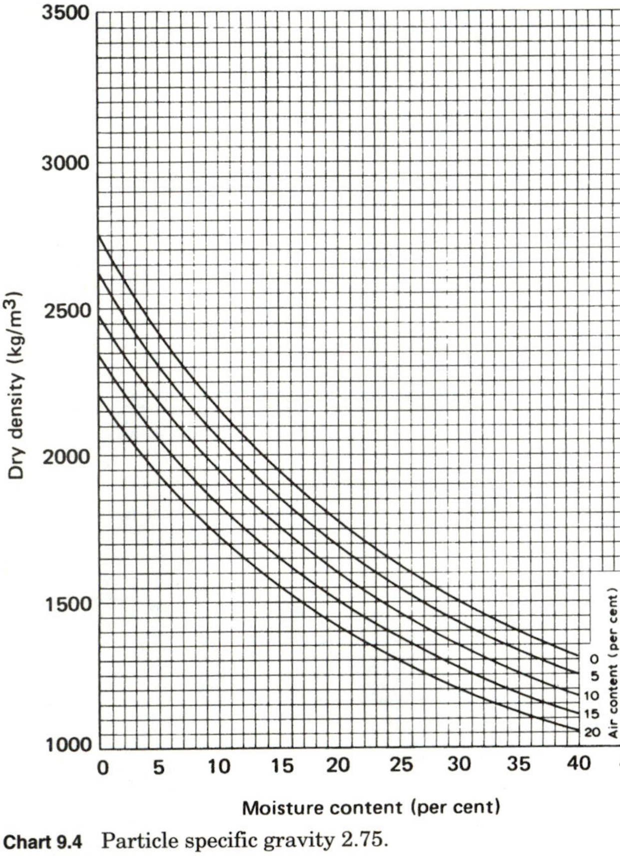

Chart 9.4 Particle specific gravity 2.75.

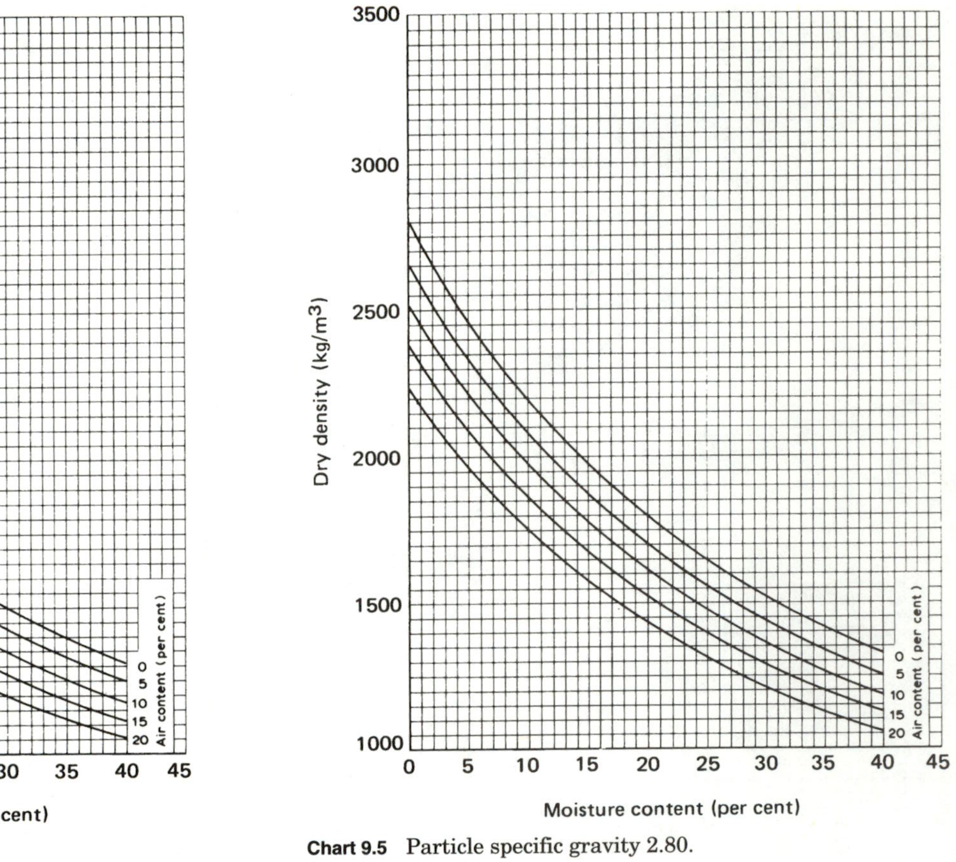

Chart 9.5 Particle specific gravity 2.80.

References

1. Croney, D., J. D. Coleman, and W. P. M. Black: Movement and Distribution of Water in Soil in Relation to Highway Design and Performance, *Highway Research Board Special Report* 40, Water and Its Conduction in Soils. National Academy of Science—National Research Council, Washington, D.C., pp. 226–252, 1958.
2. Parsons, A. W.: The Rapid Measurement of the Moisture Condition of Earthwork Material, *Transport and Road Research Laboratory Report* LR750, TRRL, Crowthorne, 1976.
3. Parsons, A. W., and J. B. Boden: The Moisture Condition Test and Its Potential Applications in Earthworks, *Transport and Road Research Supplementary Report* SR522, TRRL, Crowthorne, 1979.
4. Parsons, A. W., and A. F. Toombs: The Precision of the Moisture Condition Test, *Transport and Road Research Laboratory Research Report* 90, TRRL, Crowthorne, 1987.
5. Russam, K.: Sub-soil Drainage and the Structural Design of Roads, *Transport and Road Research Laboratory Report* LR110, TRRL, Crowthorne, 1967.
6. Russam, K.: The Distribution of Moisture in Soils at Overseas Airfields, Road Research Technical Paper 58, HMSO, London, 1962.
7. Scofield, R. K.: The pF of the Water in Soil, *Trans. 3rd Int. Congr. Soil Science,* vol. 2, pp. 37–48, 1935.
8. Croney, D., and J. C. Jacobs: The Frost Susceptibility of Soils and Road Materials, *Transport and Road Research Laboratory Report* LR90, TRRL, Crowthorne, 1967.
9. Taber, S.: Freezing and Thawing of Soils as Factors in the Destruction of Road Pavements, *Publ. Rds Wash.,* vol. 11, no. 6, pp. 113–132, 1930.
10. Sherwood, P. T., and P. G. Roe: Winter Air Temperatures in Relation to Frost Damage in Roads, *Transport and Road Research Laboratory Research Report* 45, TRRL, Crowthorne, 1986.
11. Bishop, A. W., and N. Morgenstern: Stability Coefficients for Earth Slopes, *Geotechnique, Lond.,* vol. 10, no. 4, pp. 129–150, 1960.
12. Symons, I. F.: The Application of Residual Shear Strength to the Design of Cuttings in Over Consolidated Fissured Clays, *Transport and Road Research Laboratory Report* LR277, TRRL, Crowthorne, 1968.
13. Perry, J.: A Survey of Slope Conditions on Motorway Earthworks in England and Wales, *Transport and Road Research Laboratory Research Report* 199, TRRL, Crowthorne, 1989.
14. Davis, E. H.: The California Bearing Ratio Method for the Design of Flexible Roads and Runways, *Geotechnique, Lond.,* vol. 1, no. 4, pp. 249–263, 1949.
15. Black, W. P. M.: The Calculation of Laboratory and *in situ* Values of California Bearing Ratio from Bearing Capacity Data, *Geotechnique, Lond.,* vol. 11, no. 1, pp. 14–21, 1961.
16. Black, W. P. M.: A Method of Estimating the California Bearing Ratio of Cohesive Soils from Plasticity Data, *Geotechnique, Lond.,* vol. 12, no. 4, pp. 271–282, 1962.
17. Parsons, A. W.: Moisture Condition Test for Assessing the Engineering Behaviour of Earthwork Material, *Conference on Clay Fills,* Institution of Civil Engineers, London, 1979.
18. Westergaard, H. M.: Stresses in Concrete Pavements Computed by Theoretical Analysis, *Publ. Rds Wash.,* vol. 7, no. 2, pp. 23–25, 1926.
19. Stratton, J. H.: Construction and Design Problems. Military Airfields, a Symposium. *Proc. Amer. Soc. Civ. Engrs,* vol. 70, no. 1, pp. 28–54, 1944.
20. Jones, R.: Non-destructive Testing of Roads and Structures, *Publ. Wks Munic. Services Congress, 1956, Final Report,* pp. 450–473.
21. Jones, R.: In situ Measurement of the Dynamic Properties of Soil by Vibration Methods, *Geotechnique, Lond.,* vol. 8, no. 1, pp. 1–21, 1958.
22. Grainger, G. D., and N. W. Lister: A Laboratory Apparatus for Studying the Behaviour of Soils under Repeated Loading, *Geotechnique, Lond.,* vol. 12, no. 1, pp. 3–14, 1962.

23. Croney, D., and J. D. Coleman: Soil Structure in Relation to Soil Suction (pF), *J. Soil Sci.,* vol. 5, no. 1, 1954.
24. Croney, P.: The Structural Design of Road and Airfield Pavements Using Modern Analytical Technique, London University thesis, 1975.
25. Hicks, R. G., and E. N. Finn: Analysis of Results from the Dynamic Measurements Programme on the San Diego Test Road, *Proc. Assn Ashp. Pav. Technology, Michigan,* vol. 39, pp. 153–185, 1970.
26. Chaddock, B. C. J.: Repeated Triaxial Loading of Soil: Apparatus and Preliminary Results, *Transport and Road Research Laboratory Supplementary Report* 711, TRRL, Crowthorne, 1982.

Chapter

10

Preparation and Testing of the Subgrade—Capping Layers

Introduction

10.1. The soil immediately below formation level is generally referred to as the subgrade. Specifications for the compaction of earthworks usually distinguish between the soil above and below a depth of 600 mm. It is convenient therefore to regard the subgrade as the upper 600 mm of the soil foundation.

10.2. Under a well-designed pavement, the stresses in the subgrade, induced by the passage of heavy wheel loads, decrease only marginally with depth. The engineer is therefore more concerned with the average strength than with surface measurements liable to be influenced by the prevailing weather conditions.

10.3. Major rural roads are today built to a maximum gradient of about 1:25, so that a steady flow of mixed traffic can be maintained. This involves considerable amounts of cut and fill and it is not unusual for construction to be evenly divided between cut, fill, and existing ground level.

Preparation of the Formation

10.4. Final preparation of the formation should be delayed as far as possible until the contractor is ready to lay the subbase or capping layer (if required by the specification). Ideally, the two processes should be closely linked to minimize the period when the formation is left exposed to the weather. Most specifications contain a clause requiring the contractor to keep earthworks protected from rain at all times, but this is quite impracticable in a wet climate. At best, earthworks can be scheduled for the drier periods of the year and left as smooth as is feasible so that much of the rainfall will run off. The use of plastic sheeting on major contracts is not really feasible on the scale necessary, and bituminous seals have been shown to be largely ineffective.

10.5. The drawings will show the finished levels required for the road surface at the two edges of the carriageway. The corresponding levels of the formation are obtained by subtracting the nominal thicknesses of the pavement layers. As with the other pavement interfaces, the specification will define tolerances to be applied to the nominal levels. For the formation, a tolerance of about $\pm$ 25 mm is normal. The procedure adopted by the contractor to trim the formation must be decided by the contractor. However, the engineer should develop a systematic approach to checking levels. In the United Kingdom the method recommended is to install level pegs on each side of the formation at intervals of about 10 m. Transverse level measurements are made at 2-m intervals between each pair of pegs. If the pegs are left in position (releveled as necessary) then measurements can be made at the same points after each successive pavement layer is completed so that thicknesses can be checked.

10.6. The repeated passage of heavily loaded scrapers and other earthmoving equipment over the ground connecting cuttings and embankments is very likely to cause intense disturbance and shearing of the soil. The consequent rise in the suction of cohesive soils (see Pars. 9.135 and 9.136) will cause the rapid absorption of surface water and progressive softening of the soil to a considerable depth. This is a very common cause of localized early failure of flexible pavements. It is advisable therefore as part of the final preparation and shaping of the subgrade to remove the soil in these areas to a depth of about 500 mm and to replace it with freshly dug soil at the natural moisture content. The compaction will be that specified for the subgrade.

10.7. In clay cuttings the release of overburden will give rise to a progressive increase in moisture content of the underlying soil (see Par. 9.84), causing apparent heaving of the formation. This is a common reason for dispute between contractors and engineers when levels passed as satisfactory are subsequently rejected as being too high. To economize in pavement materials, some contractors tend to work to the upper tolerance on subgrades and subbases. This is particularly unwise in clay cuttings where uplift of 50 mm or more is not uncommon.

The Testing of Subgrades

General

10.8. In road construction projects the structural design, i.e., the thickness of the various layers, particularly in the case of flexible pavements, is based on the strength of the soil. The engineer, having had various tests carried out as part of the site investigation for the project, will have used them to estimate this strength. If, as is likely to be the case, an empirical design procedure has been used, based, for example, on the California Bearing Ratio (CBR) test, then the engineer will have, or should have, studied the relationships between CBR value and the density and moisture content of the soil or soils present at the site. On the basis of this laboratory work selections will have been made for the maximum moisture content and minimum dry-density values which will ensure the CBR value used in the design. Alternatively, if the engineer elects to use a method specification for compacting the subgrade, it will be assumed that the requirements have been so framed as to produce the CBR value on which the design has been based. In either case, if the contractor carries out the compaction requirements of the specification correctly and a lower strength is achieved, the responsibility is the engineer's and not the contractor's. The engineer must then either modify the design and accept any financial penalties involved or proceed according to the original design in the knowledge that the pavement may have a shorter life than was intended. The same arguments would apply if the engineer were using a fundamental design procedure based on the elastic properties of the soil. It would not in general be feasible for the engineer to specify a minimum CBR value or a minimum modulus of elasticity for the soil, since this would transfer to the contractor all the responsibility involved in organizing and overseeing the site investigation.

10.9. The above discussion does emphasize the need, when a method specification is used for the preparation of the subgrade, for the engineer to pay close attention to the compaction plant and the way in which it is operated, i.e., the number of passes and the layer thickness being used. The engineer should at the outset check that the plant on site and its operators are sufficient to provide adequate compaction, bearing in mind the rate of progress of the work, and should have enough inspectors on site to make frequent checks that the requirements of the specification are being completely met. There may well be a case, where a method specification is being used for the construction of the embankments, to change to an end-product specification based on moisture content and dry density for the compaction of the subgrade.

Moisture Content and Dry Density Tests

10.10. Where the required state of compaction of the subgrade is defined in the contract in terms of moisture content and dry density, both contractor and engineer should carry out continuous testing of those properties. The contractor for a major road construction project should have on site a well-equipped and completely staffed laboratory. Long experience shows that disputes and delays are much less likely to occur when both contractors and engineers maintain efficient site laboratories between which there exists understanding and mutual respect. The assistant in charge of each laboratory should be made responsible for keeping records of every test made and its date and location. This information will be called for in the event of arbitration or litigation.

10.11. If in the contract the density of the subgrade is expressed as a percentage of the maximum dry density obtained in one of the standard laboratory compaction tests (see Par. 9.57) then the appropriate laboratory test should be carried out at least once daily by both the engineer and the contractor, using soil taken from the subgrade. In the early stages of the work frequent determinations of dry density and moisture content should be made using the sand replacement method (AASHTO Designation T191-86 or BS 1377:1975 Test 15), again by both the engineer and the contractor. During this stage a correlation should be established with the density measured by a nuclear density meter of the backscatter type (AASHTO Designation T238-86, Method A). This equipment gives the wet (or bulk) density, and a separate measurement of moisture content at the same location is required to deduce the equivalent dry density. Once a reliable correlation between the two methods has been obtained, increasing reliance can be placed on the nuclear tests. However, a revised calibration will be needed if a significant change of soil type occurs in the subgrade.

10.12. If the specification for the subgrade is in terms of a maximum moisture content and maximum air content, in addition to the moisture content and dry density, the specific gravity of the soil particles will need to be measured (AASHTO Designation T100-86 or BS

1377:1975 Test 6). The maximum air voids content can then be read off the appropriate chart given in the appendix to Chap. 9.

The Strength of the Subgrade

10.13. Empirically designed pavements, whether flexible or concrete, will in general employ the CBR test to define the strength of the subgrade. This test has been described in detail in Chap. 9 and is covered by AASHTO Designation T193-81 and BS 1377, Test 16. In designing a pavement using the CBR value the engineer will, on the basis of laboratory testing at the site investigation stage, have defined the required moisture content and dry density of the finished subgrade in such a way that the design CBR will be achieved in the properly constructed subgrade.

10.14. In checking the subgrade the engineer will need to carry out further laboratory CBR tests on soil taken from the prepared subgrade and remoulded in the standard CBR mould. It will be remoulded to the moisture content and dry density measured in the finished subgrade at the point from which the soil was taken. The CBR determined in this way should be equal to or marginally greater than the value used in the design of the pavement. Compliance will in general be more critical for flexible pavements than is the case for concrete construction. Several CBR determinations should be made in this way each day, and they would be coupled with the dry-density and moisture content determinations. The tests are of necessity slow, giving results with a delay of 24 hours. It is usual, therefore, to adopt supplementary CBR measurements in situ which can be carried out more quickly. Because of the mould restraint factor, laboratory CBR values tend to be greater than measurements in situ at the same density and moisture content. The engineer must therefore carry out checks to establish the relationship between laboratory tests and tests in situ for the soils in question. Some comparative tests from U.K. and U.S. sources are shown in Table 10.1. For heavy and medium clays the results in situ are only a little lower than laboratory values, but for less cohesive soils with low air voids content the difference is larger and for coarse granular soils they can be very large. CBR tests can be carried out in situ from a rig attached to the back of a truck (Fig. 10.1). For weaker soils, where the reaction required is smaller, a light pickup vehicle can be used to avoid damage to the subgrade.

10.15. A valuable tool for assessing the uniformity of subgrades in both the horizontal and the vertical direction is the hand-held soil

TABLE 10.1 Comparison of Laboratory (Remoulded) and in situ CBR Values

Soil type	Source of data	Dry density, lb/ft^3 (kg/m^3)	Moisture content, %	CBR, %	
				Remoulded	In situ
Heavy clay (LL 69 PL 27)	Transport and Road Research Laboratory	95 (1522)	24.8	8.9	7.9
Clay (LL 59 PL 22)	Transport and Road Research Laboratory	96 (1538)	25.1	3.9	3.0
Silty clay (LL 37 PL 23)	U.S. Waterways Experimental Station	109 (1746)	19.5	2.0	12
		107 (1714)	19.0	5.0	11
		104 (1666)	16.1	22	22
Sandy clay (LL 30 PL 18)	Transport and Road Research Laboratory	95 (1522)	19.2	2.2	3.1
Clayey sand	U.S. Waterways Experimental Station	116 (1858)	12.2	14	7.0
		114 (1826)	12.6	10	9.0
		109 (1746)	10.0	12	18
Single-size sand	Transport and Road Research Laboratory	98 (1570)	8.0	24	7.5
Crushed slag	Transport and Road Research Laboratory	140 (2243)	4.8	412	44

Figure 10.1 Equipment for in situ measurement of CBR.

assessment cone penetrometer shown in Fig. 10.2. Two scales, corresponding to different sizes of cone, give the soil strength in terms of a "cone index" or the equivalent CBR in situ over the range 0 to 15 percent. The readings correlate fairly closely with the CBR in situ on fine-grained soils, but not for coarse soils. For all soils, calibration against laboratory or in situ CBR tests is necessary. With an extended shaft the instrument can be used satisfactorily to examine the varia-

Figure 10.2 Soil assessment cone penetrometer.

tion of the CBR value with depth, for soils in the range CBR 0 to 5 percent. This application is useful in exploring the depth of "soft spots." The cone is pushed at constant rate into the soil and the steady reading observed at the different depths scribed on the shaft. Proof rolling to locate soft spots is strongly to be deprecated. Very heavy rollers or vehicles cause unnecessary damage to the subgrade without giving any quantitative information. On weak soils of CBR <5 percent an unloaded panel truck or a private car can sometimes be used to show weak areas by the depth of tire marks.

10.16. When pavements are designed using multilayer elastic theory, the "property" criterion for the subgrade is its modulus of elasticity, corresponding to its moisture and density condition after compaction. The determination of the modulus of elasticity (or resilient modulus, as it is called in the United States) is much more complicated than the determination of the CBR value. Various methods have already been considered in Chap. 9. Basically, it involves carrying out repeated-loading triaxial tests on the soil and determining the ratio of stress and strain for the range of stress conditions generated by traffic loading.

10.17. To carry out this complex test procedure on soil with a wide range of density and moisture content conditions is impracticable outside a research institution. The engineer is advised therefore to proceed as if designing the pavement empirically using the CBR design procedure, i.e., to follow the steps outlined in Par. 10.8 to derive an appropriate level of CBR for design purposes based on a maximum moisture content and a minimum dry density. Using these values of moisture content and dry density, remoulded cylindrical triaxial test samples would be prepared and subjected to repeated loading tests using increasing deviator stresses and associated cell pressures to simulate the vertical and radial stresses induced by wheel loads. For each such combination of stresses the effective elastic modulus after several hundred stress applications would be calculated and the minimum value used as input into the computer program adopted to calculate the stresses in the pavement layers. This subject is discussed in detail in Chap. 23. With this approach the information to be derived from the site investigation would be the same for both the empirical and structural design procedures.

10.18. The equipment and test procedure for measuring resilient modulus is described in AASHTO Designation T274-82 (1986). However, current research is directed toward relating the elastic properties of soils to other simpler tests, such as the index properties, shear strength, and CBR. This has been discussed in Chap. 9.

The Use of Subgrade Capping and Geotextile Fabrics in Earthworks

10.19. When the soil at formation level is weak (CBR 2 to 4 percent), 400 to 600 mm of subbase will be required in the construction of heavily trafficked flexible pavements. Under these circumstances it is a common cost-cutting practice to subdivide the subbase layer, the upper part normally being an angular crushed rock and the lower part a naturally occurring gravel or gravel sand. In the United Kingdom these are termed type 1 and type 2 subbases, and their specifications are discussed in Chap. 11.

10.20. There is naturally a temptation to believe that if this process of gradation of materials is extended to a third layer between the formation and the road base material, then considerable savings in cost might be made. To provide such a third layer the concept of "capping" the subgrade has been introduced. Whether it does in fact prove to be economic is very questionable in many cases, and there is no doubt that attempts by contractors in the United Kingdom to meet specifications for capping layers have led to expensive confrontations between clients and contractors. This is in part due to inadequate specifications.

10.21. If a soil is abnormally weak (CBR <2 percent), owing either to its particle size distribution or to a high natural moisture content, tipping and rolling in granular subbase material eventually produces a measure of mechanical stabilization, but the effectiveness of much of the granular material will be lost as a result of contamination by the wet soil. The local use of geotextile fabrics to separate the soil and the granular material has been shown to be effective in reducing subsequent deformation under construction traffic.[1] Table 10.2 shows the results of tests carried out by the TRL, where a crushed stone layer 200 mm thick was laid over a heavy clay soil of in situ CBR 2 percent with and without separation of the materials by a nonwoven polypropylene–nylon fabric with a specified strength of 10.5 kN/m and weight of 450 g/m^2. The two forms of construction were trafficked by repeated passes of a loaded truck with front wheel loading of 2.25 t and rear-wheel loading of 4.55 t (carried on a dual wheel assembly). There was no initial difference in elastic deflection of the two sec-

TABLE 10.2 The Influence on Deformation of a Geotextile Fabric Separation Layer between the Subgrade and a Crushed Stone Base

	Permanent deformation at the surface, mm	
Number of passes of truck	Without fabric	With fabric
50	28	20
100	37	25
200	47	29
300	53	32

tions, but it is possible that such a difference would have developed with time. The difference in deformation was therefore largely attributable to the influence of the fabric in preventing intermixing of clay and stone.

10.22. In the United States subgrade improvement by the addition of cement and lime has for many years been used to provide a capping for weak subgrades. The much smaller road construction programs in Europe have not favored similar developments, although specialist subcontractors with the necessary single- and multiple-pass processing equipment are now available to encourage this type of subgrade treatment when it is applicable, and where it can be shown to be economic.

10.23. Some broad conclusions can be drawn from the experience gained from capping subgrades in the United Kingdom, and there seems no reason why they should not apply generally. They are as follows:

1. Capping or subgrade improvement should be considered only when the in situ CBR of the formation when properly compacted is likely to be less than 4 percent.
2. With prepared subgrades of in situ CBR less than 2 percent, any imported capping material or subbase should be separated from the underlying soil by a geotextile membrane.
3. Capping with material other than normal subbase should be required by the contract documents only when the engineer is satisfied that granular materials with relevant properties between those of the subgrade and those of normal subbase are available and that their use as capping will be economically advantageous to the client.
4. Forming a capping layer by processing the soil with cement, lime, or other chemical additives will be required or permitted only when the engineer or the contractor have carried out laboratory or field trials to show the method can be economically and advantageously used.

10.24. The U.K. Department of Transport Specification for Highway Works (1986) permits the use of four capping materials or processes.[2] These include two unbound materials, designated class F1 and class F2, and two stabilized materials with very wide grading envelopes. The unbound materials, class F1 and class F2, must conform to the gradings shown in Table 10.3. The material suitable for cement stabilization must conform to the grading limits for class 6E materials and the material suitable for lime stabilization to the limits for class 7E materials also included in Table 10.3.

10.25. The grading limits for class F1 materials are virtually identical to those for type 2 subbase. The only difference appears to be that softer aggregates such as chalk and soft limestone can be used and presumably unspecified waste materials which would not be permitted as subbase. The grading for the class F2 material overlaps with the grading requirements of both type 1 and type 2 subbase. The maximum size of aggregate is greater for class F2 material, and it is presumably intended for use where greater thicknesses of capping are specified.

10.26. For both class F1 and F2 materials the specification permits the engineer to require that a stated CBR value is achieved in the compacted material. The value commonly selected is 14 percent, although it is often not made clear to the contractor whether this is a laboratory-determined CBR or whether it is intended to be an in situ value. Few engineers or contractors have any clear idea how a CBR 14 percent material will behave under construction traffic, and frequent disputes arise when the resident engineer attempts to use deformation as a performance criterion.

10.27. The stabilization alternatives for capping are primarily intended for use where the soils at the site conform to the gradings shown in Table 10.3 for class 6E and 7G materials. The importation of soil to the site for this purpose would never be economic. The performance criterion used for lime- and cement-stabilized capping is the laboratory CBR value, which must be set by the engineer, who must therefore carry out sufficient laboratory testing to ensure that the required strength will be obtainable economically. In general, once work has started contractors would also be required to carry out site trials to validate their methods of working. Cement stabilization is unlikely to be a cheap option. For soils with a grading close to the finer boundary of the grading envelope, cement contents in excess of 10 percent by weight could be needed to meet the strength requirement.

10.28. There is little experience with lime stabilization of soils in the United Kingdom, beyond laboratory and small-scale field trials. The process is intended primarily for treating wet, heavy clays and its effect is gradually to reduce the plasticity of the clay and thereby increase its strength. A method of assessing the suitability of a clay

TABLE 10.3 Grading Requirements for Capping Materials

	BS sieve size															
	mm													μm		
Class	125	90	75	37.5	28	20	14	10	6.3	5	3.35	2	1.18	600	150	63
Granular materials:																
6 F1			100	75–100				40–95		30–85				10–50		<15
6 F2	100	80–100	65–100	45–100				15–60		10–45				0–25		0–12
Cement stabilized material:																
6 E	100	85–100						25–100						10–100		0–10
Lime stabilized material:																
7E			100		95–100											15–100

for lime stabilization is given in AASHTO Designation M216-84 (ASTM Designation C977-83A). The method is based on the measurement of the pH of the soil-lime mixture. The determination of the strength of soil-lime mixtures is covered in AASHTO Designation T220-66 (1984).

References

1. Potter, J. F., and E. W. H. Currer: The Effect of a Fabric Membrane on the Structural Behaviour of a Granular Road Pavement, *TRRL Report* LR 996, TRRL, Crowthorne, 1981.
2. Department of Transport: Specification for Highway Works, Part 2, HMSO, London, Clauses 613–615, 1986.

Chapter

11

Unbound Subbases and Road Bases

Introduction

11.1. The subbase of a flexible pavement is the layer between the subgrade (or the capping layer if one is used) and the base. In a concrete pavement it is the layer separating the concrete slab from the subgrade. The base, or road base, is the layer immediately below the bituminous surfacing of a flexible pavement.

11.2. Subbases can be constructed from unbound compacted granular material, or similar material bound with a small amount of cement. Bases can also be made from unbound crushed rock, but more usually the aggregate is bound with either cement or bitumen.

11.3. This chapter deals exclusively with unbound subbases and bases. Subsequent chapters deal with cement-bound subbases and bases and with bituminous-bound bases.

11.4. The subbase has three main functions, as follows:

1. It is a structural layer which will accept greater compressive stress than the subgrade, and it thus reduces the deformation of the pavement under traffic loading.
2. It provides a working platform over which the construction plant can operate when the layer above is being placed.
3. If thick enough, it prevents frost from penetrating into frost-susceptible subgrades.

11.5. The main function of the base is to reduce the vertical compressive stress induced by traffic, in the subbase and the subgrade, to a level at which no unacceptable deformation will occur in these layers. Unbound bases perform this function solely by virtue of their thickness and state of compaction. Bound bases perform the same function by a combination of thickness and stiffness.

U.S. Specifications for Unbound Subbases and Bases

11.6. Both AASHTO and ASTM have issued specifications for unbound subbase and base materials. These are briefly discussed below.

11.7. In AASHTO Designation M147-65 (1980) are specified the six gradings, A to F, shown in Table 11.1 and Fig. 11.1. Grading A is used primarily for bases and gradings B to D refer to subbase materials. Gradings E and F are used as top courses for unsurfaced roads. Requirements in relation to plasticity of fines and aggregate durability are shown below Table 11.1.

11.8. Two gradings, one for subbase and the other for base, are specified in ASTM Designation D2940-74 (reapproved 1985), and are shown in Table 11.2 and Fig. 11.2. Again, requirements in relation to plasticity of fines and durability of aggregate are shown under Table 11.2.

TABLE 11.1 Grading Requirements for Unbound Subbases and Base Materials to AASHTO Designation M147-65 (1980)

	Grading: percentage passing					
Sieve size	A	B	C	D	E	F
50 mm	100	100				
25 mm		75–95	100	100	100	100
9.5 mm	30–60	40–75	50–85	60–100		
4.75 mm	25–55	30–60	35–65	50–85	55–100	70–100
2 mm	15–40	20–45	25–50	40–70	40–100	55–100
425 μm	8–20	15–30	15–30	25–45	20–50	30–70
75 μm	2–8	5–20	5–15	5–20	6–20	8–25

Other requirements:

1. Coarse aggregate (>2 mm) to have a percentage wear by Los Angeles test not more than 50.
2. Fraction passing 425 μm sieve to have a liquid limit not greater than 25 percent and a plasticity index not greater than 6 percent.

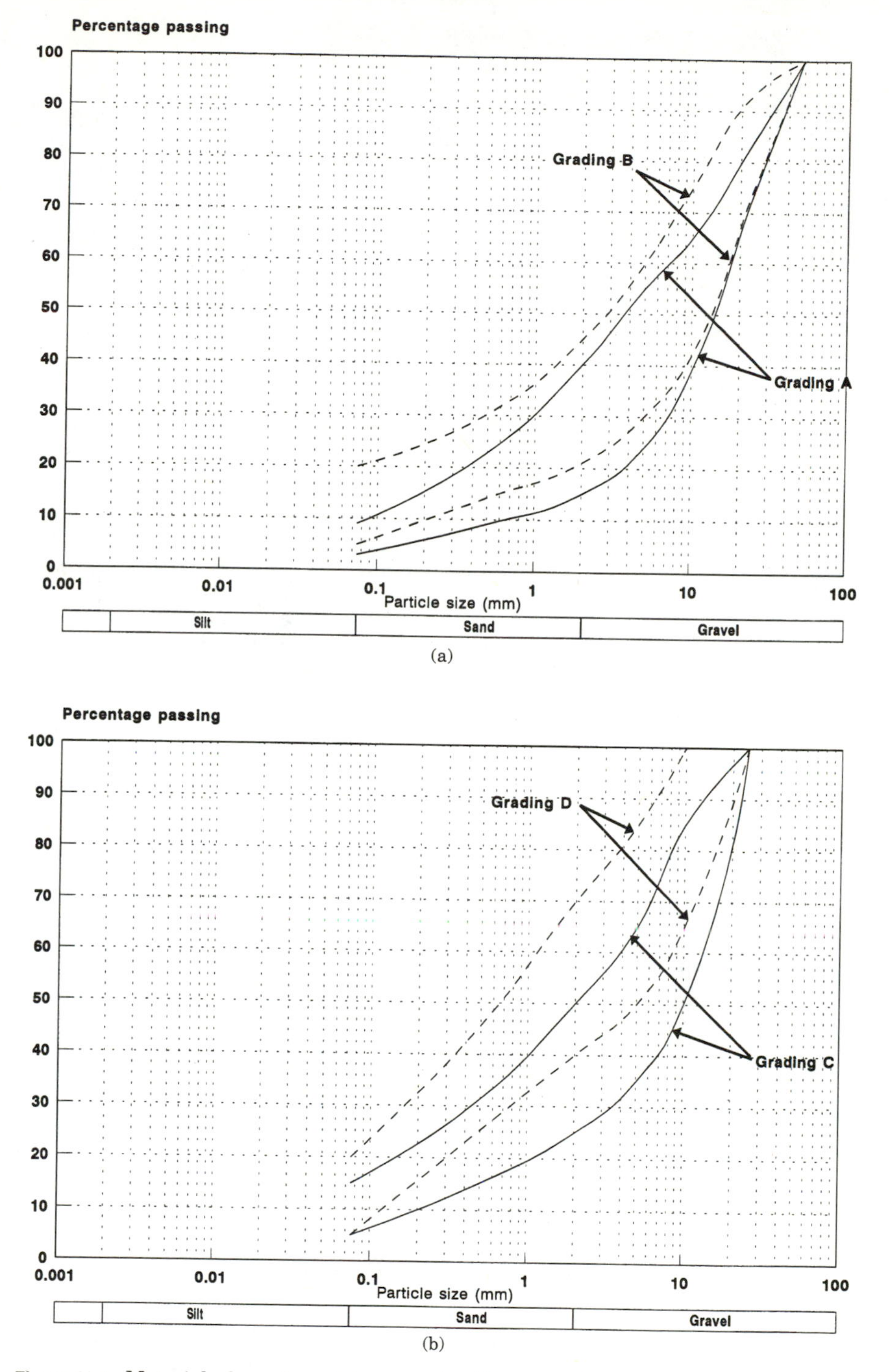

Figure 11.1 Materials for unbound bases and subbases. AASHTO Designation M147-65 (1980): (*a*) gradings A and B, (*b*) gradings C and D.

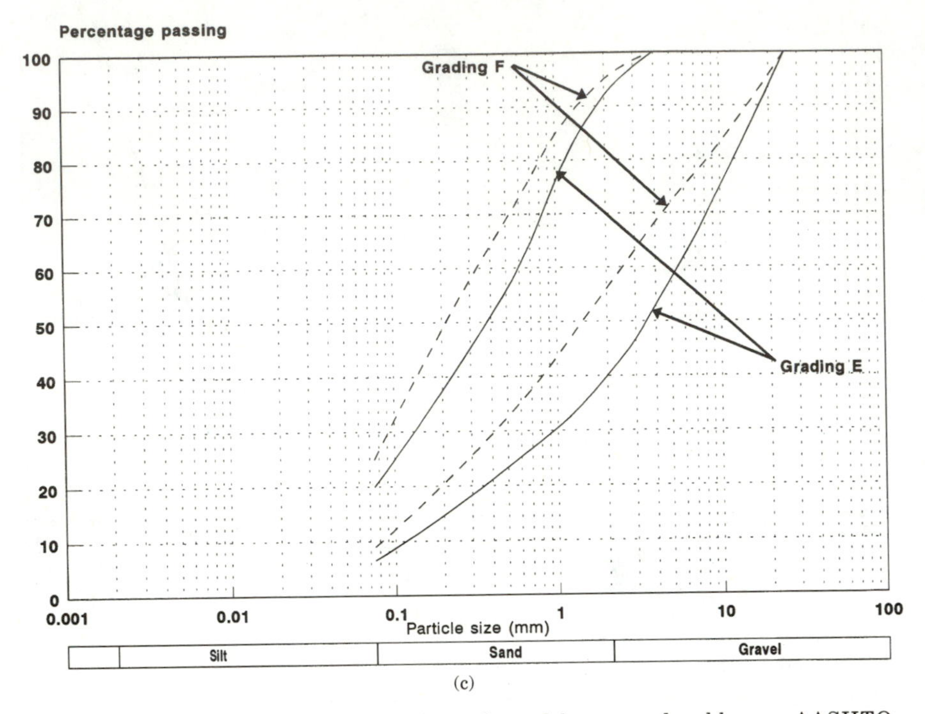

(c)

Figure 11.1 *(Continued)* Materials for unbound bases and subbases. AASHTO Designation M147-65 (1980): (*c*) gradings E and F.

TABLE 11.2 Grading Requirements for Bases and Subbases for Highways and Airports to ASTM Designation D2940-74 (Reapproved 1985)

	Grading: percentage passing	
Sieve size	Bases	Subbases
50 mm	100	100
37.5 mm	95–100	90–100
19 mm	70–92	
9.5 mm	50–70	
4.75 mm	35–55	30–60
600 μm	12–25	
75 μm	0–8	0–12

Other requirements:
1. Coarse aggregate to be hard and durable.
2. Fraction passing the 75-μm sieve not to exceed 60 percent of the fraction passing the 600-μm sieve.
3. Fraction passing the 425-μm sieve shall have a liquid limit no greater than 25 percent and a plasticity index not greater than 4 percent.

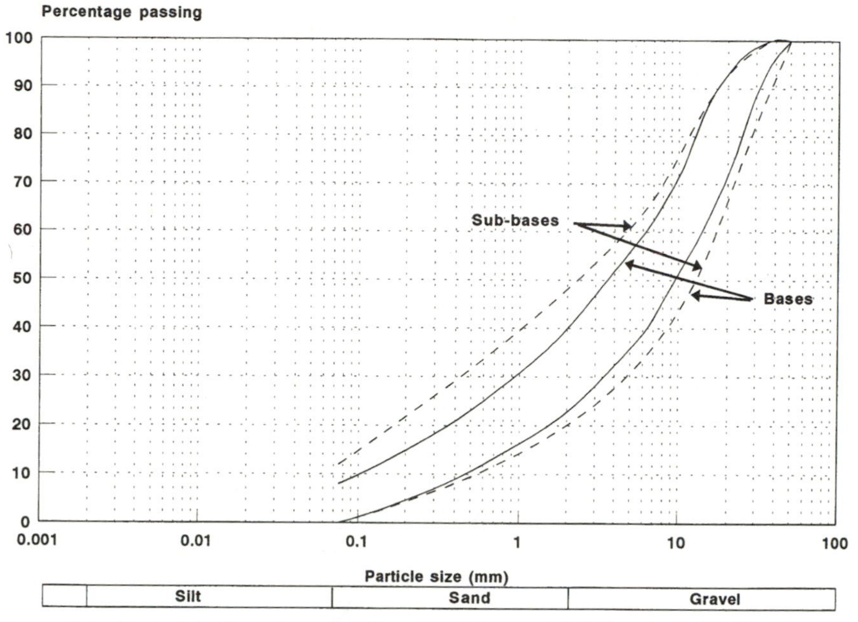

Figure 11.2 Materials for unbound subbase and base—ASTM Designation D2940-74.

11.9. Neither the AASHTO nor the ASTM specifications give a strength criterion for the compacted materials, but the Asphalt Institute Thickness Design Manual[1] requires in its Table V-3 a CBR value of 20 percent for subbase and 80 percent for base material. These are laboratory test results carried out at the appropriate moisture content and density conditions, and tested after 4 days' soaking.

British Specifications for Unbound Subbases and Road Bases

11.10. Types 1 and 2 subbases. The U.K. Department of Transport Specification for Highway Works, Part 3, Clauses 801–804, defines two unbound subbase materials, type 1 and type 2. The gradings, determined by wet sieving, are shown in Table 11.3 and the grading envelopes in Fig. 11.3.

11.11. The aggregate for type 1 subbase is restricted to crushed rock, crushed concrete, or well-burnt, nonplastic colliery shale, and the material passing the 425-μm sieve should be nonplastic. The aggregate for type 2 subbase additionally includes natural sand and gravel, and the plasticity index of the material passing the 425-μm sieve is limited to 6 percent. For both types, to ensure that the aggregate is

TABLE 11.3 Grading Limits for Type 1 and Type 2 Subbases

	Percentage by mass passing	
BS sieve size	Type 1	Type 2
75 mm	100	100
37.5 mm	85–100	85–100
10 mm	40–70	45–100
5 mm	25–45	25–85
600 μm	8–22	8–45
75 μm	0–10	0–10

sufficiently hard, the 10 percent fines value determined with the aggregate saturated but surface-dry is required to be not less than 50 kN.[2]

11.12. Unbound road bases. Two types of unbound road base are in use in the United Kingdom. These are wet-mix macadam and dry-bound macadam. The first is defined in the U.K. Department of Transport Specification for Highway Works, Part 3, Clause 805. The second is no longer included in the specifications but is widely used in various forms for the less heavily trafficked rural roads.

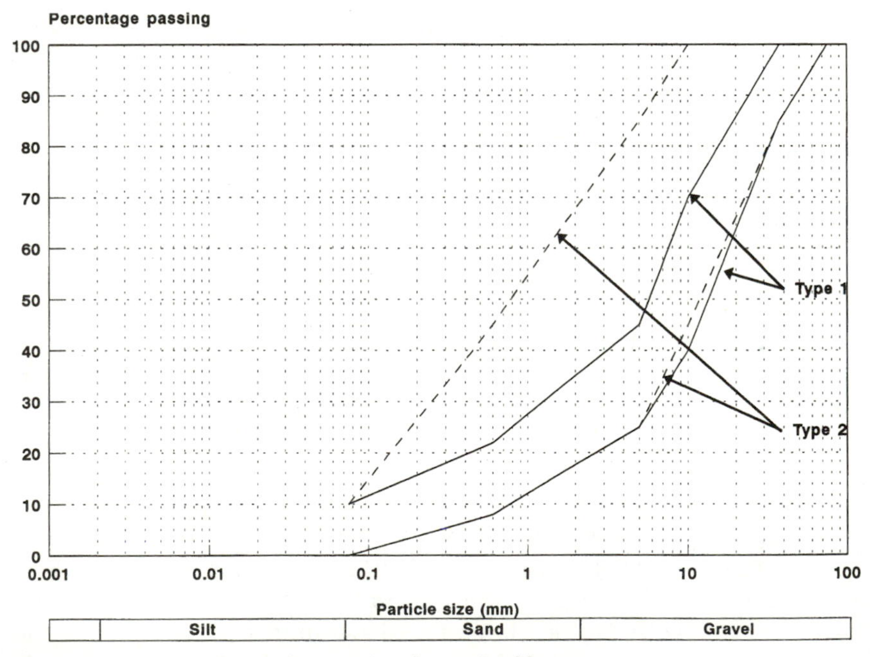

Figure 11.3 Grading limits for type 1 and type 2 subbases.

TABLE 11.4 Grading Limits for Wet-Mix Macadam

BS sieve size	Percentage of mass passing
50 mm	100
37.5 mm	95–100
20 mm	60–80
10 mm	40–60
5 mm	25–40
2.36 mm	15–30
600 μm	8–22
75 μm	0–8

11.13. Wet-mix macadam. The grading of the aggregate for wet-mix macadam is given in Table 11.4 and Fig. 11.4. The aggregate consists of nonflaky crushed rock or crushed slag.

11.14. Dry-bound macadam. In this process a layer of single-size crushed stone or crushed slag of nominal size 50 or 37.5 mm is uniformly laid to a thickness between 75 and 100 mm. The use of a box spreader or a bituminous paving machine is recommended. The spread material is given two passes of a smooth-wheel roller of weight

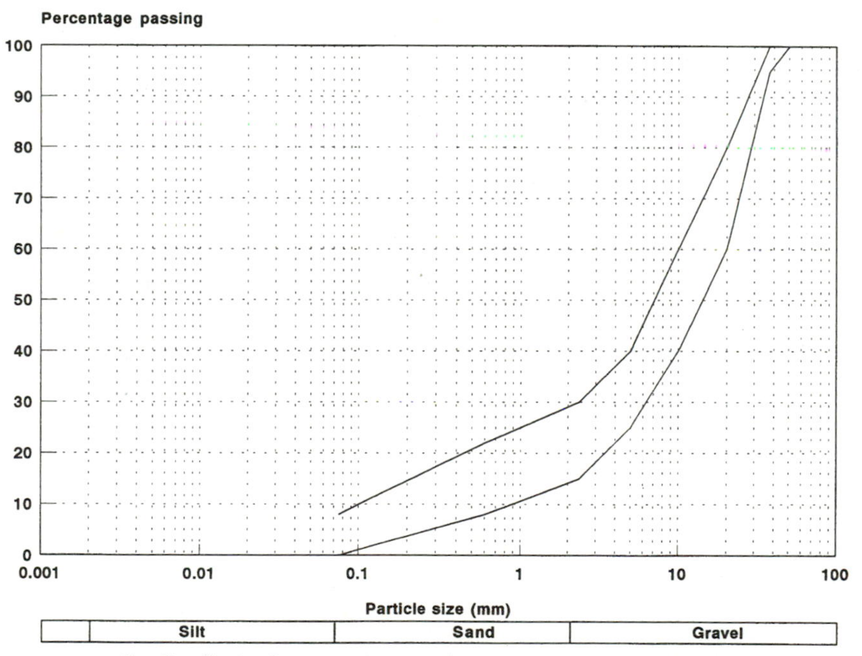

Figure 11.4 Grading limits for wet-mix macadam.

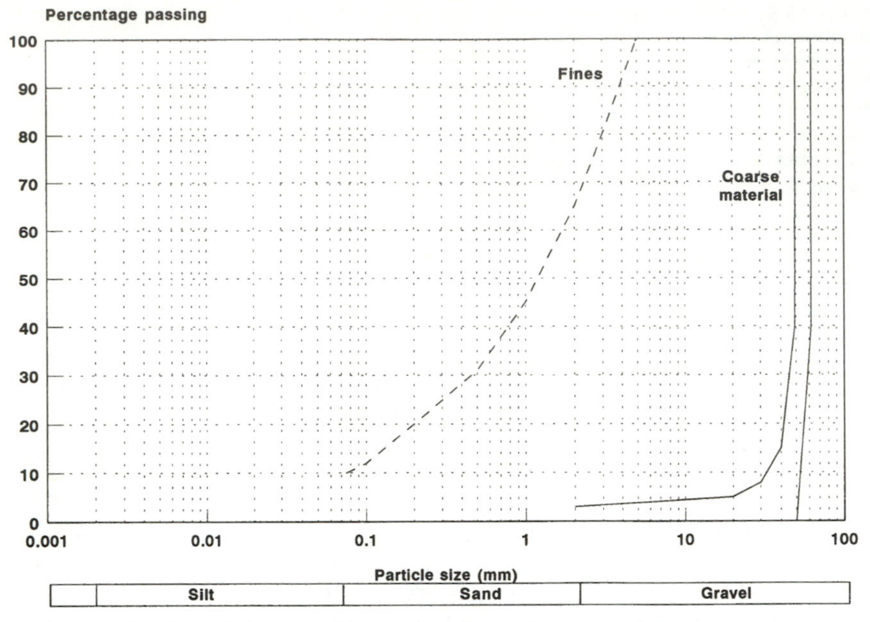

Figure 11.5 Dry-bound stone road bases—gradings of coarse and fine aggregates.

8 to 10 t. The stone is "blinded" by progressive applications of fine aggregate of maximum size 5 mm and having less than 10 percent passing the 75-μm sieve. This material is also crushed rock or crushed slag. Each application of fine aggregate is vibrated into the voids of the larger stone using a vibrating plate compactor having a static pressure under the base plate of at least 13.8 kN/m^2 or a vibrating roller having a static force per 100-mm width of at least 1.76 kN. The blinding operation is continued until no more fines are accepted, and the surface is then brushed off. Acceptable gradings for the coarse and fine aggregates are shown in Fig. 11.5. This is essentially a dry-weather process, and it is most important that the fines are kept dry before and during their application and vibration. The dry density of a dry-bound road base should be very similar to that obtained in a wet-mix road base using the same aggregate type, and the performance of the two types of road base has been shown to be very similar.

Compaction of Unbound Subbases and Bases

11.15. All unbound subbases and bases laid in the United Kingdom are required to be non-frost-susceptible. The moisture content at the

TABLE 11.5 Compaction Requirements for Granular Materials Types 1 and 2 and Wet-Mix Macadam

Type of compaction plant	Category	Number of passes for layers not exceeding compacted thickness		
		110 mm	150 mm	225 mm
	Mass per m width of roll:			
Smooth-wheeled roller (or vibratory roller operating without vibration)	2700–5400 kg	16	u/s	u/s
	Over 5400 kg	8	16	u/s
	Mass per wheel:			
Pneumatic-tired roller	4000–6000 kg	12	u/s	u/s
	6000–8000 kg	12	u/s	u/s
	8000–12,000 kg	10	16	u/s
	Over 12,000 kg	8	12	u/s
	Mass per m width of roll:			
Vibratory roller	700–1300 kg	16	u/s	u/s
	1300–1800 kg	6	16	u/s
	1800–2300 kg	4	6	10
	2300–2900 kg	3	5	9
	2900–3600 kg	3	5	8
	3600–4300 kg	2	4	7
	4300–5000 kg	2	4	6
	Over 5000 kg	2	3	5
	Mass per sq m of plate:			
Vibrating-plate compactor	1400–1800 kg/m^2	8	u/s	u/s
	1800–2100 kg/m^2	5	8	u/s
	Over 2100 kg/m^2	3	6	10
	Mass:			
Vibrotamper	50–65 kg	4	8	u/s
	65–75 kg	3	6	10
	Over 75 kg	2	4	8
	Mass:			
Power rammer	100–500 kg	5	8	u/s
	Over 500 kg	5	8	12

time of laying should be close to the optimum for the type of compaction plant to be used. Compaction should be in accordance with Table 11.5.

Comparison of British and U.S. Specifications for Unbound Base and Subbase Materials

11.16. The extremes of the grading envelopes of materials B, C, and D in the AASHTO specification correspond quite closely to the British type 2 subbase material. Material A of the AASHTO specification is also very similar to that of type 1 subbase and wet-mix road base used in the United Kingdom. The subdivisions used in the AASHTO specification cover materials of different maximum sizes, and this may have advantages when considering layer thickness.

Structural Properties of Unbound Materials

11.17. Hicks and Monismith have published the results of comprehensive studies of the modulus of elasticity (resilient modulus) of a wide range of unbound granular materials and crushed rocks. Figures 11.6 and 11.7 are reproduced from their paper.[3]

11.18. Figure 11.6 shows the results from repeated loading triaxial tests on dry crushed gravel of a type suitable for subbases. The samples were conditioned by about 1000 applications of a medium-stress regime before the commencement of each series of tests. It was found possible to characterize the resilient modulus after 50 to 100 applications of stress. As would be expected, the modulus varied widely with confining stress. Figure 11.7 shows that plotted on a loglog basis the relationships between resilient modulus and confining pressure and

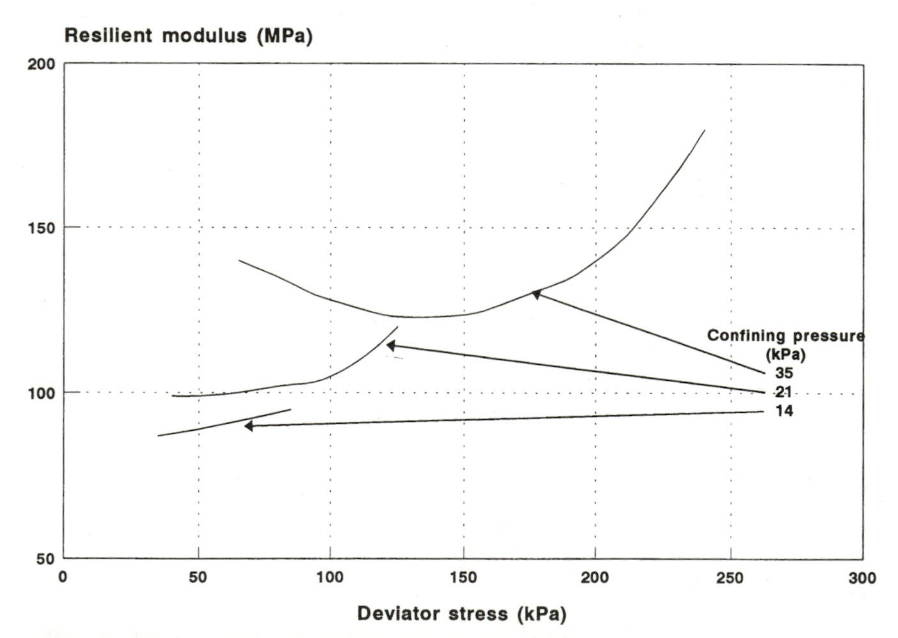

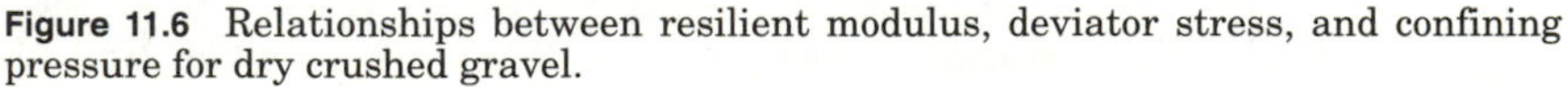
Figure 11.6 Relationships between resilient modulus, deviator stress, and confining pressure for dry crushed gravel.

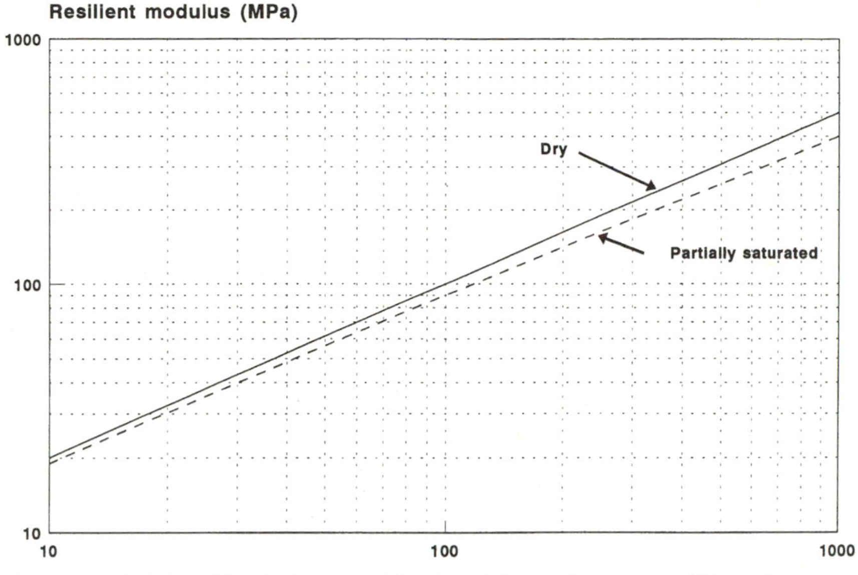

Figure 11.7 Relationships between resilient modulus and stress conditions for granular materials.

between resilient modulus and the sum of the principal stresses are linear, and moisture content has little influence on the relationships. The tests showed that the grading and state of compaction of crushed rock and crushed gravel have only a small influence on the modulus.

11.19. The results given above provide useful target values for the elastic properties of subbase and unbound base materials to be adopted in the structural analysis of pavements. An initial modulus of 150 MPa with a Poisson ratio of 0.3 is usually selected for the unbound base and a modulus of 100 to 150 MPa for the subbase. Except when the subbase is unusually stiff, the use of such high modulus values leads to the prediction of significant tensile stresses in the unbound material (>10 kPa). A stress-dependent modulus must therefore be used so that tensile stresses in excess of 10 kPa are not predicted.

References

1. Thickness Design—Pavements for Highways and Streets, Asphalt Institute Manual Series no. 1 (MS-1), Asphalt Institute, College Park, Md., 1984.
2. British Standards Institution: Methods for Sampling and Testing of Mineral Aggregates, Sands and Fillers, British Standard 812: Part 3: 1975, BSI, London, 1975.
3. Hicks, R. G., and C. L. Monismith: Factors Influencing the Resilient Response of Granular Materials, Highway Research Board Record 345, National Research Council, pp. 15–31, Washington, D.C., 1971.

Chapter

12

Cement-Treated Subgrades, Subbases, and Bases

Introduction

12.1. The procedure of mixing cement with weak soils to form a foundation on which a plant could operate was developed in the United States some 65 years ago. The material so produced is sometimes referred to as soil-cement.

12.2. During World War II the U.S. Corps of Engineers produced designs for the rapid paving of forward airfields in Europe and the Far East. The soil stabilization process was then extended to substantially clay-free materials to produce cement-bound granular materials suitable for subbases, and in stronger form for bases.

The Cement Treatment of Subgrades

12.3. A flexible pavement designed to carry 100 million standard axles would need to have a thickness of about 400 mm of granular subbase if the subgrade had a CBR value of 3 percent or less. This thickness can sometimes be reduced significantly if the soil is of a type which can be treated by the addition of a small amount of cement.

12.4. There is no doubt that the mixing of cement with a soil which is free of deleterious matter will increase the strength and render the soil more capable of carrying construction traffic. The question the engineer must answer is whether the improvement is economic when considered in relation to:

1. The cost of additional subbase material necessary to effect the same increase in pavement strength
2. The cost of the cement and the processing operations

12.5. When the soil-cement process is used for improving the subgrade or providing a subbase, the field operations must be preceded by a program of laboratory testing. This will be aimed at determining the cement and water contents to be used to obtain the required strength. The latter is measured using 150-mm cubes. To determine the optimum moisture content to be used, the soil is compacted in three layers in the cube mold, without any cement. The wet density is obtained for a range of moisture contents and the optimum moisture content for maximum density deduced. A moisture content of about 1 percent higher than this value is used for the mixing of the soil-cement. A number of cubes are then made using this moisture content and a range of cement contents, the same compaction being used. After 7 days' curing in a saturated atmosphere, crushing strengths are determined and the cement content which gives the strength requirement is selected.

12.6. Where the process is used to enable a plant to move effectively when laying subbase material, a strength of about 2.5 N/mm^2 at 7 days should be sufficient. If, however, it is used to replace subbase, the strength required should be increased to 4 N/mm^2.

12.7. Cohesive soils are difficult to process even in paddle-type mixers, and in practice a mix-in-place method will be adopted, comprising (1) scarifying the soil, (2) cement spreading, (3) mixing to the depth required, and (4) compaction. In the United States, single-pass trains to carry out these operations in sequence are sometimes used.

12.8. In Europe interest in the soil-cement process has waned in recent years, although it is sometimes used to stabilize weak pockets of soil located during major construction projects.

Cement-Bound Granular Material

12.9. This is exclusively a plant-mixed material, using a paddle or pan-type mixer, but otherwise it can be regarded as a rather stronger form of soil-cement made with granular (crushed rock or gravel) aggregate. The grading should fall between the limits shown in Table 12.1. The 7-day crushing strength requirement is between 5 and 7 N/mm^2.

TABLE 12.1 Grading Limits for Cement-Bound Granular Material

BS sieve size	Percentage by mass passing
50 mm	100
37.5 mm	95–100
20 mm	45–100
10 mm	35–100
5 mm	25–100
2.36 mm	15–90
600 μm	8–65
300 μm	5–40
75 μm	0–10

12.10. The preliminary laboratory testing to determine the optimum moisture content and the strength and density is exactly the same as that described for soil-cement in Par. 12.5.

12.11. When used as a subbase for flexible pavements this material can be hand-laid, but when it provides the base for such a pavement, or the subbase for concrete roads, it should be laid by paver.

Lean Concrete

12.12. This material is normally made from batched coarse and fine aggregates, although naturally occurring all-in washed aggregate can also be used. The aggregate will be crushed rock, gravel, or crushed air-cooled blast furnace slag or a combination of these materials. The grading limits are shown in Table 12.2, which caters for nominal maximum sizes of 40 and 20 mm.

TABLE 12.2 Grading Limits for Lean Concrete

	Percentage by mass passing nominal maximum size	
BS sieve size	40 mm	20 mm
50 mm	100	
37.5 mm	95–100	100
20 mm	45–80	95–100
5 mm	25–50	35–55
600 μm	8–30	10–35
150 μm	0–8*	0–8*

*0–10 for crushed rock fines.

12.13. The U.K. specification provides for normal-strength lean concrete with a 7-day compressive strength of 6 to 10 N/mm^2 and a higher-strength material of 10 to 15 N/mm^2. The preparatory laboratory work to determine the moisture content and cement content is the same as that described in Par. 12.5. Some adjustment to the moisture content may be necessary to improve workability. Compaction should be in accordance with Table 11.5.

12.14. Lean concrete can be mixed in drum mixers as well as the pan and paddle types. On larger jobs it should always be laid by paver. Preliminary field density measurements should be made within 24 hours of compaction, by the sand replacement method. These tests can be used to calibrate the nuclear density meter if one is being used for control purposes.

Influence of Sample Dimensions on the Measured Strength of Cemented Materials

12.15. Laboratory samples prepared in connection with the design of cement-stabilized materials are sometimes made in cylindrical moulds or, in the case of fine-grained soils, in 100-mm cubical moulds. In addition, to establish the actual strength of cured cemented subbase and base materials it is common practice to test cores cut through the full depth of the material. To equate the results to strengths based on 150-mm cubes, it is necessary to know the relationship between the various types of test. Figure 12.1 shows the results of relevant tests carried out in the United Kingdom.

U.S. Practice in Relation to Cemented Subbases and Bases

12.16. In the United States cemented subbase and road-base materials are not subdivided in terms of grading limits as is the case in Britain. Cement contents are recommended for the various soil types classified under AASHTO Designation M145-82. These cement contents, on a mass basis, vary from 3.5 to 7 percent for A1 to A3 soils (granular materials) to 7 to 10 percent for A4 to A7 soils (silt-clay materials). They are expected to give 7-day strengths of not less than 2 MN/m^2, and provide a starting point for the laboratory work necessary to establish the strength required.

12.17. Much more use is made in the United States of in situ soil stabilization than is the case in Europe. A wide range of multiple and single-pass plants has been developed and this has led to cost savings from this process which cannot in general be realized in smaller countries.

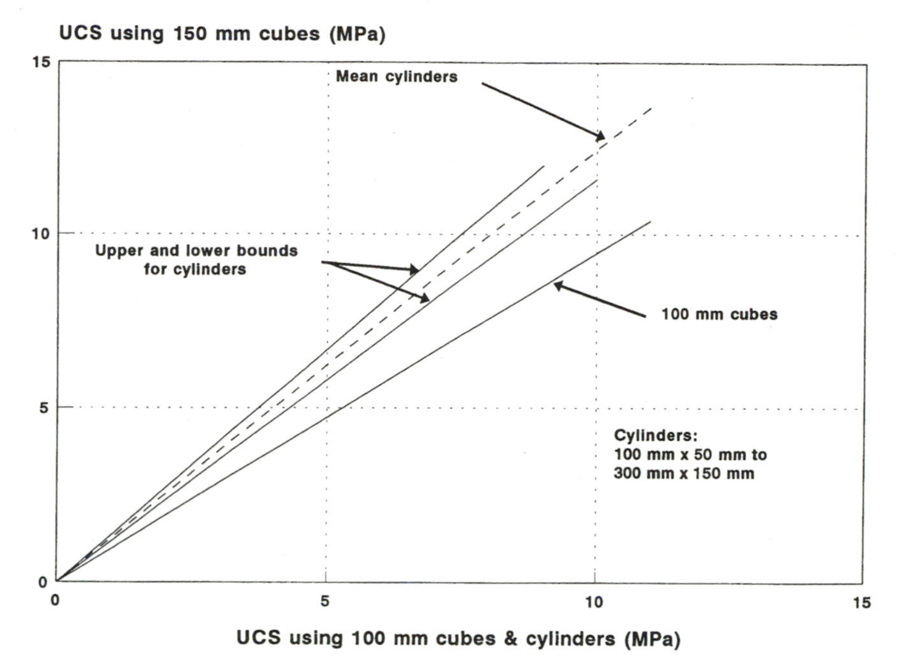

Figure 12.1 Relationship between crushing strength and sample size and shape.

The Structural Properties of Cemented Base, Subbase, and Capping Materials

12.18. Increase in compressive strength with age. The increase in compressive strength with age for cemented materials with a wide range of compressive strengths has been studied over periods up to 10 years in the United Kingdom.[1,2] For lean concrete and soil-cement the average relationship is given in Fig. 12.2. For pavement-quality concrete the rate of increase of strength with age decreases slightly with increasing compressive strength. This is shown in Fig. 14.1.

12.19. Modulus of elasticity. The modulus of elasticity of cemented materials can be measured statically by loading 150-mm diameter cylinders fitted with extensometers, or dynamically using electrodynamic excitation on long beams of 150-mm square section. The static method involves strains greater and more sustained than those experienced in pavements under the action of traffic. With the dynamic method the strains are less than occur under traffic, and of higher frequency. The comparative studies which have been made show the dynamic modulus to be consistently higher than the static value over a wide range of compressive strengths, as is indicated in Fig. 12.3. (The zones referred to in this figure are shown in Fig. 12.4.) For weak

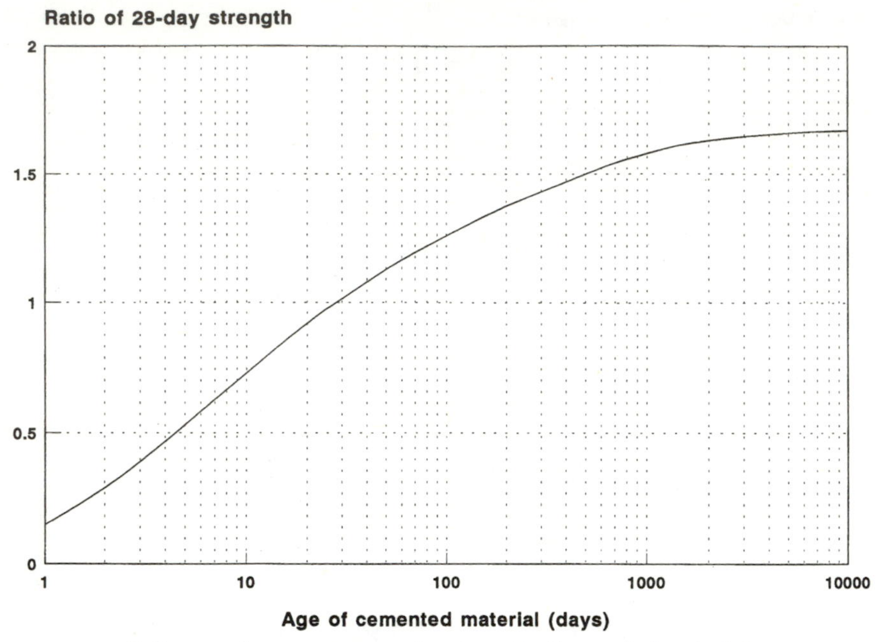

Figure 12.2 Influence of age on compressive strength of cemented materials.

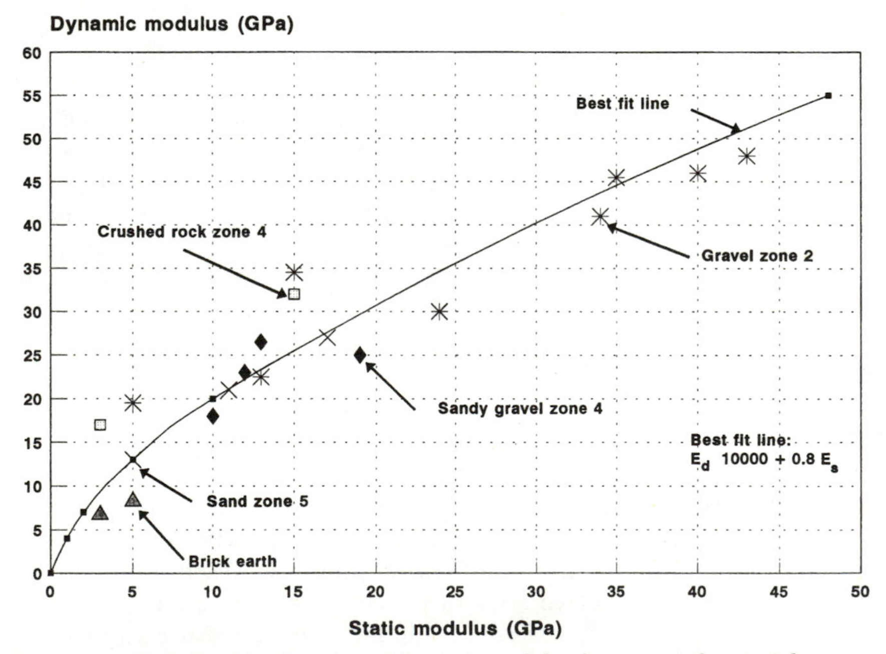

Figure 12.3 Relationship of static and dynamic modulus for cemented materials.

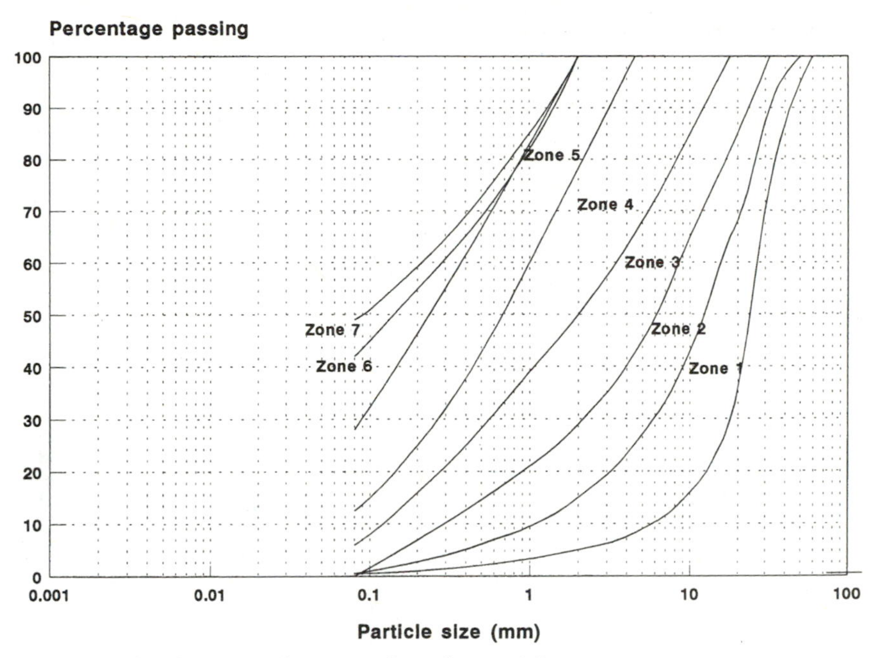

Figure 12.4 Grading zones for cement-bound materials.

cemented materials the ratio dynamic *E*:static *E* can be as high as 2:1, but for concretes it is much smaller.

12.20. Which value should be used for the structural analysis of pavements is not at present clear. Further research is required using dynamic loading tests which model traffic loading. For the time being it appears reasonable to use the mean between the two methods of measurement. Figure 12.5 shows the results of dynamic modulus tests carried out on a range of materials conforming to the grading zones shown in Fig. 12.4. Grading zones 2, 3, and 4 approximately cover the grading limits for cement-bound granular material, and grading zones 1 and 2 approximate to the grading limits for lean concrete. Grading zone 2 and finer correspond to the requirements for soil-cement and capping-layer material.

12.21. Figure 12.5 also includes measurements on two soil-cements made with cohesive soils (gradings 6 and 7). These would correspond to stabilized subgrade material. For a more cohesive capping material of plasticity index 33 percent, it has been found, using static tests, that the elastic modulus in tension was only about 10 percent of the value in compression.[3] This is considered to be due to the develop-

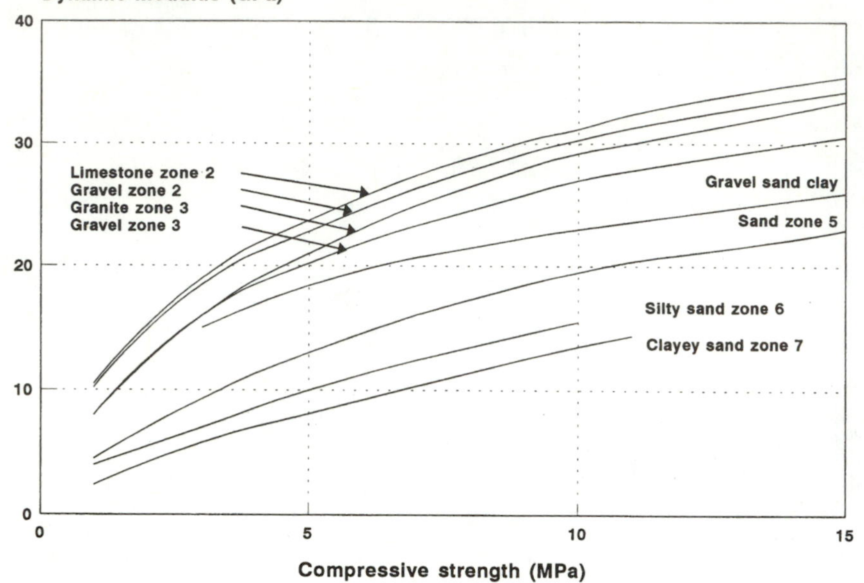

Figure 12.5 Relationship between dynamic modulus and compressive strength at 28 days for cemented materials.

ment of microshrinkage cracks during curing. In computing tensile stresses in such materials a modulus of elasticity of not more than 25 percent of the measured dynamic value would probably represent a reasonable estimate.

12.22. From the information given in Fig. 12.5 and on the basis of the discussion in Pars. 12.19 and 12.20, Table 12.3 has been prepared.

TABLE 12.3 Proposed Values of Elastic Modulus of Cemented Materials for Use in Structural Analyses

Stabilized material	Modulus of elasticity, GPa		
	Dynamic	Static	Mean
Soil-cement			
Granular soils	18	10	14
Silty soils			
PI<10	7	4	5
Clay soil PI>10	1	0	0.5
Cement-bound granular	23	13	18
Normal lean concrete	27	19	23
Stronger lean concrete	30	23	27

12.23. Poisson's ratio. Kolias and Williams have studied Poisson's ratio for various cemented materials ranging from stabilized cohesive soil to lean concrete.[4] Static values close to 0.15 were found at an age of 28 days. Anson and Newman found a similar value for dense concrete.[5] It appears that, as with elastic modulus, dynamic values are rather higher, but it is suggested that this value should be used, since structural analyses are not sensitive to small changes in this factor.

12.24. Flexural and fatigue strength. Figure 12.5 shows the relations between compressive strength and dynamic modulus for a wide range of cemented materials with gradings within the limits for soil-cement, cement-bound material, and lean concrete.

12.25. In the testing of concrete beams in flexure, the extreme fiber stress measured at the surface of the test beam is generally greater than the flexural strength as recorded by the standard test procedure. This extreme fiber stress is often referred to as the modulus of rupture. In the structural analysis of concrete pavements, the modulus of rupture, rather than the flexural strength, is used to denote the onset of cracking. Sufficient information is not available for weaker cemented materials to allow this differentiation, and the flexural strength is used to define failure by cracking.

12.26. Symons has made a study of the fatigue strength of several cement-bound materials using the equipment shown in Fig. 12.6.[6] The frequency of oscillation was determined by the fundamental flexural resonance of the beams used, which was about 140 Hz, although this varied slightly with the cement content. The results are shown in Fig. 12.7. Galloway et al. have used a similar but more sophisticated loading procedure applied to concrete beams, but they have included some tests on lean concrete of 7-day compressive strength 12 MN/m^2. The frequency of loading used was 20 Hz, and tests were conducted on samples of different ages between 13 weeks and 2 years. The results of this work also are included in Fig. 12.7. Both sets of measurements indicate a failure stress of approximately 75 to 80 percent of the flexural strength at 10^5 stress applications. This follows the generally accepted behavior of pavement concretes.

12.27. In the above tests the failure was induced in a comparatively short period. The figure of 10^5 applications in the Symons tests was achieved in about 12 min, and in the Galloway tests in 80 min. The effects of aging did not therefore influence the results. In the case of a road pavement, the incidence of very heavy wheel loads could be limited to about 10 per day, and 10^5 applications would be spread over a

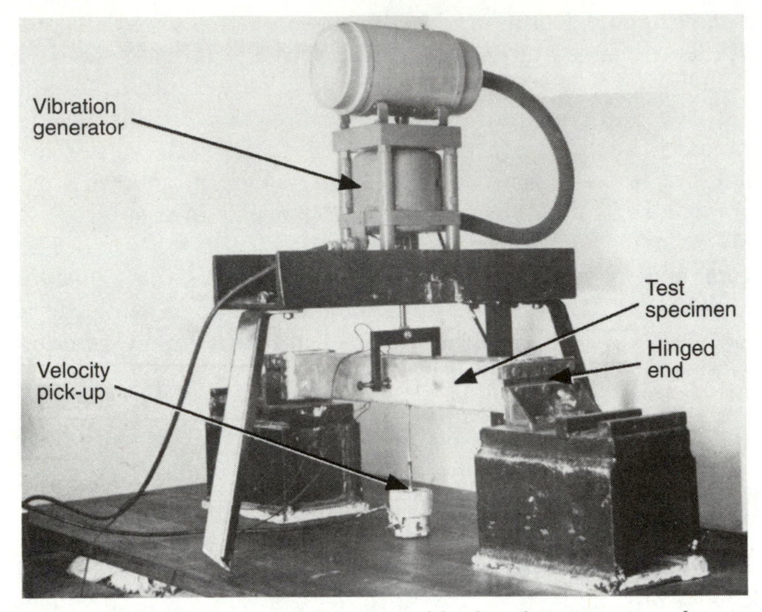

Figure 12.6 Equipment used for repeated-loading fatigue tests on beams of cemented materials.

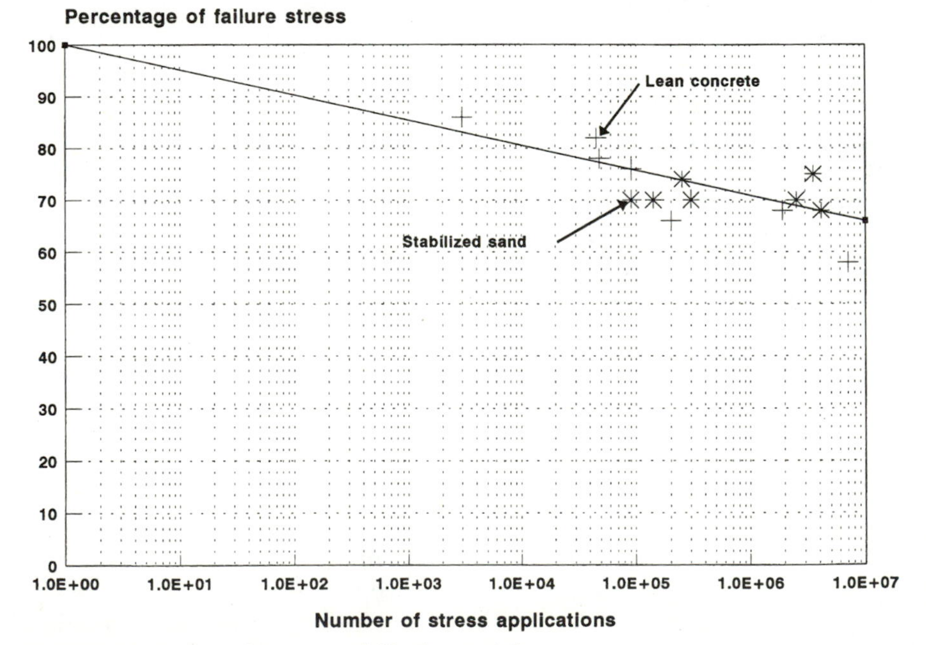

Figure 12.7 Fatigue of cement-stabilized materials.

period in excess of 20 years; also, during the early life of the road the increase in flexural strength would tend to counteract the influence of fatigue. This matter is dealt with more fully in Chap. 22.

References

1. Road Research Laboratory: *Concrete Roads—Design and Construction,* HMSO, London, p. 58, 1955.
2. Galloway, J. W., H. M. Harding, and K. D. Raithby: Effects of Age on Flexural Fatigue and Compressive Strength of Concrete, *Transport and Road Research Laboratory Report* LR 865, TRRL, Crowthorne, 1979.
3. Bofinger, H. E.: The Measurement of the Tensile Properties of Soil-cement, *Transport and Road Research Laboratory Report* LR 365, TRRL, Crowthorne, 1970.
4. Kolias, S., and R. I. T. Williams: Research on the Tensile Properties of Cement Stabilised Materials, University of Surrey.
5. Anson, M., and K. Newman: The Effect of Mix Proportions and Methods of Testing on Poisson's Ratio for Mortars and Concretes, *Magazine of Concrete Research,* vol. 18, no. 56, pp. 115–130, 1966.
6. Symons, I. F.: A Preliminary Investigation to Determine the Resistance of Cement-stabilised Materials to Repeated Loading, *Transport and Road Research Laboratory Report* LR 61, TRRL, Crowthorne, 1967.

Chapter

13

Bituminous Bases and Surfacings

Introduction

13.1. The approaches to the design of bituminous surfacings and road bases in the United States and in the United Kingdom (and in Europe generally) are different. In the United States emphasis is placed on the stability, particularly of wearing and binder courses, as indicated by the Marshall test procedure. In the United Kingdom recipe specifications relating aggregate grading and binder content are preferred to ensure that wearing course and binder courses as laid will be impermeable to water. This difference in approach has probably arisen mainly because of the large temperature range encountered in the United States and its influence on stability.

The Marshall Test Procedure

13.2. The stability and flow of an asphaltic concrete mix are defined in terms of the empirical Marshall test. This is described in AASHTO Designation T245-82 (1986),[1] or more fully in BS 598:Part 3:1985.[2] Briefly, the test is carried out on compacted samples of the mixture prepared in a steel mould 101.6 mm in diameter. Approximately 1100 g of the material at the appropriate mixing temperature is placed in the heated mould and compacted by a rammer with a circular foot of 98.5 mm diameter acted upon by a mass of 4585 g falling freely through a distance of 457 mm. Fifty blows are applied to each face of the sample at a rate of approximately one per second. After compaction the sample is stored at 20°C for not more than 8 hours prior to testing.

13.3. Before testing, the weight of the dry sample in air is determined to an accuracy of 0.1 g. It is then weighed submerged in water at 20°C to the same accuracy. The volume of the sample is numerically equal in cm³ to the difference of the mass in air and water. The relative density of the specimen S_M is equal to the mass in air divided by volume, and the aggregate density S_A is given by

$$S_A = \frac{S_M \times 100 - W_B}{100}$$

where W_B is the percentage by mass of the binder in the specimen.

13.4. For testing, the sample at a temperature maintained at 60 ± 0.5°C is placed between the jaws of a testing head, shown in section in Fig. 13.1. The testing head is mounted in a compression-testing machine capable of operating at a constant strain rate of 50 ± 3 mm/min. The applied load-time curve is observed beyond the point where the maximum load has been recorded, in the manner indicated in Fig. 13.2.

13.5. From Fig. 13.2 the maximum load is 4.8 kN, and this is defined as the stability. The time taken to reach this load is 4.5 s. At a strain rate of 50 mm/min this corresponds to a deformation of (50 × 4.5)/60, or 3.8 mm. This is termed the flow.

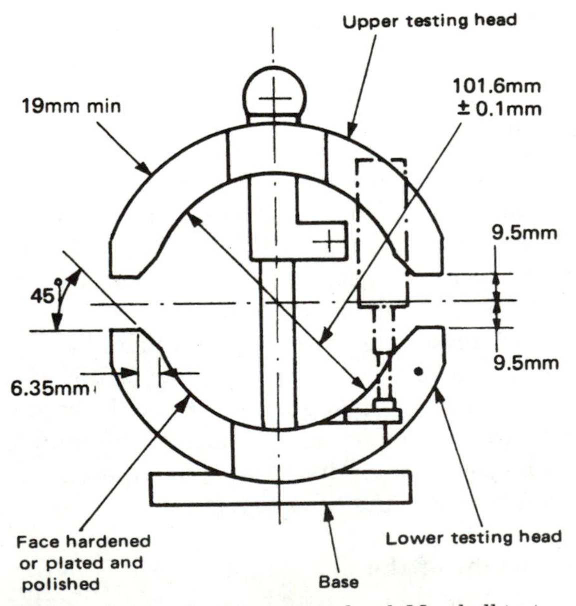

Figure 13.1. Details of the testing head, Marshall test.

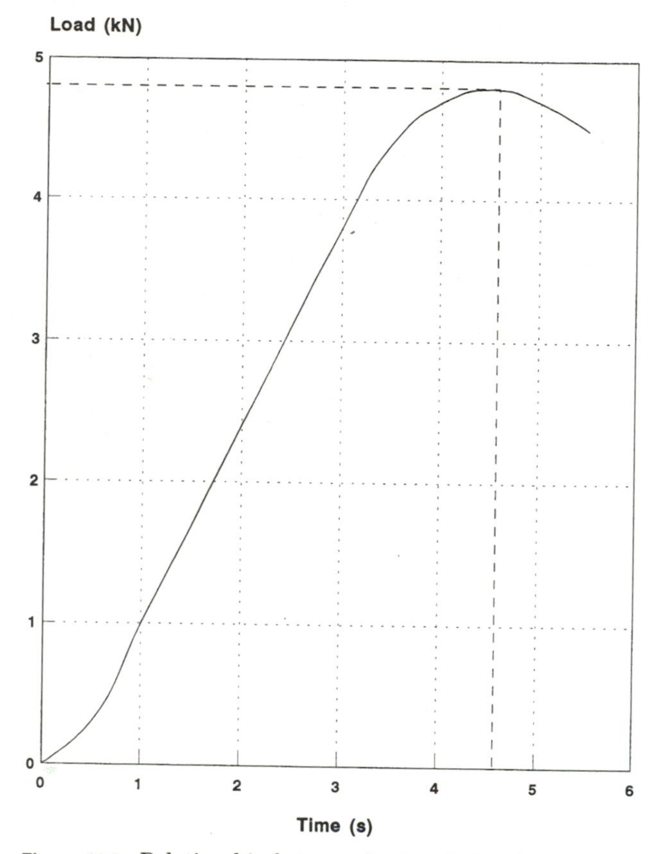

Figure 13.2 Relationship between load and time for Marshall test.

13.6. The design procedure involves making up and testing a number of samples of the bituminous material with different binder contents using the grade of bitumen intended for use. The results are plotted in the form shown in Fig. 13.3 (plots 1–4), proposed in BS 598:Part 3:1985. The binder contents corresponding to maximum stability, maximum mix density, and maximum compacted aggregate density are meaned and the stability and flow at this mean binder content are recorded. This information provides the basis for the minimum stability and maximum flow values included in a contract specification. If the stability obtained is considered to be too low, then an adjustment must be made to the grading or, more particularly, to the penetration grade of the binder used. These tests will normally be conducted using bin aggregates. The contractors will subsequently have to carry out similar tests using the aggregate they propose to

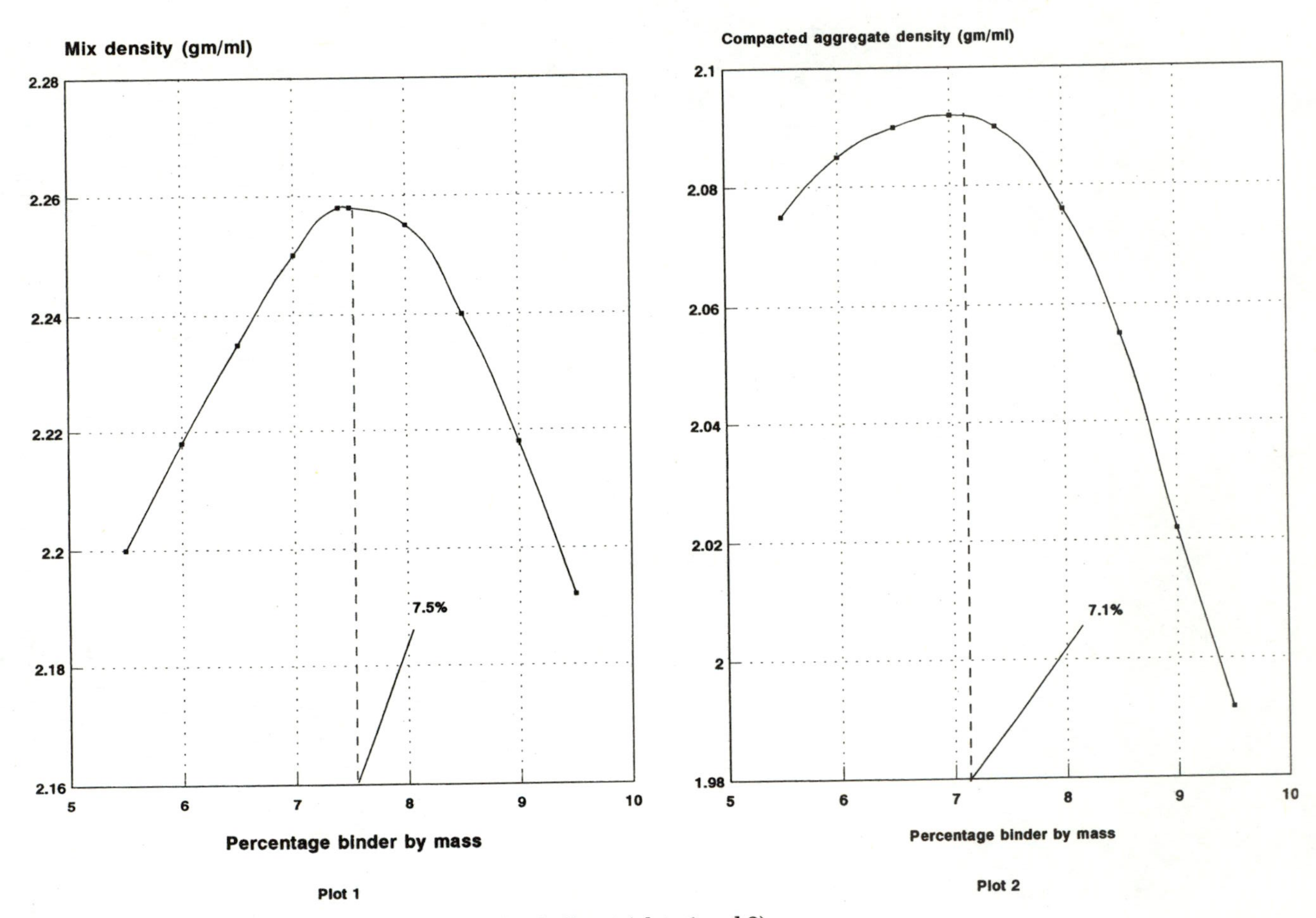

Figure 13.3 Examples of mix design based on Marshall test (plots 1 and 2).

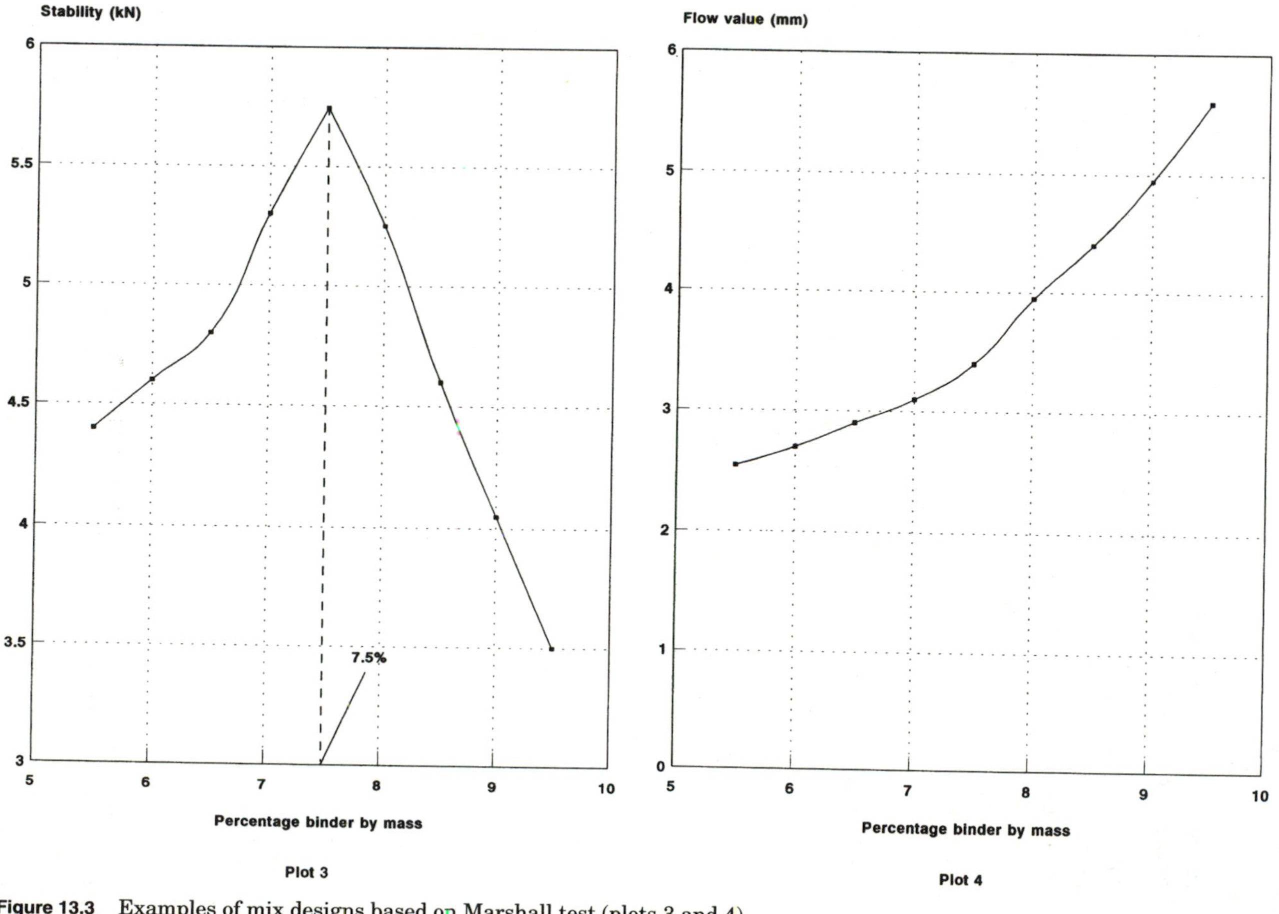

Figure 13.3 Examples of mix designs based on Marshall test (plots 3 and 4).

use to establish a job mix for the contract, and periodically checks on the stability will also need to be made.

U.S. Specifications for Asphaltic Concrete Materials

13.7. Typical grading limits for the aggregate used in American wearing and binder courses are shown in Fig. 13.4. The gradings are continuous but are normally compounded from fine and coarse constituents. As a consequence of the design procedure used, asphaltic concretes as laid are not in general impervious to water, although they may become so under traffic.

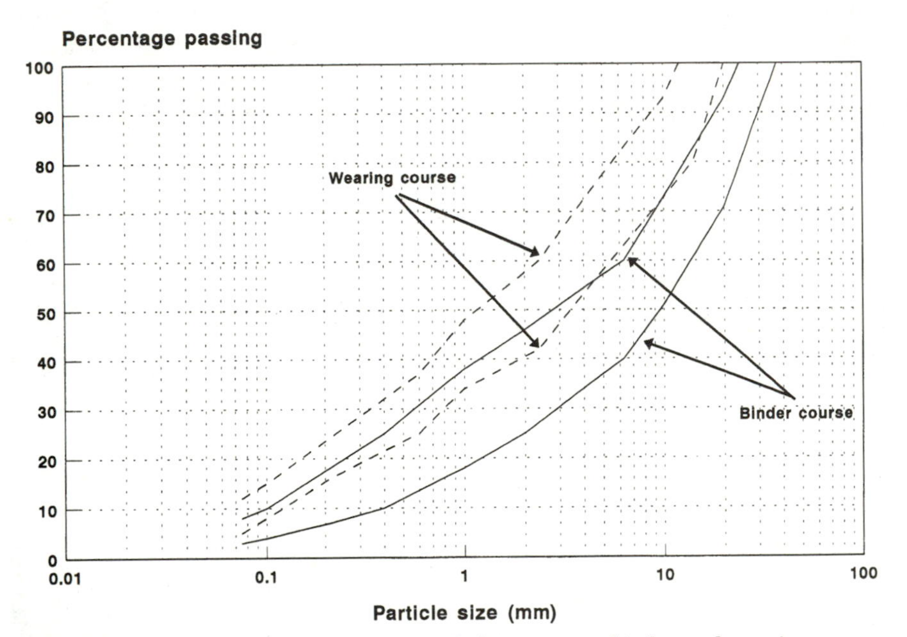

Figure 13.4 Typical grading curves for asphaltic concrete binder and wearing course materials.

U.K. Specifications for Materials with Bituminous Binders

13.8. These materials are defined as (1) rolled asphalts normally using 50 penetration bitumen, and (2) dense bitumen macadams using 100 penetration or softer binders. The former are covered by BS 594[3] and the latter by BS 4987.[4]

Rolled Asphalt Materials

13.9. These are made with gap-graded aggregates compounded from coarse and fine constituents. The gap grading permits the use of comparatively high binder contents without the danger of high pore pressures being developed during rolling.

13.10. Grading envelopes for wearing courses, binder courses, and road bases are shown, respectively, in Figs. 13.5, 13.6, and 13.7.

13.11. For wearing course material (Fig. 13.5) the proportion of coarse aggregate and the maximum size of the aggregate depend on the thickness of the layer. For a 36-mm layer the percentage of coarse aggregate would be 30 and the maximum size 10 mm. The binder content would be close to 8 percent. Precoated chippings would be added to the surface during final rolling to give adequate skid resistance.

13.12. For the base course (binder course) Fig. 13.6 shows the combined grading. The maximum size of aggregate increases from 10 to 20 mm as the thickness of layer increases from 25 to 80 mm. The binder content for crushed rock aggregate is 6.5 percent.

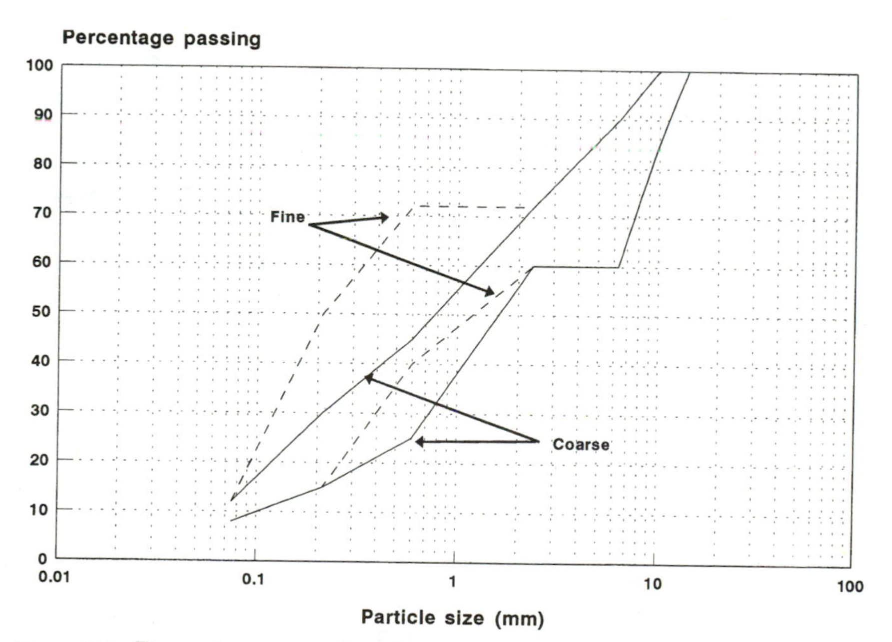

Figure 13.5 Fine and coarse gradings for recipe specification for rolled asphalt for 35-mm (nominal) wearing courses.

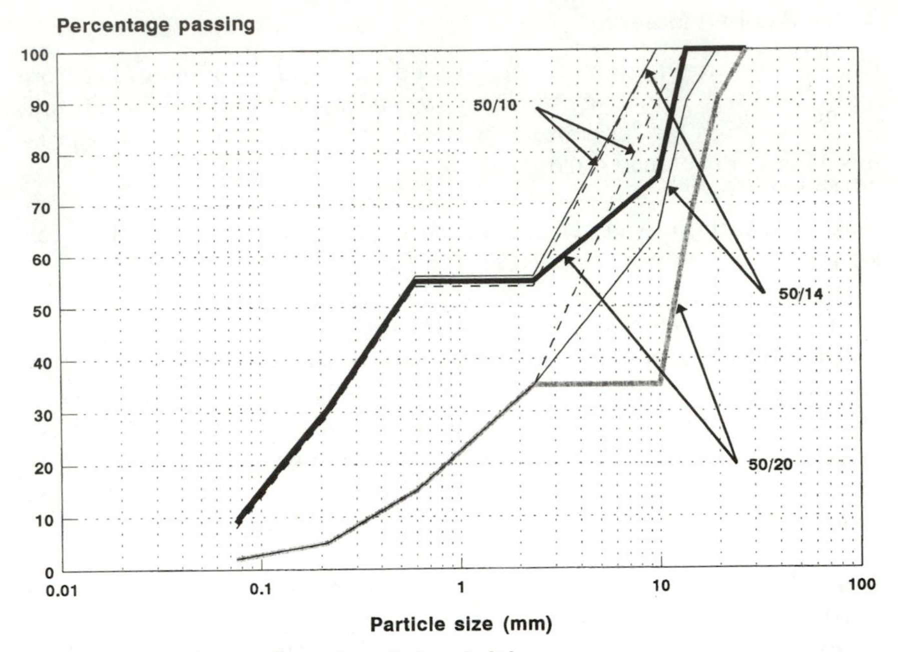

Figure 13.6 Grading envelopes for rolled asphalt base course.

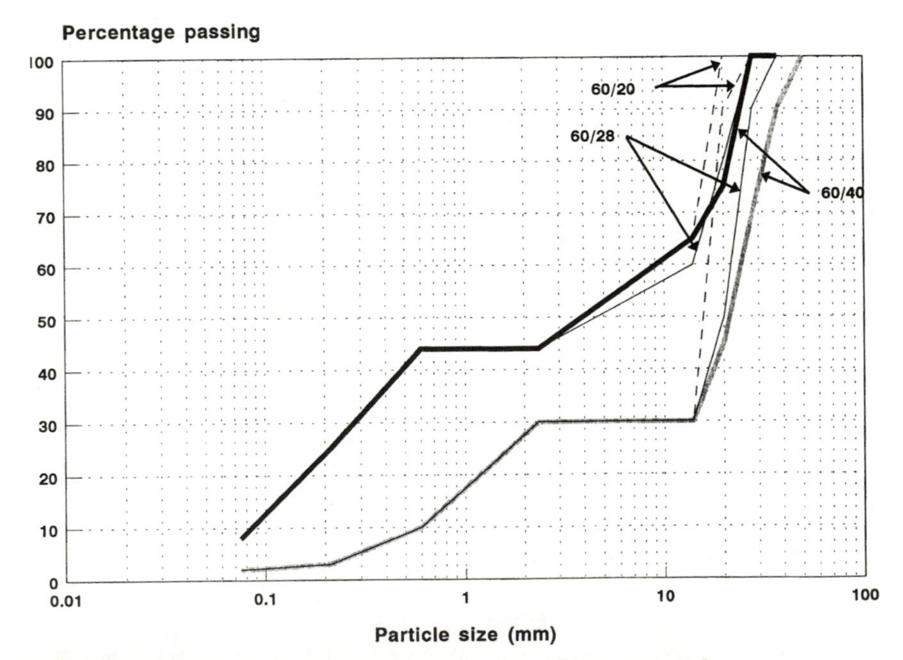

Figure 13.7 Grading envelopes for rolled asphalt road-base materials.

13.13. For road base material (Fig. 13.7) the maximum size of material increases from 20 to 37 mm as the thickness of layer increases from 45 to 150 mm. The binder content used is 5.7 percent for crushed rock aggregate.

Dense Coated Macadams

13.14. These materials, specified in BS 4987, are primarily used for binder courses and road bases. Although the specification includes a wearing course material, because of its permeability its use is confined to very lightly trafficked roads. The grading envelopes for binder course and road base materials are shown in Figs. 13.8 and 13.9. Reference to Fig. 13.4 shows that the binder course material (Fig. 13.8) is very similar with reference to grading to U.S. asphaltic concrete binder course. The binder content used in the United Kingdom is approximately 4.5 percent using either 100 or 200 penetration bitumen.

13.15. The dense coated macadam road base material (Fig. 13.9) is very widely used in Britain for heavily trafficked trunk roads and freeways. The binder content used is between 3.5 and 4 percent for crushed rock aggregates.

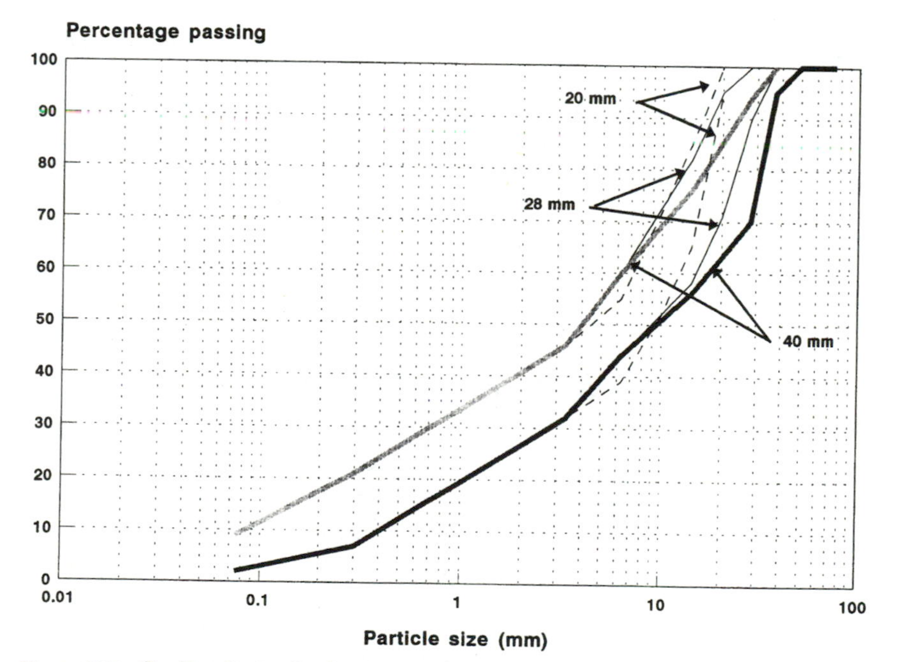

Figure 13.8 Grading limits for dense coated macadam base courses.

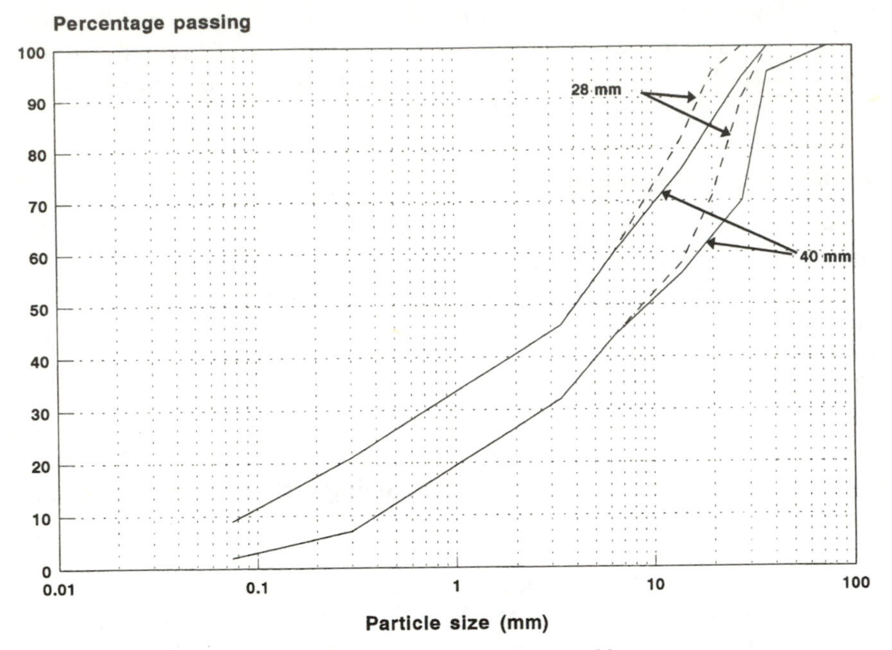

Figure 13.9 Grading limits for dense coated macadam road bases.

Pervious Wearing Courses

13.16. Increasing use is now being made of pervious wearing courses with sufficient storage capacity and transverse drainage to reduce spray from fast-moving heavy vehicles during periods of heavy rain. They contribute little to the strength of the pavement and are normally laid on impervious asphalt wearing course material. In the United Kingdom aggregates to the gradings shown in Fig. 13.10 are commonly used. For the material of maximum size 10 mm a binder content of 5 percent is used, and for the 20-mm material this is reduced to 4 percent.

The Elastic Properties of Bituminous Materials

13.17. Any bituminous mixture acts as a viscoelastic material in which the strain caused by the application of a stress is out of phase with the stress. This complicates any assessment of the elastic modulus of such materials and has led to the adoption of a complex modulus generally denoted by E^*.

13.18. If J_1 is the component of the strain in phase with the applied stress expressed as a ratio of the stress, and J_2 is the component of

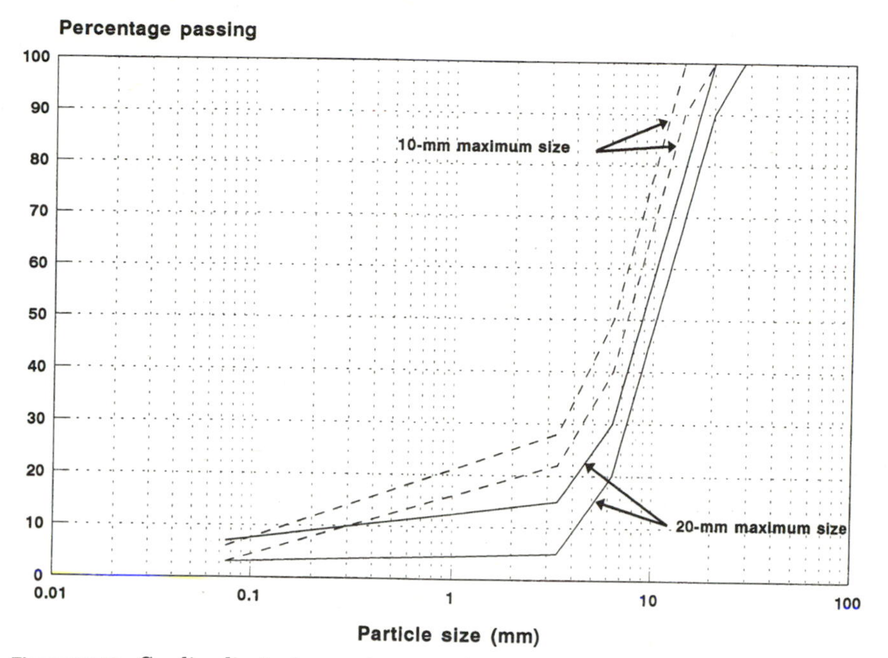

Figure 13.10 Grading limits for pervious wearing course materials.

the strain at 90° to the stress also expressed as a ratio of the stress, then the complex modulus is defined as

$$E^* = \frac{1}{\sqrt{J_1^2 + J_2^2}}$$

13.19. Shook and Kallas conducted an extensive research into the elastic modulus of asphaltic concretes using various frequencies of sinusoidal loading.[5] Samples 100 mm in diameter and 200 mm long were tested in unconfined compression using loading frequencies of 1, 4, and 16 cycles per second. They investigated the effects of asphalt content and state of compaction, as defined by air voids content. The aggregates used were continuously graded, and in this respect the mixtures resembled dense bitumen macadams as used in Britain and asphaltic concretes in the United States. The binder used was generally of about 100 penetration, but some tests were made with a harder binder of about 50 penetration. Figure 13.11 shows the relationship between E^*, air voids, and temperature for asphaltic concretes made with the 50 and 100 penetration binders at a loading frequency of 4 Hz. The harder binder produced a significantly higher modulus, and for both materials a reduction of air voids content (i.e., an increase in

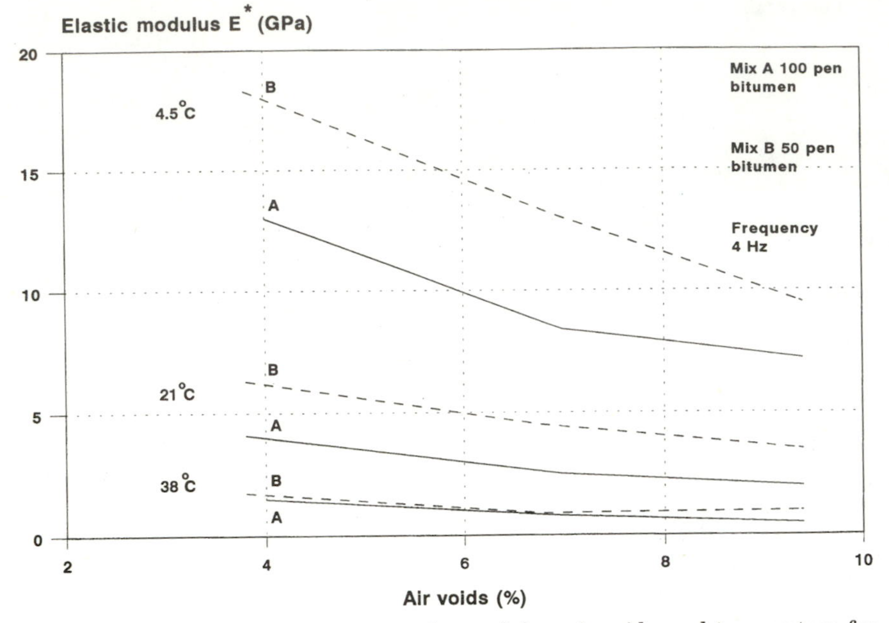

Figure 13.11 Relationship between complex modulus, air voids, and temperature for asphaltic concretes.

compacted density) resulted in an increase in modulus. The large effect of temperature on modulus is apparent. Increasing the temperature from 4 to 38°C reduces the modulus by a factor of about 12. Figure 13.12 shows the effect of binder content on elastic modulus. Increasing the binder content by about 4 percent decreases the modulus for all the temperatures investigated. A further important finding was that the modulus in tension and compression were very similar over the practical stress range.

13.20. From their work Shook and Kallas produced several statistical equations relating complex elastic modulus of bituminous materials with the physical and mechanical properties of the mixes. The equation most applicable to British materials as they are currently specified is

$$\log_{10}E^* = 3.12197 + 0.0248722\,(X_1) - 0.0345875\,(X_2) - 9.02594\,(X_4)^{0.19}/(X_6)^{0.9} \qquad (13.1)$$

where E^* = dynamic modulus, 10^5 lb/in^2 (4 cycles/s loading frequency)

X_1 = percentage of total aggregate finer than 75 μm

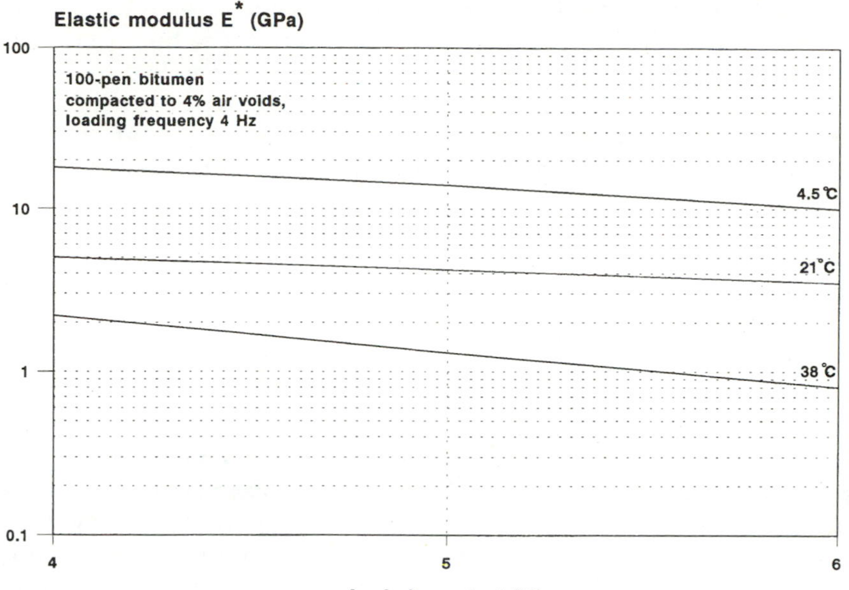

Figure 13.12 Relationship between complex modulus, asphalt content, and temperature for asphaltic concrete.

X_2 = percentage air voids in compacted mix
X_3 = asphalt viscosity at 70°F, poises
X_4 = percentage of binder by weight of mix
X_6 = $\log_{10}$ viscosity of binder at test temperature, poises

In accordance with normal American practice, Eq. (13.1) uses imperial units with temperature expressed in °F. Appropriate values of X_3 and X_6 can be deduced from Table 13.1 for various penetration grades of bitumen.

TABLE 13.1 Relationship between Binder Viscosity, Temperature, and Penetration Grade of Bitumen

Temperature		Viscosity (poises) for bitumen of penetration (at 25°C)			
°C	°F	20/30	40/50	80/100	100/200
0	32	4×10^{10}	5×10^{9}	3×10^{8}	5×10^{7}
10	50	1×10^{9}	2×10^{8}	2×10^{7}	5×10^{6}
20	68	8×10^{7}	1×10^{7}	1.5×10^{6}	4×10^{5}
30	86	6×10^{6}	1×10^{6}	2×10^{5}	5×10^{4}
40	104	7×10^{5}	1.5×10^{5}	3×10^{4}	1×10^{4}
50	122	1×10^{5}	4×10^{4}	9×10^{3}	3×10^{3}

13.21. The values obtained for E^* depend to some extent on the type of test adopted. A variety of test procedures have been used by research workers modeling to different degrees the practical condition in the pavement. Brown has made a useful summary of the information available[6] and has used it to prepare average relationships between elastic modulus and loading time at temperatures of 0 and 20°C for dense bitumen macadam made with 100-penetration binder and for rolled asphalt made with 50-penetration binder. These are modifications of curves previously prepared by the Shell Laboratories. Curves derived from Brown's data are included in Fig. 13.13 which defines the relationship between modulus and loading time for dense bitumen macadam and for rolled asphalt.

13.22. Bituminous materials in British roads operate for about 10 percent of the year at temperatures above 20°C and within the upper 40 mm they operate for about 5 percent of the year at temperatures above 30°C. The results from full-scale road experiments show that the deformation of flexible pavements occurs mainly at road temperatures in excess of 20°C. Particularly in relation to subgrade and subbase stresses it is necessary to consider the change of elastic modulus which occurs in bituminous materials at temperatures above 20°C. Unfortunately, owing to the difficulties which arise in high-temperature testing, little information is available other than that reported by Shook and Kallas, already discussed. Their results at temperatures up to 20°C do not agree very closely with the average curves prepared by Brown, but the relationship they found between modulus and temperature between 4.5 and 38°C can be used with moderate confidence to extend Brown's curves to temperatures of 30 and 40°C, and this has been done in Fig. 13.13. Figure 13.14 presents Brown's data as relationships between modulus and temperature.

13.23. The data given in Fig. 13.13 for both dense bitumen macadam and rolled asphalt and in Fig. 13.14 for dense bitumen macadam relate to freshly made materials. The data shown in Fig. 13.14 for rolled asphalt were obtained on cores cut from a full-scale experiment after at least 10 years. Recent research carried out by the TRL based on long-term deflection studies shows that the stiffness of bituminous materials increases markedly with age. Although the process continues for 20+ years it is most pronounced in the first 5 years. This is accompanied by an equally marked decline in temperature sensitivity. Deflection measurements made at 1-hour intervals over a 24-hour period at the Nately Scures full-scale experiment (see Chaps. 18 and 20) at several times of the year also indicated that for mature bituminous materials there is an approximately 4-hour lag between temper-

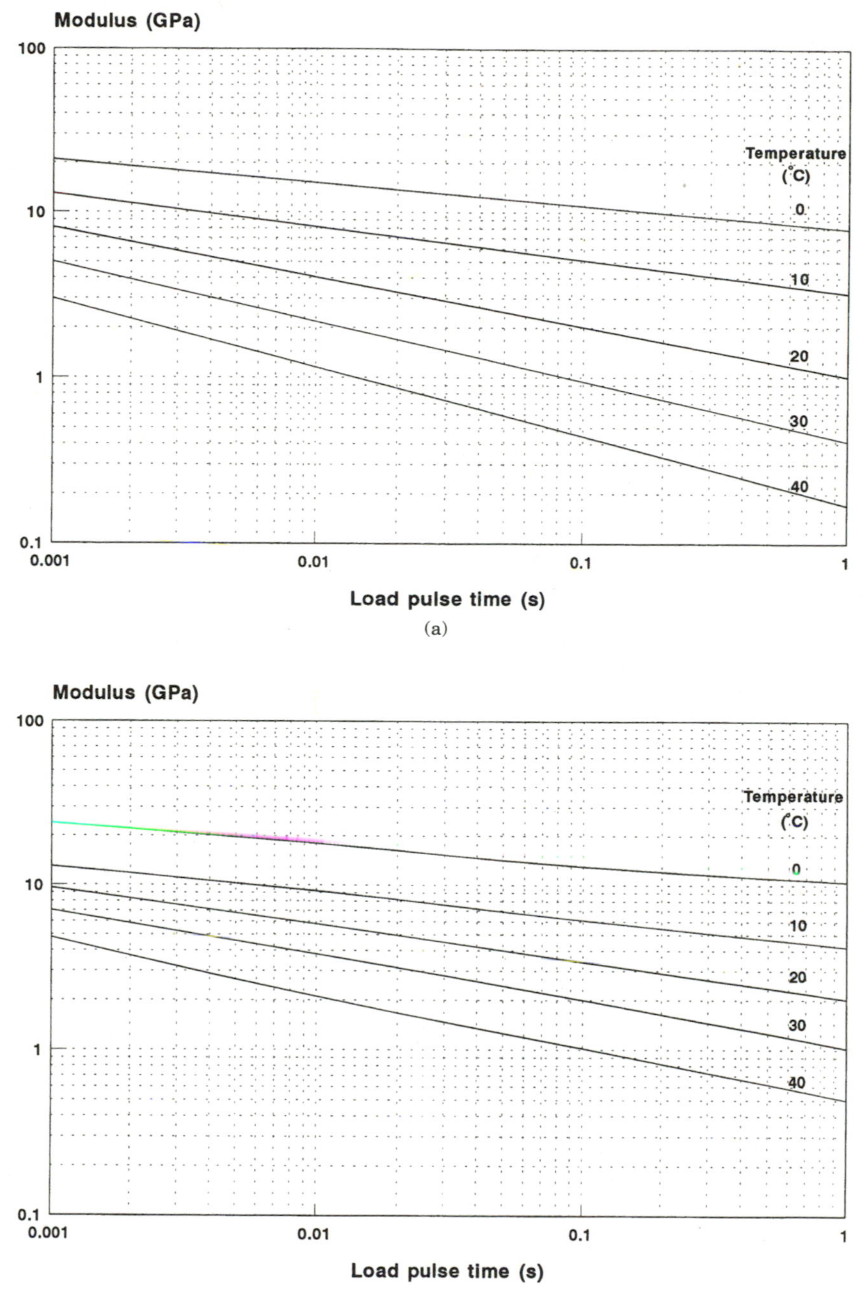

Figure 13.13 Relationship between modulus and loading time for (*a*) dense bitumen macadam and (*b*) rolled asphalt. (*After Brown.*[6])

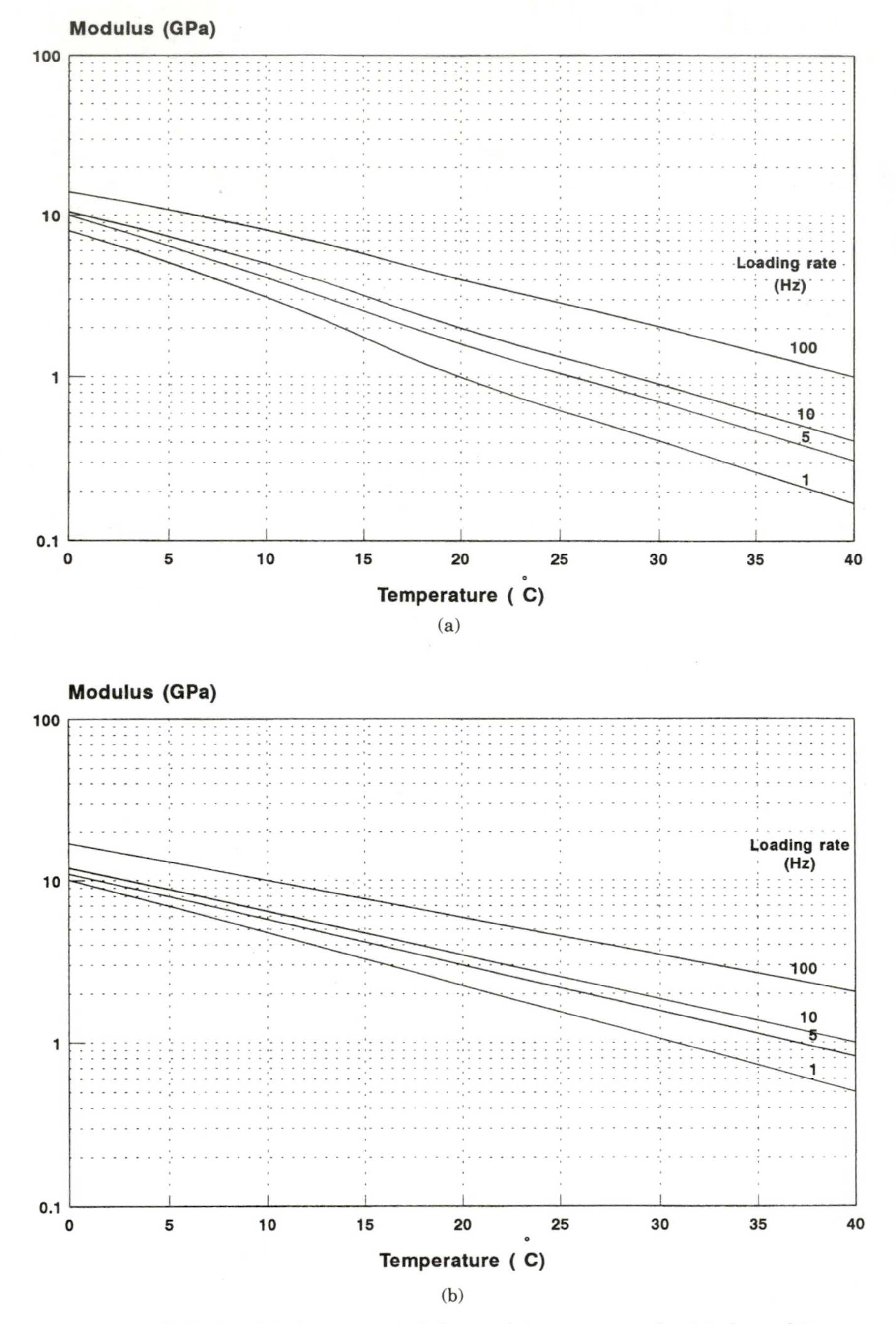

Figure 13.14 Relationship between modulus and temperature for (*a*) dense bitumen macadam and (*b*) rolled asphalt. (*After Brown.*[6])

ature and measured deflection, i.e., the modulus is lowest about 4 hours after the maximum temperature. It was concluded that measured deflections correlated well with the mean temperature measured in the bituminous materials over the 24-hour period.

13.24. Based on a back analysis of a large number of sections from full-scale experiments with rolled asphalt and dense bitumen macadam bases, Fig. 13.15 has been produced showing the modulus temperature relationships for rolled asphalt surfacings on rolled asphalt bases and for rolled asphalt surfacings on dense bitumen macadam bases. From this data, together with the data shown in Figs. 13.13 and 13.14, Fig. 13.16 has been prepared which shows, for U.K. conditions, modulus temperature relationships for use in structural analysis for prediction of stresses, strains, and deflections under both slow-moving (deflection beam) vehicles and normal traffic.

13.25. The effect of aggregate type on the elastic modulus of bituminous materials has not been studied as a primary variable. The evidence available from tests in which it was a secondary variable suggests that it is not a very important factor.

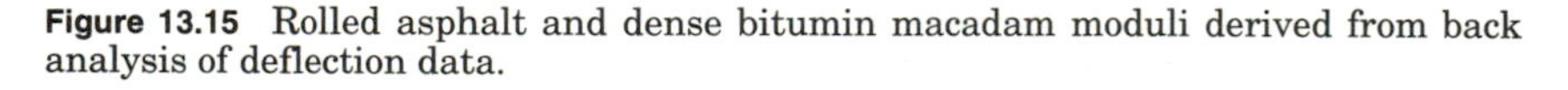

Figure 13.15 Rolled asphalt and dense bitumin macadam moduli derived from back analysis of deflection data.

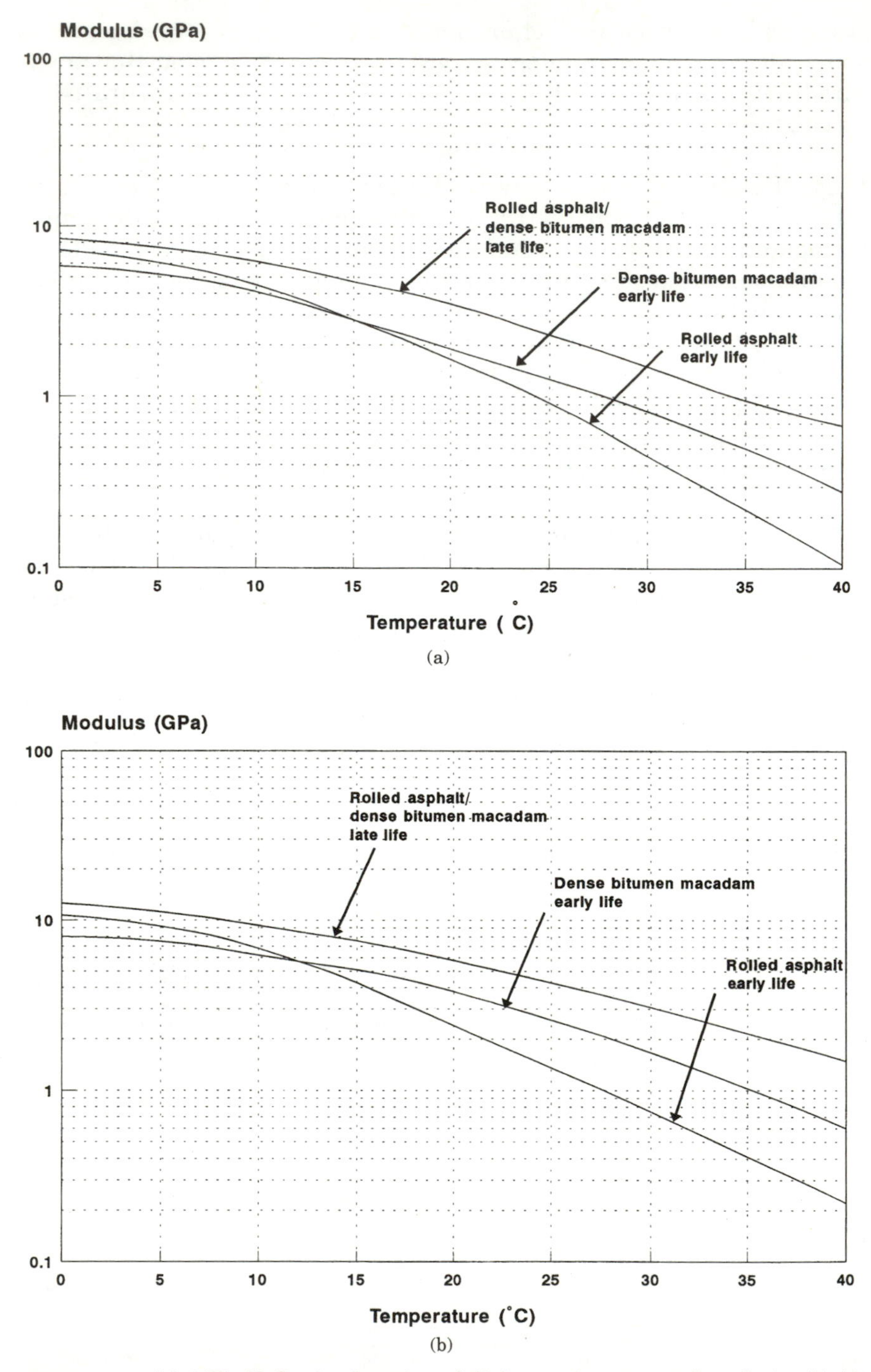

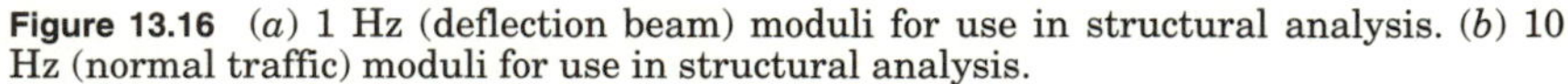
Figure 13.16 (*a*) 1 Hz (deflection beam) moduli for use in structural analysis. (*b*) 10 Hz (normal traffic) moduli for use in structural analysis.

Poisson's Ratio of Bituminous Materials

13.26. The value of Poisson's ratio for bituminous road materials is generally close to 0.4. Monismith and Secor show[7] that for extreme conditions of loading and temperature the value for asphaltic concrete mixes could vary between 0.3 and 0.5.

The Fatigue of Bituminous Materials under Repeated Loading

13.27. The complicated nature of the relationship between stress and strain for bituminous materials—its dependence on the magnitude of the stress, the frequency of loading, and the duration of rest periods between load applications—means that the fatigue life measured depends to a large extent on the test procedure used.

13.28. As with cemented materials, various methods of testing bituminous materials for fatigue have been used. These include beam and cylindrical flexural tests. Comprehensive data relating fatigue life with stress and strain levels have been published by Cooper and Pell,[8] Epps and Monismith,[9] and Kirk.[10]

13.29. Because the rise and decline of stress and strain in bituminous materials are time dependent, accelerated test procedures inevitably give much shorter fatigue lives than are found under normal traffic loading. Typical fatigue curves relating strain and fatigue life are shown in Fig. 13.17. The main conclusions reached by Cooper and Pell are that binder content and binder type are primary factors affecting the fatigue performance on the basis of applied strain. For all binders, increasing the binder content increases the fatigue life. This applies even over the range where increasing the binder content decreases the elastic modulus. Decreasing the air voids content by compaction or by modifying the aggregate grading increases the stiffness of a bituminous material and thus decreases the strain for a given applied stress. However, it has little direct effect on the fatigue life of a bituminous material subjected to repetitions of a constant strain.

13.30. An important recent finding is that rest periods between the applications of stress (such as will occur in practice between the passage of axles and of vehicles) have an important influence on fatigue life. This would be expected from the nature of the stress-strain relationship. Raithby and Sterling, using direct tensile and compressive loading on sawn beams of rolled asphalt surfacing, showed that rest

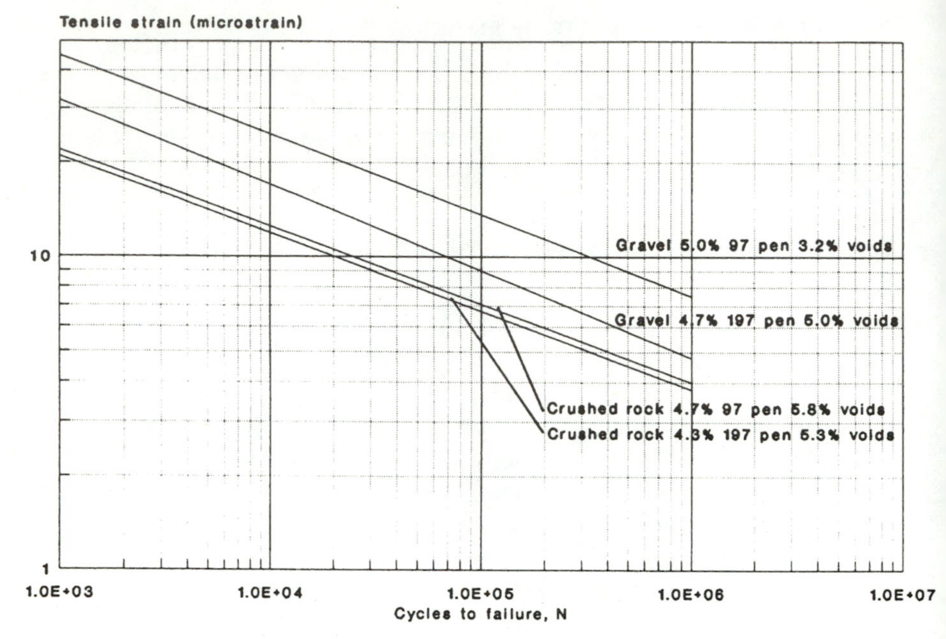

Figure 13.17 Strain/life relationships for dense bitumen macadam.

periods of 0.3 s between applications could increase the fatigue life by as much as 25 times in the temperature range 10 to 25°C.[11] The rest periods between the passage of the axles of a large commercial vehicle over a particular point on the road surface will be of the order 0.1 to 0.3 s and between the axles of different vehicles will exceed 0.3 s. The fatigue life determined at a frequency of loading of 15 to 20 Hz will therefore underestimate the fatigue life under practical conditions. To cover this point and take into account other differences between laboratory and field loading, Brown has suggested that laboratory-determined fatigue lives should be arbitrarily increased by a factor of 100 and on this basis has proposed the use of the fatigue relationships shown in Fig. 13.18 for the bituminous materials used in Britain for surfacings and road bases.[12] More recently, the Transport and Road Research Laboratory has issued a modified version of these curves shown in Fig. 13.19 which includes the effect of temperature.[13] Unless or until the fatigue lives of these materials can be more closely defined, estimates of flexible pavement lives based on fatigue tests must remain very conjectural.

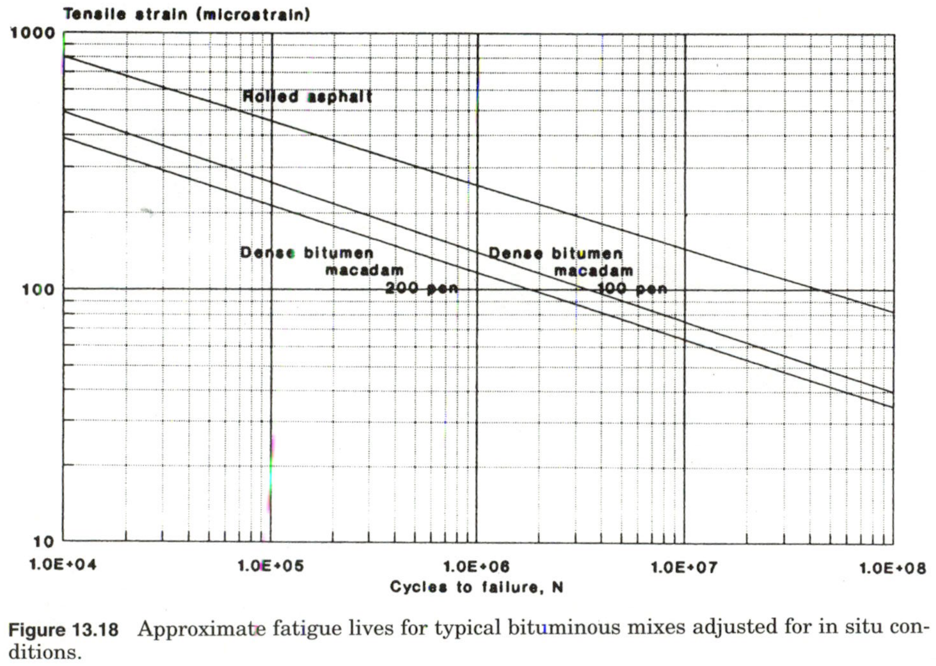

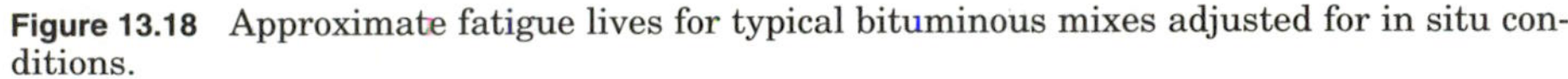

Figure 13.18 Approximate fatigue lives for typical bituminous mixes adjusted for in situ conditions.

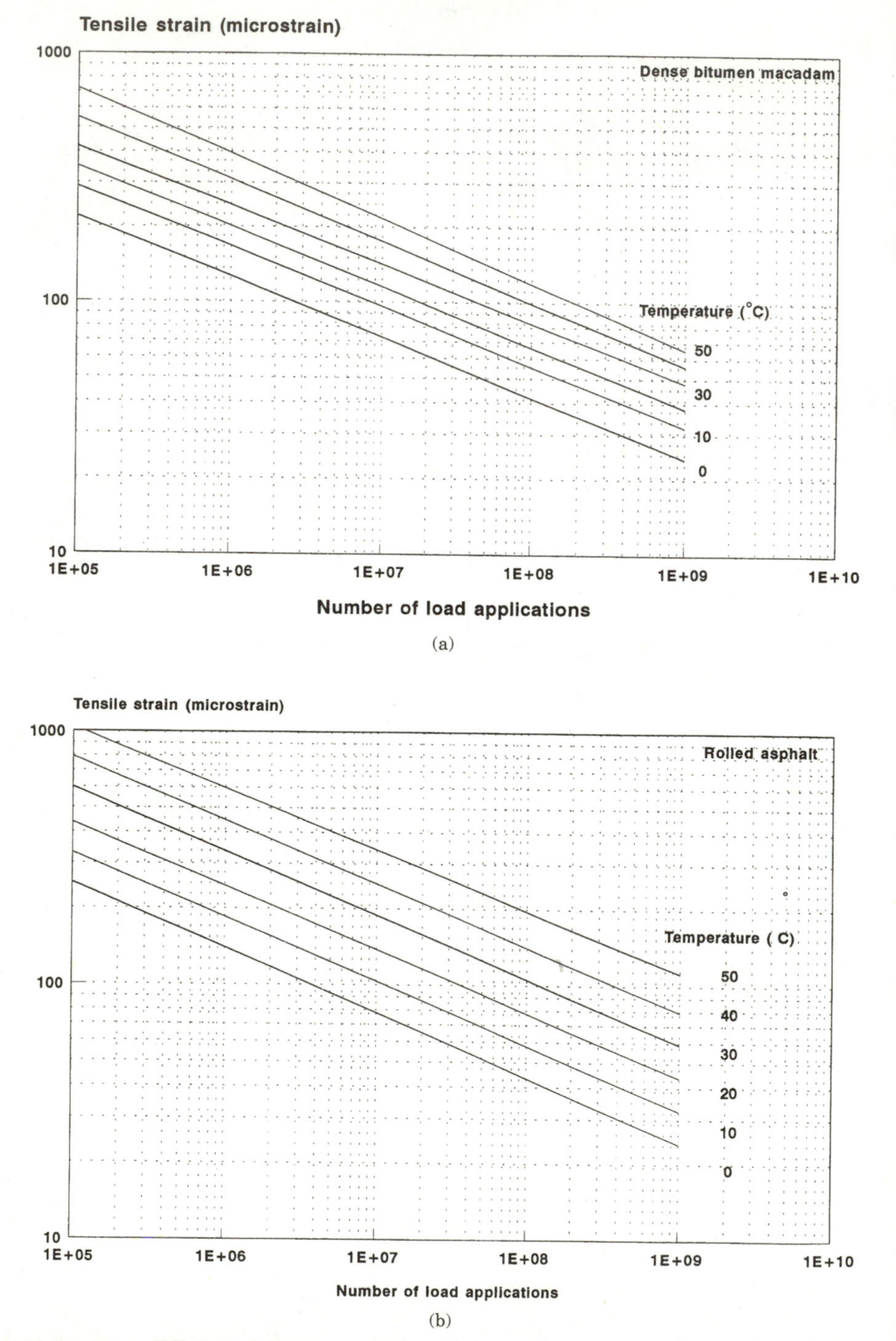

Figure 13.19 Effective fatigue curves for (*a*) dense bitumen macadam and (*b*) rolled asphalt.

References

1. American Association of State Highway and Transportation Officials: *AASHTO Materials:* Part II: Tests, 14th ed., Designation T245-82 (1986), AASHTO, Washington, D.C., p. 1006, 1986.
2. British Standards Institution: Sampling and Examination of Bituminous Mixtures for Roads and Other Paved Areas, British Standard 598:Part 3:1985, BSI, London, 1985.
3. British Standards Institution: Hot Rolled Asphalt for Roads and Other Paved Areas, British Standard 594:1985, BSI, London, 1985.
4. British Standards Institution: Coated Macadam for Roads and Other Paved Areas, British Standard 4987:1988, BSI, London, 1988.
5. Shook, J. F., and B. F. Kallas: Factors Influencing Dynamic Modulus of Asphalt Concrete, *Proc. t. Sess. Ass. Asph. Pav. Technol., Los Angeles,* vol. 33 (February), pp. 140–178, 1969.
6. Brown, S. F.: Determination of Young's Modulus for Bituminous Materials in Pavement Design, Highway Research Record no. 431, National Research Council, Washington, D.C., pp. 38–49, 1973.
7. Monismith, C. L., and K. E. Secor: Viscoelastic Behaviour of Asphalt Concrete Pavements, *Proc. 1st Int. Conf. on the Structural Design of Asphalt Pavements, Ann Arbor, Michigan, 1962*, University of Michigan, Ann Arbor, 1963.
8. Cooper, K. E., and P. S. Pell: The Effect of Mix Variables on the Fatigue Strength of Bituminous Materials, *Transport and Road Research Laboratory Report* LR 633, TRRL, Crowthorne, 1974.
9. Epps, J. A., and C. L. Monismith: Influence of Mixture Variables on the Flexural Fatigue Properties of Asphalt Concrete, *Proc. t. Sess. Asph. Pav. Technol., Los Angeles,* vol. 38 (February), pp. 423–464, 1969.
10. Kirk, J. M.: Relations between Mix Design and Fatigue Properties of Asphaltic Concrete, *Proc. 3rd Int. Conf. on the Structural Design of Asphalt Pavements, London, 1972,* University of Michigan, Ann Arbor, pp. 241–247, 1972.
11. Raithby, K. D., and A. B. Sterling: Laboratory Fatigue Tests on Rolled Asphalt and Their Relation to Traffic Loading, *Rds and Rd Constr., London,* vol. 50, pp. 596–597, 1972.
12. Brown, S. F.: A Simplified Fundamental Design for Bituminous Pavements, *The Highway Engineer* (*J. Instn. Highw. Engrs.*), vol. 21, nos. 8–9, pp. 14–23, 1974.
13. Fatigue Resistance of a Bituminous Roadpavement Design for Very Heavy Traffic, *Transport and Road Research Laboratory Report* 1050, TRRL, Crowthorne, 1982.

Chapter

14

Pavement-Quality Concrete

Introduction

14.1. Although short lengths of experimental concrete road were constructed in Britain in the later part of the 19th century, serious use of the material as a riding surface for major roads dates from the twenties. By that time concrete was being widely used in the United States and early British practice was based on American experience. Over the past 70 years the performance of concrete roads has been rather unpredictable. Either they have performed excellently or they have given trouble from a very early stage. The main problem appears to have been noncompliance with the specification, and particularly the concrete strength requirement.

Concrete Mix Design

14.2. Concrete for highway pavements is traditionally made using a combination of coarse and fine aggregates, although if suitable all-in gravel aggregates are available these can be used.

Coarse Aggregate

14.3. AASHTO Designation M43-82 gives 19 gradings for coarse aggregate varying in maximum size from 100 down to 10 mm. In practice the materials commonly used have a maximum size in the range 10 to 40 mm.

14.4. In the United Kingdom the choice of gradings included in BS 882[1] is more limited, as indicated in Table 14.1. Over this limited range, however, the U.S. and U.K. gradings are very similar.

TABLE 14.1 Grading of Coarse Aggregate

	Percentage by mass passing BS sieves for nominal sizes							
Sieve size, mm	Graded aggregate			Single-sized aggregate				
	40–5 mm	20–5 mm	14–5 mm	40 mm	20 mm	14 mm	10 mm	5 mm*
50.0	100			100				
37.5	90–100	100		85–100	100			
20.0	35–70	90–100	100	0–25	85–100	100		
14.0			90–100			85–100	100	
10.0	10–40	30–60	50–85	0–5	0–25	0–50	85–100	100
5.0	0–5	0–10	0–10		0–5	0–10	0–25	45–100
2.36							0–5	0–30

*Used mainly in precast concrete products.

Fine Aggregate

14.5. AASHTO Designation M6-81 defines the grading of fine aggregates as indicated in Table 14.2. To ensure that the fine aggregate is well graded, the U.S. specification uses the fineness modulus concept.

14.6. The U.K. specification for fine aggregate is shown in Table 14.3. The overall gradings closely follow those of Table 14.2, but subdivisions of coarse, medium, and fine types are introduced to ensure that the material used is well graded.

Mixing

14.7. The coarse and fine aggregates are mixed with cement and water in an approved plant, in proportions necessary to produce the required compacted strength and the required levels of workability and water-cement ratio. These requirements are based on laboratory testing prior to the start of any concrete pavement construction. The following requirements apply to concrete pavement construction in the United Kingdom.

TABLE 14.2 Grading of Fine Aggregate

Sieve, mm	Mass, % passing
9.5	100
4.75	95–100
1.18	45–80
0.300	10–30
0.150	2–10

TABLE 14.3 Grading of Fine Aggregate

	Percentage by mass passing BS sieve			
		Additional limits for grading		
Sieve size	Overall limits	C	M	F
10.00 mm	100			
5.00 mm	89–100			
2.36 mm	60–100	60–100	65–100	80–100
1.18 mm	30–100	30–90	45–100	70–100
600 μm	15–100	15–54	25–80	55–100
300 μm	5–70	5–40	5–48	5–70
150 μm	0–15*			

*Increased to 20 percent for crushed rock fines, except when they are used for heavy-duty floors.

NOTE: Fine aggregate not complying with Table 14.2 may also be used provided that the supplier can satisfy the purchaser that such materials can produce concrete of the required quality.

Water-Cement Ratio and Workability

14.8. The water content should be the minimum necessary to maintain the required workability, but for pavement-quality concrete the water-cement ratio by weight must not exceed 0.50.

14.9. The workability of the concrete at the time of placing should enable the concrete to be fully compacted and finished without undue flow. The compacting factor test should be used when the aggregate is crushed rock or gravel, and the target value should be 0.8 to 0.85. The VeBe test is applicable where slag or pulverized fuel ash are used in the mix, when the target level should be 6 s. In the compacting factor test,[2] freshly mixed concrete is allowed to fall under gravity in a controlled manner into a steel mould, and the density obtained is expressed as a ratio of the fully compacted density. The VeBe test[3] is a modified form of slump test in which vibration is used. The VeBe is the time taken for the slump formed in a standard manner to collapse to the horizontal condition under vibration.

Compressive Strength

14.10. The compressive strength of concrete is usually specified as a cube strength at an age of 7 or 28 days. The current U.K. Department of Transport requirement for pavement concrete is a compressive strength of 31 N/mm^2 at 7 days (44 N/mm^2 at 28 days). It is now normal practice to measure the strength at 7 days and 28 days by a series of preliminary laboratory tests at both ages using the cement and the aggregate to be used in the contract. In fact, the ratio of the 7-day and 28-day compressive strengths is not very significantly affected by the type of aggregate, although it is obviously affected by the rate of hardening of the cement. A ratio of 0.7:1 is normal for pavement-quality concrete.

14.11. Figure 14.1 shows the long-term variation of compressive strength with time for pavement-quality concrete of various 28-day compressive strengths. The measurements cover a period of 5 years. After that period the change of compressive strength was comparatively small but by no means insignificant or negligible in relation to performance.

The Relationship between the Compressive Strength and Modulus of Rupture of Concrete

14.12. Whether a concrete slab cracks under an applied tensile stress depends on the modulus of rupture of the concrete at the time the tensile stress is applied. The modulus of rupture is the stress which first initiates cracking. It is determined by the constitution of the con-

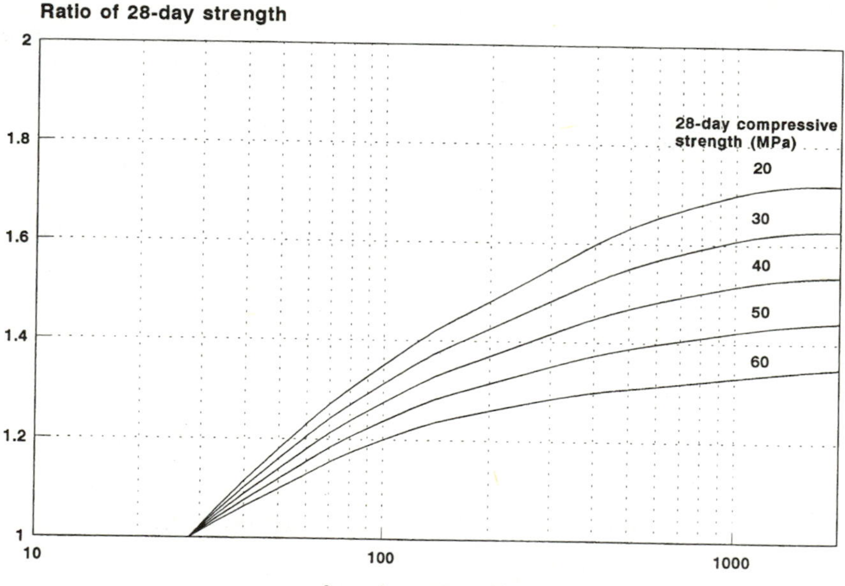

Figure 14.1 Relationship between age and compressive strength for pavement-quality concrete.

crete and its age and stress history in relation to fatigue. Research carried out in the early fifties showed that the modulus of rupture determined by slow flexural tests could be related to the compressive strength measured on cubes by relationships which were to an extent influenced by the aggregate used in the concrete.[4] The equations for gravel aggregates and for crushed rock aggregates are given below, where the compressive strength F_e and the modulus of rupture M_R are expressed in N/mm^2:

For gravel aggregate:

$$M_R = 0.49 \times F_e^{\,0.55} \tag{14.1}$$

For crushed stone aggregate:

$$\mathrm{M_R} = 0.36 \times F_e^{\,0.7} \tag{14.2}$$

14.13. These equations have been used in Table 14.4 to calculate the modulus of rupture for concrete made with crushed rock and gravel aggregates having a crushing strength of 10 to 60 N/mm^2. For a given crushing strength, the modulus of rupture of concrete made with crushed rock aggregate is considerably greater than that of gravel concrete. The lower cost of gravel may well at medium strengths justify on economic grounds the use of gravel concrete with a higher cement content and higher compressive strength.

The Relationship between Elastic Modulus, Compressive Strength, and Age

14.14. The elastic modulus of concrete is measured using both dynamic and static methods.[5,6] As with bituminous materials, the static method gives a lower value. The strains involved in static tests are greater than those which occur under wheel loading, while with the dynamic method they are smaller.

14.15. Table 14.5 shows the variation of dynamic modulus measured on two concretes, with compressive strength and age. The PQ1 material was made with gravel aggregate and PQ2 with crushed stone. The value increases with both age and strength, but the variation is small. Comparison between static and dynamic moduli measured on the same concretes indicates a ratio of about 0.8. Structural analyses carried out on concrete pavements show that computed stresses are not very sensitive to small changes in elastic modulus. It is proposed, therefore, that a constant modulus should be used in such analyses

TABLE 14.4 Effect of Age on the Compressive Strength and the Modulus of Rupture of Concrete Made with Gravel and Crushed Rock Aggregates (All Units N/mm²)

Age of		M_R			M_R			M_R			M_R			M_R			M_R	
concrete	F_e	CR*	G*	F_e	CR	G	F_e	CR	G	F_e	CR	G	F_e	CR	G	F_e	CR	G
28 days	10.0	1.80	1.74	20.0	2.93	2.55	30.0	3.89	3.18	40.0	4.76	3.73	50.0	5.51	4.21	60.0	6.32	4.66
40 days	11.0	1.93	1.84	22.1	3.14	2.69	32.9	4.15	3.35	43.5	5.05	3.90	53.8	5.86	4.39	63.6	6.59	4.81
3 months	13.3	2.20	2.03	26.5	3.57	2.97	38.4	4.63	3.64	50.4	5.59	4.23	61.8	6.46	4.73	71.1	7.12	5.11
6 months	14.7	2.36	2.15	29.4	3.84	3.15	42.5	4.97	3.85	55.0	5.95	4.44	65.8	6.75	4.90	75.6	7.43	5.29
1 year	15.9	2.50	2.24	31.8	4.06	3.28	45.8	5.23	4.01	58.6	6.22	4.60	69.0	6.97	5.03	78.3	7.62	5.39
2 years	16.6	2.57	2.30	33.3	4.19	3.37	48.0	5.41	4.12	61.0	6.39	4.70	71.3	7.14	5.12	79.5	7.70	5.44
3 years	17.0	2.62	2.33	34.0	4.25	3.41	48.6	5.46	4.15	61.6	6.44	4.73	72.0	7.19	5.15	79.8	7.67	5.45
4 years	17.2	2.64	2.35	34.4	4.28	3.43	48.7	5.47	4.15	61.8	6.46	4.73	72.3	7.21	5.16	80.1	7.74	5.46
5 years	17.3	2.65	2.35	34.5	4.29	3.44	48.9	5.48	4.16	62.0	6.47	4.74	72.3	7.21	5.16	80.1	7.74	5.46

*CR = crushed rock; *G = gravel.

TABLE 14.5 Dynamic Modulus of Elasticity, Density, and Compressive Strength

		Dynamic modulus, kN/mm²			Density, kg/m³			Equivalent cube strength, N/mm²		
Concrete	Age, weeks	Mean	No. of results	c.v.,* %	Mean	No. of results	c.v., %	Mean	No. of results	c.v., %
PQ1	4	43	20	1.8	2344	20	0.5	41.20	20	3.3
	13	45	21	1.1	2341	21	0.4	48.80	21	3.1
	26	45	47	1.5	2331	47	0.5	49.83	40	4.1
	39	45.5	20	1.0	2332	20	0.4	52.10	20	5.0
	52	46.5	27	1.6	2344	32	0.4	54.23	31	4.9
	104	46	23	0.9	2335	25	0.4	53.73	22	2.8
	156	47	20	1.1	2348	20	0.4	56.75	19	3.3
	260	47.5	53	1.5	2351	53	0.5	61.21	47	4.4
	520	48	10	1.1	2350	10	0.35	61.95	10	4.6
PQ2	13	39	20	1.4	2374	20	0.3	39.26	20	3.1
	26	39.5	54	1.2	2367	55	0.4	43.09	50	2.6
	39	40.5	20	3.8	2376	20	0.4	44.40	19	3.6
	52	40	20	1.1	2373	20	0.5	44.43	20	4.8
	104	41	19	1.7	2370	19	0.8	45.48	14	9.5

*c.v. = coefficient of variation.

and that it should be the mean value between the two methods of determination. A value is 35 to 40 $\times$ 10^3 N/mm^2 is appropriate.

Poisson's Ratio of Pavement-Quality Concrete

14.16. The value of Poisson's ratio for concrete has been discussed in Par. 12.23. The appropriate value is 0.15. There appears to be little change with the compressive strength of the material or with the type of aggregate used.

Fatigue of Concrete Pavements

14.17. Concrete pavements are subject to fatigue under repeated loading, in the sense that they may crack under repeated applications of a stress less than the modulus of rupture. As with most brittle materials there is a linear relation between the applied tensile stress and the logarithm of the number of applications of that stress which will cause cracking. The fatigue of concrete pavements appears to have been first studied in detail by Kesler in the United States.[7] Using a repeated loading flexural test, operated at a frequency designed to obviate the effect of aging on the tensile strength, he showed that the stress to cause failure after 10^5 applications was close to 0.74 times the modulus of rupture at 28 days. This compares with the value of about 0.77 shown in Fig. 12.7 referring to lean concrete and cement-bound materials.

14.18. Kesler used concrete beams approximately 1.5 m long and of 150 × 150 mm cross section. These were freely supported near the ends and loaded between the third points as in the normal flexural strength test. The load was applied through a lever and cam system to give loading frequencies of 70, 230, and 440 cycles per minute. Typical results are shown in Fig. 14.2. Even with the lowest loading frequency the tests were completed within 24 hours. Kesler used the flexural strength measured on the free ends of the same beams to determine the static strength.

14.19. More recently Galloway, Harding, and Raithby have carried out fatigue tests on two concretes,[8] using the dynamic loading machine shown in Fig. 14.3. Tests were made on a concrete PQ1 of 28-day compressive strength 45 N/mm^2, and a concrete PQ2 of 28-day compressive strength 33 N/mm^2. The tests were conducted with the materials at various ages between 28 days and 5 years. Figure 14.4 shows the results. It is clear that the strength of the concrete and its

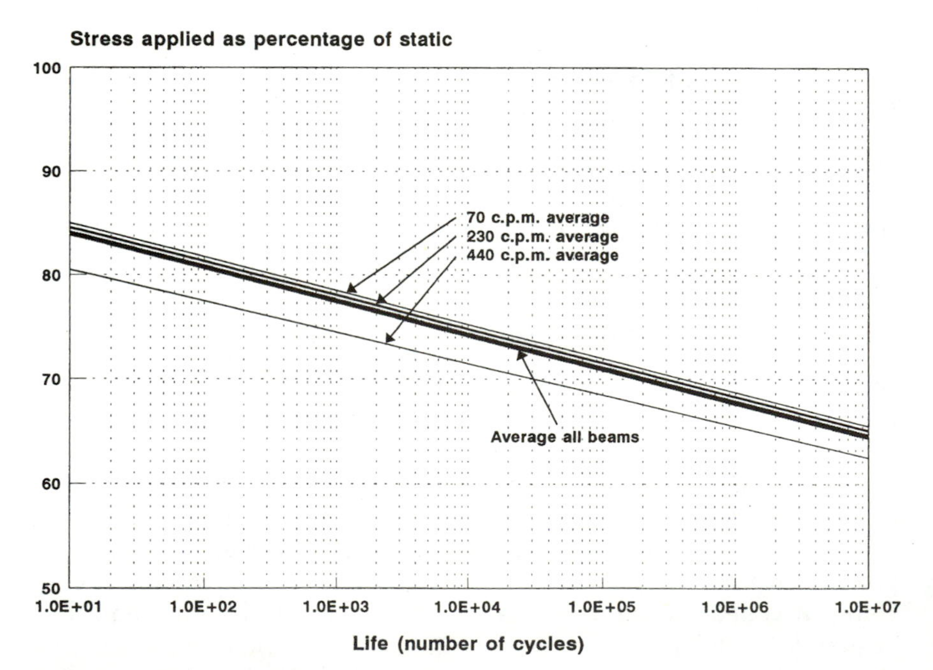

Figure 14.2 Effect of frequency of loading on fatigue of concrete.

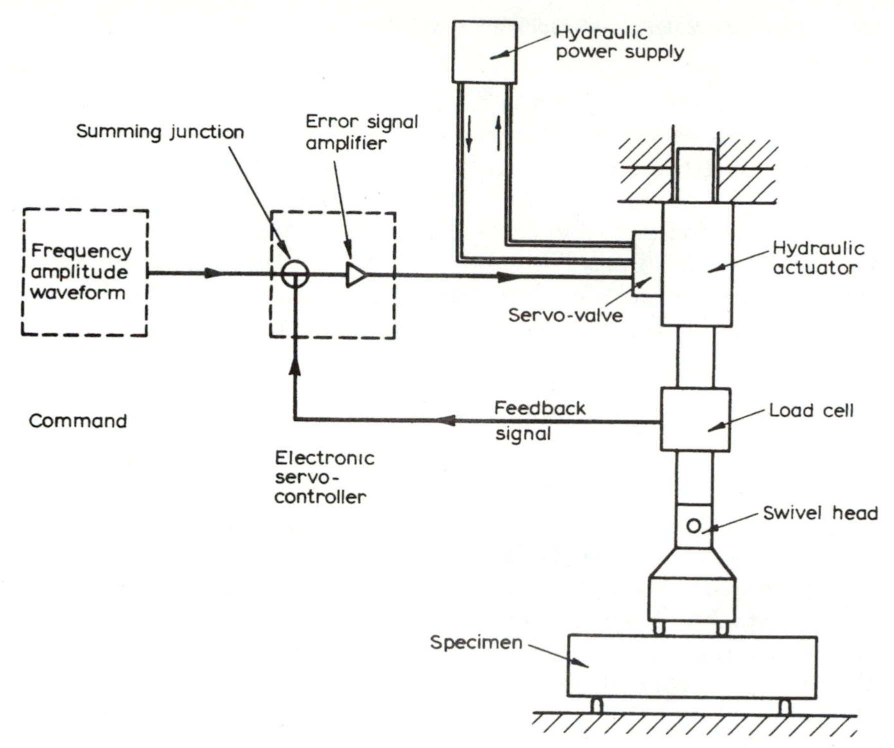

Figure 14.3 Closed-loop electrohydraulic testing system.

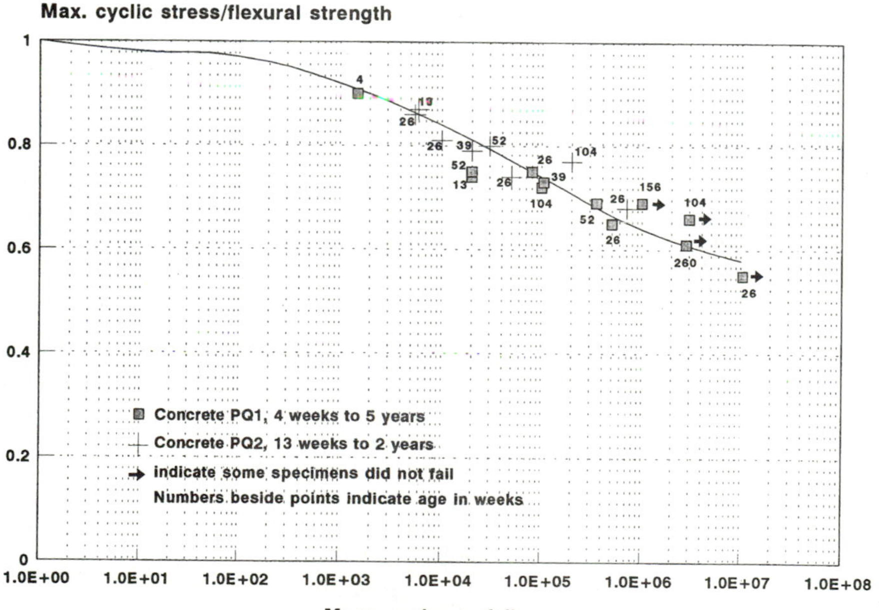

Figure 14.4 Fatigue performance related to flexural strength.

age have little influence on the fatigue life expressed as a percentage of the flexural strength at the time of testing. The percentage of the flexural strength required to cause fatigue at 10^5 stress applications is close to 75. The relationship is less linear than that from the Kesler tests. This is due partly to the fact that a number of the tests were not continued to failure. There is likely to be little error involved in assuming a linear relationship with fatigue failure due to a stress of 80 percent of the flexural strength at 10^5 applications.

Relationship between Fatigue and Age for Concrete

14.20. Linear fatigue relations of this type can be drawn for any concrete at ages between 28 days and 5 years. The information given in Table 14.4 and in Par. 14.19 will enable a family of such relations to be prepared. The rupture envelope will cross the fatigue lines in a manner which depends solely on the rate at which the stress applications are repeated. This must be taken into account in the structural design of concrete pavements, and it is considered in detail in Chap. 24.

References

1. British Standards Institution: Aggregates from Natural Sources for Concrete, BS 882:1983, BSI, London, 1983.
2. British Standards Institution: Testing Concrete. Method of Determining the Compacting Factor, BS 1881:Part 103:1983, BSI, London, 1983.
3. British Standards Institution: Testing Concrete. Method of Determination of VeBe Time, BS 1881:Part 104:1983, BSI, London, 1983.
4. Road Research Laboratory: *Concrete Roads—Design and Construction,* HMSO, London, p. 74, 1955.
5. British Standards Institution: Testing Concrete: Recommendations for Measurement of Velocity of Ultrasonic Pulses in Concrete, BS 1881:Part 203:1983, BSI, London, 1983.
6. British Standards Institution: Testing Concrete: Method of Determination of Static Modulus of Elasticity in Compression, BS 1881:Part 121:1983, BSI, London, 1983.
7. Kesler, C. E.: Effect of Speed of Testing on Flexural Fatigue Strength of Plain Concrete, *Proc. Highw. Res. Bd., Wash.,* vol. 32, pp. 251–258, 1953.
8. Galloway, J. W., H. M. Harding, and K. D. Raithby: Effects of Age on Flexure, Fatigue and Compressive Strength of Concrete, *Transport and Road Research Laboratory Report* LR865, TRRL, Crowthorne, 1979.

Chapter

15

Block Pavements

Introduction

15.1. Concrete and ceramic block pavements have enjoyed a considerable degree of popularity in recent years. They undoubtedly look attractive, particularly when new. In situations such as shopping malls and pedestrian precincts they can provide a pleasing architectural feature. An advantage frequently stressed is that they can be lifted when necessary to restore regularity, or to provide access to underground services. However, it is to be deprecated that there is a growing tendency to infill areas which have been raised or damaged with concrete or bitumen as shown in Fig. 15.1. Because of changes in the shape and color of blocks any authority using this form of construction should order sufficient spare blocks to permit adequate maintenance.

15.2. Hard-selling advertising claims that block pavements spread loads more effectively than continuous concrete and bituminous surfacings, or that 30 percent savings can be made by their use. These claims need to be treated with caution. There is no long-term full-scale evidence to support such claims.

History of Block Pavements

15.3. Block paving is far from a new concept. Stone setts, laid in a very similar manner to modern blocks, have been a feature of many European cities since medieval times. In the early 19th century interlocking wood blocks, generally hexagonal in shape, were introduced into the City of London, as a quieter alternative to stone setts. They were laid on a cemented base at least as thick as the blocks. In various forms, wood blocks remained a very popular form of paving in

Figure 15.1 Bituminous infill in block pavement.

London until after World War II. Later they were surfaced dressed with hot tar and chippings to provide adequate skid resistance.

15.4. Brick pavements were used very widely in the United States in the last two decades of the 19th century, particularly in Ohio and Illinois. The bricks were fired at a high temperature to provide a vitreous appearance and high crushing strength. They were generally 9 $\times$ 4 $\times$ 3 in deep with an alternative size for lightly trafficked situations where the depth was 2 in.

In 1899, a U.S. commission was set up to report on brick pavements. A major conclusion was that the bricks should be set on a bed of sharp sand 1 in thick, over a concrete base. It was recommended that the joints should be filled with a cement slurry.

15.5. In western Holland, which includes the cities of Amsterdam, Rotterdam, and The Hague, and where the subsoil is a well-compacted sand, brick pavements have been widely used for more than 90 years. The bricks are laid directly on the sand and experienced gangs are employed to relay the surface as deformation occurs. The frequency of the relaying operation depended on the type and constitution of the traffic. After World War II the increase in the loading of commercial traffic made the continued use of this form of construction on major roads uneconomic.

Modern Block Pavements

15.6. The blocks being used today for this type of paving are of either pressed concrete or fired clay. Thicknesses range from 50 to 100 mm and they are either rectangular or shaped to provide lateral interlock. The overall size is such that, when laid, each square meter contains about 40 blocks. Present thinking appears to prefer the rectangular shape as being less liable to crack during service. The blocks are produced in a variety of colors.

15.7. The manufacturers' recommendation is to use a base and subbase thickness appropriate to an asphalt pavement designed for the traffic to be carried. In place of the asphalt, the blocks are laid on a bedding of sand of 50 mm compacted thickness. The sand, which should have a maximum size of 5 mm and less than 3 percent by weight finer than 0.06 mm, is laid to an appropriate uncompacted thickness. After placement the blocks are surface-vibrated using an appropriate vibrating-plate machine. During compaction, sand is applied to the surface in an attempt to fill the spaces between the blocks.[1] The blocks are normally required to have a crushing strength in excess of 50 N/mm^2 at 28 days.

Load Spreading by Block Pavements

15.8. The manufacturers' recommendations as stated above imply that the load-spreading properties of block pavements are equal to those of an uncracked two-course asphalt pavement of equal total thickness. This would be possible only if the friction between the vertical faces of the blocks was sufficient to ensure load transfer without differential sliding between the blocks. This is a most unlikely situation. Seddon, working in New Zealand, appears to have been the first to investigate this matter using the blocks as part of a road structure, as distinct from laboratory tests in shallow tanks. He assessed the load-spreading ability by a process of back calculation from measured deflection under a heavily loaded wheel. He concluded that the effective modulus of elasticity of the combination of blocks and their bedding was similar to that of a compacted unbound material.[2]

15.9. Clark,[3] following earlier similar work carried out by the Cement and Concrete Association,[4] attempted to assess the load-spreading ability of block pavements using tank tests and a synthetic subgrade of expanded polystyrene. The blocks were laid on a 60-mm uncompacted sand bedding over 300 mm of well-compacted type 1 crushed stone subbase (see Chap. 11). The subbase was instrumented with pressure cells; five were in the surface of the subbase, close to the axis of loading, and one was in the lower surface of the subbase. The last was on the axis of loading at the middepth of the subbase. Loading was by a 300-mm-diameter plate and a hydraulic ram. The report states that a number of the pressure cells failed to operate satisfactorily, and as a result the observations were restricted to some measurements of compressive stress in the surface of the subbase. These indicated a stress level of about 40 to 50 percent of the applied stress at the surface. The Boussinesq stress analysis for circular loading indicates that this is consistent with a surfacing having an effective modulus of elasticity very similar to the subbase material. The report states:

> In the tests, the surface deformation of the paving had been confined solely to the blocks in direct contact with the loading plate and virtually no movement had been detected in adjacent blocks. This suggests that the level of load transfer across the boundary between the loaded and unloaded blocks was small.[3]

This indicates that there was relatively unimpaired vertical sliding between the loaded and unloaded blocks. A photograph in the paper shows this clearly.

15.10. A most comprehensive study of the load spreading of block pavements has recently been published by the TRRL.[5] Various combinations were used. These included (1) brick paver blocks with sand bedding laid on 135 mm of type 1 subbase over a deep clay fill of CBR

5 to 8 percent, compared with a dense bitumen macadam road base material laid on the same subbase and to a thickness equal to that of the blocks and their bedding, and (2) a similar comparison, but with an additional layer of dense bitumen base material 60 mm thick laid beneath both the blocks and the bituminous material of equal thickness. Wheel loads of 40 to 80 kN were applied at speeds of 2.5 and 20 km/h to give cumulative traffic up to 8000 standard axles. The sections were instrumented for the measurement of vertical subgrade strain and the horizontal strain under the base.

15.11. The measured vertical strain in the subgrade beneath the block pavement was about three times that under the equivalent thickness of bituminous road base. The horizontal strain under the base was 2 to 3 times greater under the block pavement. Elastic deflection measurements were made on the pavements using the falling weight deflectometer (see Chap. 20). These showed deflections of the block pavements three to five times greater than those observed on the bituminous pavements. When the permanent deformation had produced a rut depth of 10 mm, there was no indication from the strain measurements that a "lock-up" effect had occurred.

15.12. These measurements appear to confirm Seddon's conclusion that the effective modulus of brick pavements is similar to that of an equivalent thickness of crushed stone, and that it does not compare in this respect with two-course rolled asphalt.

15.13. The inference from all the research work referred to above is that, to be successful, blocks should be laid on a foundation which itself is strong enough to resist both deformation and cracking under the envisaged traffic loading.

15.14. Confirmation that block pavements spread traffic loads less adequately than asphalt of equivalent thickness is shown in Figs. 15.2 and 15.3. These refer to the performance of blocks laid on a heavily trafficked A-class road. The original asphalt surfacing was removed and replaced by blocks on a sand bedding to restore the original surface level. Considerable deformation is shown in Fig. 15.2 some 2 years after the work was completed. Figure 15.3 shows the tendency of the blocks to move in the direction of traffic flow in an area subject to braking and acceleration forces.

Figure 15.2 Examples of deformation in block pavement laid on a heavily trafficked road.

Figure 15.3 Displacement of blocks in an acceleration and braking situation.

References

1. Cement and Concrete Association: *Concrete Block Paving: Model Specification Clauses for Roads Subject to Adoption,* C & CA, Wexham Springs, Slough, Bucks., 1978.
2. Seddon, P. A.: The Behaviour of Interlocking Concrete Block Paving at the Canterbury Test Track, *Tenth Australian Road Research Board Conference,* August 1980, ARRB, Vermont South, Victoria 3133.

3. Clark, A. J.: Further Investigations into the Load-spreading of Concrete Block Paving, *Cement and Concrete Association Technical Report* 545, C & CA, Wexham Springs, Slough, Bucks., 1981.
4. Knapton, J.: The Design of Concrete Block Roads, *Cement and Concrete Association Technical Report* 415, C & CA, Wexham Springs, Slough, Bucks., 1976.
5. Addis, R. R., R. G. Robinson, and A. R. Halliday: The Load-spreading Properties of Clay Brick Pavements, *Transport and Road Research Technical Report* 234, TRRL, Crowthorne, 1989.

Part

Pavement Performance Studies and Empirical Design

Chapter

16

The Approaches to Pavement Design

Introduction

16.1. Chapters 9 to 15 discuss the materials with which the road engineer has to construct pavements. The combination of these materials needs to be equated to the traffic to be carried in the most cost-effective manner to meet the design life requirements. This process, generally referred to as pavement design, is the subject of the next three chapters.

16.2. The most reliable approach to pavement design is by full-scale experiments constructed on heavily trafficked routes, in which a variety of surfacing, base, and subbase materials are laid to a range of thicknesses. Observations of deformation, cracking, and the constitution and volume of the traffic carried out over a long period will enable the most economical design to be formulated. This approach has been adopted in the United Kingdom for more than 50 years, as is discussed in Chaps. 18 and 19. Current British designs are based on the results of such experiments.

Accelerated Testing Procedures

16.3. As would be expected, more rapid methods of collecting design information, particularly in relation to new materials, are constantly being sought. In 1912, the road testing machine shown in Fig. 16.1 was constructed at Teddington in the southern outskirts of London.[1] The load on each of the six wheels was adjustable to give the required wheel load spectrum, as is indicated in Fig. 16.2. The wheels were replaced with a more modern type with pneumatic tires several years

Figure 16.1 Teddington road testing machine (1912).

later. The machine was operated in a temperature-controlled building. The maximum wheel load was 2800 lb, or 1270 kg, with a coverage of 80,000 wheel coverages per 24-hour period.

16.4. The main use of this machine was to develop specifications for stable bituminous mixtures for the surfacing of existing water-bound

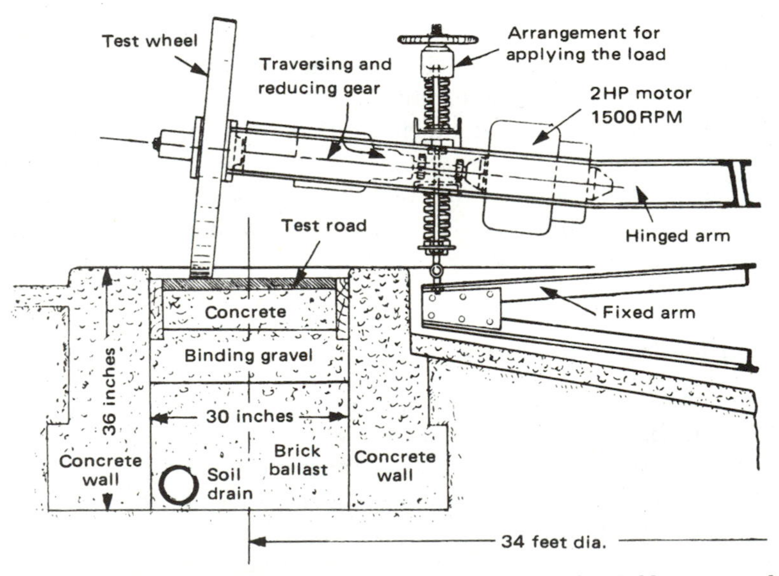

Figure 16.2 Details of wheel load adjustment used on the Teddington road testing machine (1912).

Figure 16.3 Bureau of Public Roads test track (1925).

macadam roads. Early British specifications for such materials were developed in this way.

16.5. In 1925, the Bureau of Public Roads in the United States had constructed the circular test track shown in Fig. 16.3.[2] This open-air facility was used to test a range of bituminous material under heavily loaded truck traffic.

16.6. The circular road machine shown in Fig. 16.4 was constructed in 1933 at the original site of the U.K. Road Research Laboratory. Experimental road sections were trafficked by a tethered commercial vehicle using electric propulsion. The machine was subsequently modified to have a very heavily loaded central wheel assembly as shown in Fig. 16.5. The machine was primarily used to study the stresses generated in pavement structures.

16.7. In 1985, a machine working on the linear principle was constructed at the Crowthorne site of the U.K. Transport Research Laboratory. The test pit is 25 m long and 10 m wide, which can accommodate several test pavements instrumented for the measurement of stress and deflection. The design of the machine is shown in Fig. 16.6.

16.8. Although machines of the type discussed above are useful in exploring the stress regime in pavement structures, they do not pro-

Figure 16.4 Road machine no. 3 as originally constructed.

Figure 16.5 Arrangement of test wheel of road machine no. 3.

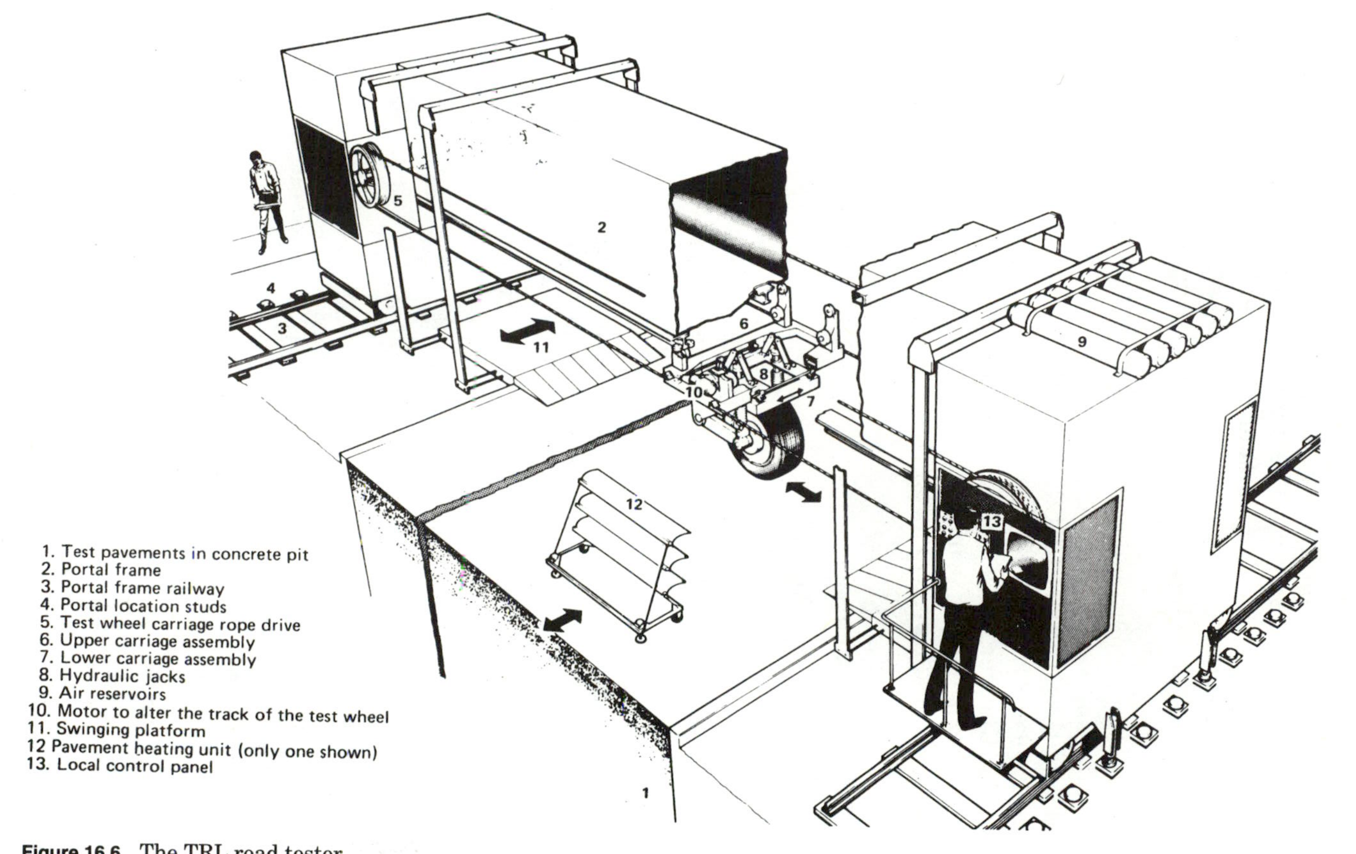

Figure 16.6 The TRL road tester.

vide a significant advantage in relation to speed of pavement testing compared with full-scale road experiments. It has been shown in Chaps. 13 and 14 that both bituminous and cemented materials increase in strength as they age. In this respect any form of accelerated testing is likely to lead to misleading conclusions.

16.9. As a compromise the WASHO and AASHO experiments carried out in the United States gave a modest degree of acceleration combined with the objective of establishing the relative damaging effect of a wide range of axle loads. These two experiments are described in Chap. 17. Chapters 18 and 19 summarize the conclusions from full-scale road experiments using flexible and concrete pavements carried out in the United Kingdom during the past 50 years.

The Theoretical Approach

16.10. Since the late 1920s when Westergaard published his papers on the structural design of concrete pavements, a great deal of effort has been devoted to applying structural theory to the design of both flexible and concrete pavements. A great many papers have, for example, been published in the proceedings of the Conferences on the Structural Design of Asphalt Pavements which have been sponsored by the University of Michigan at 5-year intervals since 1962, and more recently by the International Society for Asphalt Pavements.

16.11. The present position with regard to asphaltic pavements is reviewed in Chap. 23, and with regard to concrete pavements in Chap. 24.

References

1. Boulnois, H. P.: *Modern Roads,* Edward Arnold, London, 1919.
2. Woodrow, J. H., and J. Y. Welborn: Development of Asphalt Tests and Specifications in the United States, *Pub. Rds., Wash.,* vol. 39, no. 1, pp. 7–15, 1975.

Chapter

17

The AASHO and WASHO Road Tests

Introduction

17.1. These two pavement test projects were designed, respectively, by the American Association of State Highway Officials and the Western Association of State Highway Officials in the 1950s. Both were administered by the Highway Research Board of America. Although the testing work of the WASHO test took place about 2 years earlier than the AASHO experiment, the latter was much more comprehensive, and it will be considered first to enable the conclusions from the two experiments to be compared more easily.

Purpose of the AASHO Road Test*

17.2. The principal objective of the test was to determine the significant relationship between the number of repetitions of specific axle loads of different magnitude and arrangement, and the performance of different thicknesses of uniformly designed and constructed asphaltic concrete and reinforced portland cement concrete surfacings on different thicknesses of base and subbase when laid on a basement soil of known characteristics.[1]

17.3. Planning the project continued until 1955 so that lessons learned from the WASHO test could be incorporated in the proposals.

*The AASHO test was conducted and analyzed in terms of imperial units, and these have been retained in this chapter.

Site Details

17.4. The test roads were built at Ottawa, Ill., about 100 km southwest of Chicago, where the climatic and soil conditions were typical of large areas of the northern United States. The construction was entirely on embankment, the top 3 ft of which were a uniform sandy clay (liquid limit 30 percent and plastic limit 13 percent). The average depth of frost penetration during the two winters of the testing was 35 in. The in situ CBR of the soil as placed was about 4 percent, but this fell to about 2 percent after the spring thaw. The level of the water table was between 2 and 7 ft below the finished road surface.

Layout of the Experiment

17.5. The experiment consisted of 6 loops (1 to 6). Loops 2 to 6 were trafficked, but loop 1 was used for climatic and other observations. Each traffic loop had two traffic lanes 12 ft wide which were independently trafficked. The two straight tangential lengths of each loop were used for experimental flexible and concrete pavements. Test traffic operated in rigidly enforced lanes at 35 mi/h, for about 19 h per day and for a little more than 2 years. The total number of axle loads over each experimental section was over 1.1 million. The axle loads ranged from 2000 lb on single axles to 48,000 lb carried on tandem axles. (In AASHO road test reports axle loads are quoted in kips, i.e., in units of 1000 lb.) The axle loads and the loops and lanes on which they operated are shown in Table 17.1.

Thickness Combinations and Materials Used for Flexible Pavements

17.6. The axle loads and combinations of surfacing, road base, and subbase used on the flexible lengths of each of the loops 2 to 6 are shown in Table 17.1.

TABLE 17.1 Axle Loads and Pavements Used; Main Flexible Sections

	Loop 2		Loop 3		Loop 4		Loop 5		Loop 6	
Axle load, lb	Lane 1 2000 S*	Lane 2 6000 S	Lane 1 12,000 S	Lane 2 24,000 T*	Lane 1 18,000 S	Lane 2 32,000 T	Lane 1 22,400 S	Lane 2 40,000 T	Lane 1 30,000 S	Lane 2 48,000 T
Subbase, in	0 and 4		0, 4, and 8		4, 8, and 12		4, 8, and 12		8, 12, and 16	
Road base, in	0, 3, and 6		0, 3, and 6		0, 3, and 6		3, 6, and 9		3, 6, and 9	
Surfacing, in	1, 2, and 3		2, 3, and 4		3, 4, and 5		3, 4, and 5		4, 5, and 6	

*S = carried on single axles; T = carried on tandem axles.

17.7. Each loop comprised a complete factorial design in the sense that each thickness of surfacing, base, and subbase was used in combination with every other thickness. The sections were placed in statistically random order around the loops, but juxtaposition of very thick and very thin sections was avoided.

17.8. The subbase used was a local sandy gravel, modified by plant mixing with a fine sand and a slightly cohesive soil. The road base was a wet-mix crushed limestone. The gradings of the road base and the subbase are shown in Fig. 17.1 compared with British specifications.

17.9. On loops 3 to 6 a subsidiary experiment was included to compare the performance of crushed limestone wet mix with cement-bound and bitumen-bound road bases. A sloping subgrade and subbase were used to give a varying thickness of base over these sections, often referred to as the "wedge sections." These experiments used only one thickness of surfacing in each loop as indicated in Table 17.2. The grading of the crushed-stone wet-mix base material was the same as that used in the main factorial experiment. The bitumen and cemented road bases were made by mixing binder with the subbase material used for all the sections. The binder content for the bituminous base material was 5.2 percent and the cement content for the cemented road base was 4 percent, giving a 7-day (cylinder) compressive strength of 840 lb/in^2, corresponding to a 28-day compressive strength of 1450 lb/in^2. Details of the surfacing materials used for all the flexible sections are given in Fig. 17.2.

Thickness Combinations and Materials Used for Concrete Pavements

17.10. The axle loads used on the concrete pavements were identical with those on the flexible pavements since both types of construction were on the same loops. The axle loads, together with the slab and base thicknesses used, are shown in Table 17.3.

17.11. The experiment was again fully factorial, giving 20 sections in each lane for loop 2 and more than 28 sections in the other loops. Reinforced sections were 240 ft long with contraction joints at 40-ft spacing. Unreinforced sections were 120 ft long with contraction joints at 15-ft intervals. All joints were doweled.

17.12. The subbase was the same material as used for the flexible sections. The mix design for the concrete was 6 bags per cubic yard

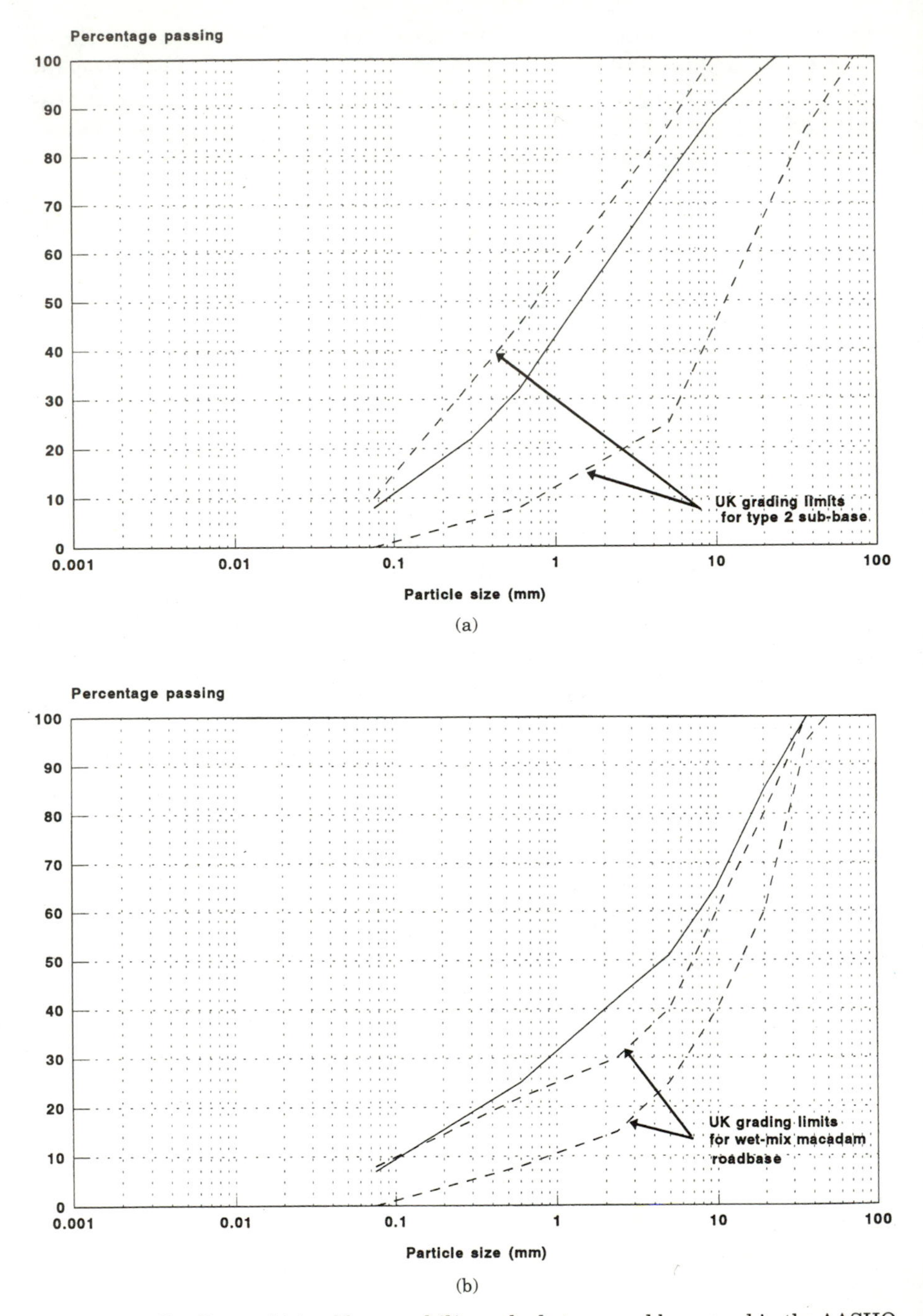

Figure 17.1 Gradings of (*a*) subbase and (*b*) crushed stone road base used in the AASHO road test.

TABLE 17.2 Axle Loads and Pavements Used for the Wedge Sections

	Loop 3		Loop 4		Loop 5		Loop 6	
Axle load, lb	Lane 1 12,000 S*	Lane 2 24,000 T*	Lane 1 18,000 S	Lane 2 32,000 T	Lane 1 22,400 S	Lane 2 40,000 T	Lane 1 30,000 S	Lane 2 48,000 T
Subbase, in	0		4		4		4 and 8	
Road base, in	2–14		2–16		3–18		3–19	
Surfacing, in	3		3		3		4	

*S = carried on single axles; T = carried on tandem axles.

with a maximum water content of 5 gal per bag of cement. The ratio of sand to total aggregate was 1:3. The average 28-day flexural and compressive strengths of the concrete were 725 and 4450 lb/in^2 (5 and 30.5 MN/m^2, respectively). For the reinforced sections the weight of reinforcement ranged from 21 to 81 lb/100 ft^2 for the thickness range 2.5 to 12.5 in for the slabs, and was approximately proportional to the thickness. All longitudinal and transverse joints were doweled.

The Concept of Present Serviceability

17.13. The analysis procedures used and the results of the AASHO road test were first unveiled at a special 3-day meeting of the

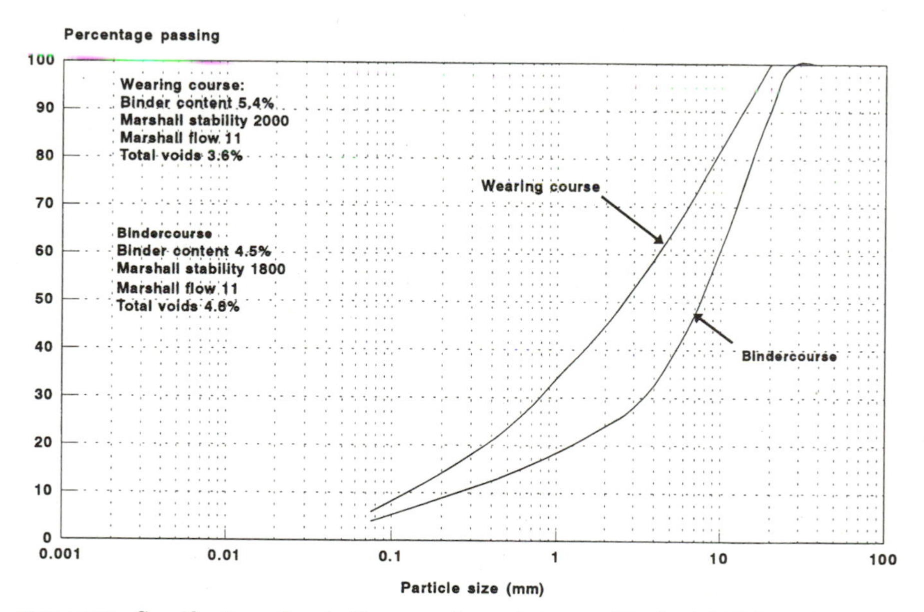

Figure 17.2 Specification of asphaltic concrete surfacing used in the AASHO road test.

TABLE 17.3 Axle Loads and Pavements Used in the Concrete Experiment

	Loop 2		Loop 3		Loop 4		Loop 5		Loop 6	
Axle load, lb	Lane 1 2000 S*	Lane 2 6000 S	Lane 1 12,000 S	Lane 2 24,000 T*	Lane 1 18,000 S	Lane 2 32,000 T	Lane 1 22,400 S	Lane 2 40,000 T	Lane 1 30,000 S	Lane 2 48,000 T
Subbase, in	0, 3, and 6		3, 6, and 9		3, 6, and 9		3, 6, and 9		3, 6, and 9	
Slab thickness (unreinforced), in	2½, 3½, and 5		3½, 5, 6½, and 8		5, 6½, 8, and 9½		6½, 8, 9½, and 11		8, 9½, 11, and 12½	
Slab thickness (reinforced), in	2½, 3½, and 5		3½, 5, 6½, and 8		5, 6½, 8, and 9½		6½, 8, 9½, and 11		8, 9½, 11, and 12½	

*S = carried on single axles; T = carried on tandem axles.

Highway Research Board held in St. Louis in May 1962. Those of us present soon began to wonder if the work was entirely in the hands of statisticians and where the engineers had been.

17.14. Before the start of the road test it was apparently decided to use the concept of "present serviceability" to quantify the condition of each experimental section of pavement, in addition to the observations of deformation and cracking normally used.

17.15. This concept is based on the assumption that road users are not interested in the extent of structural deterioration but are concerned solely with the quality of ride which they think they experience. To assess riding quality, a subjective assessment panel was constituted of drivers of private and commercial vehicles, and subjective observations were made on a total of 99 selected lengths of road in the states of Illinois, Minnesota, and Indiana. The selected sites were equally divided between flexible and concrete construction. The panel was asked to rate the serviceability of the pavement sections, each member using his or her judgment of what was meant by serviceability, using a scale of 0 to 5 as defined in Fig. 17.3. The members were also asked to give an overall judgment on whether the section was acceptable or not for further service. In this way a level of acceptability on the rating scale was to be established. The mean rating of the panel and the mean opinion on acceptability were used to define the present serviceability rating (PSR) of each section. A conclusion was that a PSR value of 2.5 represented the critical condition likely to require attention in the near future, and a value of 1.5 represented the condition of a pavement unfit to carry further traffic.

17.16. Armed with the present serviceability ratings for the 99 state highway sections referred to above, it became the responsibility of the

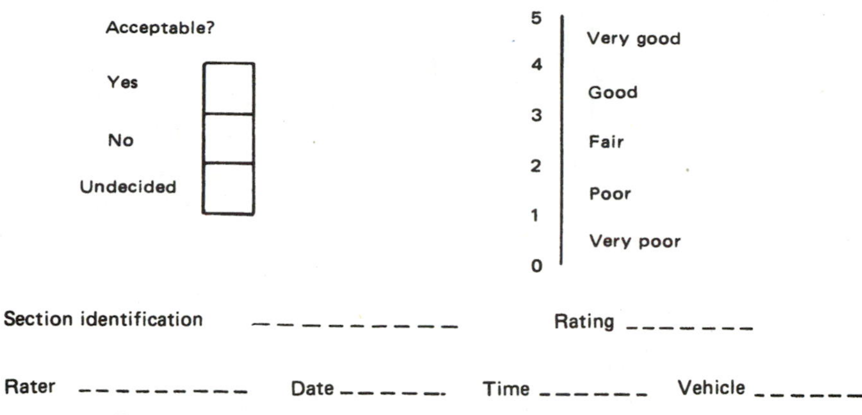

Acceptable?

Yes

No

Undecided

5

Very good

4

Good

3

Fair

2

Poor

1

Very poor

0

Section identification __________ Rating ________

Rater __________ Date ______ Time ______ Vehicle ______

Figure 17.3 Present serviceability rating form used in the AASHO road test.

statisticians and engineers involved in the road test to produce equations relating riding quality, deformation, cracking, and patching to give a present serviceability index (PSI) which matched the PSR values of the observation panels.

The Application of Present Serviceability to the Flexible Pavements

17.17. The equation finally used to evaluate the PSI of the flexible pavements included in the main factorial design using crushed-stone bases was

$$\text{PSI} = 5.03 - 1.91 \log (1 + \text{SV}) - 1.38\text{RD}^2 - 0.01 \sqrt{C + P} \qquad (17.1)$$

where RD = rut depth measured in inches over a 4-ft span embracing each wheel track (average for both wheel tracks)

SV = slope variance $\times 10^6$ (average of both wheel tracks)

C = cracking, expressed as the area of pavement in square feet exhibiting grid-pattern cracking or other cracking leading to the breakout of the bituminous surfacing, measured over an area of 1000 ft^2

P = area of patching per 1000 ft^2

Slope variance was measured by the CHLOE profilometer in which the pavement slope over a 9-in baseline was measured with respect to the average slope of the pavement at intervals of 1 ft. (Since the slope is generally less than $\pm$ 3° the angle in radians is approximately equal to the slope.) The slope variance is defined as follows:

$$SV = \left\{ \frac{\left[\sum_{i=1}^{i=n} X_i^2 - \frac{1}{n} \left(\sum_{i=1}^{i=n} X_i \right)^2 \right]}{n-1} \right\} \times 10^6 \tag{17.2}$$

where X_i is the ith slope measurement and n is the total number of measurements made.

17.18. In Eq. (17.1) the value of SV is much more important than the other factors in determining the level of PSI. For example, two pavements with the same value of SV, one of which showed no cracking or patching and the other with a surface covered by cracking or patching, would vary in PSI by only 0.3. Similarly, a difference in rut depth between 0 and 0.5 in would affect the PSI by only 0.4. Because of the statistical nature of the equations this does not mean that cracking and rutting are not important, since they will react on the value of SV. However, it does mean that Eq. (17.1) cannot be directly used in comparing the AASHO road test results with those from other experiments in which rutting and cracking have been used as the criterion of structural performance, except where the pavements are close to failure.

Evaluation of the Performance of Flexible Pavements

17.19. During the period of trafficking the PSI of each of the sections was measured at 14-day intervals and the relationship between axle-load applications and PSI obtained. Because of the seasonal effects, and particularly the weakening of the subgrade after the winter thaw, load applications at certain times of year were more damaging than at others. A concept of "weighted" load applications was developed to enable deterioration to be assessed in terms of "average" load applications. Coefficients above and below unity were derived from observations made at regular intervals of the elastic deflection of the pavements under a standard wheel load. These were applied to the performance data before analysis.

17.20. The shape of the curves PSI and W (the number of weighted applications of the axle loads for all the flexible sections in the main factorial experiment) was analyzed statistically in relation to the coordinates at the various levels of PSI and a model to fit the data was sought. The model finally chosen was as follows:

$$\text{PSI} = 4.2 - 2.7\left(\frac{W}{\rho}\right)^{\beta} \tag{17.3}$$

where β and ρ are functions of the design variables D_1, D_2, and D_3 (thickness of surfacing, road base, and subbase, respectively) and the load variables L_1 (axle load in kips) and L_2 (unity for single axles and 2 for tandem axles).

The equations β and ρ for the materials tested and the range of axle loads used were

$$\beta = 0.4 + \frac{0.081\,(L_1 + L_2)^{3.23}}{(D + 1)^{5.19} \cdot L_2^{3.23}} \tag{17.4}$$

$$\rho = \frac{10^{5.93}(D + 1)^{9.36} \cdot L_2^{4.33}}{(L_1 + L_2)^{4.79}} \tag{17.5}$$

and

$$D = 0.44D_1 + 0.14D_2 + 0.11D_3 \tag{17.6}$$

D is defined as the thickness index and suggests that unit thickness of surfacing was three times as effective as the same thickness of wet-mix road base and four times as effective as the same thickness of gravel subbase in improving pavement performance.

For any value D_1, D_2, and D_3 the thickness index D is calculated from Eq. (17.6) and, using the appropriate values of L_1 and L_2, 10 discrete values of β and ρ are obtained from Eqs. (17.4) and (17.5). These are substituted in Eq. (17.3) and for any value of PSI between 1 and 5 an appropriate value of W is obtained. Thus for any value of PSI, 10 curves relating D and W are obtained. These relationships are shown in Fig. 17.4 for terminal PSI values of 2.5 and 1.5.

17.21. From Fig. 17.4, equivalence factors can be obtained relating the damage caused by applications of one axle load to the corresponding damage caused by applications of a "standard" axle. The standard axle load normally adopted is 18,000 lb, so that the equivalence factor is the ratio of the number of applications of an 18,000-lb axle to the number of applications of the test axle to give the same terminal PSI value. Equivalence factors of this type, derived from the AASHO road test, were first published by Liddle,[2] and they are reproduced in Table 17.4. It will be noticed that as the thickness index increases the equivalence factor tends to decrease, particularly for the higher axle loads. The equivalence factor for tandem axles is rather less than twice the factor for half the total axle load, suggesting that the use of

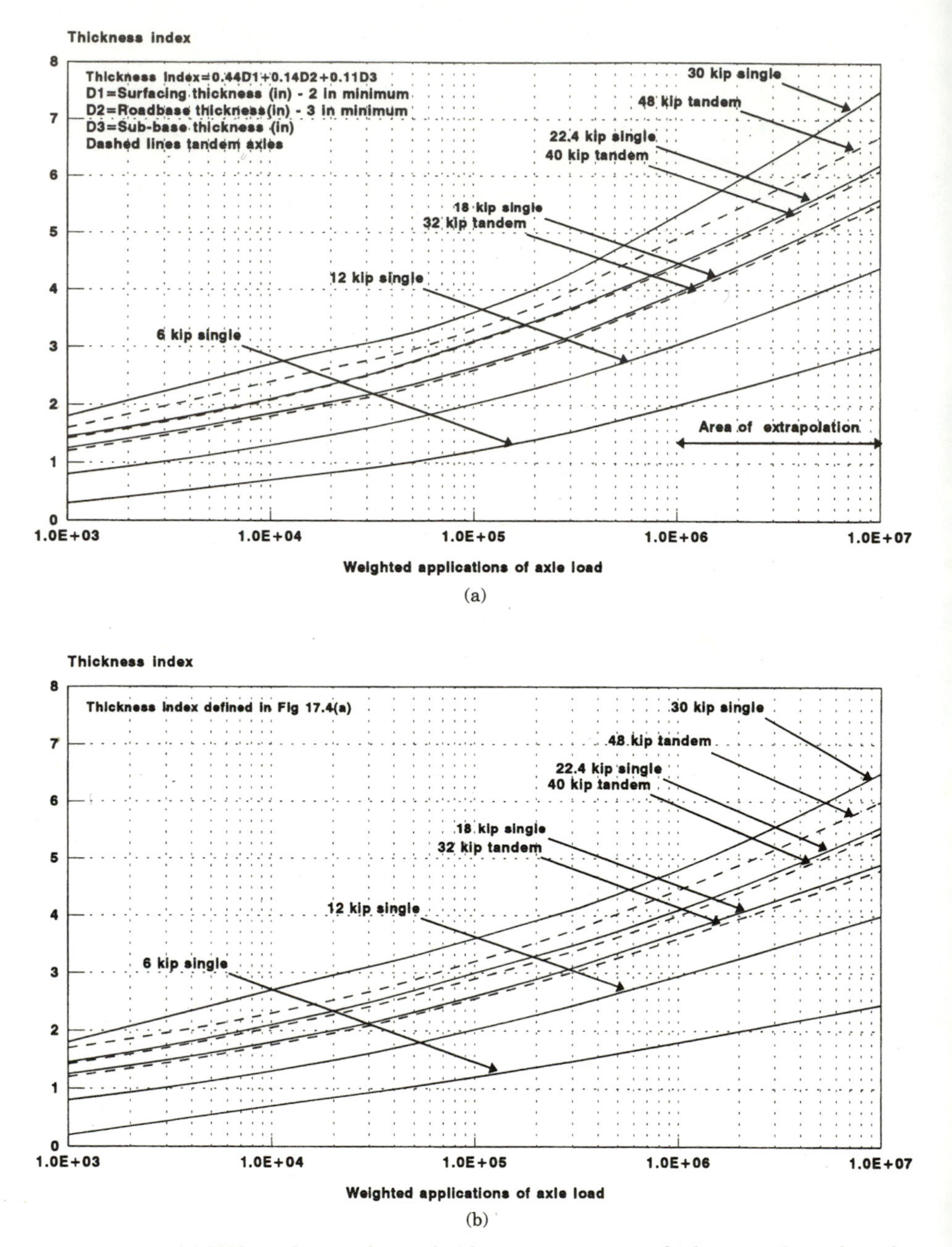

Figure 17.4 AASHO road test relationship between pavement thickness and number of load applications for *(a)* PSI 2.5 and *(b)* PSI 1.5.

TABLE 17.4 Axle Load Equivalence Factors for Flexible Pavements

Single axles, PSI = 2.0							Tandem axle sets, PSI = 2.0						
	Equivalence factors							Equivalence factors					
Gross axle load, kips*	Thickness index						Gross axle load, kips	Thickness index					
	1	2	3	4	5	6		1	2	3	4	5	6
2	0.0002	0.0002	0.0002	0.0002	0.0002	0.0002	10	0.01	0.01	0.01	0.01	0.01	0.01
4	0.002	0.003	0.002	0.002	0.002	0.002	12	0.01	0.02	0.02	0.01	0.01	0.01
6	0.01	0.01	0.01	0.01	0.01	0.01	14	0.02	0.03	0.03	0.03	0.02	0.02
8	0.03	0.04	0.04	0.03	0.03	0.03	16	0.04	0.05	0.05	0.05	0.04	0.04
10	0.08	0.08	0.09	0.08	0.08	0.08	18	0.07	0.08	0.08	0.08	0.07	0.07
12	0.16	0.18	0.19	0.18	0.17	0.17	20	0.10	0.12	0.12	0.12	0.11	0.10
14	0.32	0.34	0.35	0.35	0.34	0.33	22	0.16	0.17	0.18	0.17	0.16	0.16
16	0.59	0.60	0.61	0.61	0.60	0.60	24	0.23	0.24	0.26	0.25	0.24	0.23
18	1.00	1.00	1.00	1.00	1.00	1.00	26	0.32	0.34	0.36	0.35	0.34	0.33
20	1.61	1.59	1.56	1.55	1.57	1.60	28	0.45	0.46	0.49	0.48	0.47	0.46
22	2.49	2.44	2.35	2.31	2.35	2.41	30	0.61	0.62	0.65	0.64	0.63	0.62
24	3.71	3.62	3.43	3.33	3.40	3.51	32	0.81	0.82	0.84	0.84	0.83	0.82
26	5.36	5.21	4.88	4.68	4.77	4.96	34	1.06	1.07	1.08	1.08	1.08	1.07
28	7.54	7.31	6.78	6.42	6.52	6.83	36	1.38	1.38	1.38	1.38	1.38	1.38
30	10.38	10.03	9.24	8.65	8.73	9.17	38	1.76	1.75	1.73	1.72	1.73	1.74
32	14.00	13.51	12.37	11.46	11.48	12.17	40	2.22	2.19	2.15	2.13	2.16	2.18
34	18.55	17.87	16.30	14.97	14.87	15.63	42	2.77	2.73	2.64	2.62	2.66	2.70
36	24.20	23.30	21.16	19.28	19.02	19.93	44	3.42	3.36	3.23	3.18	3.24	3.31
38	31.14	29.95	27.12	24.55	24.03	25.10	46	4.20	4.11	3.92	3.83	3.91	4.02
40	39.57	38.02	34.34	30.92	30.04	31.25	48	5.10	4.98	4.72	4.58	4.68	4.83

*1 kip = 1000 lb.

TABLE 17.4 Axle Load Equivalence Factors for Flexible Pavements (Continued)

Single axles, PSI = 2.5							Tandem axle sets, PSI = 2.5						
	Equivalence factors							Equivalence factors					
Gross axle load, kips*	Thickness index						Gross axle load, kips	Thickness index					
	1	2	3	4	5	6		1	2	3	4	5	6
2	0.0004	0.0004	0.0003	0.0002	0.0002	0.0002	10	0.01	0.01	0.01	0.01	0.01	0.01
4	0.003	0.004	0.004	0.003	0.003	0.002	12	0.02	0.02	0.02	0.02	0.01	0.01
6	0.01	0.02	0.02	0.01	0.01	0.01	14	0.03	0.04	0.04	0.03	0.03	0.02
8	0.03	0.05	0.05	0.04	0.03	0.03	16	0.04	0.07	0.07	0.06	0.05	0.04
10	0.08	0.10	0.12	0.10	0.09	0.08	18	0.07	0.10	0.11	0.09	0.08	0.07
12	0.17	0.20	0.23	0.21	0.19	0.18	20	0.11	0.14	0.16	0.14	0.12	0.11
14	0.33	0.36	0.40	0.39	0.36	0.34	22	0.16	0.20	0.23	0.21	0.18	0.17
16	0.59	0.61	0.65	0.65	0.62	0.61	24	0.23	0.27	0.31	0.29	0.26	0.24
18	1.00	1.00	1.00	1.00	1.00	1.00	26	0.33	0.37	0.42	0.40	0.36	0.34
20	1.61	1.57	1.49	1.47	1.51	1.55	28	0.45	0.49	0.55	0.53	0.50	0.47
22	2.48	2.38	2.17	2.09	2.18	2.30	30	0.61	0.65	0.70	0.70	0.66	0.63
24	3.69	3.49	3.09	2.89	3.03	3.27	32	0.81	0.84	0.89	0.89	0.86	0.83
26	5.33	4.99	4.31	3.91	4.09	4.48	34	1.06	1.08	1.11	1.11	1.09	1.08
28	7.49	6.98	5.90	5.21	5.39	5.98	36	1.38	1.38	1.38	1.38	1.38	1.38
30	10.31	9.55	7.94	6.83	6.97	7.79	38	1.75	1.73	1.69	1.68	1.70	1.73
32	13.90	12.82	10.52	8.85	8.88	9.95	40	2.21	2.16	2.06	2.03	2.08	2.14
34	18.41	16.94	13.74	11.34	11.18	12.51	42	2.76	2.67	2.49	2.43	2.51	2.61
36	24.02	22.04	17.73	14.38	13.93	15.50	44	3.41	3.27	2.99	2.88	3.00	3.16
38	30.90	28.30	22.61	18.06	17.20	18.98	46	4.18	3.98	3.58	3.40	3.55	3.79
40	39.26	35.89	28.51	22.50	21.08	23.04	48	5.08	4.80	4.25	3.98	4.17	4.49

*1 kip = 1000 lb.

tandem axles is less damaging than uncoupled, more widely spaced, axles. The equivalence factors also tend to increase with decreasing terminal PSI level.

17.22. The equivalence factors contained in Table 17.4 can be used to convert the weighted applications of each axle load shown in Figs. 17.4*a* and *b* to equivalent applications of standard 18-kip standard axles. As would be expected, when this is done the curves become closely coincident to form a single relationship between thickness index and the number of standard axles to the particular level of PSI being considered. Figure 17.5 shows such relationships for PSI values of 2.5 and 1.5 (the small scatter indicated in the figure indicates the accuracy of the equivalence values shown in Table 17.4).

17.23. The wedge sections included crushed-stone bases to the same grading specification as those incorporated in the main factorial sections. The relationship between thickness index and cumulative standard axles for those wedge sections is indicated by the points shown in Fig. 17.6 which refer to PSI 2.5. There is general agreement below one million standard axles but poorer agreement between 1 and 10 million standard axles. This could be due to some ambiguity in the

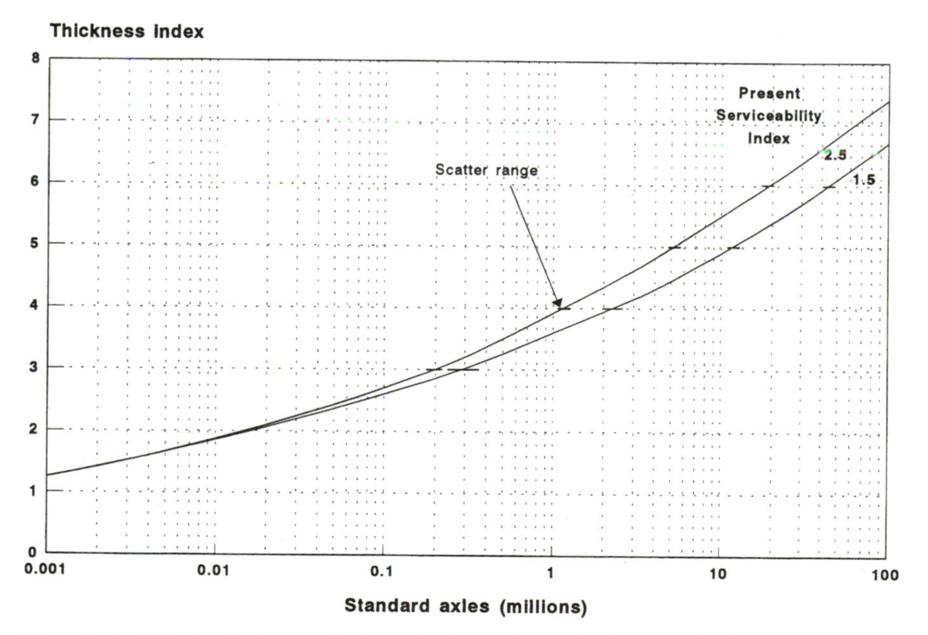

Figure 17.5 Relation between thickness index and cumulative standard axles for main factorial sections.

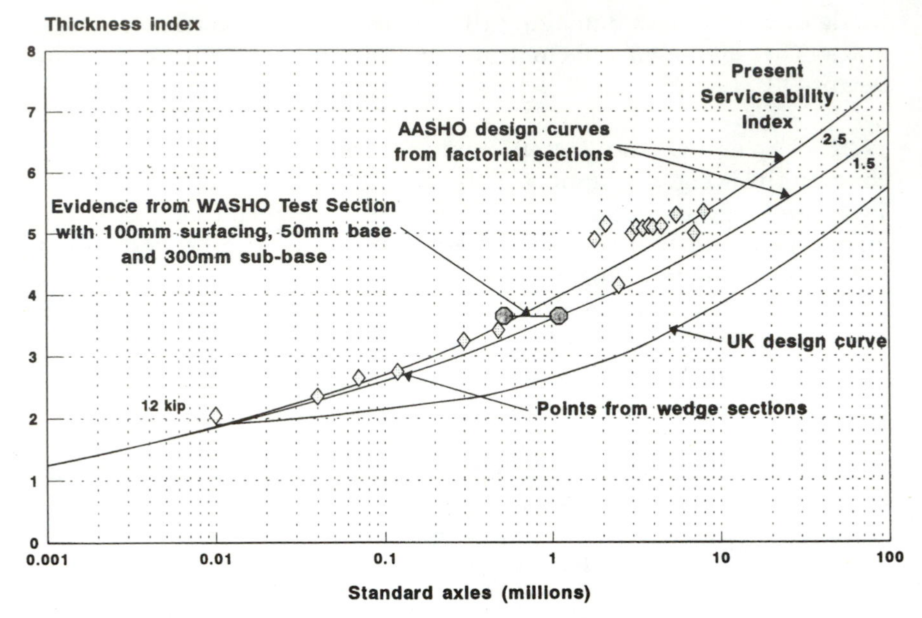

Figure 17.6 Comparison between AASHO and U.K. design standards for pavements with crushed-stone bases.

surfacing thickness used for the wedge sections of loop 3. Table 2 of the AASHO Report 61E[1] shows the thickness of the asphalt surfacing to be 4 in, whereas Fig. 36 of the same report indicates 3 in. The points between 2 and 10 million standard axles are more consistent with the 4-in thickness.

17.24. The relationship between base thickness and cumulative number of standard axles for the wedge sections with the cement-bound base is shown in Fig. 17.7. To avoid confusion, the points shown are for the most damaging and the least damaging axle loads. The points for the other axle loads follow the same relationship.

17.25. For the wedge sections using bituminous-bound bases the relationship between base thickness and cumulative standard axles is shown in Fig. 17.8. The scatter of points referring to the various axle loads was rather greater, and the envelope containing the points is shown rather than the individual points.

17.26. Figures 17.6, 17.7, and 17.8 include relationships between base thickness and cumulative standard axles derived for base materials of very similar specification to those incorporated in the AASHO

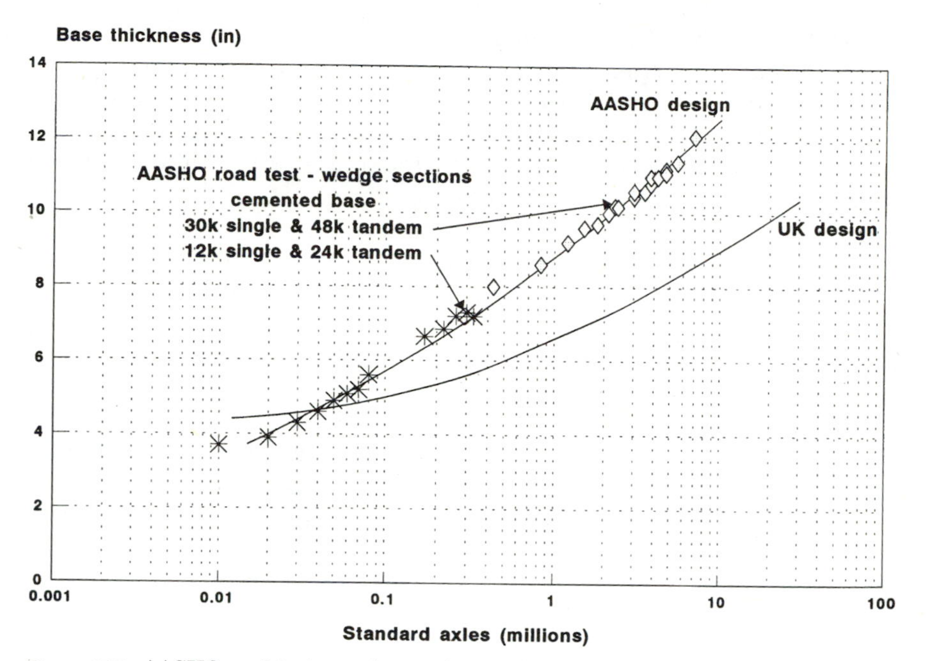

Figure 17.7 AASHO road test—wedge sections with cemented base.

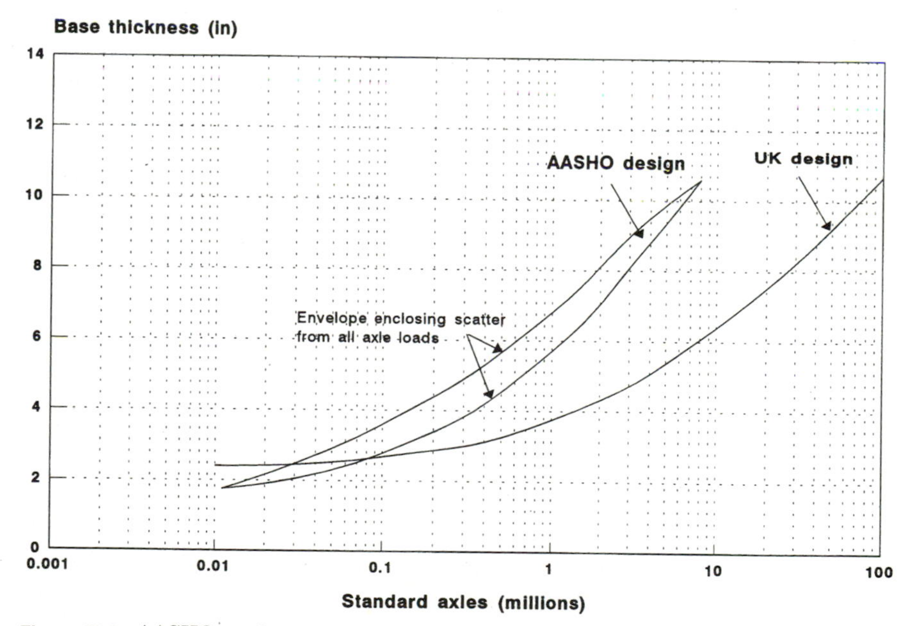

Figure 17.8 AASHO road test—wedge sections with bituminous base.

experiment, which have been incorporated in U.K. full-scale pavement design experiments. These experiments are on heavily trafficked commercial trunk roads for which the axle load spectra are recorded. The observations have been made over periods of 20 to 30 years, and based on rut depth and cracking, the terminal condition approximates a PSI value of about 2.

17.27. It will be noted that the lives derived from the U.K. experiments are several times greater than those for the AASHO pavements. The probable reason for this lies in the rapidity of loading used in the AASHO road test, and the consequent absence of long-term hardening of the bitumen (see Par. 16.8).

17.28. A properly designed flexible pavement carrying the traffic for which it was intended becomes progressively stiffer with the passing of time owing to the hardening of the bitumen in the surfacing (and in the base where a bitumen-bound base is used). This hardening, which makes an important contribution to the life of the pavement, may continue for many years, until the materials cannot accept the increasing tensile stresses developed by the hardening process. It is at this point that cracking is initiated. The early stages of hardening will generally be accompanied by some deformation in the wheel tracks, particularly in hot weather.

17.29. It follows that any form of accelerated testing of either road materials or complete pavements is likely to neglect or curtail the influence of time-dependent hardening and as a consequence underestimate the lives which will be obtained under normal traffic conditions.

17.30. In the AASHO road test, the 30-kip single-axle load and the 48-kip tandem axle loads dominate the performance of the pavements for traffic in excess of 0.5 million standard axles. Such axle loads are about 30 percent above the maximum loads permitted in most of the states of the United States. In Britain axle load spectra measured on the slow traffic lanes of industrial trunk roads and freeways show that less than 0.15 percent of commercial axle loads exceed 30 kips and that 1 million of such loads would involve a time span of about 70 years. If this also applies to roads in the United States it would appear that the road test condensed into 2 years the possible *overloads* of 30 times that period.

17.31. For axle loads of less than 18 kips the AASHO traffic was less than would be expected on the slow lane of industrial roads in the United Kingdom, and less difference between the AASHO and British design curves would be expected. It seems probable therefore that the differences apparent in Figs. 17.6, 17.7, and 17.8 arise largely from the concentration of axle loading. Although some compensation for frost action was made in the AASHO test, it is possible that the markedly different frost penetrations at the AASHO site and those affecting British roads may be partly responsible for the difference between U.K. and U.S. designs.

The Application of Present Serviceability to Concrete Pavements

17.32. The equation used to evaluate the PSI of the concrete pavements included in the AASHO road test was

$$\text{PSI} = 5.41 - 1.78 \log (1 + \text{SV}) - 0.09 \sqrt{C + P} \qquad (17.7)$$

In this equation C has a different meaning from the same symbol used in Eq. (17.1). C is here defined as the total linear footage of class 3 and class 4 cracks (cracks opened or spalled to a width of $\frac{1}{4}$ in or more) per 1000 ft^2 of lane area. On some of the concrete sections of the AASHO experiment it was not possible to use the CHLOE profilometer, so to determine SV the Bureau of Public Roads Roughometer was employed at a speed of 10 mi/h and the alternative formula given below was used to determine PSI:

$$\text{PSI} = 5.41 - 1.8 \log (0.4R - 33) - 0.99 \sqrt{C + P} \qquad (17.8)$$

where R is the roughness index in inches per mile.

17.33. As for flexible pavements the value of PSI obtained from Eq. (17.7) or (17.8) is determined almost entirely by the riding quality of the road and the values of C and P are secondary, except insofar as they influence riding quality. The criterion used in Britain to assess the performance of experimental concrete roads is based entirely on the amount and severity of cracking, and to relate British experience with the results of the AASHO road test it is necessary to examine more closely the relationship between PSI and cracking. In the preliminary tests on state roads from which Eq. (17.7) was established (see Par. 17.15) measurements of crack length were made, and this enables the relationships between degree of cracking and PSI to be examined directly. Figure 17.9 shows the mean relationship between the total length of cracking per 100 ft of traffic lane and PSI derived from Eq. (17.7).

17.34. The "failure" criterion used for experimental reinforced concrete roads in Britain is 250 m of total crack length per 100 m of traf-

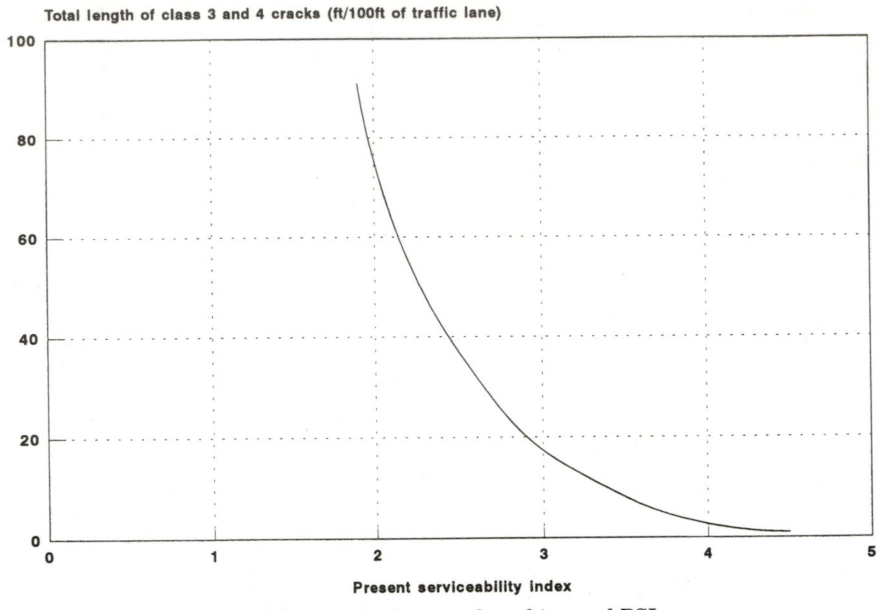

Figure 17.9 Relationship between the degree of cracking and PSI.

fic lane. Experience shows that when such pavements are approaching the end of their lives about one-third of the total cracking falls into the class 3 and class 4 categories as defined in relation to the AASHO road test. It follows from Fig. 17.9 that the British failure condition would correspond to a PSI level of about 2.

Evaluation of the Performance of Concrete Pavements

17.35. The present serviceability index was measured at 14-day intervals using the equation already discussed and curves relating PSI and the number of applications were obtained for each section. No weighting factor for the seasonal effect was applied to the curves obtained in the case of the concrete pavements. The model used for representing the results statistically was of a form similar to that discussed for the flexible sections, but the constants and coefficients were different. The model equation was

$$\mathrm{PSI} = 4.5 - 3\left(\frac{W}{\rho}\right)^{\beta} \tag{17.9}$$

where β and ρ are given by Eqs. (17.10) and (17.11)

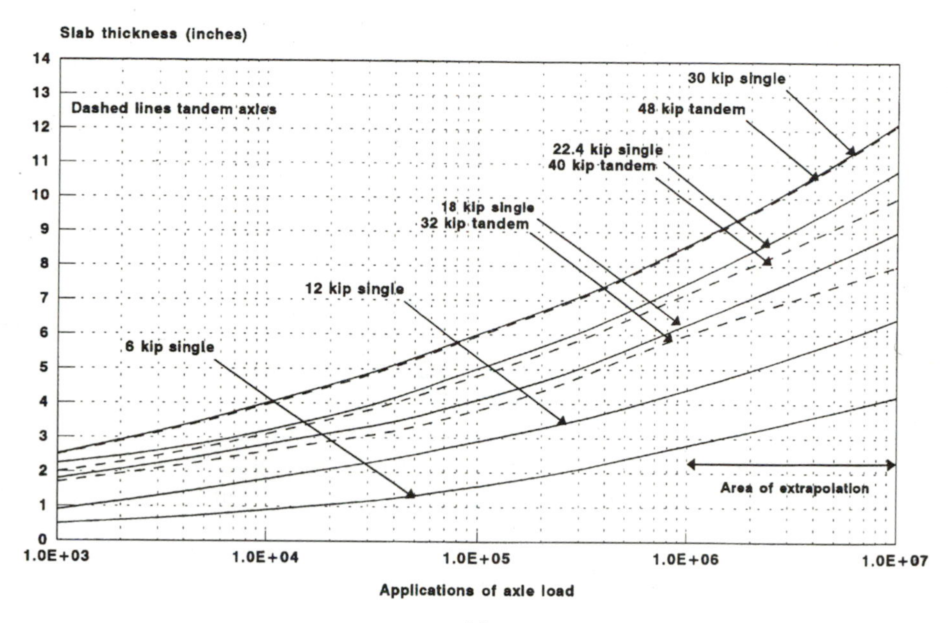

(a)

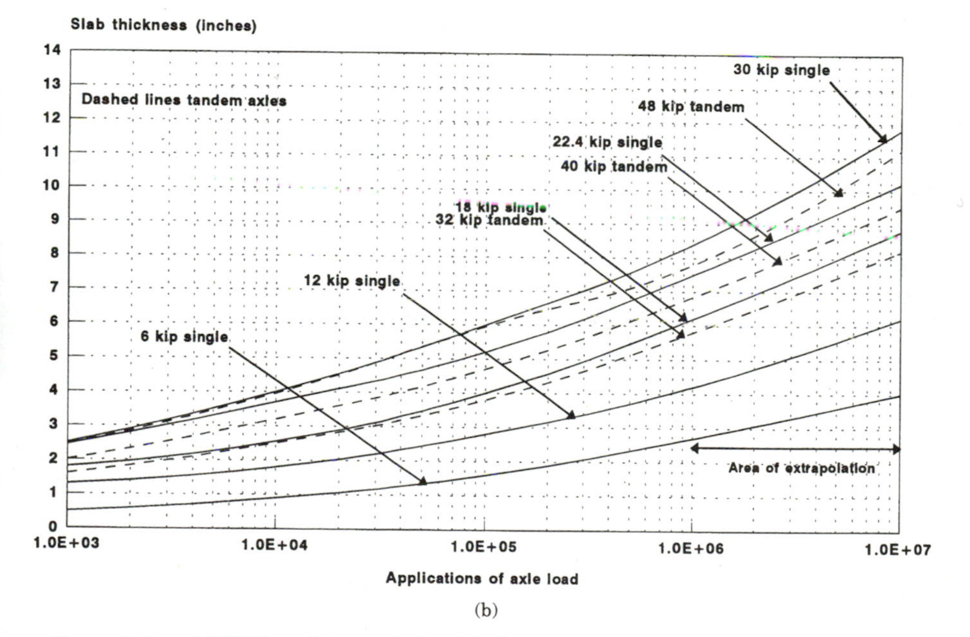

(b)

Figure 17.10 AASHO road test relationship between pavement thickness and number of load applications for (*a*) PSI 2.5 (*b*) PSI 1.5 (concrete).

$$ß = 1 + \frac{3.63(L_1 + L_2)^{5.20}}{(D_2 + 1)^{8.46} \cdot L_2^{3.52}} \tag{17.10}$$

and

$$\rho = \frac{10^{5.85}(D_2 + 1)^{7.35} \cdot L_2^{3.28}}{(L_1 + L_2)^{4.62}} \tag{17.11}$$

It will be noted that the thickness of subbase and degree of reinforcement do not appear in these equations, since they were found not to be significant variables in determining performance. In Eqs. (17.10) and (17.11), D_2 is the thickness of slab and L_1 and L_2 are the same load factors as were defined in connection with the flexible sections. By combining Eqs. (17.9), (17.10), and (17.11) the slab thickness can be related to the number of applications of the axle load for any level of terminal PSI. As for the flexible pavement experiment, the results are presented as families of curves related to specific values of PSI. The curves for terminal PSI levels of 2.5 and 1.5 are shown in Fig. 17.10 (*a*) and (*b*).

17.36. The conclusion that reinforcement had no influence on the performance of the concrete pavements is not surprising in view of the relatively short bay lengths used for the reinforced sections (40 ft). That subbase thickness had no significant influence is rather more surprising in view of the liability of the thinner concrete pavements to pumping at the AASHO site. This suggests that the subbase and subgrade materials may have been equally liable to pumping.

17.37. Equations (17.9), (17.10), and (17.11) can be used to obtain equivalence factors in terms of cumulative standard axles for all the axle loads used in the concrete pavement tests. The factors given in Table 17.5 lead to a single relationship between slab thickness and cumulative standard axles, with very little scatter of points.

17.38. Using the equivalence factors given in Table 17.5 the relationship between axle loads and applications of those axle loads contained in Fig. 17.10*a* and *b* have been converted to single curves relating slab thickness and cumulative standard axles for PSI values of 2.5 and 1.5. These curves are shown in Fig. 17.11. As was the case with the flexible pavements, the overloads represented by the 30-kip single and 48-kip tandem axles were applied some 70 times faster than would be expected on U.K. industrial roads, which means that 70 years of such traffic was condensed into the 2 years of the test. The

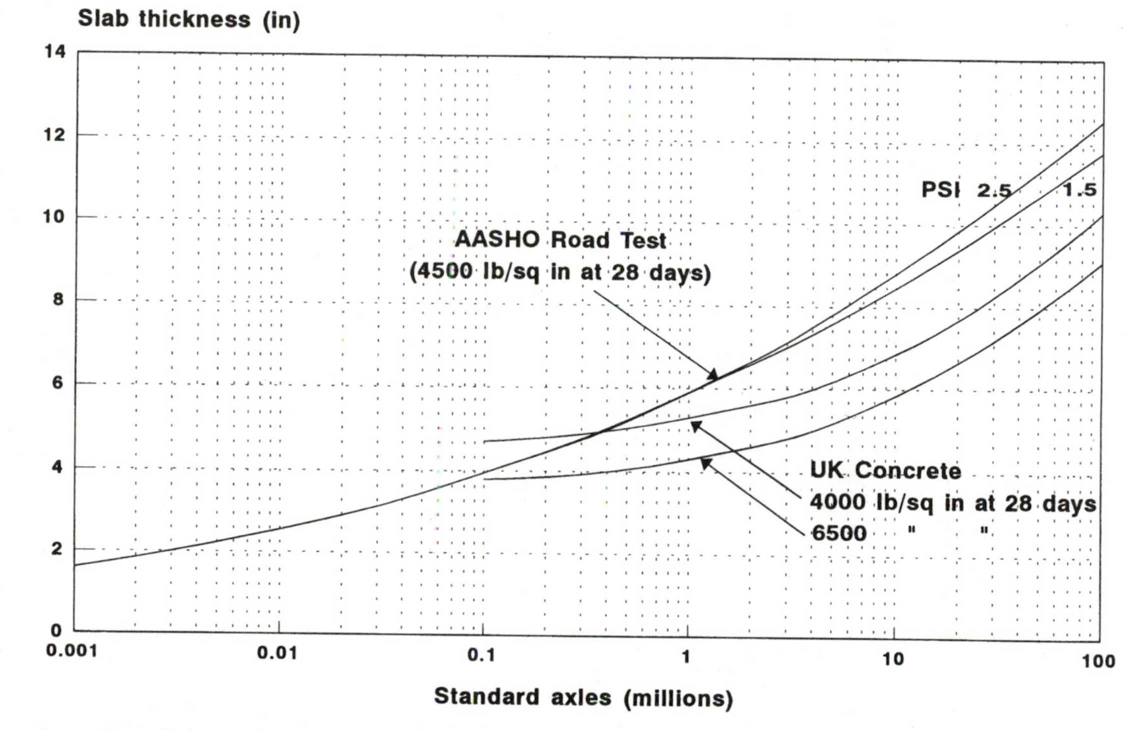

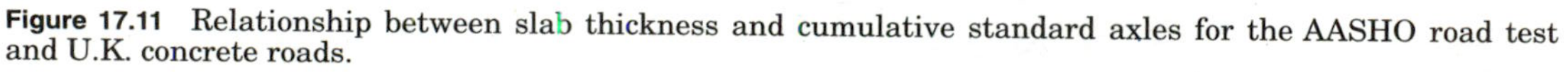

Figure 17.11 Relationship between slab thickness and cumulative standard axles for the AASHO road test and U.K. concrete roads.

TABLE 17.5 Axle Load Equivalence Factors for Concrete Pavements

Single axles		Tandem axles	
Axle load, kips	Equivalence factor	Axle load, kips	Equivalence factor
30	8.3	48	6.7
22.4	2.5	40	3.1
18	1.0	32	1.2
12	0.2	24	0.4
6	0.01		
2	0.0002		

smaller axle loads were applied at a much reduced rate compared with normal industrial roads.

17.39. U.K. curves derived from the performance of experimental concrete roads in Britain over a period in excess of 20 years are included in Fig. 17.11. They indicate much longer lives for pavements of the same thickness as those used in the AASHO road test. This is partly due to the excessive overloads combined with the accelerated rate of testing. Over a normal life span of 20 to 30 years the compressive strength of a road slab would increase by a factor of 1.7 compared with the 28-day strength; i.e., a slab with a 28-day compressive strength of 4000 lb/in^2 would increase to a compressive strength of about 6900 lb/in^2, which would be accompanied by a much greater increase in fatigue strength. The design curve for high-strength concrete shown in Fig. 17.11 illustrates the importance of aging of the concrete on long-term performance.

The WASHO Road Test*

17.40. The WASHO road test preceded the AASHO test by about 2 years. The object was to study the performance of experimental flexible road pavements constructed to a wide range of overall thicknesses, when they were trafficked by repetitions of known axles. No concrete pavements were included. The site chosen for the experiment was in the south of the state of Idaho where the subgrade was a silty clay of in situ CBR value approximately 14 percent. The rainfall was only 200 mm per year, but the winter air temperature averaged about −10°C, giving rise to subgrade freezing in December and January.

*The WASHO Road Test was conducted using metric units.

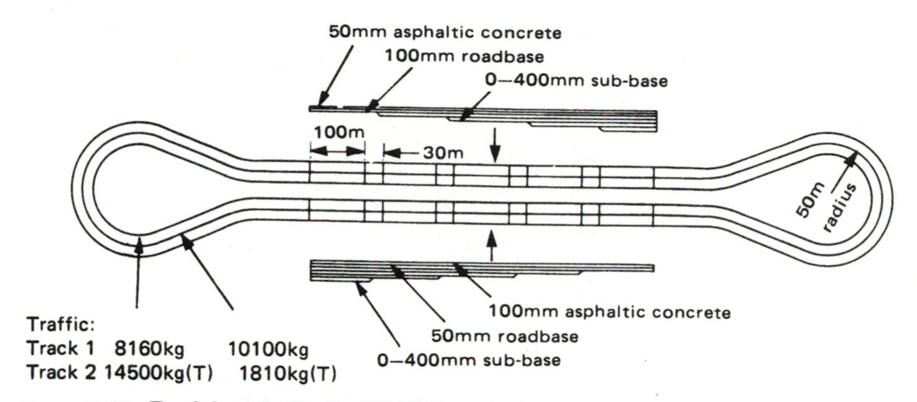

Figure 17.12 Track layouts for the WASHO road test.

17.41. As with the AASHO test the test sections were arranged around closed loops consisting of straight sections of two-lane construction connected by turnabouts. The sections were 100 m long connected by 30-m transition lengths. The forms of construction used on each of the two identical loops are shown in Fig. 17.12. The materials used for surfacing and base were very similar to those used in the AASHO test, i.e., asphaltic concrete, crushed-stone base, and granular subbase. Mean grading curves are shown in Fig. 17.13. The two lanes of each loop carried different axle loads. On one loop the single-load axles were 8160 and 10,100 kg, and on the other tandem-load axles of 14,500 and 18,100 kg were used.

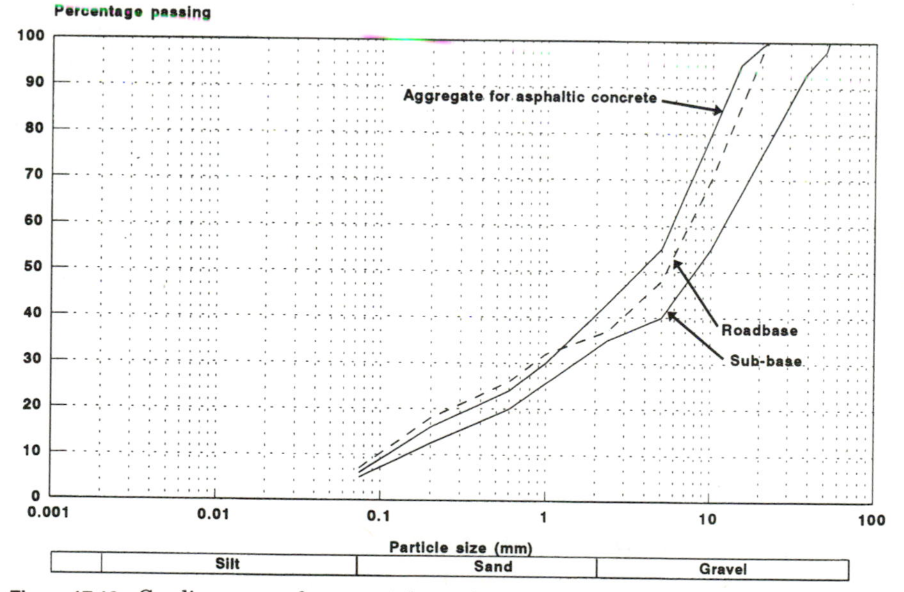

Figure 17.13 Grading curves for materials used in the WASHO road test.

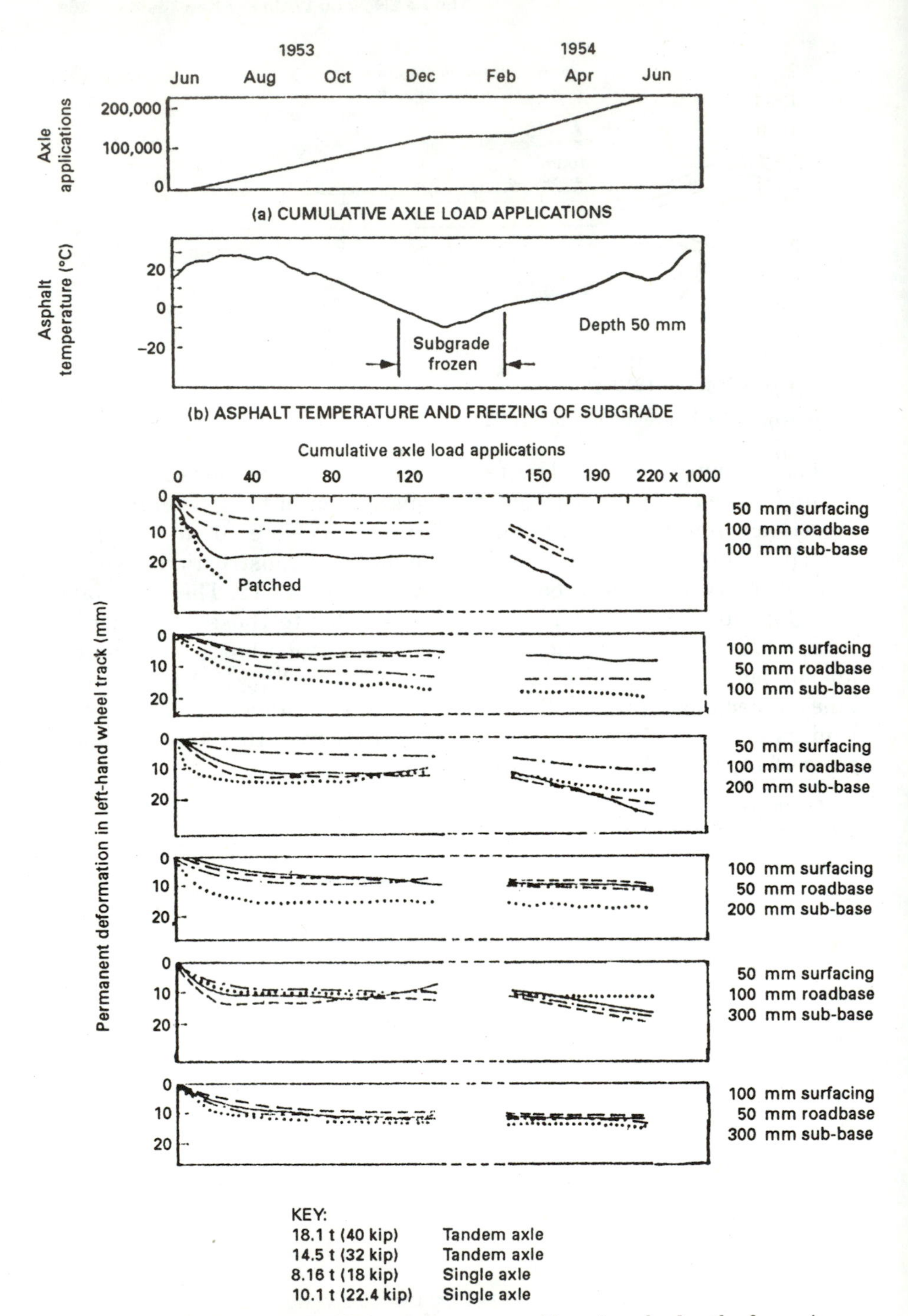

Figure 17.14 Total permanent deformation measured in outer wheel path of experimental sections of the WASHO road test.

17.42. After a very short period of traffic in November 1952 the test tracks remained untrafficked until June 1953. Trafficking at a uniform rate was commenced in mid-June and continued to the end of May 1954, with a break of 2 months between mid-December 1953 and mid-February 1954 while the subgrade was frozen (see Fig. 17.14).

17.43. The performance of the pavements was assessed mainly in terms of the permanent deformation which developed under the action of traffic. Measurements were made in the wheel tracks at monthly intervals. Observations were also made of the amount of cracking which occurred and of the elastic deflection measured by the Benkelman beam. Where excessive deformation and cracking occurred, patching was undertaken to keep the sections serviceable and to minimize sympathetic deterioration in adjacent sections.

17.44. Results from all but the thinnest sections (those without subbase) are shown in Fig. 17.14. The thinnest sections deformed rapidly under all the axle loads, and little useful information was obtained from them. The deformation curves are characterized by a steep initial portion associated with the first 30,000 applications of the axle loads, followed by comparatively little deformation during the subse-

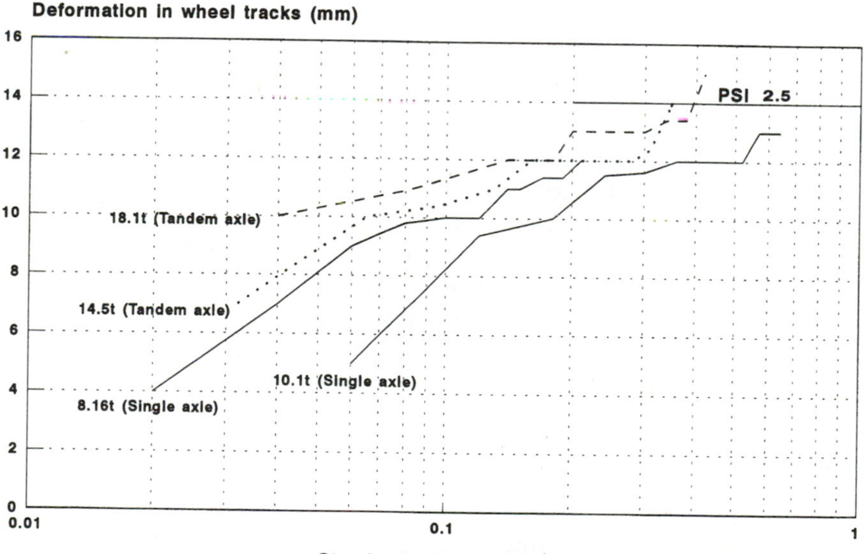

Figure 17.15 Relationship between deformation and cumulative standard axles for four-axle loads on a pavement with 100-mm surfacing on 50-mm base and 300-mm subbase—WASHO road test.

quent 100,000 applications. When the traffic was resumed after the thawing of the subgrade in February 1954 the sections with the 50-mm asphalt surfacing deformed comparatively rapidly. Under the 100-mm asphalt surfacing there was little apparent acceleration of the deformation which could be attributed to a weakened subgrade.

17.45. The initial rapid deformation under traffic coincided with the highest summer temperature conditions, when the elastic modulus of the asphalt would be expected to be low, and the stress transmitted to the base, subbase, and subgrade correspondingly higher. The performance of the sections with the 100-mm asphalt surfacing was much superior to that of the sections with the same total thickness but with a 50-mm surfacing. Increasing the thickness of the subbase in all cases reduced the permanent deformation, although the influence of subbase thickness is less under the 100-mm surfacing than under the 50-mm surfacing.

17.46. As with the AASHO road test, the traffic on the various sections can be expressed in terms of standard axles, using the equivalence factors in Table 17.4. This has been done in Fig. 17.15, which refers to the strongest section in the WASHO road test, i.e., that with a 100-mm asphalt surfacing on a 50-mm base and 300-mm subbase. Using the AASHO thickness equivalence equation (17.6), this corresponds to a thickness index of 3.36. This section was selected as being least affected by traffic during the period of thawing in the subgrade. Comparison between the performance of sections in the WASHO and AASHO tests is complicated by the different performance criteria used. However, the present serviceability rating panels referred to in Par. 17.15 concluded that a PSR value of 2.5 corresponded to a rut depth in the wheel tracks of about 14 mm. Applying this criterion to Fig. 17.15, it appears that for that section the 2.5 level of PSI occurred over a range of standard axles carried between 0.34 and 0.6 million depending on the axle load. Using the Thickness Index of 3.36 this enables a limited direct comparison between the results of the two road tests. This comparison, which is included in Fig. 17.6, shows reasonable agreement between the two tests.

Conclusions

17.47. The attempt made in the AASHO road test to combine a subjective assessment of riding quality with structural aspects of pavement performance such as deformation and cracking appears, with hindsight, to have been a serious mistake. Most engineers feel that it would have been a more cost-effective use of resources had several simpler experiments of the WASHO type been carried out in different climatic areas of the United States. The problem of evaluating riding quality with cracking and deformation could have been tackled by separate observations made on in-service highways. It is of course true that detailed cracking and deformation information is available in the massive volumes of test results. Few engineers, however, are likely to have the time to unravel this documentation, bearing in mind that it refers to one location only.

17.48. The intention was that the AASHO road test results would be extended by satellite experiments throughout the United States. So far as the authors are aware, there are no published results from such experiments, with the exception of the San Diego experiment supervised by the Asphalt Institute and reported in 1977.[3] At this site 33 experimental sections were laid using asphalt surfaces and various asphalt bases, on an in-service suburban road. The criteria of performance were deformation and cracking. The traffic was unfortunately rather light and in 5 years fewer than 2 million standard axles were carried.

17.49. The Asphalt Institute carried out a very comprehensive program of laboratory research into the materials used in the San Diego project and the results are reported in detail in the final report.[3] Any research worker proposing to apply structural theory to the design of pavements should study this report carefully.

References

1. Highway Research Board: The AASHO Road Test, Report 5: Pavement Research, Special Report 61E, National Academy of Sciences, National Research Council, *Publication* 954, Washington, D.C., 1962.
2. Liddle, W. J.: Application of AASHO Road Test Results to the Design of Flexible Pavement Structures, *Proc. Int. Conf. on the Structural Design of Asphalt Pavements, Ann Arbor, Michigan, 1962,* University of Michigan, Ann Arbor, pp. 42–51, 1962.
3. Kallas, B. F., and J. F. Shook: San Diego County Experimental Base Project, Final Report, *Research Report* 77-1, November 1977, The Asphalt Institute, College Park, Md.

Chapter

18

Performance Studies of Full-Scale Experimental Sections Incorporated in In-service Highways in the United Kingdom—Flexible Pavements

Introduction

18.1. Since the 1930s experimental flexible sections have been incorporated in public highways in the United Kingdom, and the performance studied under normal road traffic. Prior to World War II the work was largely directed toward developing specifications for durable nonskid bituminous surfacings. After the war the experiments concentrated on the development of thickness standards for flexible pavements in relation to subgrade properties and traffic intensity. The latter aspect is primarily considered in this chapter.

18.2. The experiments are too numerous to discuss individually, and it is proposed therefore to concentrate here on six of the larger experiments constructed since 1949, most of which are still under observation. Details of these experiments and their objectives are given in Table 18.1.

18.3. The experiments have all been located on heavily trafficked trunk routes, with dual carriageways separated by a central reservation. Each of the sections has a length between 60 and 80 m, extend-

TABLE 18.1 Details of Major Full-Scale Pavement Design Experiments Using Flexible Construction

Year of construction	Location	Main variables (numbered) and construction details	Subgrade type and strength	Initial traffic, commercial vehicles/day	Growth rate of commercial traffic, % per annum	Damaging effect of commercial traffic, standard axles/100 commercial axles
1949	A1 (16 km north of Boroughbridge), North Yorkshire	1. Type of road base—dry stone, tarmacadam 2. Thickness of road base—200–430 mm 3. Type of wearing course—bitumen macadam, rolled asphalt Surfacing—100 mm thick Subbase—None	Silty sand CBR 10%	1000	4	13
1957	A1 Alconbury Hill, Cambridgeshire	1. Type and thickness of surfacing—asphalt, bitumen macadam—38–100 mm thick 2. Type and thickness of road base—wet-mix, soil-cement, lean concrete, tarmacadam, rolled asphalt—75–230 mm thick 3. Thickness of sand subbase—100–350 mm	Silty clay (average) LL 57% (average) PL 21% CBR 4%	1400	5	25
1963	A30 Nately Scures (3 km west of Hook), Hampshire	1. Type of thickness of road base—wet-mix, lean concrete, dense-coated macadam, rolled asphalt—80–300 mm thick 2. Type of base course over coated macadam road bases Subbase—gravel—150 mm thick	Silty clay LL 60% PL 21% CBR 3.5%	1500	5	35

1963	A40 Wheatley bypass (12 km east of Oxford)	1. Grading and strength of cemented road base materials—200 mm thick 2. Grading and binder content of bituminous road base materials—200 mm thick Surfacing—asphalt—100 mm thick Subbase—gravel—150 mm thick	Silty clay LL 57% PL 20% CBR 5.5%	850–1300	3	20-35 (east-bound) 25-30 (west-bound)
1964	A1 Alconbury bypass, Cambridgeshire	1. Thickness of certain of the cemented and bituminous road base materials used in the Wheatley bypass experiment (see above) in the range 150–250 mm 2. Type and thickness of surfacing—rolled asphalt, bitumen macadam—100–200 mm thick Subbase—gravel—150 or 300 mm thick	Silty clay (average) LL 51% (average) PL 20% CBR 5%	1300–2200	3	25 (north-bound) 45–55 (south-bound)
1965	Al Conington, Cambridgeshire	1. Type of base course under asphalt wearing course (total thickness 100 mm), on wet-mix road base 200 mm thick 2. Type of bituminous road base 150 mm thick under rolled asphalt surfacing 100 mm thick Base course and road base—crushed rock, various gravel aggregates Subbase—gravel—150 mm thick	Silty clay LL 50% PL 20% CBR 4%	2400	3	50

ing over the full width of the carriageway, i.e., across the slow and overtaking lanes. Deformation caused by traffic is measured at three or five points along each section, depending on the length. At each point of measurement a row of dimpled metal leveling studs is set in the surfacing across the full width of the carriageway with a spacing of 300 mm.

18.4. The studs are leveled several times each year against a deep benchmark, to give the deformation induced by traffic in each traffic lane. Each experiment has a recording weighbridge set in the surface of the slow traffic lane, to enable the traffic in terms of numbers of standard axles to be calculated.

18.5. The transient deflection under the passage of a 6350-kg axle load is measured by the deflection (Benkelman) beam, close to each leveling point, in the wheel tracks of the slow traffic lane twice each year in the spring and autumn. The conclusions from these measurements are given in Chap. 20.

The Boroughbridge Experiment—1946[1]

18.6. The variables introduced into this experiment are shown in Table 18.1. The main objective was to examine the influence (1) of an impermeable asphalt wearing course and (2) of a permeable bitumen macadam surfacing when laid on bases of unbound stone and similar bases stabilized with a tar binder.

18.7. Figures 18.1 and 18.2 show the development of maximum deformation in the nearside wheel track and the transverse deformation across both traffic lanes after 6 years. The thicknesses of the sections are shown inset in the diagrams. No subbase was used in this experiment because of the relatively high CBR value of the subgrade. It will be noted that the deformation is greatest in the nearside wheel tracks of the slow traffic lane. This is characteristic of all flexible pavement experiments. The loading on the nearside wheels will be slightly increased by the crossfall, but a more likely reason is moisture migration from the verge into the subgrade.

18.8. The results show that the sections with the rolled asphalt wearing course gave a performance much superior to those surfaced with the tarmacadam wearing course. This was partly due to the superior stiffness of the asphalt, but a more important factor seems to be the impermeability of the asphalt. The coated macadam wearing course and base course allowed water to pass into the base and the subgrade.

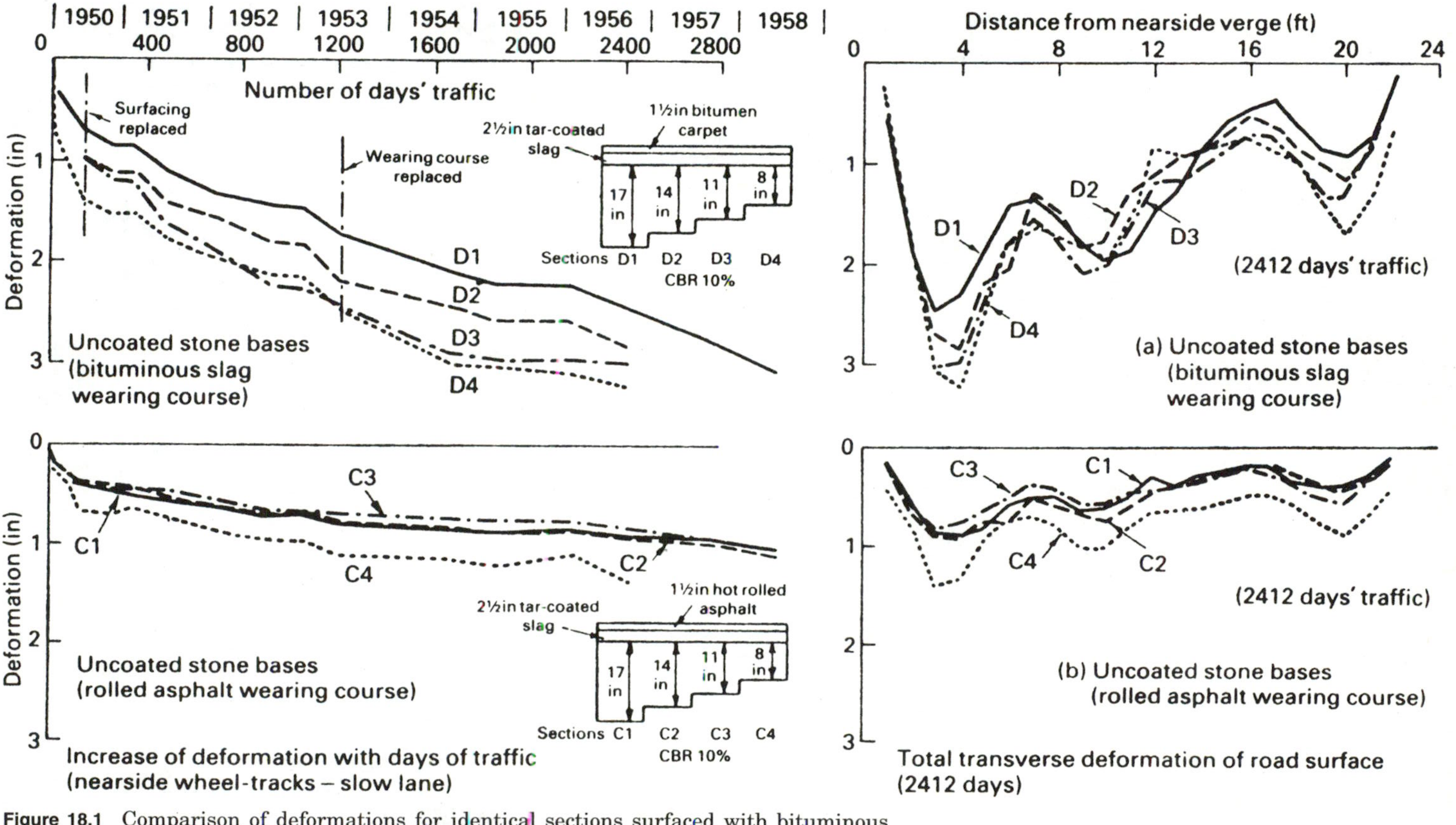

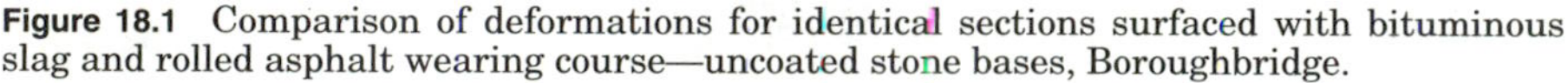

Figure 18.1 Comparison of deformations for identical sections surfaced with bituminous slag and rolled asphalt wearing course—uncoated stone bases, Boroughbridge.

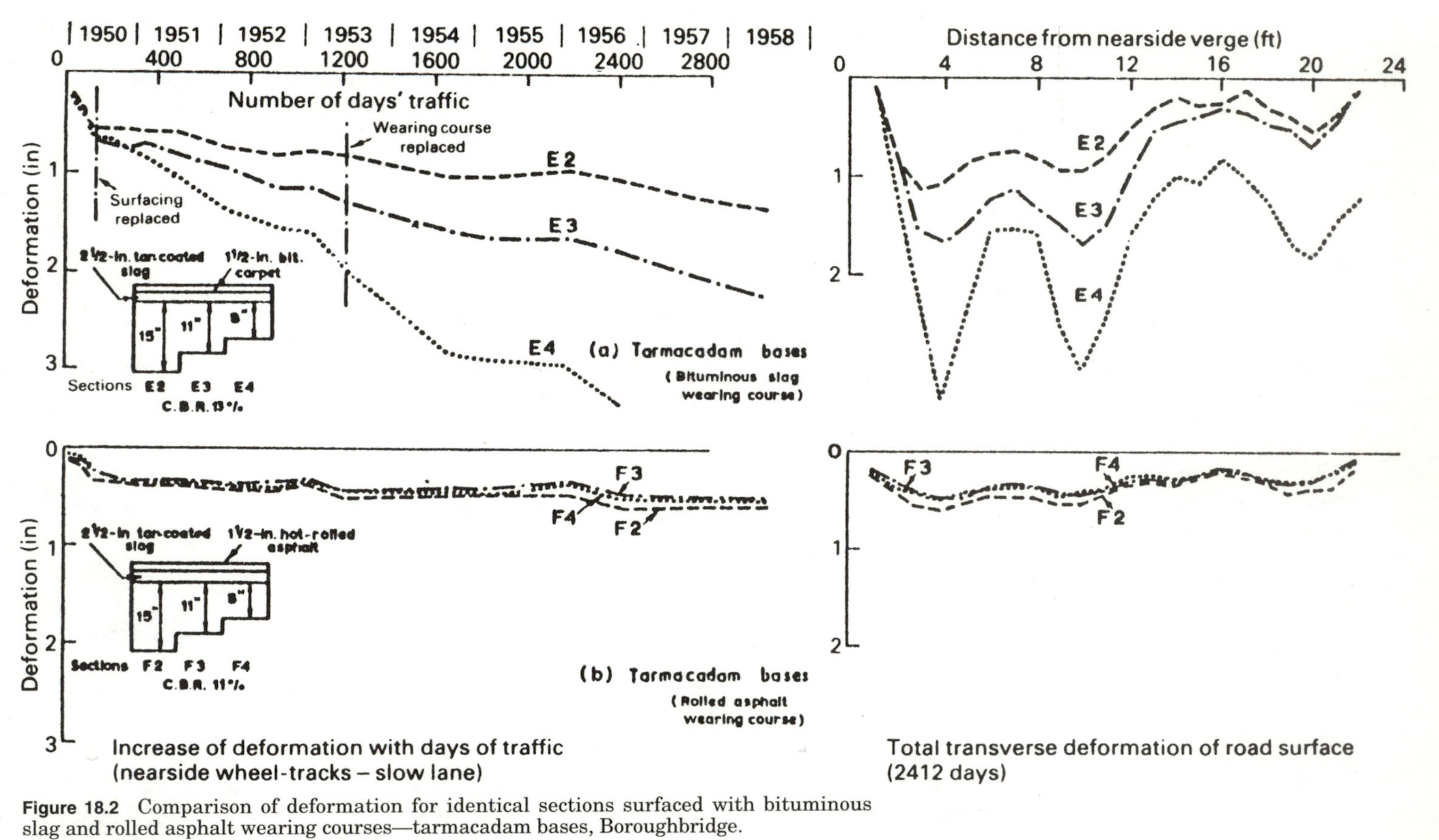

Figure 18.2 Comparison of deformation for identical sections surfaced with bituminous slag and rolled asphalt wearing courses—tarmacadam bases, Boroughbridge.

TABLE 18.2 Lives of Sections Incorporated in the Boroughbridge Experiment

Section	Life, millions of standard axles	Surfacing
C1	1.5	Asphalt w/c
C2	1.5	
C3	1.2	
C4	0.4	
D1	<0.1	Bitumen macadam w/c
D2	<0.1	
D3	<0.1	
D4	0.1	
E2	1.1	Bitumen macadam w/c
E3	0.5	
E4	0.2	
F2	>3*	Asphalt w/c
F3	>3*	
F4	>3*	

*Corresponding to the 10-year period.

18.9. Using the failure criteria discussed in Chap. 4, Table 18.2 shows the lives to failure of the various sections.

The Alconbury Hill Experiment, Flexible Sections—1957[2,3]

18.10. This experiment consisted of 33 flexible sections each 46 or 60 m in length. The main purpose was to compare the performance of wet-mix stone, open-textured tar macadam, lean concrete, rolled asphalt, and sand-cement bases laid to a range of thicknesses between 75 and 230 mm, when laid under asphalt surfacings 38 to 100 mm thick. A few duplicate sections were included with bitumen macadam surfacings 100 mm thick. The thickness of the subbase varied along each section, as shown in Fig. 18.3. A relatively weak sand subbase of CBR value 14 percent (unsoaked) was used to accentuate differences in performance of the various base materials. The soil at the site was a medium to heavy boulder clay of CBR (unsoaked) of 3 to 5 percent.

18.11. At the time the experiment was constructed in 1957 the road was one of the most heavily trafficked industrial trunk roads in Britain with an average damaging effect per commercial vehicle of about 0.9 standard axle. Since then this figure has more than doubled.

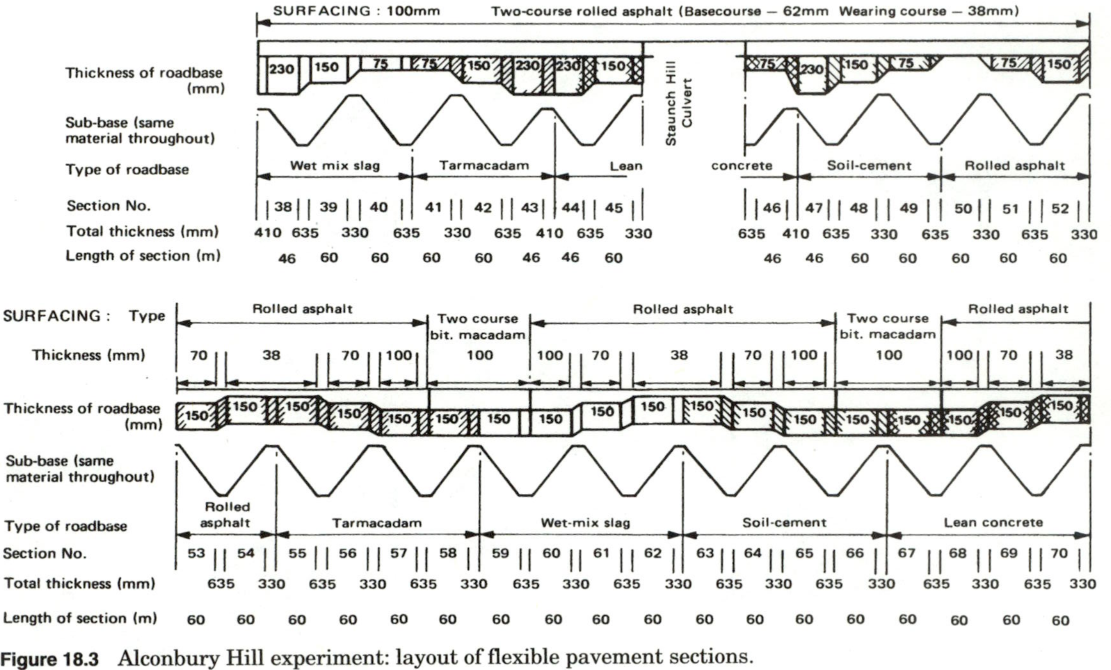

Figure 18.3 Alconbury Hill experiment: layout of flexible pavement sections.

18.12. As with the Boroughbridge experiment the performance was judged in terms of deformation and cracking in accordance with the failure criterion discussed in Chap. 4. This criterion corresponds to a PSI value close to 2. Again the maximum deformation occurred in the nearside wheel track followed by the heavy commercial vehicles. Table 18.3 summarizes the performance of the various sections. Measurements were continued for 10 years, during which time the

TABLE 18.3 Performance of Sections in the Alconbury Hill Experiment, A1 Trunk Road

Section no.	Surfacing	Base thickness, mm	Subgrade CBR, %	Life, million standard axles
Sections with wet-mix road bases				
38	100-mm RA*	230	4.5	8
39	100-mm RA	150	4.5	4.2
60	100-mm RA	150	4	0.7
40	100-mm RA	75	3.5	1.3
61	70-mm RA	150	4	2.8
62	40-mm RA	150	4	0.3
59	100-mm CM*	150	4	0.8
Sections with lean concrete bases				
44	100-mm RA	230	5.5	16
45	100-mm RA	150	6.5	10
68	100-mm RA	150	3	2.7
46	100-mm RA	75	5	1.6
69	70-mm RA	150	5	0.6
70	40-mm RA	150	5	1
67	100-mm CM	150	3	2.6
Sections with tarmacadam bases				
43	100-mm RA	230	4	14
42	100-mm RA	150	4	5
41	100-mm RA	75	5	3.5
56	70-mm RA	150	5.5	3.2
55	40-mm RA	150	5.5	2
58	100-mm CM	150	4	2.5
Sections with rolled asphalt bases				
52	100-mm RA	150	4	23
51	100-mm RA	75	4.5	8
53	70-mm RA	150	4	9
54	40-mm RA	150	4	10
Sections with soil-cement bases				
47	100-mm RA	230	5	5
48	100-mm RA	150	4	2.5
65	100-mm RA	150	3.5	1.2
49	100-mm RA	75	3.5	0.4
64	70-mm RA	150	3.5	0.5
63	40-mm RA	150	4	0.1
66	100-mm CM	150	4	0.2

*RA = rolled asphalt; CM = coated macadam (surface-dressed)

road had carried about 6 msa. Where lives greater than this are shown in Table 18.3 they have been obtained by extrapolation.

18.13. For all the sections the lives for each base material were largely determined by the thickness of base and surfacing, although where replicate sections were laid there was some evidence that the CBR value of the subgrade and the thickness of the subbase were important.

18.14. The sections with rolled asphalt bases gave the best performance. The lives were determined principally by the combined thickness of base and surfacing. When open-textured tarmacadam was used in place of rolled asphalt, the lives were much reduced. When the combined thickness of base and surfacing was reduced below 330 mm, the lives of the sections were radically reduced.

18.15. The lives of the sections with lean concrete bases decreased rapidly as the combined thickness of base and surfacing was reduced below 350 mm; for the thinner sections cracking through the surfacing was the main indication of failure.

18.16. The sections with wet-mix stone and sand-cement gave a poor performance, and for the traffic carried by this experiment, a combined thickness of base and surfacing of 350 mm would have been required.

The Nately Scures Experiment—1963[4]

18.17. This experiment consists of 21 sections laid on one carriageway of a dual carriageway trunk road carrying moderately heavy commercial traffic. The objective was to compare the performance of dense coated macadam base materials of various thicknesses with unbound and cemented bases. Details of the layout and of the materials used are given in Fig. 18.4. The average CBR of the subgrade was 5 percent, and a 150-mm granular subbase was used.

18.18. The performance of the sections is summarized in Table 18.4. The route was duplicated in 1972 by a new freeway, which removed a large proportion of the heavy commercial traffic. At that time the experiment had carried approximately 5 msa. Lives in excess of that figure shown in Table 18.4 have been deduced by extrapolation of the measured deformation and degree of cracking.

18.19. The results, which confirm the superiority of bound over unbound bases, are reasonably consistent with those from the

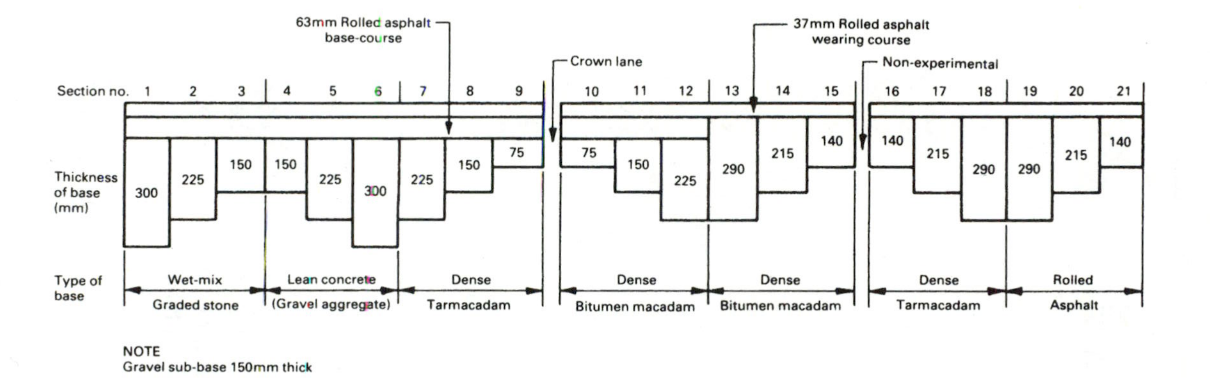

NOTE
Gravel sub-base 150mm thick

Figure 18.4 Nately Scures experiment: layout and description of the experimental sections.

TABLE 18.4 Performance of Sections in the Nately Scures Experiment, A30 Trunk Road

Section no.	Surfacing	Base thickness, mm	Life, million standard axles	Condition, 1988
Sections with wet-mix road bases				
1	100-mm RA*	300	4.5	Deformed and cracked
2	100-mm RA	225	3	Overlaid
3	100-mm RA	150	2	Overlaid
Sections with lean concrete bases				
4	100-mm RA	150	20	Satisfactory
5	100-mm RA	225	>20	Satisfactory
6	100-mm RA	300	>20	Satisfactory
Sections with dense tarmacadam bases				
7	100-mm RA	225	>20	Satisfactory
8	100-mm RA	150	>20	Satisfactory
9	100-mm RA	75	4	Cracked
16	37-mm RA	140	2.5	Reconstructed
17	37-mm RA	215	>20	Satisfactory
18	37-mm RA	290	>20	Satisfactory
Sections with dense bitumen macadam bases				
10	100-mm RA	75	5	Deformed and cracked
11	100-mm RA	150	>20	Satisfactory
12	100-mm RA	225	>20	Satisfactory
13	37-mm RA	290	>20	Satisfactory
14	37-mm RA	215	15	Satisfactory
15	37-mm RA	140	2.5	Reconstructed
Sections with rolled asphalt road bases				
19	37-mm RA	290	>20	Satisfactory
20	37-mm RA	215	>20	Satisfactory
21	37-mm RA	140	2	Reconstructed because of local foundation slip

*RA = rolled asphalt.

Alconbury Hill site. The lean concrete bases have performed rather better than those laid at Alconbury, probably because the traffic at that site contains a greater proportion of very heavy axle loads likely to cause early cracking in lean concrete.

The Wheatley Bypass and Alconbury Bypass Experiments—1964[4]

18.20. These two large experiments were designed to be complementary. They were each laid on both carriageways of four-lane highways with a central reservation and hard shoulders. Because at each site the traffic was more damaging in one direction than in the other, and because there were intermediate interchanges, the sections were not all subjected to the same number of standard axles. A number of weighbridges were installed at each site to assess the traffic accurately over each section.

18.21. The overall objective of the experiments was to study the effects of (1) aggregate grading, (2) binder content, and (3) thickness on the performance of bases bound with bitumen and with cement. A few sections with unbound granular bases were also included. For the bituminous bases (bound with bitumen and with tar) three grading zones were selected as indicated in Fig. 18.5. For the cement-bound bases four grading zones were used as shown in Fig. 18.6. Both gravel and crushed-stone aggregates were used.

18.22. At Wheatley bypass a constant thickness of base of 200 mm was used in conjunction with 150 mm of granular subbase and a 100-mm rolled asphalt surfacing. The type of base used for each section is shown in Fig. 18.7. (After the road had been opened to traffic for a few weeks the approach alignment had to be modified and sections 1 to 8 were lost.) All the sections were 52 m in length.

18.23. The binder content for the bases bound with bitumen and tar are shown in Fig. 18.7. The cement contents for the sections with cement-bound bases were determined by prior laboratory testing on

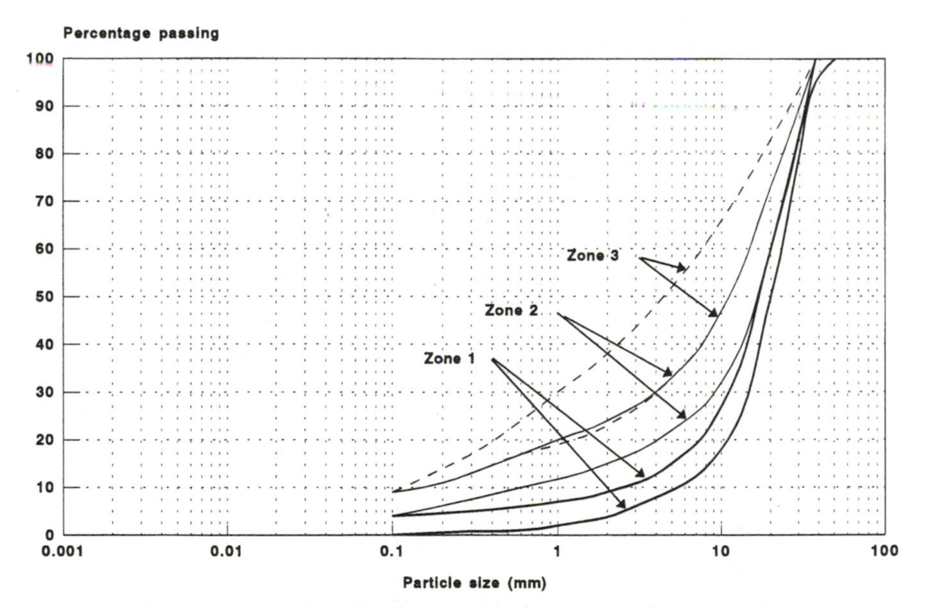

Figure 18.5 Alconbury and Wheatley bypass experiments: grading zones for aggregates used in bituminous bases.

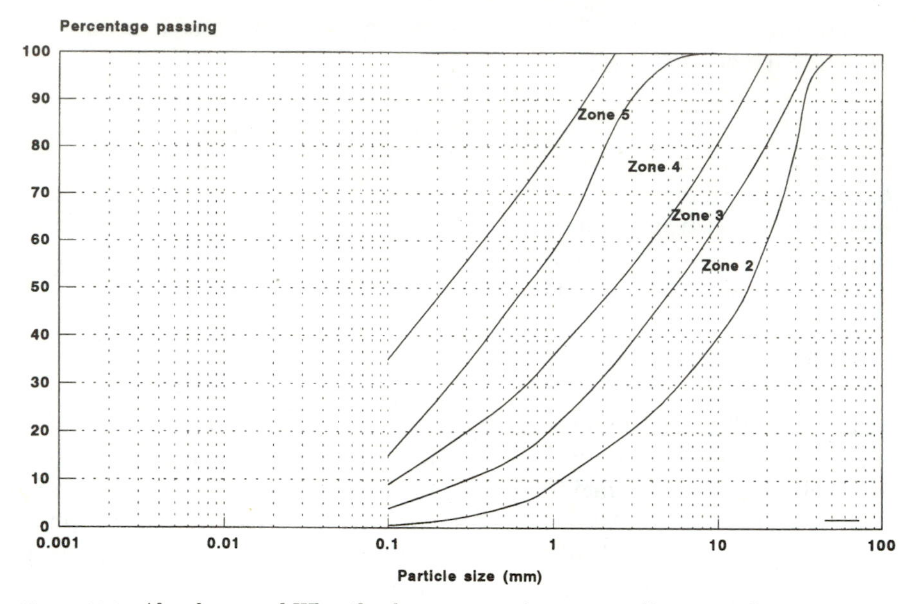

Figure 18.6 Alconbury and Wheatley bypass experiments: grading zones for aggregates used in cemented bases.

the aggregates used during construction. The cement contents were chosen to give the 28-day crushing strengths of 3.2, 7.6, and 15.2 MN/m^2 as indicated in Fig. 18.7.

18.24. The sections incorporated in the Alconbury bypass experiment are shown in Fig. 18.8. The principal variable was the thickness of the base. Using the mean binder content in the case of the bituminous sections and the mean strength in the case of the cemented bases, adopted at Wheatley bypass, bases were laid with thicknesses of 150, 200, and 250 mm, as indicated in Fig. 18.8.

18.25. As with the earlier experiments, transverse deformation caused by traffic was measured at five points along each section twice yearly and the crack patterns were recorded. Based on these measurements the condition was classified as satisfactory (S), critical (C), or failed (F). The condition after 22 years, assessed in this manner, is shown for each section in Figs. 18.7 and 18.8 together with the number of standard axles carried.

18.26. To facilitate discussion the performance of the cemented and unbound bases at the two sites is summarized in Tables 18.5 and 18.6.

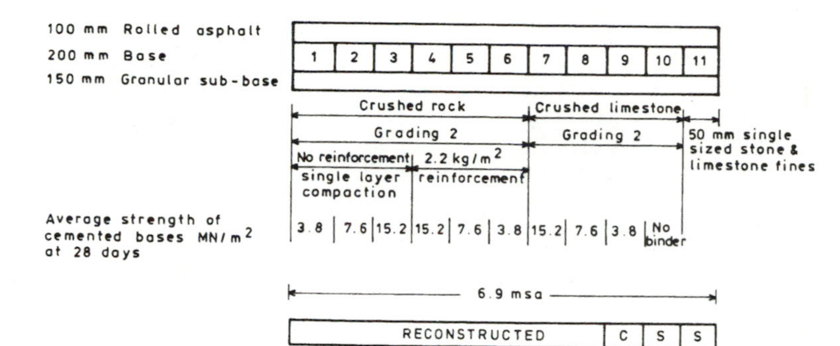

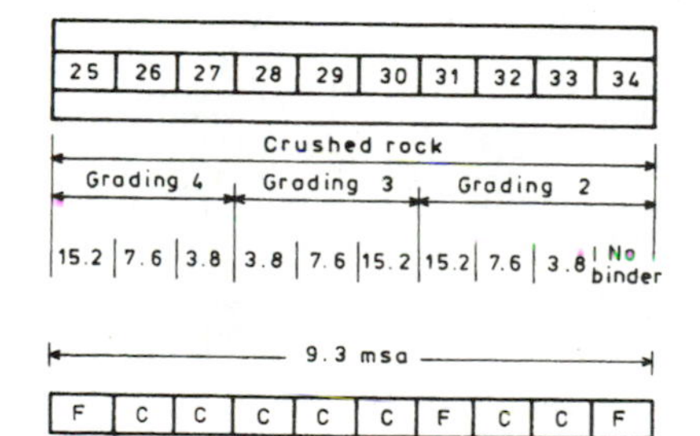

Figure 18.7 Wheatley bypass: experimental sections and condition after 22 years. The condition of the sections was assessed in 1987: S = satisfactory; C = critical; and F = failed.

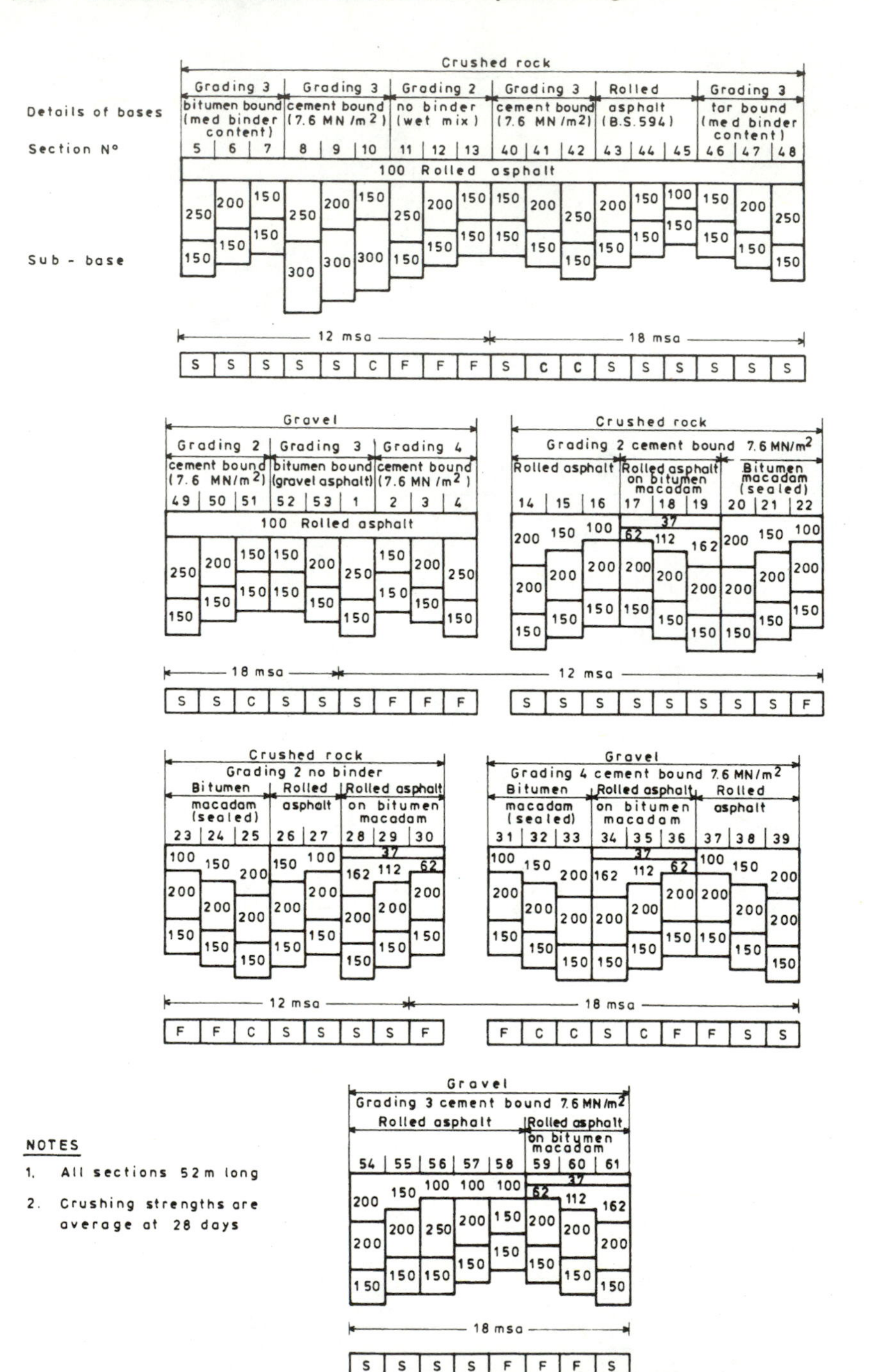

Figure 18.8 Alconbury bypass: experimental sections and condition after 22 years. The condition of the sections was assessed in 1987: S = satisfactory; C = critical; and F = failed.

TABLE 18.5 Performance of Cemented and Unbound Bases—Wheatley Bypass

Aggregate type	Grading	Strength, MN/m²	Base thickness, mm	Section no.	Condition,† 1988	msa
Gravel	2	3.8*	200	14	C	6.9
		7.6	200	13	S	6.9
		15.2	200	12	S	6.9
	3	3.8	200	16	C	6.9
		7.6	200	17	C	6.9
		15.2	200	18	S	6.9
	4	3.8	200	21	S	6.9
		7.6	200	20	C	6.9
		15.2	200	19	S	6.9
	5	3.8	200	22	F	6.9
		7.6	200	23	F	6.9
		15.2	200	24	F	6.9
	2	None‡	200	15	F	6.9
Crushed rock	2	3.8	200	33	C	9.3
		7.6	200	32	C	9.3
		15.2	200	31	F	9.3
	3	3.8	200	28	C	9.3
		7.6	200	29	C	9.3
		15.2	200	30	C	9.3
	4	3.8	200	27	C	9.3
		7.6	200	26	C	9.3
		15.2	200	25	F	9.3
	2	None‡	200	34	F	9.3

*28-day.
†S = satisfactory, C = critical, F = failed.
‡Laid as wet-mix.

TABLE 18.6 Performance of Cemented and Unbound Bases—Alconbury Bypass

Aggregate type	Grading	Strength, MN/m²	Base thickness, mm	Section no.	Condition, 1988	msa
Gravel	2	7.6	250	49	S	18
		7.6	200	50	S	18
		7.6	150	51	C	18
	3	7.6	250	56	S	18
		7.6	200	57	S	18
		7.6	150	58	F	18
	4	7.6	250	4	F	18
		7.6	200	3	F	18
		7.6	150	2	F	18
Crushed rock	2	7.6	250	None		
		7.6	200	16	S	12
	3	7.6	250	42	S	18
		7.6	150	40	C	18
	2	No binder (wet-mix)	250	11	F	12
			200	12	F	12
			150	13	F	12
			200	27	C	12

18.27. The low-strength cemented bases tended to develop alligator-type cracking under the action of traffic, and the performance became similar to that associated with unbound stone bases. Deformation was therefore the main cause of failure of these sections. With the highest-strength cemented bases wide cracks in the base were reflected through the surfacing under the action of traffic and temperature. These cracks were either transverse to the direction of traffic or longitudinal in the wheel tracks of the commercial vehicles.

18.28. In the United Kingdom cemented bases are generally made with gravel aggregates. Table 18.5 shows that the grading of the aggregate as well as the crushing strength affected the performance of the sections. The most satisfactory performance for both gravel and crushed-rock aggregates was obtained from the grading 2 aggregates and the medium crushing strength. For a 100-mm asphalt surfacing, a 200-mm base, and a 150-mm subbase the life obtained was about 10 msa. This is in line with the evidence obtained from the earlier Alconbury Hill experiment, the results of which are summarized in Table 18.3.

18.29. The soil-cement used at the Alconbury Hill experiment was made with a fine sand, much finer than grading 5 in Fig. 18.6. The 28-day crushing strength was only 2 MN/m^2. In view of the later experience it is not surprising that the soil-cement sections at Alconbury Hill failed so quickly.

18.30. Within the period of the observations there is no evidence from the Alconbury bypass experiment that increasing the thickness of the base from 200 to 250 mm had an influence on the behavior of the sections with cemented bases, although it is reasonable to assume that it would have had a longer-term effect in increasing life. Decreasing the thickness to 150 mm certainly reduced the life for both types of aggregate and for all the gradings investigated.

18.31. The effect on pavement life of changing the type and thickness of the surfacing at the Alconbury bypass experiment is shown in Table 18.7.

18.32. The conclusion which can be drawn from Table 18.7 is that the rolled asphalt surfacing was more effective than the same thickness of coated macadam in extending the lives of the sections with cemented bases. For example, 200 mm of bitumen macadam was less effective than 150 mm of rolled asphalt. This applies to bases using both gravel and crushed-stone aggregates.

TABLE 18.7 Effect of Changes to Surfacing on the Performance of Cemented Bases—Alconbury Bypass

Aggregate type	Grading	Type of sufacing	Section no.	Condition, 1988	msa
Gravel	4	100-mm R.A.*	37	F	18
		150-mm R.A.	38	S	18
		200-mm R.A.	39	S	18
		100-mm B.M.*	31	F	18
		150-mm B.M.	32	C	18
		200-mm B.M.	33	C	18
		37-mm R.A. on 162-mm B.M.	34	S	18
		37-mm R.A. on 112-mm B.M.	35	C	18
		37-mm R.A. on 62-mm B.M.	36	F	18
Crushed rock	2	100-mm R.A.	16	S	12
		150-mm R.A.	15	S	12
		200-mm R.A.	14	S	12
		200-mm B.M.	20	S	12
		150-mm B.M.	21	C	12
		100-mm B.M.	22	F	12
		37-mm R.A. on 62-mm B.M.	17	S	12
		37-mm R.A. on 112-mm B.M.	18	S	12
		37-mm R.A. on 162-mm B.M.	19	S	12

NOTE: All the sections considered had (1) a strength of 7.6 MN/m^2, (2) a thickness of base of 200 mm, and (3) a 150-mm granular subbase.

*R.A. = rolled asphalt, B.M. = bitumen macadam.

18.33. The unbound bases laid at both sites under 100 mm of rolled asphalt surfacing had failed before 7 msa had been carried, although one section at Alconbury bypass was satisfactory after 12 msa under 150 mm of rolled asphalt. This is consistent with the experience from the Alconbury Hill experiment (see Table 18.3).

18.34. The performance of the sections at Wheatley bypass using bituminous and tar-bound bases is summarized in Table 18.8.

18.35. It will be seen from this table that all the sections were still in satisfactory condition after 22 years, during which 13 to 15 msa were carried. There was no significant difference in either deformation or cracking for the range of binder contents used. This is regarded as most surprising in view of the very "hungry" appearance of the low-binder-content bases at the time of laying.

18.36. The performance of sections 7, 45, 46, and 52 at the Alconbury bypass experiment (see Fig. 18.8) shows that even when the thickness of the base was reduced to 150 mm the sections with the intermediate binder content were able to carry 18 msa without signs of distress.

TABLE 18.8 Performance of Bitumen- and Tar-Bound Bases at Wheatley Bypass

Aggregate type	Grading	Binder		Section no.	Condition, 1988	msa
		Type	Content			
Crushed rock	1	Bitumen	2.0	35	S	12.6
		Bitumen	3.5	36	S	12.6
	2	Bitumen	5.0	37	S	12.6
		Bitumen	3.5	38	S	12.6
		Bitumen	2.5	39	S	12.6
	3	Bitumen	3.0	40	S	12.6
		Bitumen	4.0	41	S	12.6
		Bitumen	5.0	42	S	12.6
	3	Tar	5.5	48	S	15.3
		Tar	4.5	49	S	15.3
		Tar	3.5	50	S	15.3
	2	Tar	2.5	51	S	15.3
			3.5	52	S	15.3
			5.5	53	S	15.3
	1	Tar	3.5	54	S	15.3
			2.0	55	S	15.3

NOTE: For all the above sections the surfacing is 100 mm of rolled asphalt, the base is 200 mm thick, and the granular subbase is 150 mm thick.

18.37. The results of these two major experiments indicate a marked superiority of bituminous bases over cemented bases of the same thickness. The excellent performance of the "lean" bituminous bases indicates an economic advantage in the use of this form of construction. Cemented bases are now very little used in the United Kingdom for major road projects.

The Conington Experiment—1965

18.38. This experiment was constructed in 1965 on the London-bound carriageway of Trunk Road A.1 close to the Alconbury Hill site already discussed. Observations of cracking and deformation were continued for rather more than 20 years, by which time the sections had carried about 32 msa. The layout of the sections and the materials used are shown in Fig. 18.9. For this experiment a crushed-rock subbase 150 mm thick was used over the boulder clay which had an unsoaked CBR of about 5 percent.

18.39. The objectives of the experiment were: (1) To compare the performance of binder courses made with various types of gravel aggregate with the performance of binder courses made with crushed-rock aggregate. For this part of the experiment a base of unbound crushed stone (wet-mix) was used. (2) To compare the performance of bitumi-

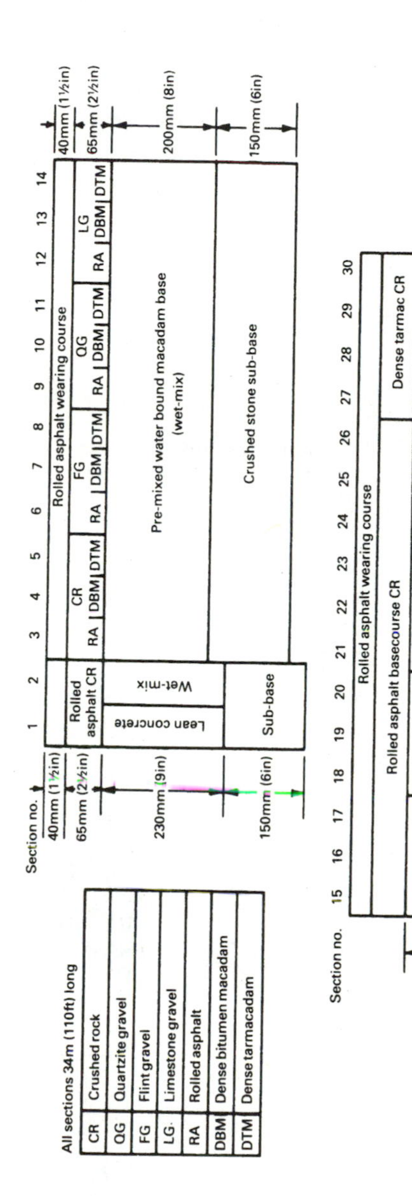

Figure 18.9 Conington experiment: construction details of sections.

nous bases made with various gravel aggregates with bases made with crushed-rock aggregate. For this part of the experiment, the wearing course and binder course were of rolled asphalt made with crushed-rock aggregate.

18.40. The lives of the various sections are shown in Table 18.9. For sections which became critical after carrying less than 30 msa the lives quoted are based on the degree of cracking and the extent of deformation. The longer lives quoted in the table are deduced by extrapolating the evidence available. The sections have been grouped so that the effect of changing the aggregate or binder can be readily seen.

TABLE 18.9 Performance of the Sections at the Conington Experiment

Section no.	Wearing course	Base course (binder course)	Base	msa
3	RA (CR)	RA (CR)	WM	27
6	RA (CR)	RA (FG)	WM	27
9	RA (CR)	RA (QG)	WM	27
12	RA (CR)	RA (LG)	WM	10
4	RA (CR)	DBM (CR)	WM	20
7	RA (CR)	DBM (FG)	WM	25
10	RA (CR)	DBM (QG)	WM	20
13	RA (CR)	DBM (LG)	WM	12
5	RA (CR)	DTM (CR)	WM	16
8	RA (CR)	DTM (FG)	WM	27
11	RA (CR)	DTM (QG)	WM	6.5
14	RA (CR)	DTM (LG)	WM	10
15	RA (CR)	RA (CR)	RA (CR)	50
18	RA (CR)	RA (CR)	RA (FG)	100
21	RA (CR)	RA (CR)	RA (QG)	25
24	RA (CR)	RA (CR)	RA (LG)	40
16	RA (CR)	RA (CR)	DBM (CR)	100
19	RA (CR)	RA (CR)	DBM (FG)	40
22	RA (CR)	RA (CR)	DBM (QG)	17
25	RA (CR)	RA (CR)	DBM (LG)	40
17	RA (CR)	RA (CR)	DTM (CR)	100
20	RA (CR)	RA (CR)	DTM (FG)	80
23	RA (CR)	RA (CR)	DTM (QG)	23
26	RA (CR)	RA (CR)	DTM (LG)	25
27	RA (CR)	DTM (CR)	DTM (LG)	25
28	RA (CR)	DTM (CR)	DTM (QG)	90
29	RA (CR)	DTM (CR)	DTM (FG)	100
30	RA (CR)	DTM (CR)	DTM (CR)	120

NOTE: CR = crushed rock, QG = quartzite gravel, FG = flint gravel, LG = limestone gravel, RA = rolled asphalt, WM = wet-mix, DBM = dense bitumen macadam, DTM = dense tarmacadam.

18.41. In the binder course, crushed rock, flint gravel, and quartzite gravel behaved similarly when the binder was bitumen. Limestone gravel, however, gave a poorer performance in bituminous mixes. When tar was used as the binder the performance of binder courses using both quartzite and limestone gravel aggregates was relatively poor.

18.42. The performance of the various bases under rolled asphalt wearing courses and binder courses (sections 15 to 26) shows very similar lives for bases made with crushed rock and flint gravel aggregate, but significantly shorter lives when quartzite and limestone gravels were used. The same pattern of behavior was found when a DTM/CR binder course was used (sections 27 to 30). The experiment showed that the use of limestone gravel in bituminous mixtures should be avoided.

Design Standards for Flexible Pavements Based on Full-Scale Road Experiments

18.43. The long-term evidence obtained from the experiments described above and from a number of smaller experiments has been used to construct Figs. 18.10 to 18.12. The curves shown in these dia-

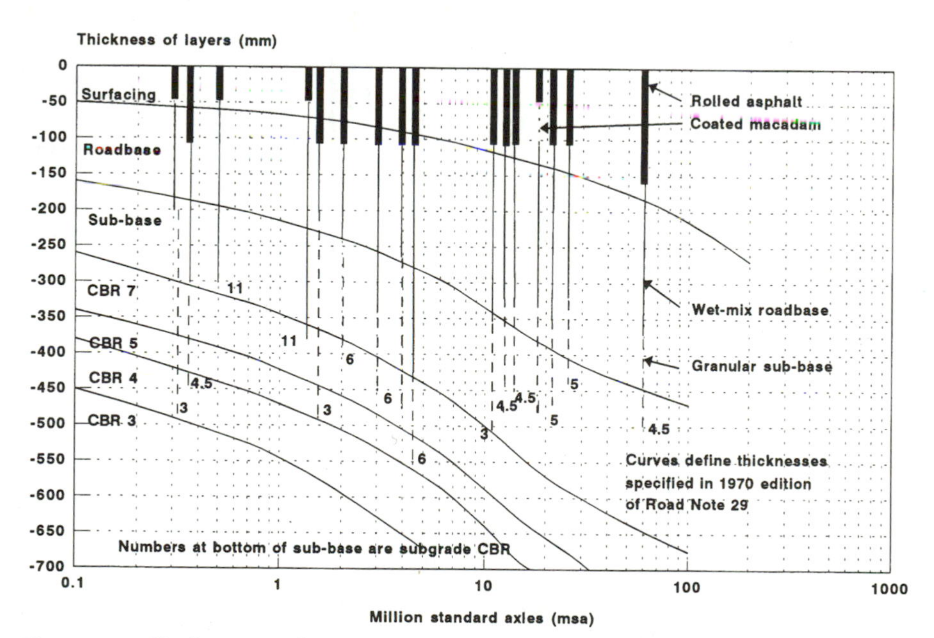

Figure 18.10 Performance of experimental sections with unbound road bases.

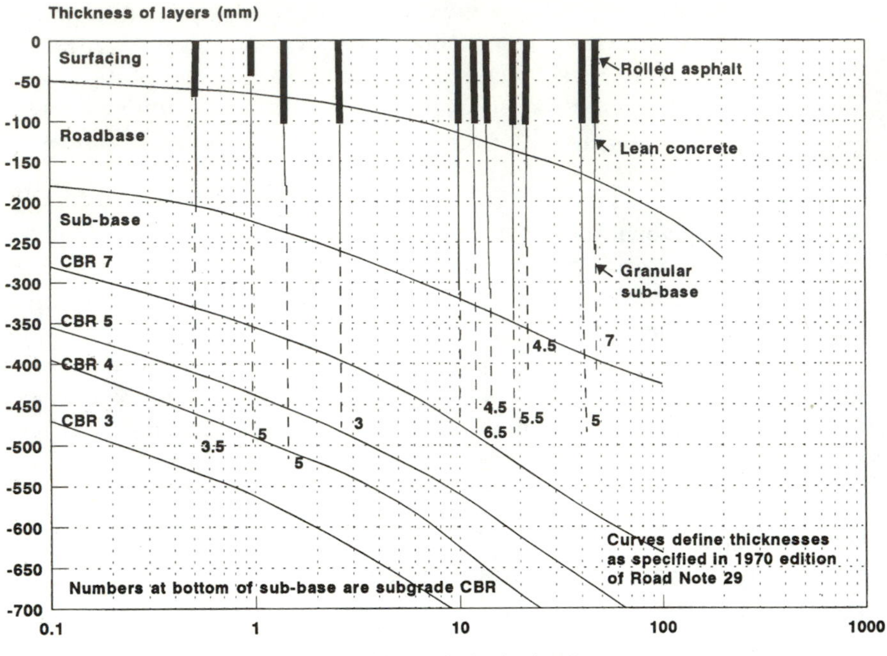

Figure 18.11 Performance of experimental sections with lean concrete bases.

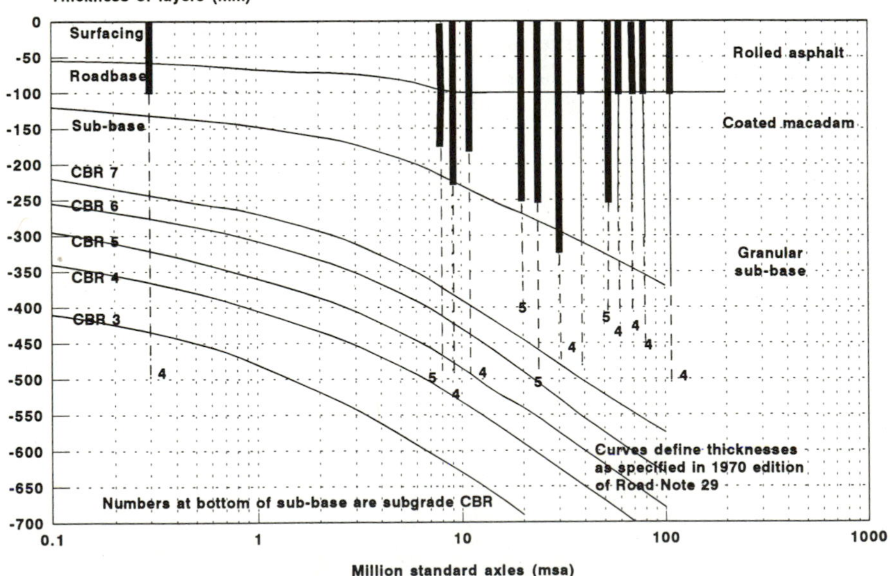

Figure 18.12 Performance of experimental sections with asphalt or coated macadam bases.

grams form the basis of the design procedures used for flexible pavements in the United Kingdom. This matter is discussed further in Chap. 21.

References

1. Lee, A. R., and D. Croney: British Full-scale Design Experiments. *Proc. 1st Int. Conf. on the Structural Design of Asphalt Pavements, Ann Arbor, 1962,* University of Michigan, Ann Arbor, 1962.
2. Croney, D., and J. A. Loe: Full-scale Pavement Design Experiments on A1 at Alconbury Hill, *Proc. Inst. Civ. Engrs.,* vol. 30 (February), pp 225–270, 1965.
3. Thompson, P. D., D. Croney, and E. W. H. Currer: The Alconbury Hill Experiment and Its Relation to Flexible Pavement Design, *Proc. 3rd Int. Conf. on the Structural Design of Asphalt Pavements, London, 1972,* University of Michigan, Ann Arbor, 1972.
4. Salt, G. F.: Recent Full-scale Pavement Design Experiments in Britain, *Proc. 2nd Int. Conf. on the Structural Design of Asphalt Pavements, Ann Arbor, Michigan, 1967,* University of Michigan, Ann Arbor, 1967.

Chapter

19

Performance Studies of Full-Scale Experimental Sections Incorporated in In-Service Highways in the United Kingdom—Concrete Pavements

Introduction

19.1. In the 1920s concrete pavement, hand tamped between forms, was widely used in the United Kingdom for housing estate developments. This form of construction was not, however, used for major heavily trafficked roads, because of lack of experience in relation to both specification and construction.

19.2. The massive road-building program in the United States, which started in the 1920s and continued through the 1930s, has been referred to in Chap. 2. Much of this work was carried out using concrete pavements, and the then Ministry of Transport in Britain sent a small study group to the United States to examine specifications and construction methods. Following this initiative a number of new "arterial" roads in concrete were constructed in the south of England. Several of these are still in use under bituminous surfacings.

19.3. Two of these roads, constructed in 1930 and 1933, were built with experimental features, and they are still under observation after more than 60 years of steadily increasing traffic. These two roads head the list of U.K. concrete pavement experiments summarized in Table 19.1. This table shows the main objectives of each experiment

TABLE 19.1 Details of Major Full-Scale Pavement Design Experiments Using Concrete Construction

Year of construction	Location	Main objectives and construction details	Subgrade type	Initial traffic, commercial vehicles/day	Growth rate of commercial traffic, % per annum	Damaging effect of commercial traffic, standard axles/100 commercial axles
1930	A316 Great Chertsey Road, Middx.	To obtain information on the design of joints and load transfer devices Slab thickness—230 mm (reinforced) Compressive strength—(28 days) 21.4 MN/m^2 Subbase—clinker—75 mm thick Slab length—6.1 m	Silty clay and gravel	500	4	20 (average)
1933	A309 Hampton Court Way, Surrey	As above, but slab length increased to 9.1 m	Deep gravel fill	300 (estimated)	4	15 (average)
1946	A6097 at Oxton, Notts.	To study the effect on performance of: 1. Thickness of slab (100–200 mm) 2. Thickness of sand-cement subbase (50–150 mm) Compressive strength—37 MN.m^2 Slab length—4.5 m (unreinforced) 9.0 m (reinforced)	Sandy gravel	200	10 (average)	25
1948	B379 Longford to Stanwell Road, Middx.	To study: 1. Performance of unreinforced concrete slabs 2. Effect of spacing of expansion joints in unreinforced concrete pavement on joint movements Slab thickness—200 mm Compressive strength—39 MN/m^2 Slab length—4.6 m Subbase—None	Gravel embankment	500	4	20

1955	A48 Llangyfelach, Glams.	To study the influence of strength of the concrete on the performance of a reinforced concrete pavement Slab thickness—200 mm Slab length—37 m Reinforcement—3.5 kg/m^2 Subbase—clinker—150 mm thick	Gravel	450	5	25
1957	A1 Alconbury Hill, Cambs.	To study the effect on performance of: 1. Slab thickness for reinforced and unreinforced slabs 2. Strength of concrete 3. Weight of reinforcement when one slab length was used 4. Thickness and type of subbase Slab thickness—125–200 mm Slab length—4.5 m (unreinforced) 37 m (reinforced) Compressive strength—(28 days) 44 MN/m^2 and 66 MN/m^2 Subbase—gravel and lean concrete 76 and 230 mm thick	Silty clay	1400	5	25
1961	A46 Winthorpe (3 km east of Newark), Notts.	To study the effect on performance of prestressing thin concrete slabs Slab thickness—125 and 178 mm Compressive strength—(28 days) 48 MN/m^2 Subbase—lean concrete—100 mm thick	Imported granular material	500	4	20
1962	A1 Grantham Bypass, Lincs.	To study the effect on performance of: 1. Type and weight of reinforcement in relation to slab length (including continuously reinforced concrete) 2. Using a sliding layer between the concrete slabs and the subbase Slab thickness—230 mm, increased to 254 mm on embankments Compressive strength—(28 days) 38 MN/m^2 Subbase—lean concrete—75 mm thick	Various—ranging from silty clay to limestone brash	1200	5	30

and gives the relevant construction details. In the space available it is not possible to discuss the pavement performance at each site in detail. This information is available in the published papers referred to in the text. The main conclusions from each experimental road are given below.

The Great Chertsey Road (1930) and Hampton Court Way (1933) Experiments[1]

19.4. These are the two roads referred to in Par. 19.3. Following American practice of the 1930s the concrete made with a gravel aggregate, had a 28-day cube strength of 28 MN/m^2 (4150 lb/in^2). Top and bottom reinforcement was used of total weight 12 lb/yd^2 (6.4 kg/m^2). The slabs were all 20 ft (6.1 m) long and 9 in (230 mm) thick. At the Chertsey Road site a clinker subbase was used and at Hampton Court Way the concrete was laid on compacted gravel fill.

19.5. At the time these roads were constructed vertical movement at joints under heavy wheel loads was a major problem in the United States. It was therefore decided to investigate various load transfer devices aimed at reducing long-term "faulting" and mud pumping at joints. Six joint designs ranging from butt joints to doweled joints were used. Details of the designs used are given in Fig. 19.1. Deflection measurements were made at the joints under the passage of an 18,000-lb (8200-kg) axle load. The results of tests made after 22 and 34 years are shown in Table 19.2. The superiority of the doweled joints is clearly shown. At Hampton Court Way the relative deflections followed the same pattern, but the magnitudes were smaller, owing to the deep gravel fill under the later experiment.

19.6. The traffic at both sites has increased by a factor of 10 to 15 since the 1930s and the Great Chertsey Road in particular now carries almost continuous heavy commercial traffic. However, less than 4 percent of the 150 slabs at that site were showing any cracking after 40 years, when a bituminous surfacing was added. The percentage of the 500 slabs laid at Hampton Court Way which showed cracking after 60 years was even smaller. This excellent performance at both sites was in part due to the comparatively short slab length and the relatively heavy reinforcement used. However, another important factor is the delay in opening the roads to traffic, which in each case was nearly 3 years. This was because each road included a new bridge over the River Thames. As a consequence of this delay the concrete increased markedly in strength before it was required to carry heavy vehicles.

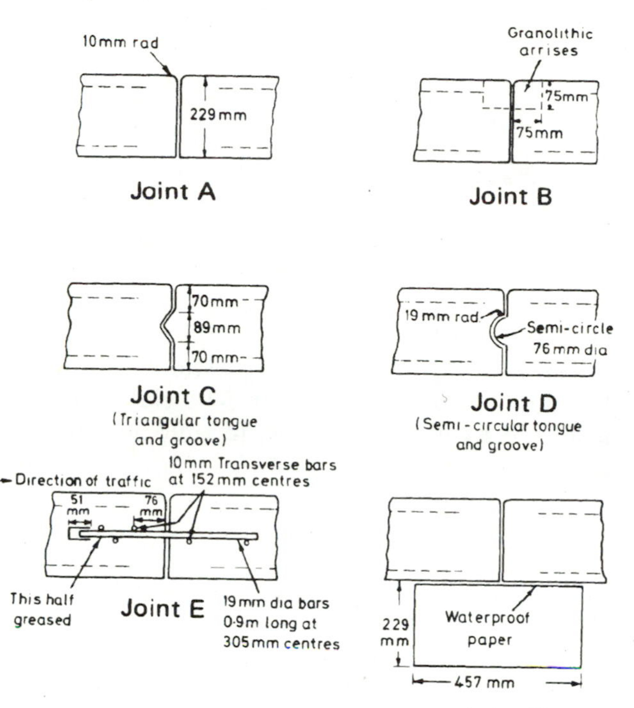

Figure 19.1 Great Chertsey Road and Hampton Court Way experiments: types of joints used.

TABLE 19.2 Vertical Deflections Observed on Six Types of Joint Used in the Great Chertsey Road, June 1952 and July 1968

Joint type (see Fig. 19.1)	Average total deflection observed, μm	
	June 1952	July 1968
A	145	284
B	185	279
C	46	198
D	5	107
E	3	25
F	572	96

19.7. When an asphalt surfacing was added to restore skid resistance, the Chertsey Road had carried about 15 msa and Hampton Court Road about 10 msa.

The Oxton Experiment—1946[2]

19.8. This was the earliest concrete pavement experiment in the United Kingdom in which the relationship between traffic intensity and slab thickness was investigated using both reinforced and plain concrete (Fig. 19.2). The concrete had a compressive strength of 37 N at 28 days and the thickness ranged from 75 to 200 mm. The reinforced slabs were 9 m long with a weight of reinforcement of 4.1 kg/m^2. The unreinforced slabs were 4.5 m long. The subgrade was granular with a CBR value of about 15 percent. The slabs were laid on a cement-stabilized sand base, which varied in thickness between 0 and 150 mm as shown in Fig. 19.2.

19.9. The performance of the sections was assessed in terms of the amount of cracking (Fig. 19.3). The lives of the sections are summarized in Table 19.3. Lives in excess of 30 years were obtained by extrapolation based on the rate at which cracking was taking place. The results show clearly for the traffic being carried the need (1) for a slab thickness of 200 mm, and (2) the advantage of using reinforcement.

TABLE 19.3 Oxton Experiment: Lives of Experimental Sections

	Life	
Slab thickness, mm	Years	msa
75U*	1.3	0.2
75R*	7	1
100U	1.5	0.3
100R	5	0.5
125U	8	1.3
125R	11	1.7
150U	9	1.4
150R	13	2.2
200U	36	12
200R	50	26

*U = unreinforced; R = reinforced

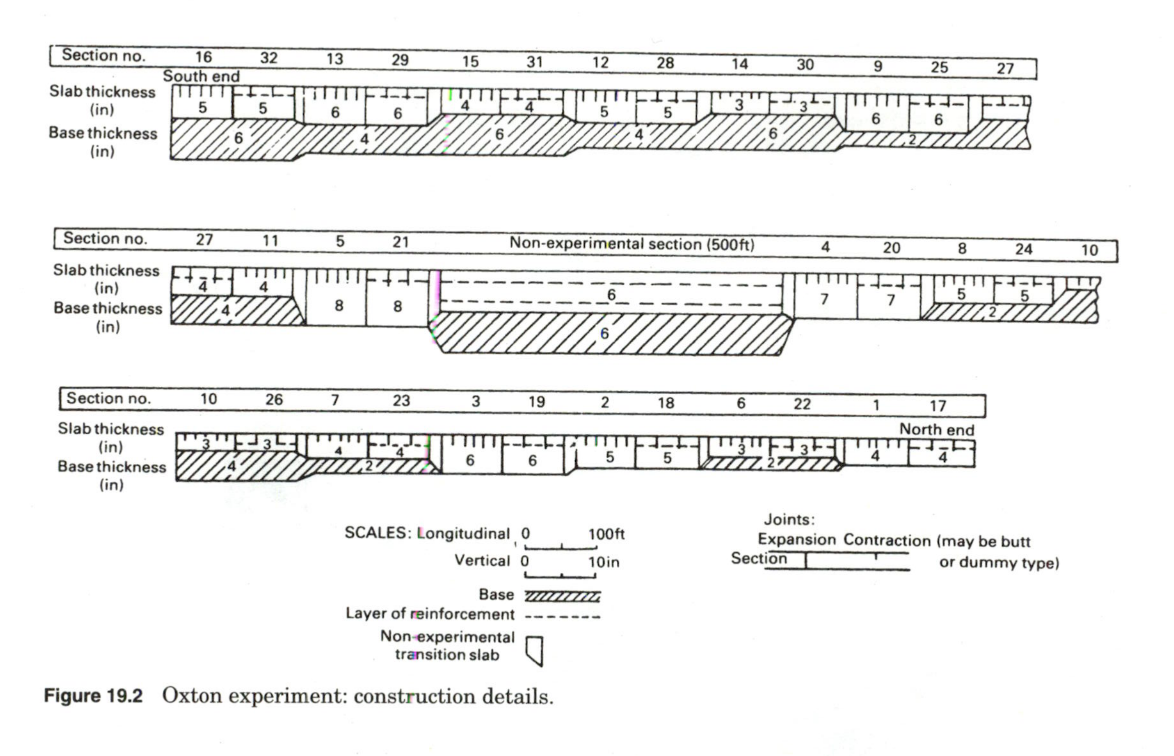

Figure 19.2 Oxton experiment: construction details.

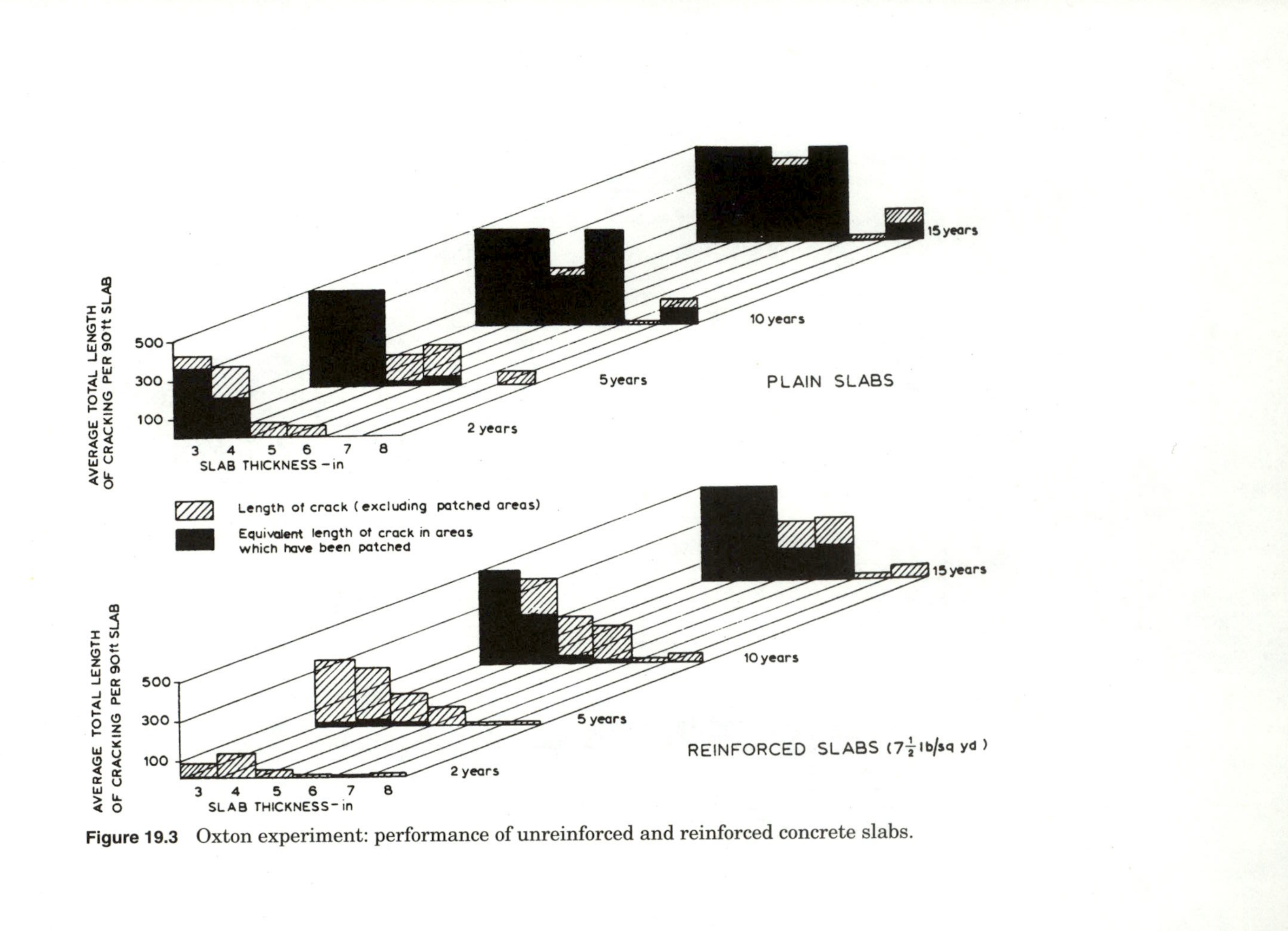

Figure 19.3 Oxton experiment: performance of unreinforced and reinforced concrete slabs.

The Alconbury Hill Experiment—1957[3]

19.10. This large experiment was sited close to the flexible pavement sections already discussed in Chap. 18. The traffic over the two experiments was therefore identical.

19.11. The purpose of the experiment was to compare the performance of normal (44.1 N/m²) and high-strength (66.1 N/m²) concrete pavements and to study the influence of reinforcement and thickness of subbase for concrete of both strengths. Details of the 35 sections are given in Fig. 19.4. As for the Oxton experiment, the development

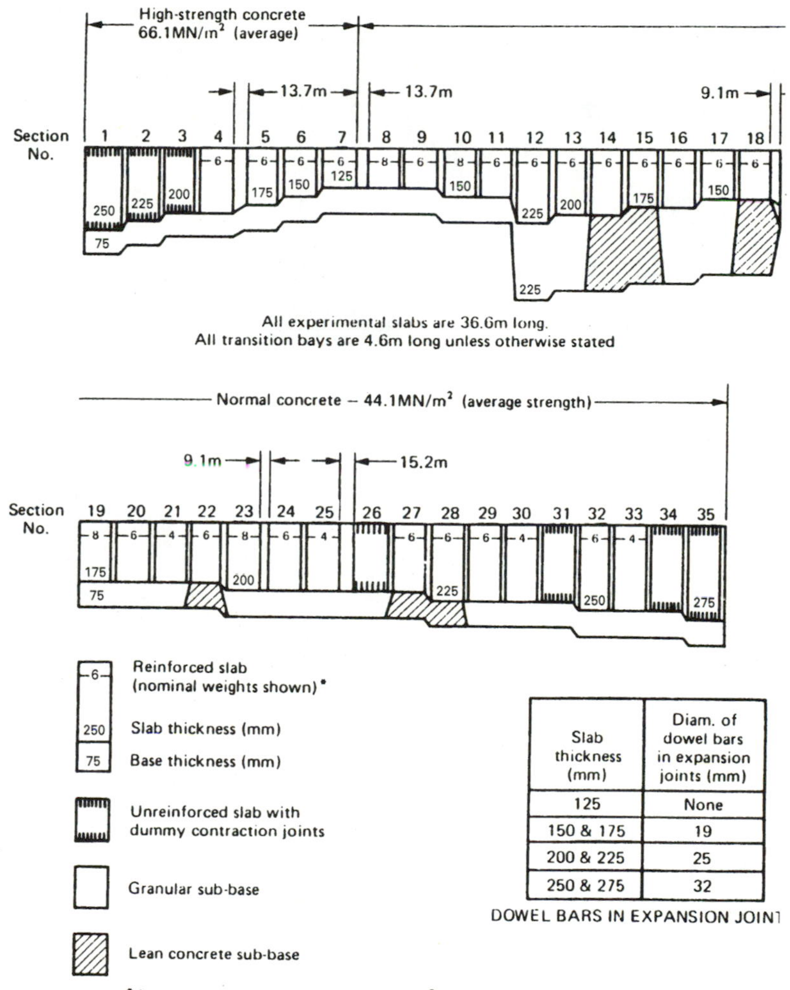

Slab thickness (mm)	Diam. of dowel bars in expansion joints (mm)
125	None
150 & 175	19
200 & 225	25
250 & 275	32

DOWEL BARS IN EXPANSION JOINT

Figure 19.4 Alconbury Hill experiment (concrete sections); details of sections.

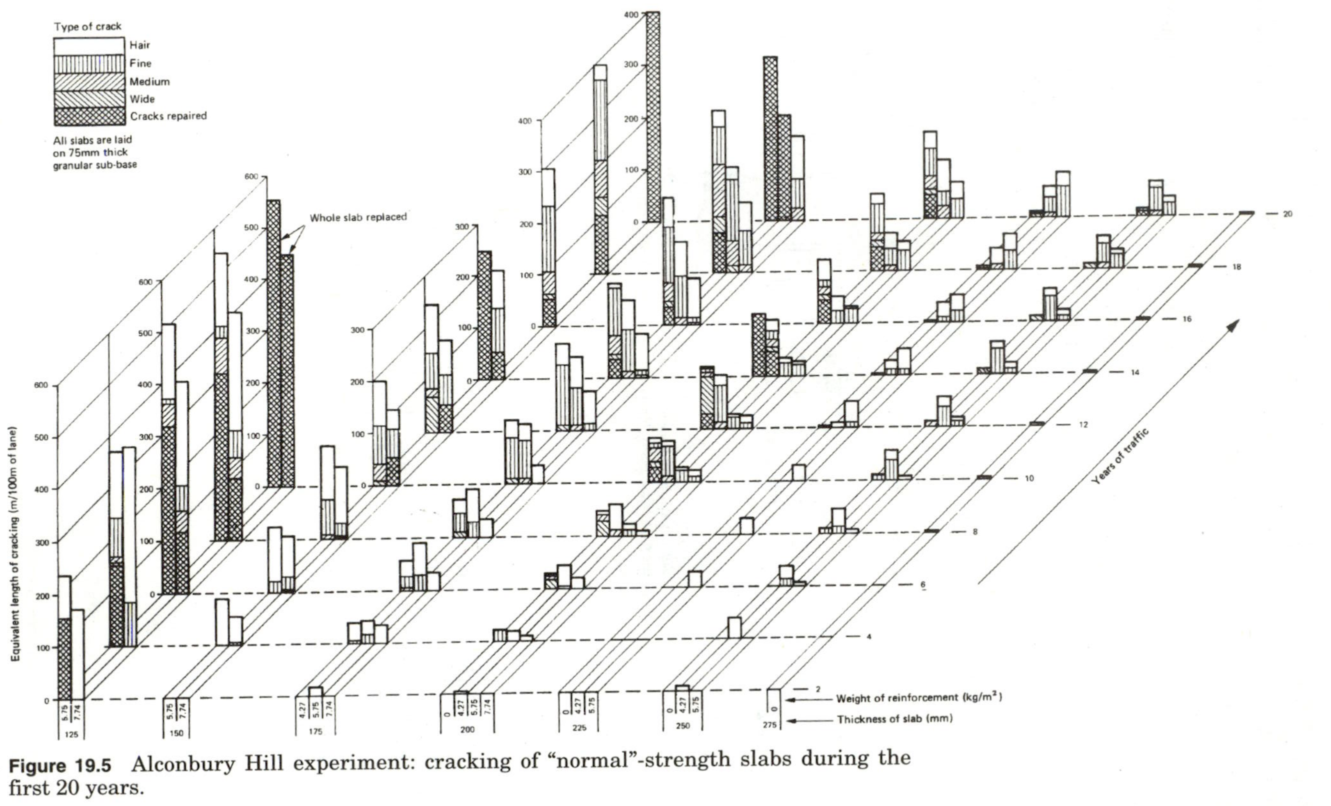

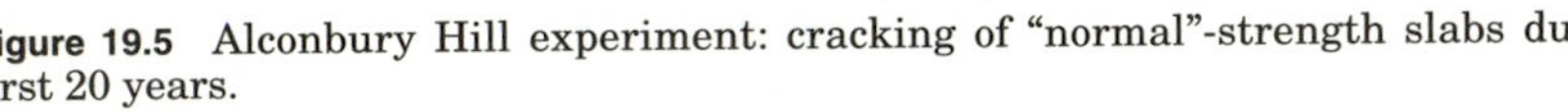

Figure 19.5 Alconbury Hill experiment: cracking of "normal"-strength slabs during the first 20 years.

and width of cracks was studied by regular visual surveys throughout the 20-year period of observations. Some of the sections are still carrying traffic after 35 years.

19.12. Figure 19.5 shows the effect on performance of the thickness of concrete and weight of reinforcement for slabs made with "normal" concrete (28-day cube strength 44.1 MN/m^2), all with a 75-mm granular subbase. To give a maintenance-free life of 20 years a minimum thickness of slab of 200 mm was required in conjunction with reinforcement of weight 5.75 kg/m^2.

19.13. Figure 19.6 compares the performance of the normal and high-strength concrete slabs and shows that 175-mm slabs of high-strength concrete gave a much better performance than 200 mm of the normal concrete. Figure 19.7 shows a marked improvement in performance when a lean concrete subbase was used in place of a granular subbase of the same thickness.

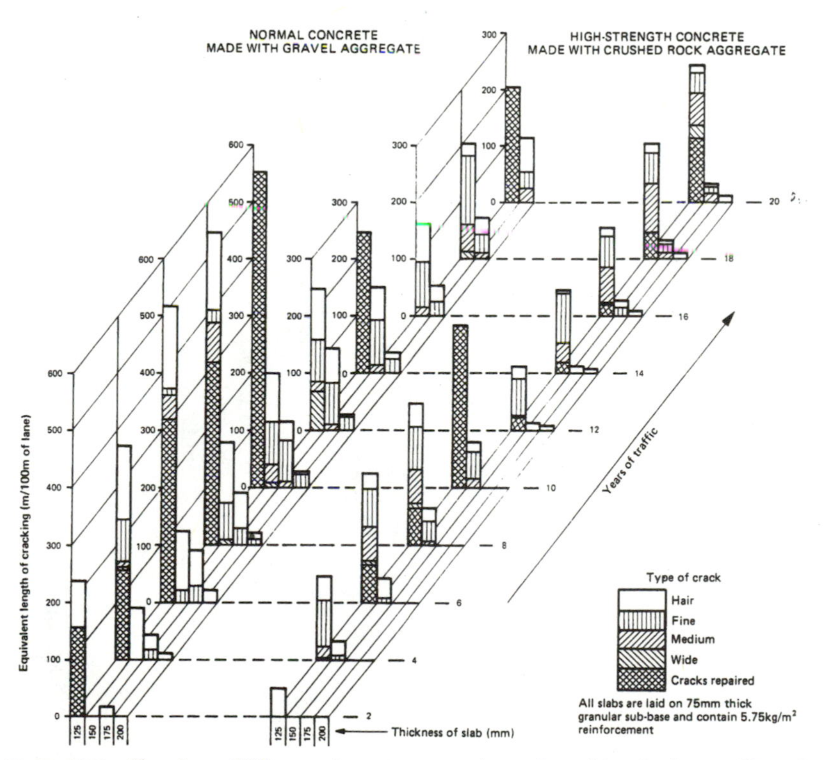

Figure 19.6 Alconbury Hill experiment: comparison of cracking in "normal"- and high-strength concrete slabs.

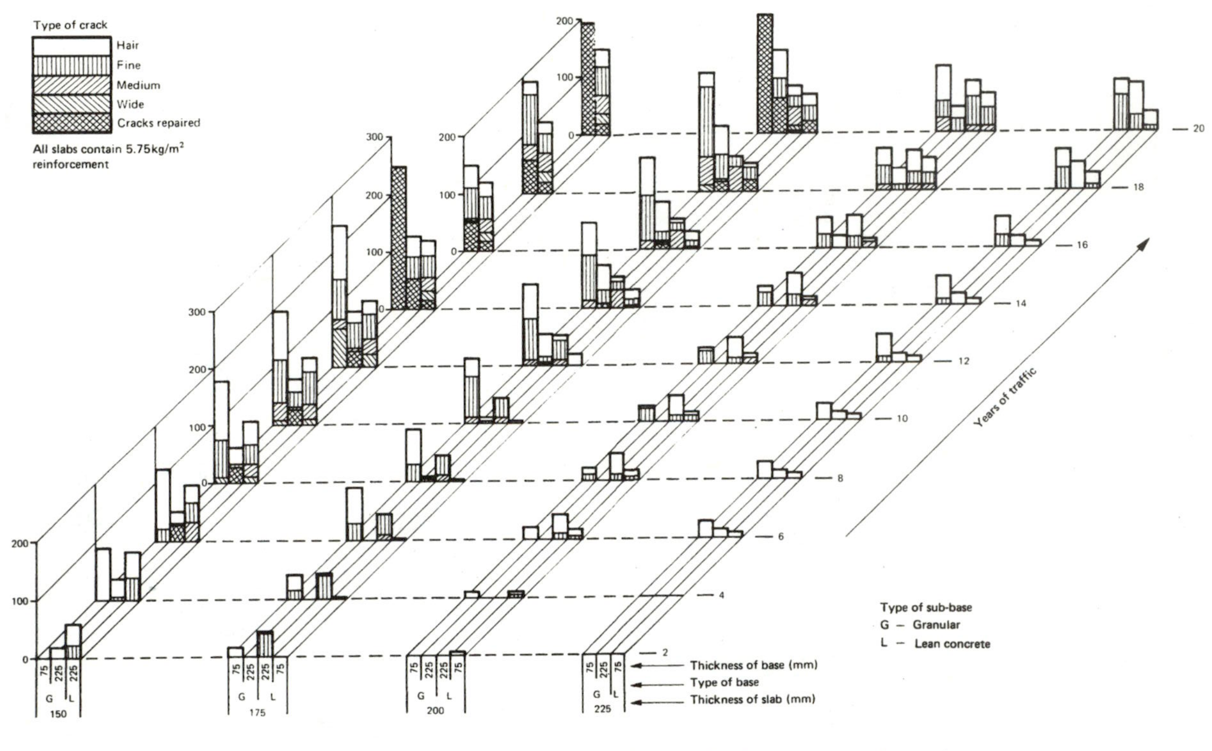

Figure 19.7 Alconbury Hill experiment: influence of type and thickness of subbase on cracking of reinforced slabs.

19.14. The lives of the various sections in years and in msa carried are given in Table 19.4. The normal-strength concrete used at Alconbury Hill (44.1 MN/m²) was stronger than the concrete laid at Oxton (37 MN/m²), and this is reflected in the longer lives at Alconbury when a comparison is made between Tables 19.3 and 19.4.

TABLE 19.4 Alconbury Hill Experiment: Lives of Concrete Sections

Section no.	Concrete strength, N/mm²	Slab thickness, mm	Reinforcement weight, kg/m²	Subbase		Life	
				Type	Thickness, mm	Years	msa
1	66.1	250	None	G*	75	40+†	40+
2	66.1	225	None	G	75	40+	40+
3	66.1	200	None	G	75	40+	40+
4	66.1	200	5.75	G	75	40+	40+
5	66.1	175	5.75	G	75	30	27
6	66.1	150	5.75	G	75	17	10
7	66.1	125	5.75	G	75	5	2.5
8	44.1	125	7.74	G	75	5	2.5
9	44.1	125	5.75	G	75	1	0.4
10	44.1	150	7.74	G	75	9	4.9
11	44.1	150	5.75	G	225	11	6.1
12	44.1	225	5.75	G	225	40+	40+
13	44.1	200	5.75	G	225	30	28
14	44.1	200	5.75	LC*	225	27	23
15	44.1	175	5.75	LC	225	22	16
16	44.1	175	5.75	G	225	18	11
17	44.1	150	5.75	G	225	8	4.5
18	44.1	150	5.75	LC	225	12	7
19	44.1	175	7.74	G	75	18	11
20	44.1	175	5.75	G	75	17	10
21	44.1	175	4.27	G	75	13	7.3
22	44.1	175	5.75	LC	75	30	28
23	44.1	200	7.74	G	75	30	28
24	44.1	200	5.75	G	75	23	18
25	44.1	200	4.27	G	75	13	7
26	44.1	200	None	G	75	8	4.3
27	44.1	200	5.75	LC	75	30	28
28	44.1	225	5.75	LC	75	40+	40+
29	44.1	225	5.75	G	75	35	35
30	44.1	225	4.27	G	75	30	28
31	44.1	225	None	G	75	24	18
32	44.1	250	5.75	G	75	30	27
33	44.1	250	4.27	G	75	30	27
34	44.1	250	None	G	75	25	20
35	44.1	275	None	G	75	40+	40+

*G = gravel; LC = lean concrete.
†40+ = probably between 40 and 100.

The Llangyfelach Experiment—1955[4]

19.15. The main purpose of this experiment was to obtain long-term information on the importance of compressive and flexural strength in concrete mixes on the performance of concrete pavements. Seven different concrete mixes A to G, and slabs made from these mixes were distributed along the ½-mile length used for the experiment, as shown in Fig. 19.8. Mixes A to C used gravel aggregate and mixes D to G crushed porphyry from North Wales. The mix proportions used are shown in the figure. The concrete was of constant thickness, 200 mm. The slabs were 37 m long and 4.6 m wide, and they were all reinforced with a weight of 3.5 kg/m^2.

19.16. The mixes as laid gave the 28-day compressive and flexural strengths shown in Table 19.5. For each slab 14 beams (50.8 × 10.2 × 10.2 cm) were cast at the time of laying; 6 were tested at 28 days and 2 each at 3 months and 1, 2, and 5 years. The tests were for flexural and compressive strength. The results of the means of these tests are included in Table 19.5, and the relationships between compressive and flexural strength for the materials made with the two aggregates are shown in Fig. 19.9.

19.17. The extent of the cracking after 12 years of traffic is shown in Fig. 19.10. It is clear that the sections with crushed-rock aggregate

To Morriston →

Slab no.	10	9	8	7	6	5	4	3	2	1
Mix	A	E	B	E	B	G	A	F	C	D
Slab no.	30	29	28	27	26	25	24	23	22	21
Mix	C	D	A	F	A	E	C	E	B	G

Slab no.	20	19	18	17	16	15	14	13	12	11
Mix	F	C	F	B	E	G	A	D	C	D
Slab no.	40	39	38	37	36	35	34	33	32	31
Mix	D	B	E	A	D	F	C	F	B	G

← To Penilergaer

	Mix	Proportions by weight
Thames Valley gravel aggregate	A	1:9.2/0.69
	B	1:7.2/0.55
	C	1:4.0/0.4
North Wales porphyry aggregate	D	1:8.7/0.66
	E	1:7.6/0.58
	F	1:6.7/0.54
	G	1:4.5/0.45

Figure 19.8 Llangyfelach experiment: details of experimental concrete mixes used.

TABLE 19.5 Compressive and Flexural Test Results for Mixes Laid at Llangyfelach

	Average strength, N/mm²									
	Flexural					Compressive				
Mix	28-day	3-month	1-year	2-year	5-year	28-day	3-month	1-year	2-year	5-year
A	2.8	3.2	3.8	4.1	4.7	27	34	40	43	45
B	3.5	4.3	4.6	4.9	5.6	39	50	57	58	62
C	4.5	4.9	5.2	5.7	6.1	55	66	69	73	71
D	3.3	3.8	4.4	4.7	4.6	29	37	44	46	49
E	3.9	4.8	5.0	5.0	5.1	37	48	54	57	57
F	4.6	5.1	5.5	5.6	5.4	45	56	65	68	67
G	5.4	5.9	6.0	6.2	6.5	57	69	78	81	79

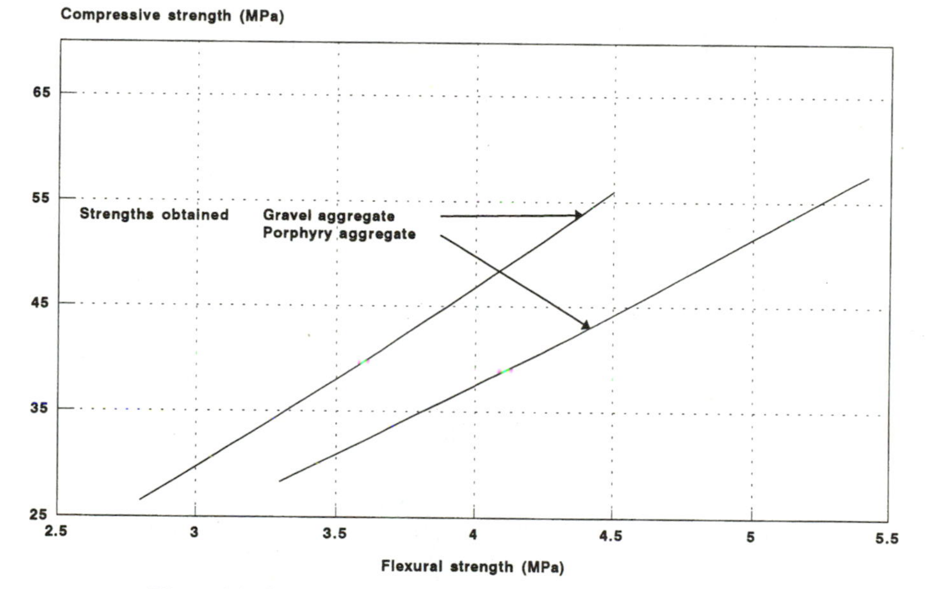

Figure 19.9 Llangyfelach experiment: relationship between compressive and flexural strengths of mixes at 28 days.

performed rather better than those using gravel aggregate, owing to their higher flexural strength. The road carried less than 2.5 msa during the first 12 years. The experiment would have been much more useful had it been constructed on a more heavily trafficked major road. The Department of Transport in the United Kingdom has always been reluctant to allow pavement design experiments on freeways and industrial trunk roads. This policy has undoubtedly con-

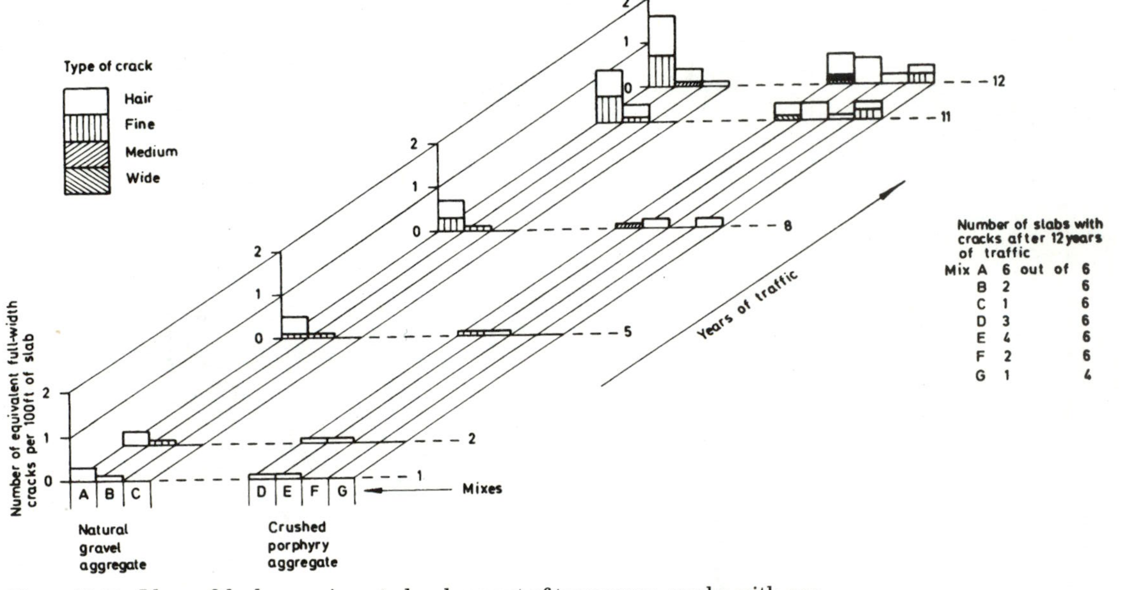

Figure 19.10 Llangyfelach experiment: development of transverse cracks with age.

tributed to expensive freeway pavement failures. In the United States each state has its own highway department and experimentation appears less stifled.

The Longford to Stanwell Road—1948[5]

19.18. The objective of this experiment was to study seasonal movements at joints in concrete roads over a long period of time. The pavement had a uniform thickness of 200 mm and a compressive strength of 40 N/mm^2. The unreinforced concrete was laid on deep gravel fill and the length of the slabs between expansion joints varied between 37 and 690 m, as shown in Fig. 19.11. Contraction joints were spaced at 4.6 m intervals. The road was initially lightly trafficked, but as it forms a new route around the western extremity of Heathrow Airport the traffic has increased steadily and now includes more than 3000 heavy commercial vehicles per day. Nevertheless after 40 years less than 2 percent of the 850 bays (between contraction joints) show any cracking, and such cracks generally originate at drainage inlets.

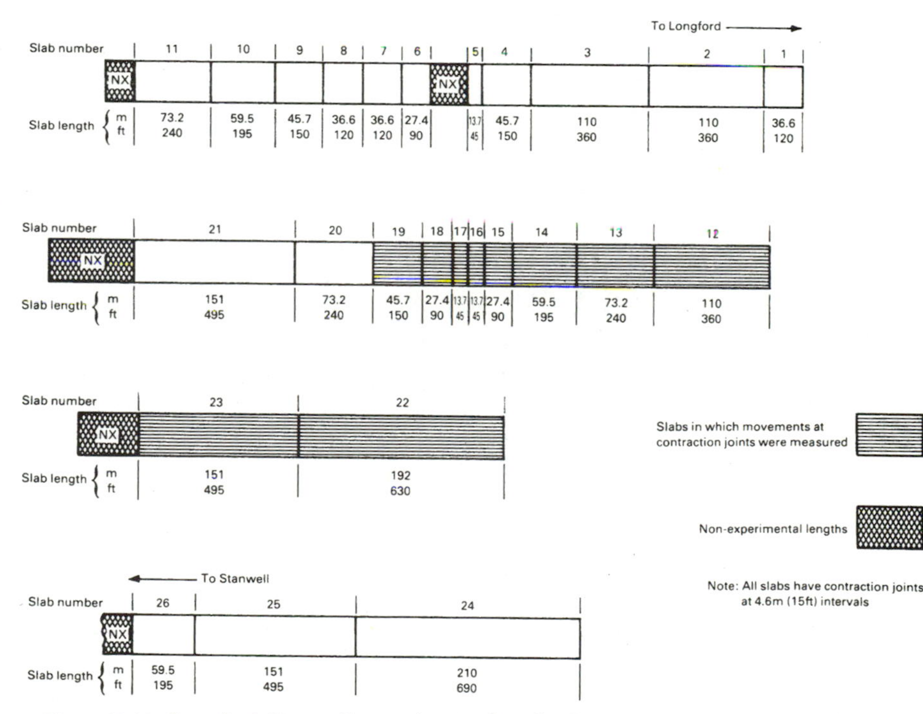

Figure 19.11 Longford-Stanwell experiment: details of experimental slabs.

19.19. Joint movements were measured between brass studs set in the concrete 150 mm on either side of each joint. Figure 19.12 shows reversible joint movements measured during diurnal summer temperature changes approximately 7 years after the road was opened to

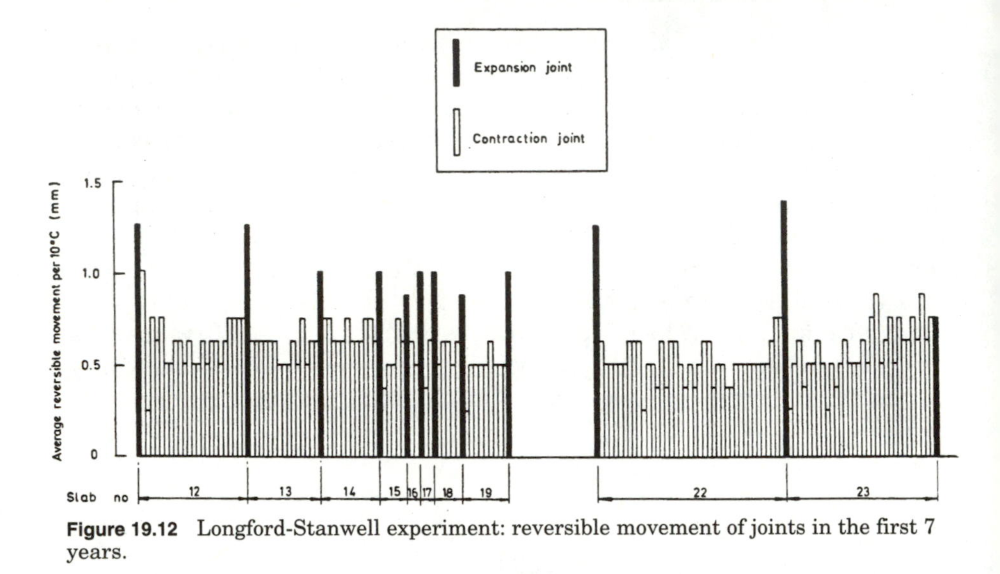

Figure 19.12 Longford-Stanwell experiment: reversible movement of joints in the first 7 years.

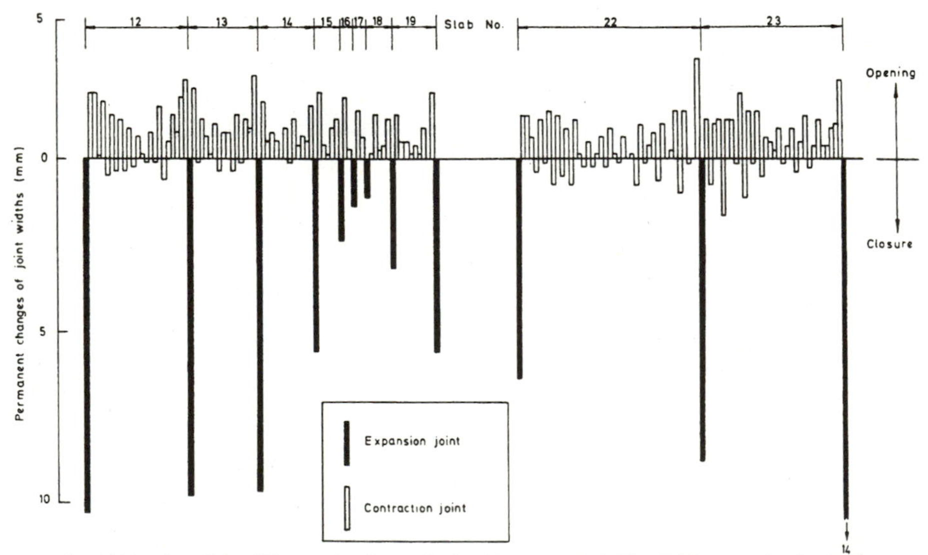

Figure 19.13 Longford-Stanwell experiment: permanent changes in the widths of joints after 7 years.

traffic. Permanent joint movements made at the same time are shown in Fig. 19.13. During the 7 years expansion joints closed progressively and the contraction joints opened to give no significant change in total joint space. Observations of the condition of the road continued for about 35 years, when a thin bituminous surfacing was added to restore skid resistance.

The Grantham Bypass Experiment—1962[6]

19.20. This was not a structural design experiment of the type constructed at Alconbury Hill, referred to above. The work was carried out as a normal road construction contract, but some design variations were introduced. The main objective was to study the relationship between slab length and weight of reinforcement for reinforced concrete construction. Secondary objectives were to compare the performance of concrete pavements reinforced with (1) mild steel reinforcement mats, (2) hard-drawn steel mats, and (3) ribbed wire mats all of the same weight.

19.21. Figure 19.14 shows the relationship which was established between weight of reinforcement (using hard-drawn steel wire rein-

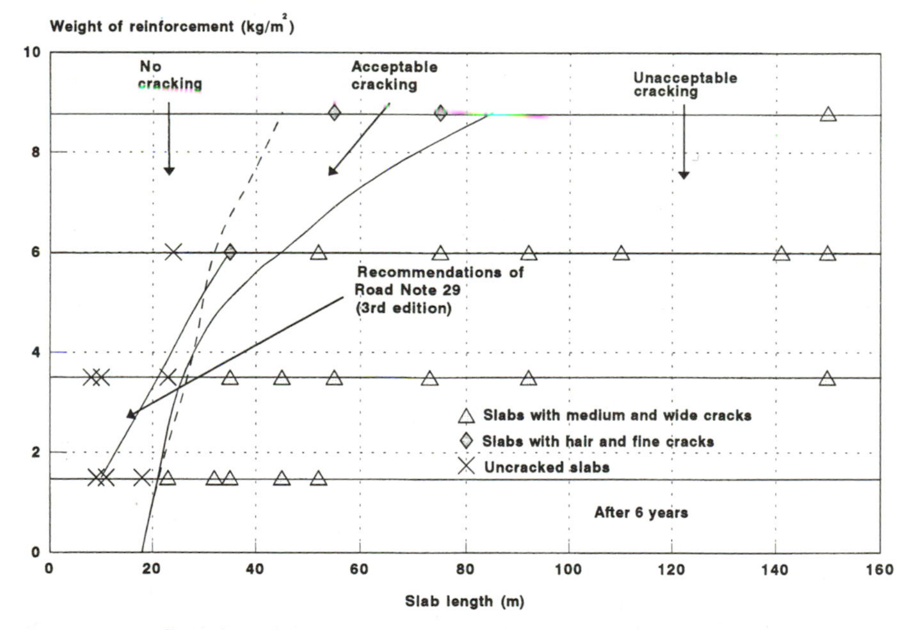

Figure 19.14 Grantham bypass experiment: cracking in slabs reinforced with plain round hard-drawn steel wire fabric.

forcement) and slab length for different degrees of cracking. These results relate to the condition of the pavement after 6 years of traffic and approximately 9 msa. Design would normally be based on the full line separating the acceptable and unacceptable cracking zones. Current recommendations in the United Kingdom are based on the rather more conservative relationship, included in the U.K. design manual, Road Note 29. This relationship is included in Fig. 19.14.

19.22. The part of the experiment in which hard-drawn and mild steel reinforcement mats, both with a weight of 5.84 kg/m^2, were compared showed after 6 years a very marked superiority in the performance of the hard-drawn mats, incorporated in slabs of length between 55 and 275 m.

19.23. There was a comparatively small difference in performance between reinforcement mats made with ribbed wire and hard-drawn steel.

19.24. This experiment also included 7 km of experimental continuously reinforced concrete construction. Two weights of reinforcement, 5.84 kg/m^2 and 8.87 kg/m^2, were used. Pretraffic cracking increased markedly with the mean slab temperature at the time of placement. The first wide cracks appeared at, or close to, construction joints as shown in Fig. 19.15. This may have been due to weak reinforcement bonds at these points or to delays during construction. After repair, further wide cracks tended to develop in the same areas over a period of 1 to 3 years. There was little difference in performance which could be related to the weight of the reinforcement. This may mean that the weights used were insufficient.

19.25. The main lesson learned from these experiments with continuously reinforced concrete construction was the importance of adequate preparation and the development of constructional skills beyond those required for jointed reinforced or unreinforced concrete. In 1974, the Department of Transport in the United Kingdom set up a small committee to examine the future for this form of construction in Britain. The conclusion was that it was unlikely to present a viable alternative to normal concrete or flexible construction. However, since 1975, certain small lengths of freeway have been constructed using continuous reinforcement. No results have been published to date.

19.26. From recent published statements it appears that the U.K. Department of Transport is currently considering continuously reinforced concrete as a very long-lasting base which could be surfaced

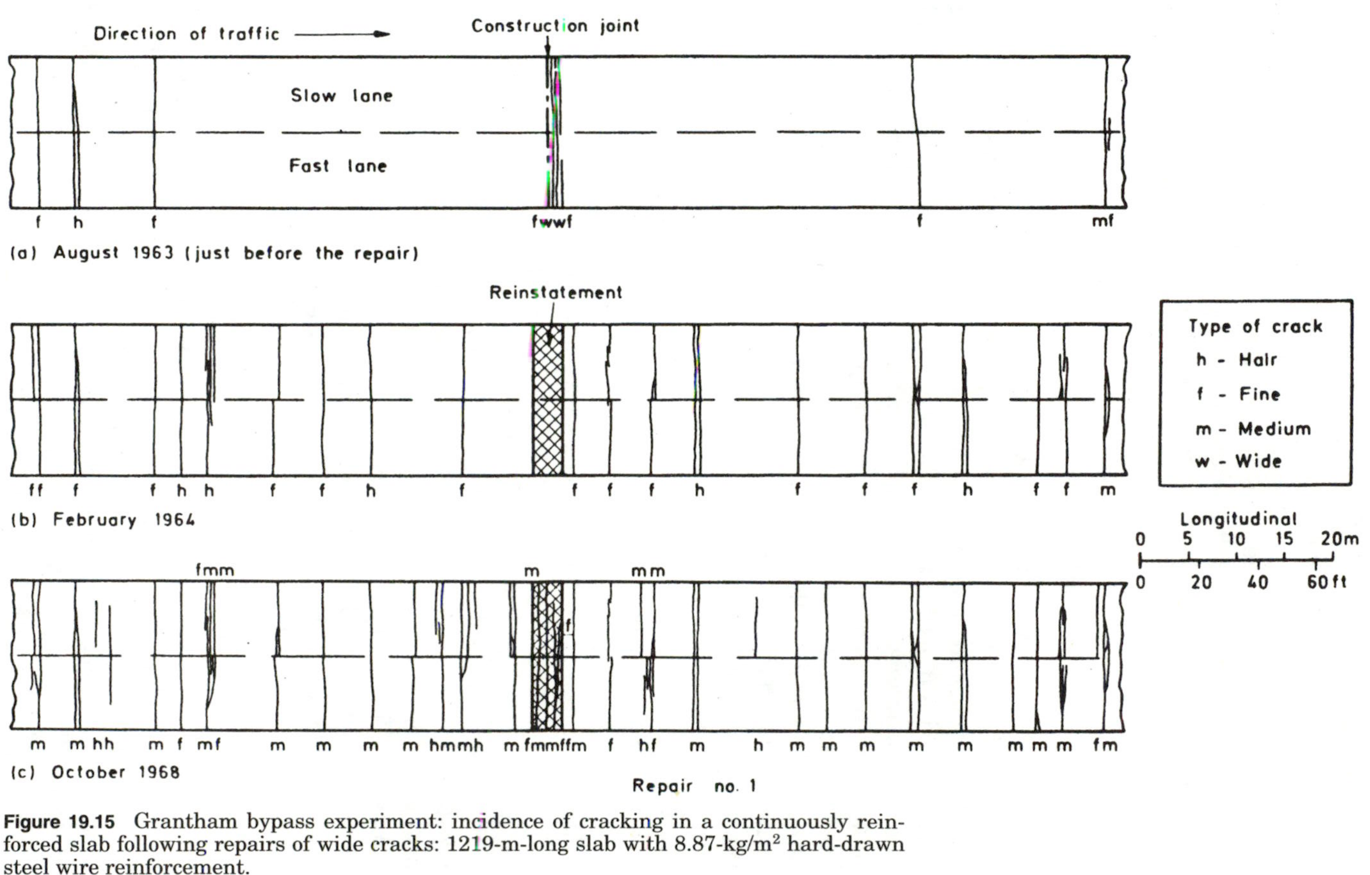

Figure 19.15 Grantham bypass experiment: incidence of cracking in a continuously reinforced slab following repairs of wide cracks: 1219-m-long slab with 8.87-kg/m² hard-drawn steel wire reinforcement.

with a thin easily replaceable asphalt wearing course, as suitable construction for new freeways. The Grantham experience illustrated in Fig. 19.15 suggests that within 5 years, medium-width cracks would appear at intervals of 3 or 4 m. It is doubtful if such cracks could be contained by a surfacing of asphalt as thin as 50 mm, and the engineer would need to think in terms of a two-course asphalt surfacing at least 100 mm thick.

The Winthorpe Experiment—1961[7]

19.27. This was an early attempt to compare the performance of thin (125 to 175 mm) unreinforced concrete slabs in a prestressed condition, with adjacent reinforced slabs of greater thickness. Prestressing was carried out using piston jacks each of 125 mm diameter and placed approximately 150 mm apart. These jacks were placed and removed in blocks of four. When the required degree of prestress had been achieved, the blocks of four jacks were removed and replaced by adjustable spacers, the degree of prestress being gradually transferred to the spacers from the jacks in this manner. After all the jacks were removed the joint was filled with concrete. The need to break out the blocking and to replace the jacks whenever an adjustment of the prestress level was necessary made stress changes a laborious and somewhat hazardous procedure.

19.28. Setting the level of prestress was complicated by the effects of temperature and creep. The prestress/temperature ratio was found to be approximately 450 kN/m^2 per °C. The minimum prestress at −10°C was designed to be 700 kN/m^2, which meant that at 10°C the prestress level needed to be set at 9 MN/m^2 (allowing 1.4 MN/m^2 to overcome frictional restraint between the slab and the lean concrete subbase). With such a level of prestress, considerable creep occurred in the concrete, and during the 5 years that the experiment was under observation the 125-mm slab decreased in length by 230 mm. To allow for creep, it was necessary to restress the slabs annually, generally in the later autumn.

19.29. Early in July 1968 there was an abnormally sudden increase in temperature at the site and the mean slab temperature rose to a maximum of 35°C. When the maximum compressive stress rose to 24 MN/m^2, the major blow-up of the 125-mm slab shown in Fig. 19.16 occurred. No damage was sustained by the 175-mm slab, but this was immediately destressed and the experiment terminated.

Figure 19.16 Blowup of 125-mm prestressed concrete pavement at Winthorpe, 1968.

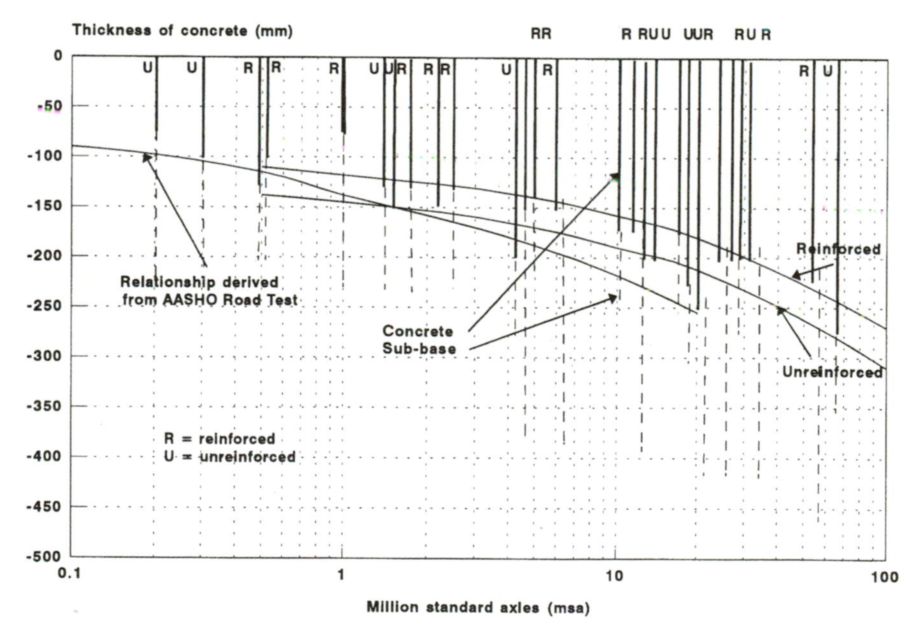

Figure 19.17 Thickness design curves for concrete roads based on experimental evidence.

19.30. The conclusion reached from this experiment was that any savings on slab thickness would be absorbed many times by the costs involved in any form of prestressing.

Thickness Design for Concrete Pavements Based on Experimental Evidence

19.31. Figure 19.17 shows the thickness design requirements for concrete pavements based on evidence from full-scale experiments. The curves relate to "normal" strength concrete with a compressive strength at 28 days of 35 to 45 N/mm^2. There is a clearly defined need for a greater thickness for unreinforced concrete, the difference being about 25 mm irrespective of the cumulative traffic. The figure also includes the relationship between thickness and standard axles deduced from the results of the AASHO road test.

References

1. Gregory, J. M.: An Experimental Concrete Road 38 Years Old: Condition and Performance of Sections of A.316, Great Chertsey Road, *Transport and Road Research Laboratory Report* LR 317, TRL, Crowthorne, 1972.
2. Loe, J. A.: The Performance during the First Five Years of the Experimental Concrete Road at Oxton, Northamptonshire, *Proc. Inst. Civ. Engrs.,* vol. 4, no. 1, part 2, pp. 137–166, 1955.
3. Croney, D., and J. A. Loe: Full-Scale Pavement Design Experiment on A.1 at Alconbury Hill, *Proc. Inst. Civ. Engrs.,* vol. 30 (February), pp. 225–270, 1965.
4. Nowak, J. R.: Influence of Strength on the Performance of Concrete Slabs. Report on the Performance after 12 Years of the Experimental Road at Llangyfelach, *Road Research Laboratory Report* LR 199, TRR66L, Crowthorne, 1968.
5. Nowak, J. R., and J. Gaunt: The Full-Scale Unreinforced Concrete Experiment on Longford-Stanwell Road, B. 379. Performance during the First 20 Years, *Road Research Laboratory Report* LR 349, TRRL, Crowthorne, 1970.
6. Nowak, J. R.: The Full-Scale Reinforced Concrete Experiment on the Grantham Bypass. Performance during the First Six Years. *Road Research Laboratory Report* LR 345, TRRL, Crowthorne, 1970.
7. Kidd, R. A., and J. P. Stott: The Construction of an Experimental Prestressed Concrete Road at Winthorpe, Nottinghamshire, *Proc. Inst. Civ. Engrs.,* vol. 36, pp. 473–498, 1967.

Chapter

20

Pavement Deflection and the Life of Asphaltic Pavements

Introduction

20.1. Flexure under passage of heavy axle loads is one of the principal causes of deterioration in asphaltic road pavements. It is reasonable, therefore, to suppose that the structural life of such pavements can be related empirically to the change of deflection with time and traffic carried. This chapter discusses information on such relationships from a 40-year study made on full-scale road experiments in the United Kingdom.

Measurement of the Deflection of Flexible Pavements

20.2. The most direct and reliable method of measuring the elastic deflection of a pavement as a heavy wheel load passes over it is that based on the Benkelman beam. This simple method was developed by A. C. Benkelman in the United States when he was Highway Research Engineer to the Bureau of Public Roads, Washington, D.C.

20.3. The version of the Benkelman beam used in the United Kingdom since the late 1940s is slightly modified to suit the trucks in common use in Britain at that time. The modifications relate principally to the width of the beam. The dimensions of the British deflection beam are shown in Fig. 20.1. Figure 20.2 is a photograph of the beam.

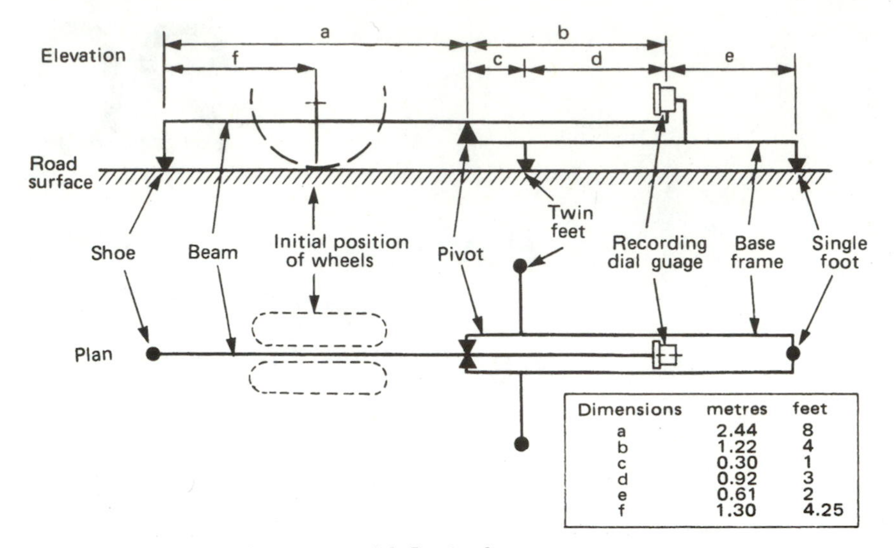

Figure 20.1 Principal dimensions of deflection beam.

20.4. Deflection measurements are normally made in the nearside wheel tracks of the road 0.9 to 1.2 m from the verge, but if two beams are available simultaneous measurements can be made in both wheel tracks. In making a measurement, a transverse line is drawn on the pavement 1.3 m behind the point at which a measurement is

Figure 20.2 The British deflection beam.

required. The truck is positioned parallel with the verge with its front wheels pointing straight ahead and the point of contact of the rear wheel directly over the line. The transverse positioning is such that when the vehicle is driven forward the gap in the rear wheel assembly will pass over the point of measurement. At a signal from the operator, the vehicle is driven forward at creep speed to a point where the rear wheels are at least 3 m beyond the test point. The maximum reading of the dial gauge is noted, together with the final reading after the rear axle has reached a point 3 m beyond the test point. The magnitude of the pavement deflection is obtained by adding the maximum reading to the difference between the maximum reading and the final reading. (This sum is not averaged because of the 2:1 length ratio of the beam arms.)

20.5. Although considerable lengths of road have been surveyed by the deflection beam, the process is slow and not conducive to close coverage. Various attempts have been made to mechanize the process while retaining the same principle. A particularly successful machine of this type called the Lacroix Deflectograph was developed in the middle 1960s by the Laboratoire Central des Ponts et Chaussees in France. A number of such machines now operate in the United Kingdom. The beam assembly (Fig. 20.3) is slung beneath the truck (Fig. 20.4) and is kept in position by guides. Between points of measurement the beam assembly is pulled forward at twice the speed of the vehicle; it then comes to rest while the rear wheels of the truck

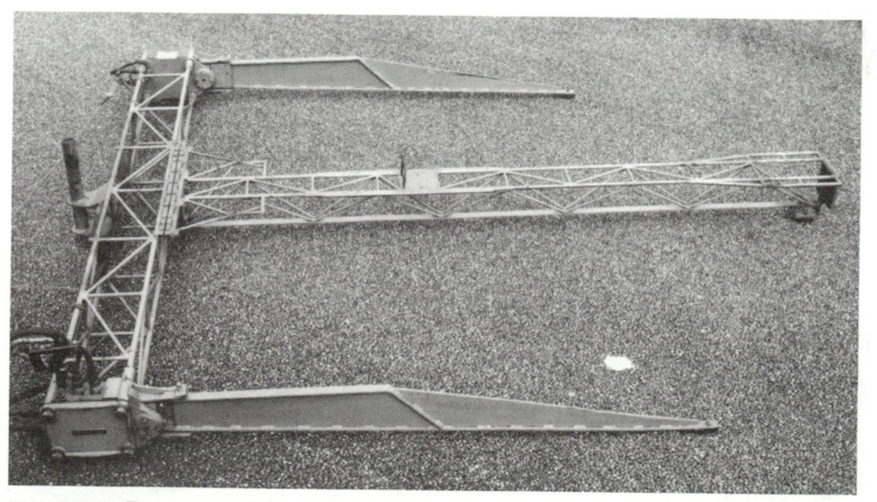

Figure 20.3 Beam assembly—deflectograph.

Figure 20.4 Lacroix Deflectograph.

advance to the point of measurement where the deflection is measured electronically and fed into the on-board computer.

20.6. Figure 20.5 shows typical tire print outlines for the rear axle twin wheel assemblies used for deflection beam and deflectograph surveys. Although the rear-axle load used for both test procedures is the same (6350 kg), the tire print areas are considerably greater for the deflectograph. This is associated mainly with the type of tires used and not with the tire pressure.

20.7. Correlation between deflections measured by the two methods can be established experimentally. Figure 20.6 shows the mean relationship obtained on a variety of pavements with different base materials. Correlation can also be obtained using analytical analyses, as discussed in Chap. 23. The experimental work described in this chapter was carried out with the deflection beam.

Deflection Studies on Full-Scale Pavement Design Experiments

20.8. Deflection studies were made on all the full-scale pavement design experiments using flexible construction already discussed in Chap. 18. On each of the sections, the deflection beam was used to monitor transient deflections under a 14,000-lb (6350-kg) axle load. Measurements in the nearside and offside wheel tracks were made on

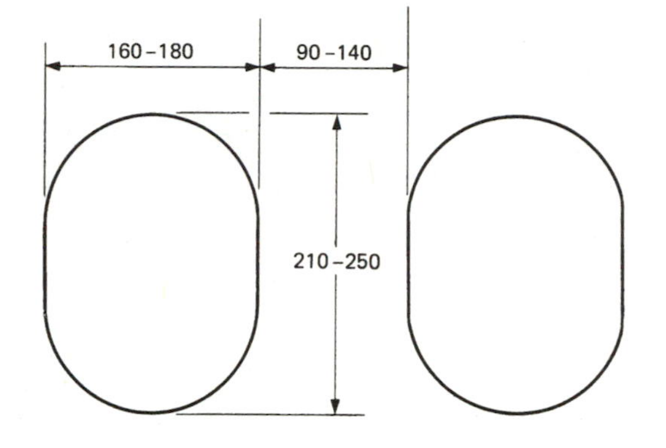

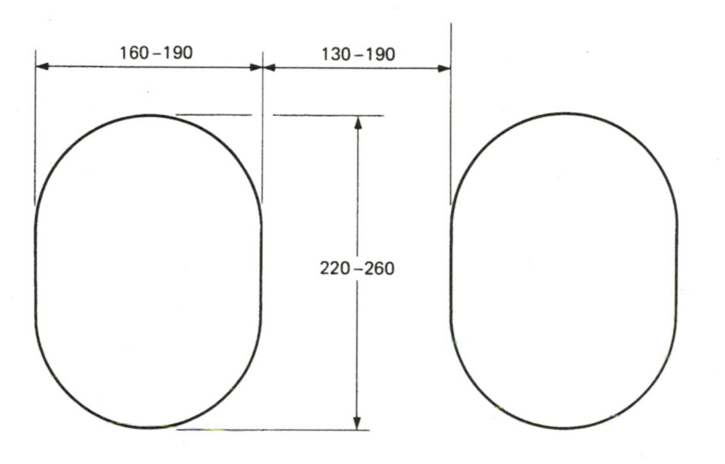

Figure 20.5 Range of tire contact dimensions of the deflection beam and deflectograph.

the slow traffic lane, at five points equally spaced along the section. Paint marks on the edge channel identified the points of measurement.

20.9. The first major experiment included in this research was that at Alconbury Hill constructed in 1957 (see Chap. 18, Pars. 18.10 to 18.16). This experiment was intentionally underdesigned to give an early comparison of the performance of various types of road base material, i.e., a similar objective to that of the AASHO Road Test in progress at much the same time. The underdesign was chiefly

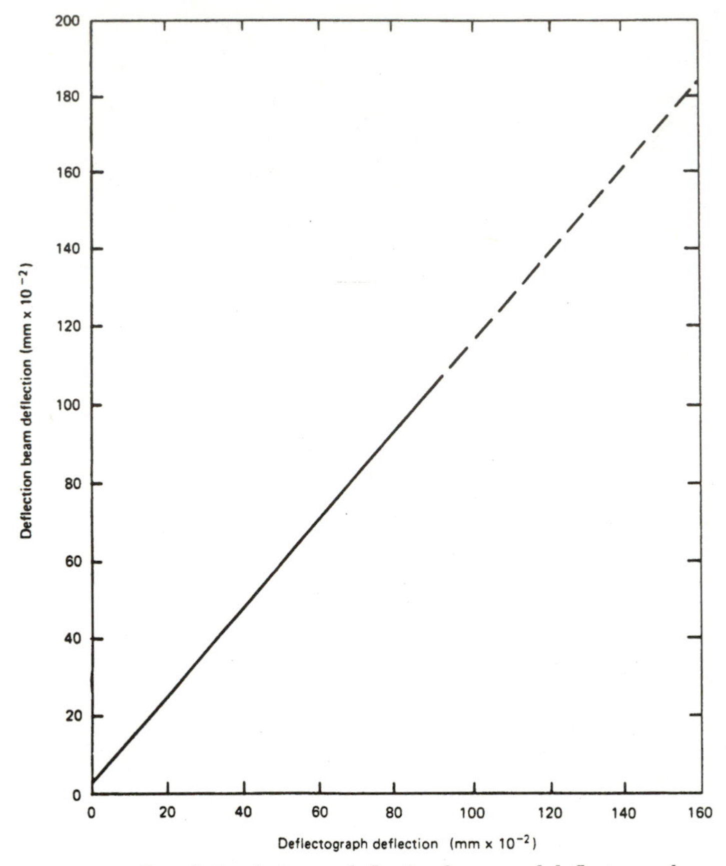

Figure 20.6 Correlation between deflection beam and deflectograph.

achieved by the use of a weak single-size sand subbase in conjunction with a high water table.

20.10. It was the measurements made at this site that provided the bulk of the information used to develop the deflection/standard axle contours included in various TRL publications.[1,2] These contours for the various base materials used are shown in Figs. 20.7 to 20.10. They have also been reproduced in the AASHTO *Guide for Design of Pavement Structures.*

20.11. Based on the early results from the Alconbury Hill experiment, relating deflection with the number of standard axles carried, it was assumed that the deflection of a pavement would inevitably increase under the action of traffic, and this is reflected in the shape

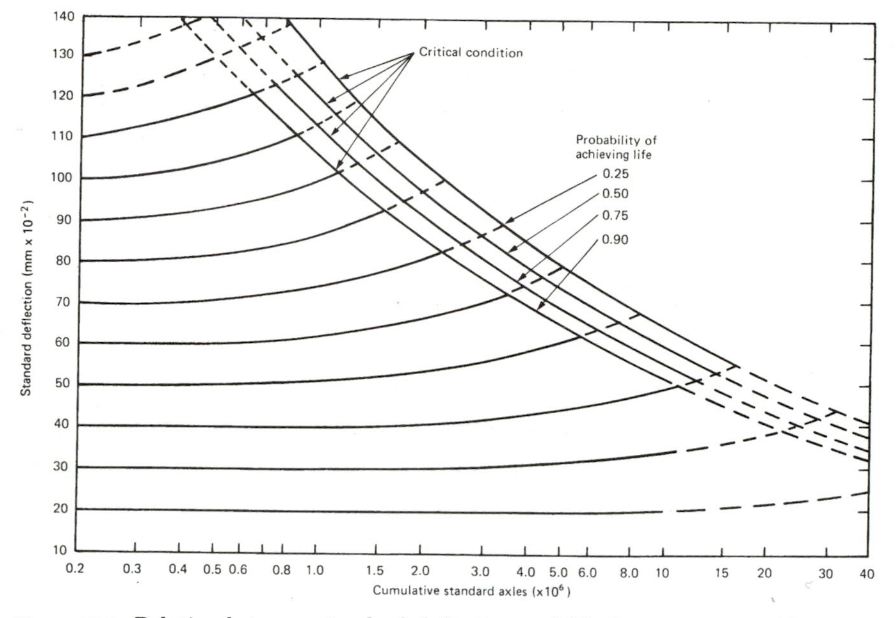

Figure 20.7 Relation between standard deflection and life for pavements with noncementing granular road bases.

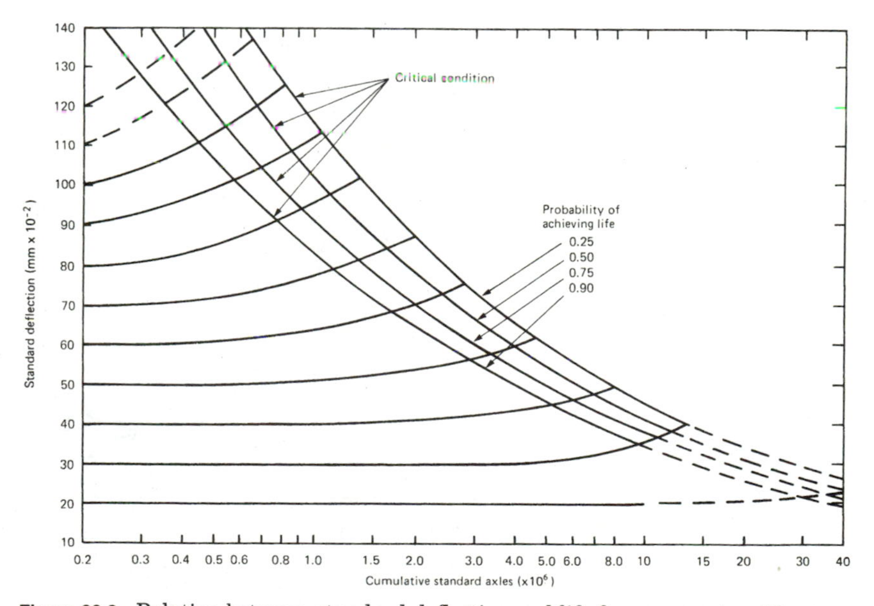

Figure 20.8 Relation between standard deflection and life for pavements with granular road bases whose aggregates exhibit a natural cementing action.

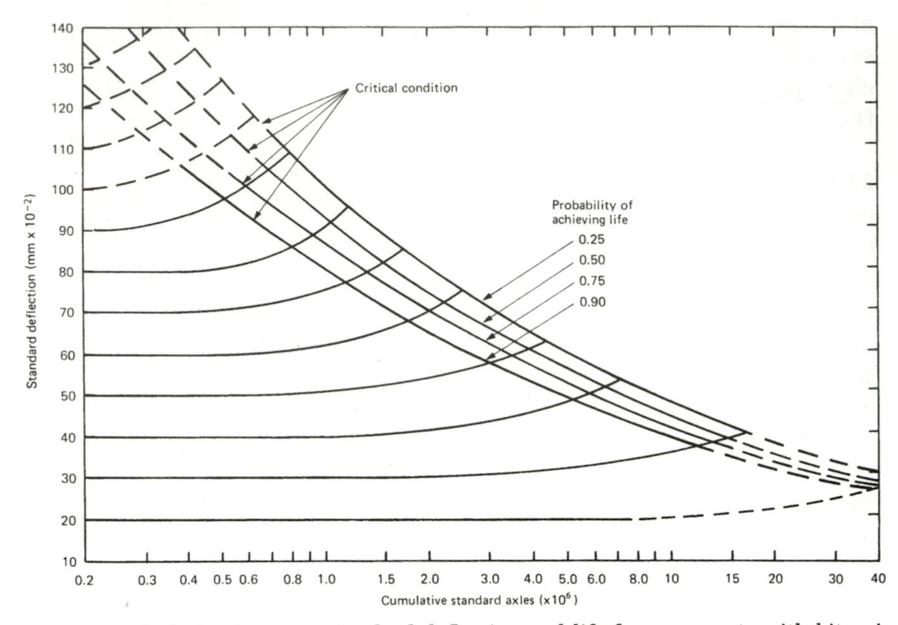

Figure 20.9 Relation between standard deflection and life for pavements with bituminous road bases.

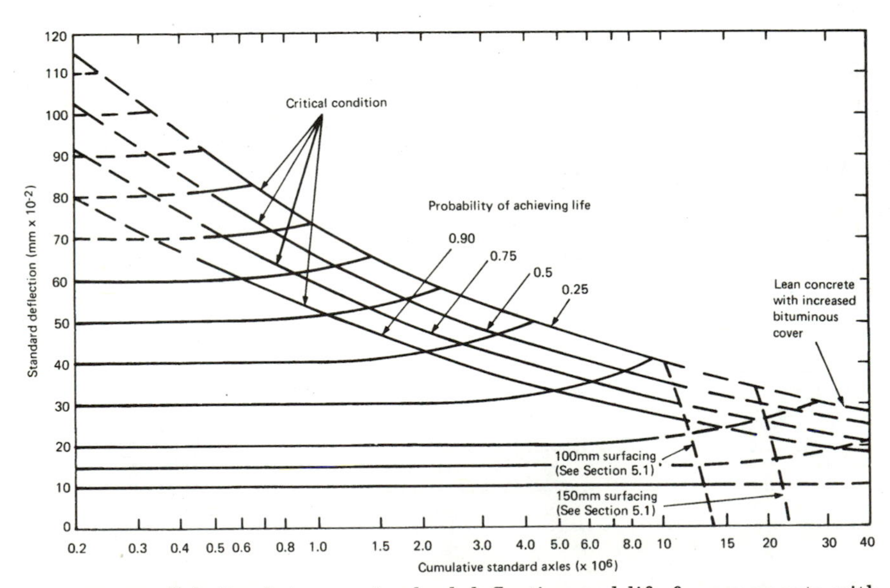

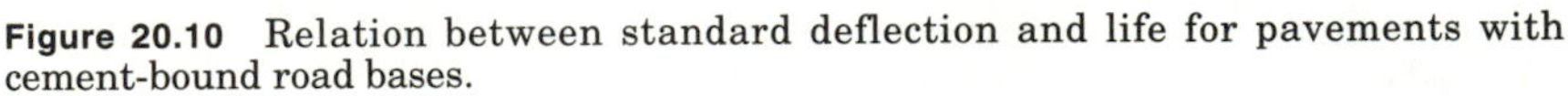

Figure 20.10 Relation between standard deflection and life for pavements with cement-bound road bases.

of the deflection contours shown in Figs. 20.7 to 20.10. For short-life sections, this may well have been a valid assumption, but it is now clear that for well-designed asphaltic pavements it is very far from true.

20.12. For such pavements designed to have a life of 20 years or more, two processes affect the deflection. The first is hardening of the bituminous elements. In the case of untrafficked pavements this will cause a continuous reduction of deflection for at least 10 years. Under traffic the development of microcracks in the structure of the bituminous materials will, on the other hand, tend to increase the deflection, and whether in practice an increase of deflection is recorded with time depends on the relative importance of the two processes in the particular pavement under consideration.

20.13. It is much to be regretted that no results have been published in connection with the vast amount of deflection data obtained from the later pavement design experiments discussed in Chap. 18. The present authors have, however, been given access to the deflection measurements made in connection with the Conington experiment (see Pars. 18.38 to 18.42) and with the Nately Scures experiment (see Pars. 18.17 to 18.19). They also have a limited amount of information relating to deflections measured at the Alconbury bypass and Wheatley bypass experiments (see Pars. 18.20 to 18.37). The information obtained from these sources is discussed below.

20.14. The Conington experiment. The layout of the sections at this site has been shown in Fig. 18.8. Sections 3 to 14, all with 200-mm wet-mix bases and a 100-mm two-course asphalt surfacing, represent a type of construction suitable for heavily trafficked trunk roads. Sections 15 to 26 with a similar two-course asphalt surfacing on a 150-mm asphaltic base would be suitable for a heavily trafficked freeway. Both types of construction would be expected to give a structural life of 20 years before resurfacing was necessary assuming crushed-rock aggregate was used in the bituminous materials. The object of the experiment was to find out how the performance was affected when various types of gravel aggregate were substituted for crushed rock in the binder course of sections 3 to 14 and in the road base of sections 15 to 26. The conclusions relating to these objectives have been given in Chap. 18. Interest here centers on the deflection measurements made at the site.

20.15. Figure 20.11 shows typical deflection histories for three of the sections with wet-mix bases, and Fig. 20.12 shows similar informa-

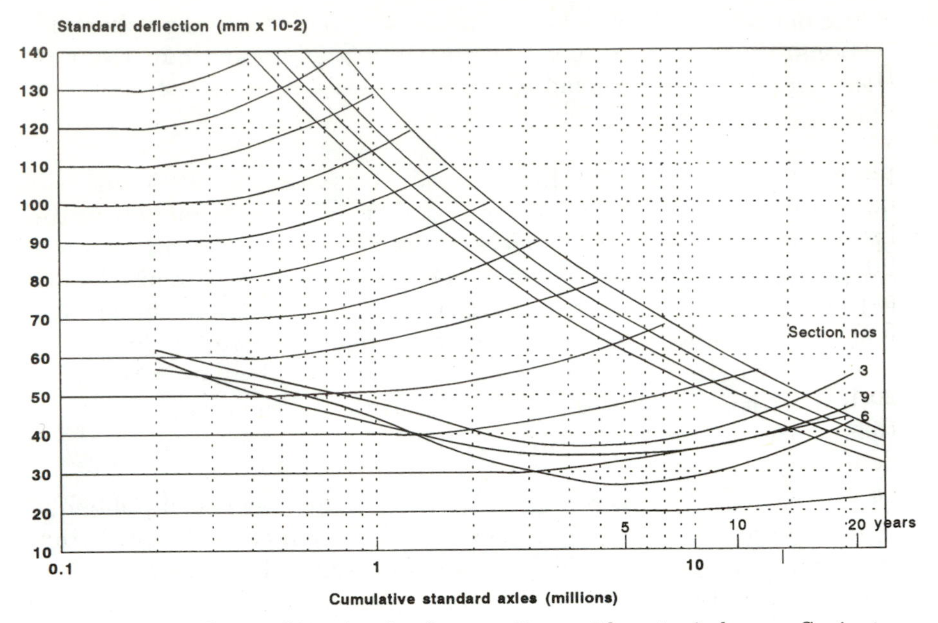

Figure 20.11 Deflection histories for three sections with wet-mix bases—Conington experiment (see Fig. 18.9).

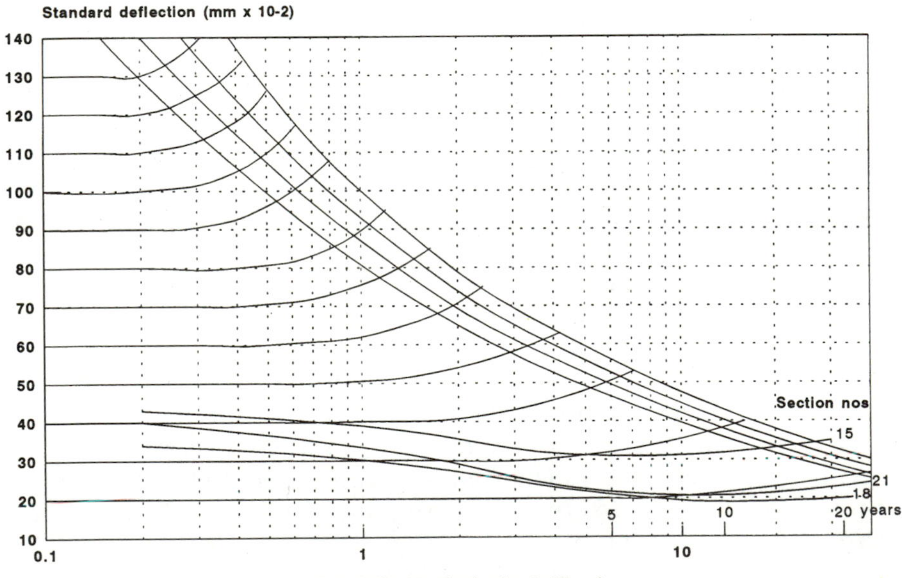

Figure 20.12 Deflection histories for three sections with asphaltic bases—Conington experiment (see Fig. 18.9).

tion for three sections with asphalt bases. For both types of base, hardening of the bitumen is the major influence on deflection for the first 5 years. The deflection then increases as the pavements approach the failure condition. The observed deflection histories are very different from the contours shown in the diagrams. As an example, the average early-life deflection of the sections with wet-mix bases (Fig. 20.11) is 60 mm $\times 10^{-2}$; this would correspond to a future life of about 3 years, whereas the actual future life was observed to be 20 years. For the sections with bituminous bases (Fig. 20.12) the average early-life deflection of 40×10^{-2} mm indicates a future life of 5 years compared with the actual life of more than 20 years.

20.16. The Nately Scures experiment. The layout of the sections at this site is given in Fig. 18.4, and their performance (either actual or deduced from the rate of deformation and cracking) is given in Table 18.4. This experiment is particularly important because all heavy commercial traffic ceased early in 1972 when the adjacent M3 freeway was opened. After that time only local traffic consisting mainly of private cars was carried.

20.17. Figure 20.13 shows the deflection histories of three sections (10, 11, and 12) having bituminous bases of thicknesses 75, 150, and

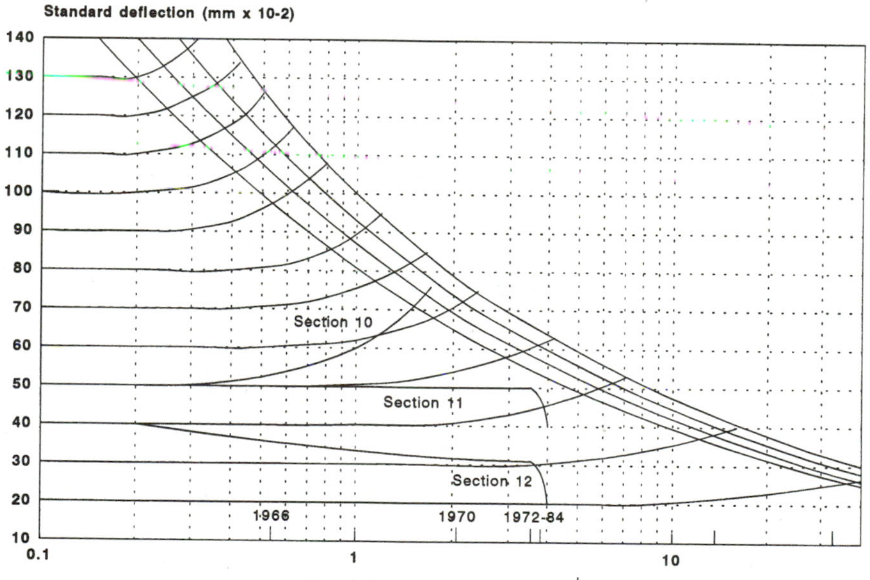

Figure 20.13 Deflection histories for three sections with asphaltic bases—Nately Scures experiment (see Fig. 18.4).

225 mm under an asphalt surfacing 100 mm thick. The thinnest section (section 10) increased in deflection rapidly until failure occurred after about 4 years. The increase was more rapid than the contours suggested, leading to an earlier failure than would be expected from the appropriate chart. In the section of intermediate thickness (section 11) the process of hardening of the bitumen and traffic deterioration were approximately matched and there was virtually no change of measured deflection when the experiment was carrying mixed traffic (1964–1971). The deflection of the thickest section (section 12) decreased over the same period owing to the increased importance of bitumen hardening.

20.18. Deflection measurements at the site were continued for a further 12 years after the commercial traffic had ceased. The measurements on sections 11 and 12 show that over this period the deflection fell markedly. This indicates that the hardening process continued throughout the 20-year life of the experiment.

20.19. Figure 20.14 shows deflection histories for sections 1, 2, and 3 of the Nately Scures experiment. These sections had wet-mix bases of thicknesses 300, 225, and 150 mm, respectively. The deflection of sec-

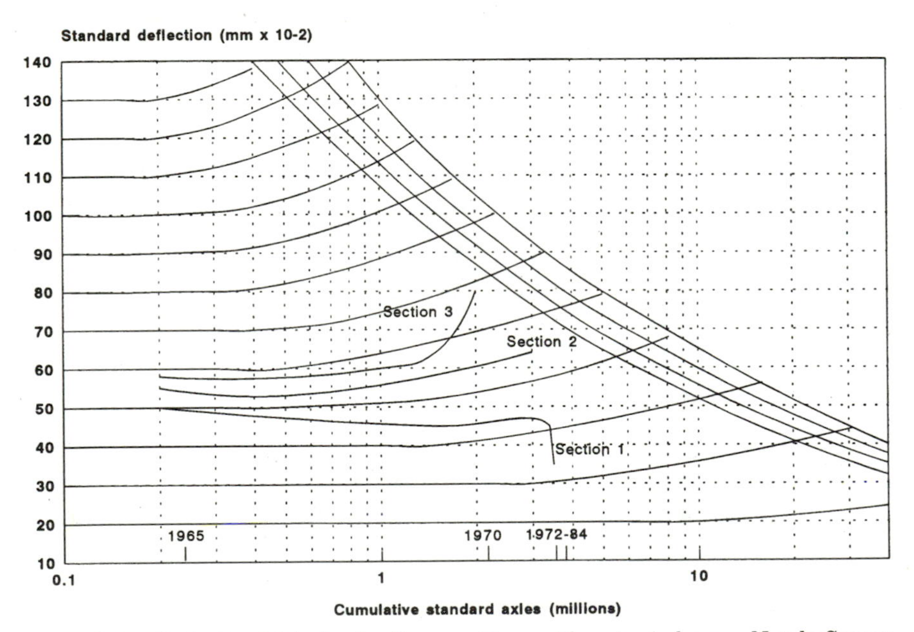

Figure 20.14 Deflection histories for three sections with wet-mix bases—Nately Scures experiment (see Fig. 18.4).

tion 3 increased rapidly, leading to failure after 5 years. Section 2 showed cracking and some deformation after 6 years, and it was resurfaced with section 3 at that time. The deflection of section 1 decreased progressively with time until the heavy traffic ceased. The deflection subsequently decreased with time owing to the hardening of the bitumen, as was the case with sections 11 and 12 already discussed.

20.20. The Alconbury bypass experiment. This large experiment has been discussed in Chap. 18 and the layout and constitution of the experimental sections are given in Fig. 18.7.

20.21. Deflection histories of sections 52, 53, and 1 at Alconbury bypass are shown in Fig. 20.15. These sections had bitumen-bound bases 150, 200, and 250 mm thick under a 100-mm asphalt surfacing. In all three cases the deflection decreased as the bitumen hardened and there was no evidence of an upturn of the deflection during the period of the observations.

20.22. Figure 20.16 shows deflection histories for sections 26 and 27 at Alconbury bypass; these have wet-mix bases 200 mm thick and surfacings of 150 and 100 mm of asphalt, respectively. The deflections in both cases show a marked decrease with time due to hardening of

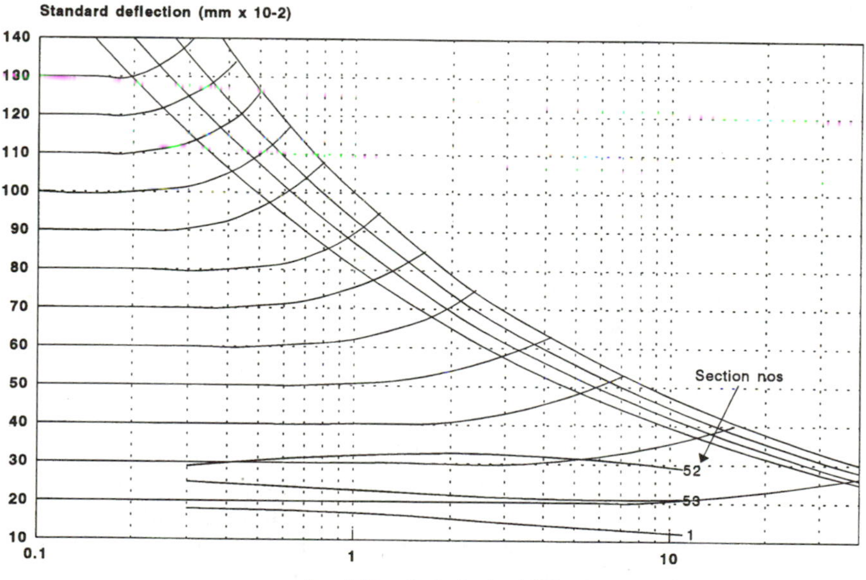

Figure 20.15 Deflection histories for three sections with asphaltic bases—Alconbury bypass experiment (see Fig. 18.8).

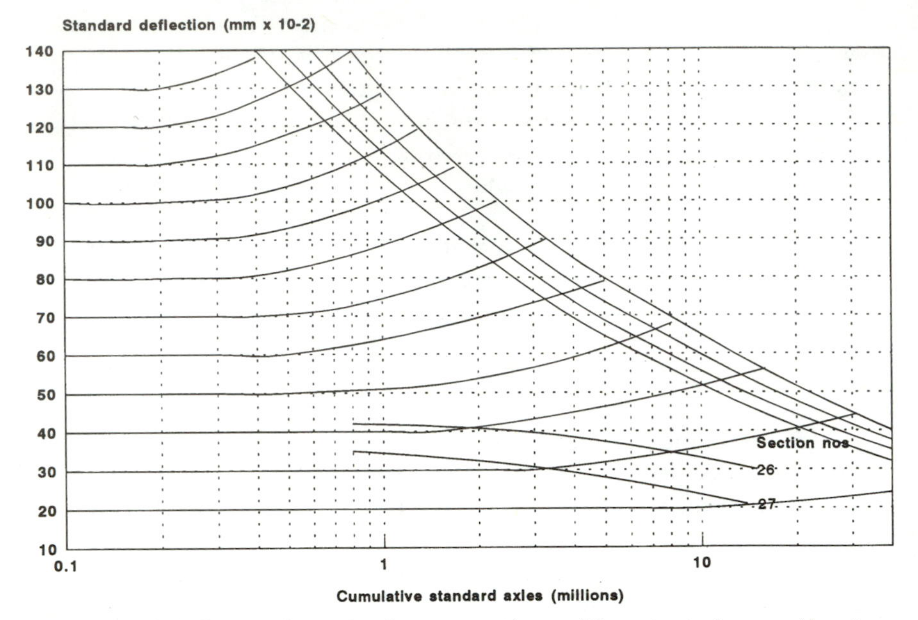

Figure 20.16 Deflection histories for two sections with wet-mix bases—Alconbury bypass experiment (see Fig. 18.8).

the bitumen. In this case the section with the thicker surfacing appears to have a higher deflection. The reason for this is not known; the numbering of the sections may be incorrect.

20.23. Sections 23, 24, and 25 had a 200-mm base of wet-mix material under, respectively, 100, 150, and 200 mm of bitumen macadam surfacing. This material is made with 100 penetration bitumen and is therefore rather less stiff than rolled asphalt made with 50 penetration bitumen. It is also slightly permeable to water, and for this reason it was surface dressed with hot bitumen at the time of construction. The deflection histories for the three sections are shown in Fig. 20.17. There is little change of deflection between laying and failure for all three surfacing thicknesses, and this suggests that any hardening of the bitumen was counteracted by the onset of internal cracking.

20.24. Figure 20.18 shows deflection histories for sections 49 and 51 of the Alconbury bypass experiment. These sections had lean concrete bases 250 and 150 mm thick under 100-mm rolled asphalt surfacings. (The intermediate section 50 with a 200-mm base had a deflection history very similar to that of section 49.) For both sections there is a small reduction of deflection with time and traffic. This is probably due more to a gain of strength in the lean concrete than to hardening

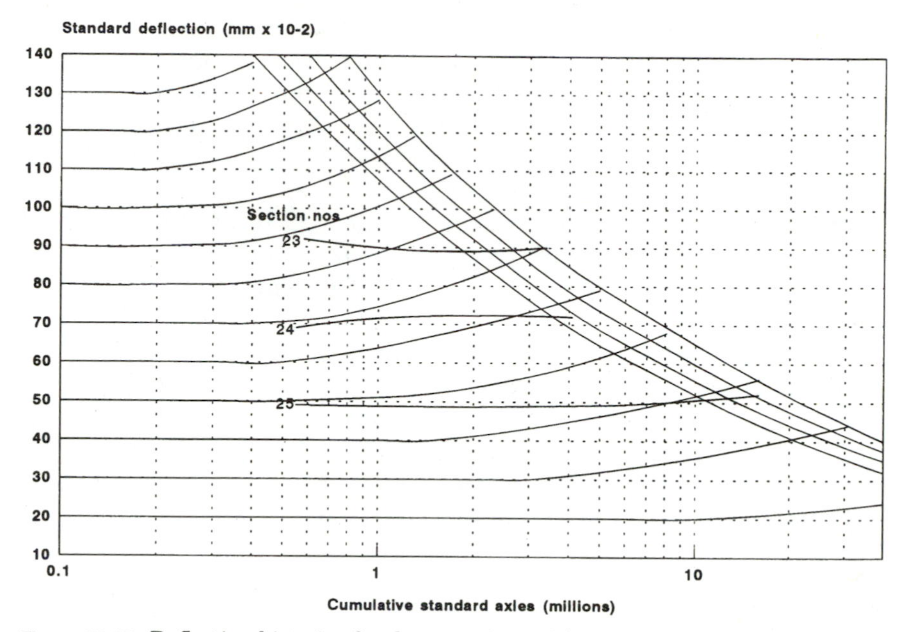

Figure 20.17 Deflection histories for three sections with wet-mix bases and bituminous macadam surface—Alconbury bypass experiment (see Fig. 18.8).

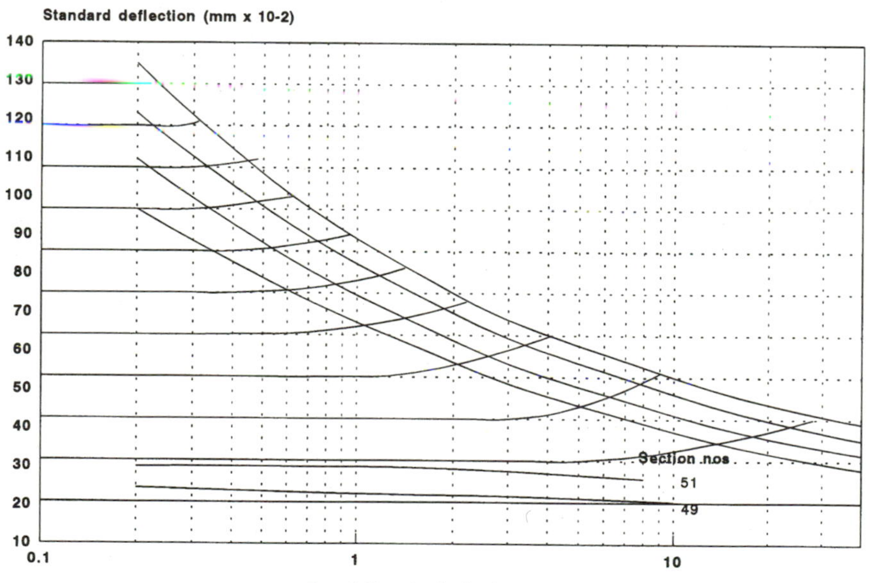

Figure 20.18 Deflection histories for two sections with cement-bound bases—Alconbury bypass experiment (see Fig. 18.8).

of the bituminous surfacing. Section 51 showed some cracking through the asphalt after 5 million standard axles, but no cracking was observed in sections 49 and 50 after 10 million standard axles.

Discussion

20.25. The relationship between deflection and life of flexible pavements is clearly much more complex than was envisaged when the current relationships for different types of pavement were issued by the Transport and Road Research Laboratory in the United Kingdom in 1972. The contours relating traffic (expressed in equivalent standard axles) with measured deflection were derived using early data obtained mainly from underdesigned pavements. The assumption made based on this early data that measured deflections would inevitably increase with time and traffic has been shown by the examples discussed in this chapter to be invalid.

20.26. In a recent article entitled Deflectograph—the Way Forward for Structural Assessment,[3] the U.K. Department of Transport included the diagram, reproduced here as Fig. 20.19. No explanation was given why the majority of the measurements show a decrease of deflection with increasing traffic, against the trend of the published

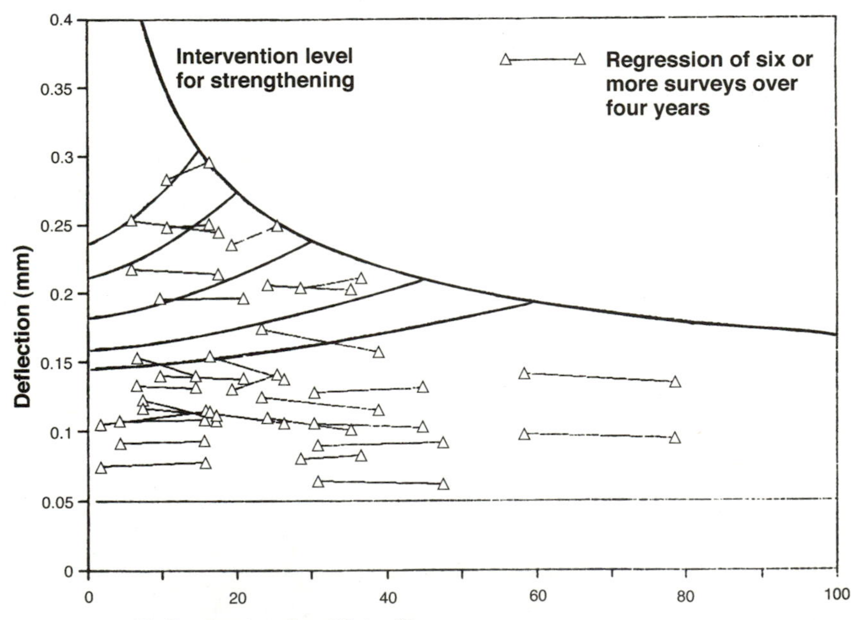

Figure 20.19 Deflection trends with traffic.

contours. The authors state that the figure "shows how the deflection trends from fifteen freeway sites being monitored compare with the present deflection model. This comparison suggests that for these strong pavements in good condition deflection shows no statistically significant change with time." This statement seems to imply that measured deflection does not give a reliable guide to residual life or maintenance requirements for the major road network. Nevertheless computer programs linked to the trend contours in Figs. 20.7 to 20.10 are still used by the U.K. Department of Transport to assess future life and maintenance requirements for the major road network of Britain.

References

1. Norman, P. J., R. A. Snowdon, and J. C. Jacobs: Pavement Deflection Measurements and Their Application to Structural Maintenance and Overlay Design, Department of Transport, *TRRL Report* LR 571, 1973.
2. Kennedy, C. K., and N. W. Lister: Prediction of Pavement Performance and the Design of Overlays, Department of Transport, *TRRL Report* LR 833, 1978.
3. Ferne, B. W., and P. K. Roberts: Deflectograph—The Way Forward for Structural Assessment, *Highways and Transportation,* vol. 10, no. 38, October 1991.

Chapter

21

Current Design Procedures for Flexible and Concrete Pavements in the United Kingdom

Introduction

21.1. Current design standards for flexible and concrete pavements in the United Kingdom are based on Road Note 29, one of a series of monographs on highway design and construction matters produced by the TRRL during the last 40 years. The Road Note is entitled *A Guide to the Structural Design of Pavements for New Roads*. The first edition was published in 1960, followed by a second edition in 1965, and a third in 1970.[1] Since then a number of changes have been made affecting the document by Department of Transport memoranda and by two recent reports issued by the TRRL.[2,3] In this chapter the various changes are brought together for the convenience of engineers, and comments on the changes are provided.

21.2. The various factors involved in formulating a design suitable for a particular situation are (1) present commercial traffic, (2) growth rate of commercial traffic, (3) design life required, (4) conversion of traffic to equivalent standard axles to be carried during the design life, and (5) a series of relationships between subbase thickness, base thickness, and surfacing thickness and cumulative standard axles for flexible construction, and similar relationships between subbase thickness, slab thickness, and weight of reinforcement for concrete pavements.

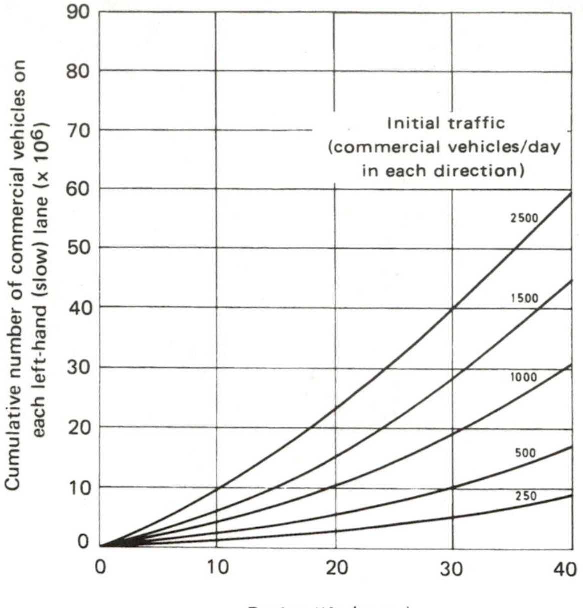

Figure 21.1 Relation between the cumulative number of commercial vehicles carried by each slow lane and design life–growth rate 4 percent per annum.

Traffic

21.3. The 1970 edition of Road Note 29 included a series of charts of the type shown in Fig. 21.1 each relating to a different growth rate of commercial traffic between 2 and 6 percent. They were constructed using the assumption that 80 percent of the commercial traffic would be carried by the slow traffic lane. Figure 8.9 shows that this would apply to carriageways carrying about 400 commercial vehicles per day. In Appendix B of LR1132 these charts have been replaced by the following formula:

$$T_n = 365F_0 \frac{(1 + r)^n - 1}{r} P \tag{21.1}$$

where T_n = total number of commercial vehicles using the slow lane over the design life, n

F_0 = initial flow in commercial vehicles per day

r = growth rate expressed as a percentage

P = proportion of the commercial vehicles in the slow traffic lane

Strictly, P will be changing as the traffic increases (see Fig. 8.9), but an average figure representing the midlife should be sufficient.

Equation (21.1) gives the same total flow as the charts in Road Note 29 so that either approach can be used where the value of P is close to 0.8.

21.4. The total number of commercial vehicles to be carried by the slow lanes during the design life was converted to cumulative standard axles by multiplying factors given in Table 2 of the Road Note. These factors were modified by the Department of Transport in 1978[4] and further modified by the TRRL in 1979.[5] These changes were made necessary by the normal development of road transport together with the increase in oil prices which occurred in the seventies and changes in the Construction and Use Regulations.[6] These factors were reflected in the returns from the TRRL weighbridges. On the basis of this information, the numbers of standard axles per commercial vehicle were increased considerably above the 1970 values as shown in Table 21.1.

21.5. To avoid the rather illogical steps in Table 21.1, the TRRL has developed an equation to relate the vehicle damaging factor to the commercial traffic flow. It gives an estimate of the vehicle damage factor D for any midterm year t based on the 24-hour flow of commercial vehicles for that year F; the base year is 1945, so that the year 1984 corresponds to $t = 39$. The equation is:

$$D = \frac{0.35}{0.93^t + 0.082} - \frac{0.26}{0.92^t + 0.082} \cdot \frac{1.0}{3.9^{F/1550)}} \tag{21.2}$$

As an example, the equation generates vehicle damage factors which vary with time and daily flow of commercial vehicles, as shown in Table 21.2.

TABLE 21.1 Vehicle Damage Factors Recommended for the Design of New Roads

Category of road (commercial vehicles per day in one direction)	Standard axles per commercial vehicle	
	Road Note 29 (1970)	1979
>2000	1.08	2.9
1000–2000	1.08	2.25
250–1000	0.72	1.25
<250	0.45	0.75

TABLE 21.2 Vehicle Damage Factors

	Daily flow of commercial vehicles, AADF*			
Year	250	1000	2000	4000
1985	0.78	1.64	2.17	2.49
1990	0.93	1.89	2.49	2.84
1995	1.18	2.12	2.76	3.14
2000	1.22	2.31	3.00	3.40
2005	1.34	2.47	3.18	3.60
2010	1.43	2.60	3.33	3.76

*Average annual daily flow.

21.6. Table 21.2 shows that extrapolation from the evidence available in 1984 indicates that by the year 2030 the *average* damaging effect of commercial vehicles will be about four standard axles per vehicle. If the eventual EC agreement on vehicle and axle loading allows either 11-t axle loads or 44-t gross vehicle weights the maximum damaging effect of commercial vehicles will rise to between five and seven standard axles. It is unlikely therefore that the *average* will rise to four, although this is not impossible in the case of some industrial freeways.

Thickness of Flexible Pavement Layers

21.7. Subbase thickness. The 1970 edition of Road Note 29 defined the thickness and type of subbase in relation to the CBR value of the soil and the cumulative traffic in standard axles. The relevant requirements for all types of flexible construction are shown in Fig. 21.2 reproduced from Road Note 29. The thicknesses are largely based on the early experience gained from the full-scale road experiments described in Chap. 18. However, in drawing up these thickness standards consideration was also given to the ability of the subbase/subgrade layers to carry construction traffic, particularly during the laying of the base. For this reason, thicknesses were increased above those shown to be necessary by structural requirements. In Figs. 18.10, 18.11, and 18.12 the subbase thickness requirements of Fig. 21.2 have been superimposed on the experimental results and the measure of the overdesign intended to protect the subgrade during construction can be seen. This latest examination of the performance of the full-scale experiments does not indicate that any change of subbase thickness is required. As is discussed in Chap. 11, the concept of introducing a capping layer between low-CBR-value earthworks and the subbase has been introduced recently by the Department of Transport in the United Kingdom. The onus of finding

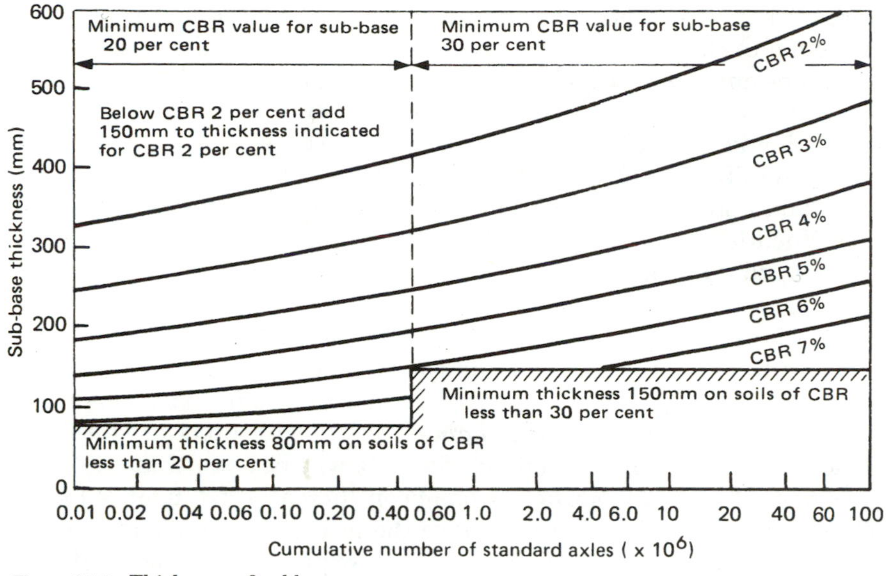

Figure 21.2 Thickness of subbase.

a suitable material of CBR in the region of 10 percent is thus passed to the contractor. This is a common cause of dispute, and whether any money or resources are saved is doubtful.

Pavements with Bituminous Road Bases

21.8. Bituminous materials are now widely used for road bases in Europe and the United States. They are generally more costly than unbound and cemented bases, and overdesign should be avoided, particularly because with this form of construction overlaying is simple if allowed for in the initial design. Figure 18.12 has shown the performance of bituminous surfacing—base combinations used in the British full-scale pavement design experiments discussed in Chap. 18. A variety of coated macadam bases have been used including rolled asphalt, dense coated macadam, and the lean bitumen and tar-coated materials included in the Alconbury and Wheatley bypass experiments. The thicknesses of surfacing and dense coated macadam bases recommended in the 1970 edition of Road Note 29 are included in Fig. 18.12. These thicknesses were based on the results then available from the full-scale road experiments.

21.9. The later evidence does not appear to justify any significant changes to the thicknesses recommended in Road Note 29. It does appear that when rolled asphalt is used as the base material some reduction in thickness (with respect to coated macadam) can be made, as was permitted in the 1970 edition of the Road Note. However, it is not now recommended that substantial thicknesses of rolled asphalt bases should be used on economic grounds. There does not appear to be any significant difference between the performance of dense coated macadam and the leaner bituminous bases adopted in the Alconbury and Wheatley bypass experiments. This is an area where significant savings could be made.

21.10. Figure 21.3 compares the latest British recommendations relating to the total surfacing and base thicknesses for pavements with rolled asphalt surfacings and coated macadam bases given in *Transport and Road Research Laboratory Report* LR1132[2] with the Road Note 29 (1970) recommendations. It will be seen that the total thickness of base and surfacing has been increased by about 40 mm. The reason for this is not clear, since the latest recommendations are understood to be based on the full-scale experiments. For traffic in excess of 10 msa, some extrapolation is inevitable. In compiling Fig. 18.12 this extrapolation has been based on the development of rutting and cracking. The extrapolation adopted in LR1132 was based partly on rutting but in addition future lives were estimated from deflection

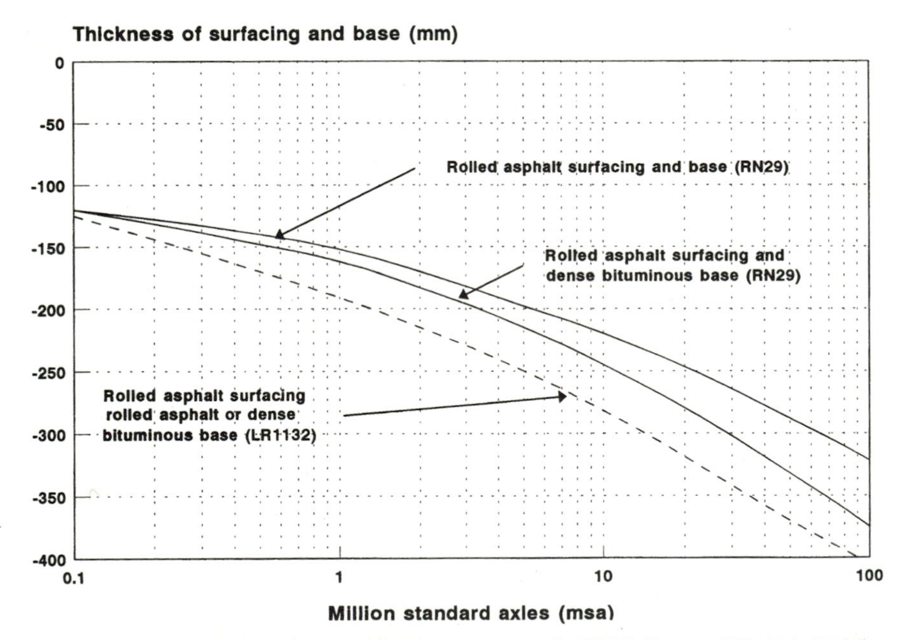

Figure 21.3 Comparison of Road Note 29 (1970) with TRRL Report LR1132 (1984)—asphalt surfacing and bituminous base.

measurements. This procedure has been shown in Chap. 20 to underestimate the lives of adequately designed pavements by a factor which may be as high as 3 or 4. This would have the effect of increasing the apparent thickness required. At 100 msa, where extrapolation is greatest, there may be a case for a small increase in thickness, but in the middle range >20 msa, which covers most trunk roads, an increase of thickness of 40 mm seems excessive and will inevitably make fully flexible pavements less competitive.

Pavements with Lean Concrete Road Bases

21.11. As with the bituminous bases considered above, Fig. 18.11 shows the performance of full-scale experiments using lean concrete and various types of cemented bases. The thicknesses of surfacing and base recommended in Road Note 29 (1970) are also included in the figure. The later information does not suggest that any modification to the 1970 recommendations is necessary. Figure 21.4 compares the Road Note 29 thicknesses with the amended recommendations in Report LR1132. For traffic >20 msa the thickness of surfacing has been increased to 200 mm, presumably to contain reflected cracking from the base. This seems a reasonable precaution. Above 6 msa the thickness of the base has also been increased by as much as 60 mm. This may again

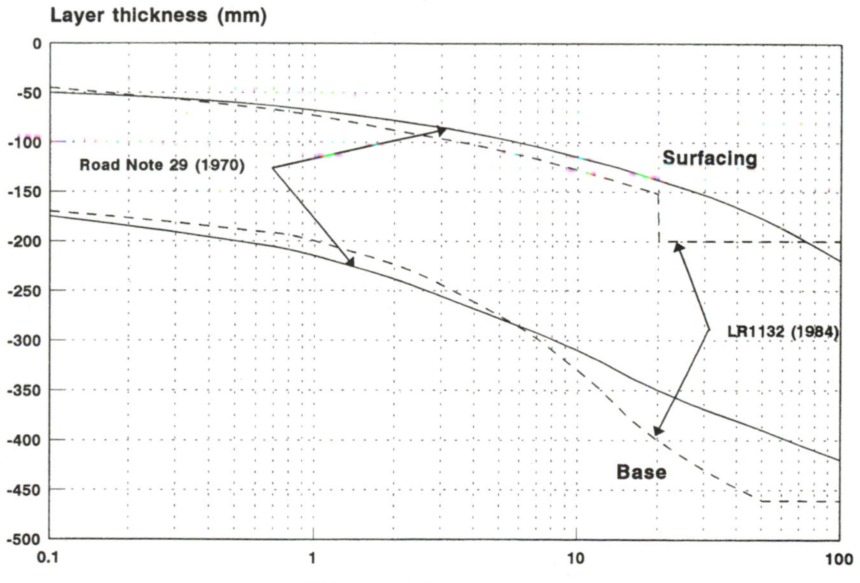

Figure 21.4 Comparison of Road Note 29 (1970) with the latest TRRL recommendations for pavements with lean concrete bases.

be associated with the method of extrapolation used. The reason for the small decrease in base thickness below 6 msa is not obvious.

Pavements with Unbound Road Bases

21.12. Figure 18.10 shows the performance of full-scale experiments using unbound road bases. With this type of base material, the similar strengths of the base material and the subbase make it difficult to define the thickness necessary for the base. However, it does not appear that the later information now available requires changes to the recommendations of Road Note 29 (1970). Figure 21.5 shows that Report LR1132 recommends an increase in the thicknesses of both surfacing and base. In effect it has also limited this form of construction to traffic below 20 msa. The later seems a wise precaution. In practice, the use of this form of construction for roads intended to carry more than 10 msa is not advisable. The experimental evidence does not appear to justify the large increases in base and surfacing thickness, in a form of construction very suited to overlaying when necessary.

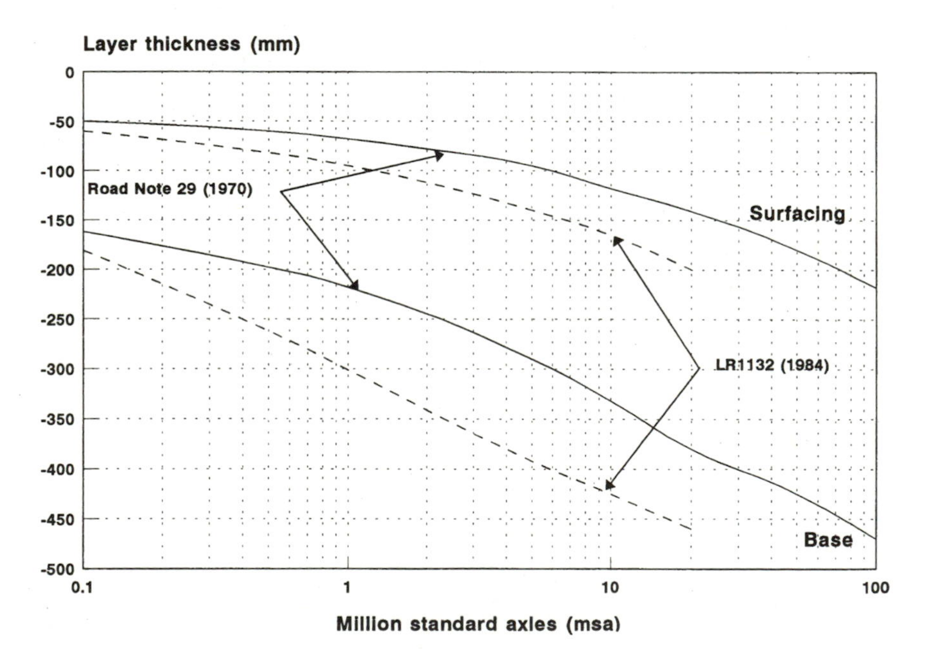

Figure 21.5 Comparison of Road Note 29 (1970) with the latest TRRL recommendations for pavements with crushed-stone bases.

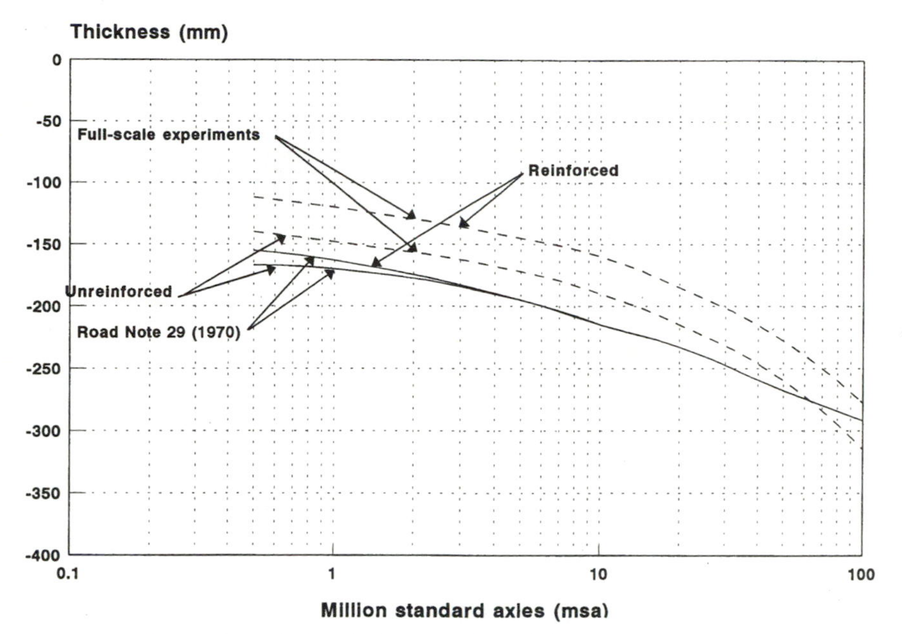

Figure 21.6 Comparison of slab thicknesses from full-scale experiments with the recommendations of Road Note 29.

Concrete Pavements

21.13. Slab thickness. Figure 19.17 shows the minimum slab thickness derived from British full-scale road experiments using concrete construction, related to cumulative standard axles. In Fig. 21.6 these minimum slab thicknesses are compared with the recommendations given in Road Note 29 (1970). When the Road Note was drafted, a conscious decision was made to place a lower limit on thickness of concrete roads, largely because of the difficulty of laying thin slabs without an undue risk of early thermal and traffic cracking. A small difference was made in the thickness of reinforced and unreinforced slabs for lightly trafficked roads. The latest evidence shows that the Road Note recommendations are safe except perhaps for very heavy traffic conditions, where the number of cumulative standard axles to be carried exceeds 50 msa.

21.14. The Road Note 29 recommendations have been recently reviewed in *Transport and Road Research Laboratory Report* RR87.[3] Based on experimental evidence, thickness curves for reinforced and unreinforced concrete pavements are given, each with a "probability of survival rate" of 50 to 90 percent. These curves are reproduced in

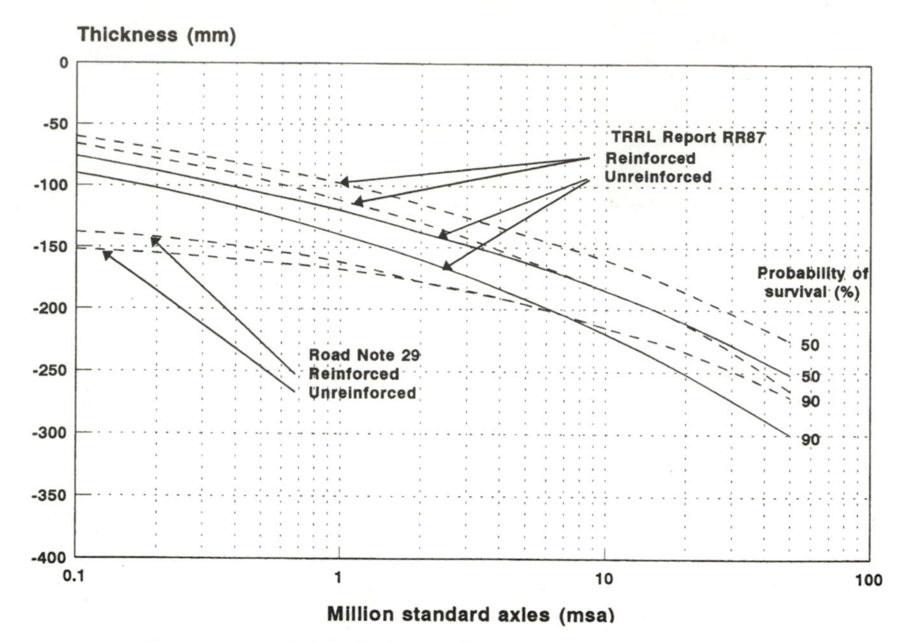

Figure 21.7 Comparison of slab thickness from TRRL Report (1987) with the recommendations of Road Note 29 (1970).

Fig. 21.7. The curves which correspond to a survival probability of 50 percent correspond closely, for both reinforced and unreinforced concrete, with those shown in Fig. 21.6. The low survival rate compared with the evidence in Fig. 19.17 may arise from the use of different methods of failure prediction. Comparison with the recommendations of Road Note 29 indicates a considerable reduction of thickness for pavements designed to carry less than 5 msa. For the reason given in Par. 21.13 the reduction of concrete thickness below 125 mm is not recommended for public roads.

21.15. The latest evidence does not suggest that there could be a greater thickness differential (about 30 mm) between designs for reinforced and unreinforced concrete pavements.

Slab Length, Weight of Reinforcement, and the Use of Dowel Bars

21.16. The slab length between contraction joints is determined largely by the weight of reinforcement used, which in turn is determined by the cumulative traffic to be adopted in the design. Figure

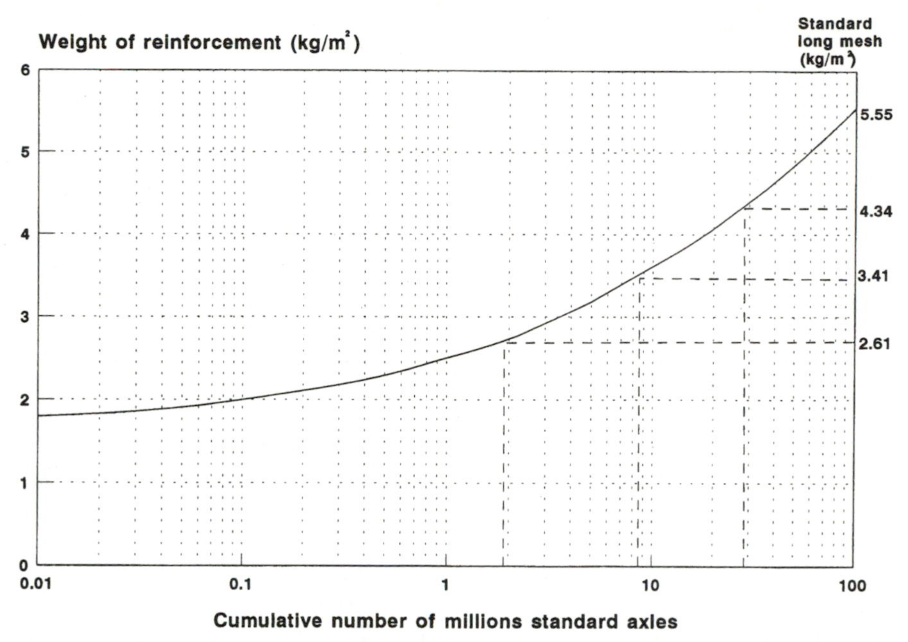

Figure 21.8 Relationship between cumulative standard axles and weight of reinforcement.

21.8 shows the relationship adopted in Road Note 29 (1970) between cumulative standard axles and the weight of reinforcement; the weights of typical long-mesh reinforcement mats available in the United Kingdom are shown in the figure. Figure 21.9 shows the relationship between the weight of reinforcement and the spacing of joints recommended in Road Note 29. In the case of reinforced concrete pavements every third joint should be an expansion joint, and the rest contraction joints. For unreinforced concrete pavements the spacing of the joints adopted is 5 m, with expansion joints at 60-m intervals for slabs <200 mm thick and 40 m for thinner slabs. The Road Note permits the omission of expansion joints in pavements constructed during the summer months, but there is now evidence that this can lead to expansion failures in abnormally hot weather. In all concrete construction, tied longitudinal joints are provided so that the slabs are no more than 4.5 m wide.

21.17. All joints in slabs thicker than 150 mm are recommended to be provided with sliding dowel bars in accordance with Table 21.3. The Road Note also contains details concerning joint sealing.

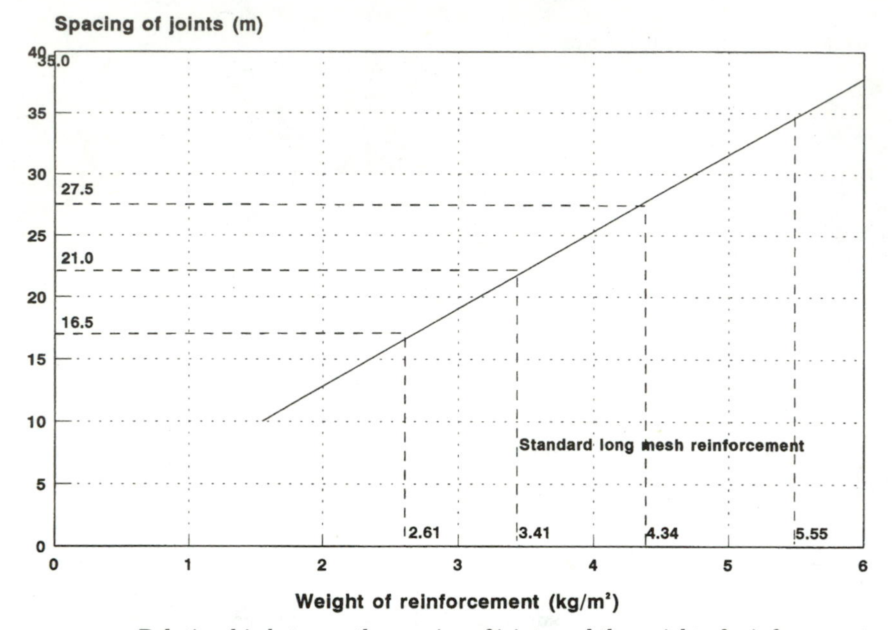

Figure 21.9 Relationship between the spacing of joints and the weight of reinforcement.

TABLE 21.3 Dimensions of Dowel Bars for Expansion and Contraction

	Expansion joints		Contraction joints	
Slab thickness, mm	Diameter, mm	Length, mm	Diameter, mm	Length, mm
150–180*	20	550	12	400
190–230	25	650	20	500
240 and over	32	750	25	600

*Dowel bars are not recommended for slabs thinner than 150 mm.

Subbase for Concrete Pavements

21.18. Road Note 29 (1970) classified subgrades for concrete roads as weak, normal, and very stable, and thicknesses of type 1 subbase for each class of subgrade were defined as shown in Table 21.4.

This requirement was unsatisfactory, particularly with regard to normal subgrades, where the 80 mm of type 1 subbase was often inadequate to permit the passage of construction traffic. It was also insufficient to contain "mud pumping," i.e., the slurrying of fines in the subgrade at joints and cracks. In 1978, the Department of

TABLE 21.4 Classification of Subgrades for Concrete Roads and Minimum Thickness of Subbase Recommended (1970)

Type of subgrade	Definition	Minimum thickness of subbase required, mm
Weak	All subgrades of CBR value 2% or less	150
Normal	Subgrades other than those defined by the other categories	80
Very stable	All subgrades of CBR value 15% or more This category includes undisturbed foundations of old roads	0

Transport introduced into its specification the concept of capping layers already referred to in Par. 21.7. The requirement for concrete pavements was that a subbase of type 1 material 130 mm thick should be used for all subgrades and traffic levels. Where the subgrade had a CBR value of 2 percent or less a capping layer having a CBR value of 7 percent or more is additionally required. It is not clear whether this is intended to be a laboratory or in situ CBR value. Such a material would in general be hard for the contractor to find, outside the type 2 subbase grading limits.

Design of Hard Shoulders

21.19. Hard shoulders are an important feature of freeways and some trunk roads. They are provided to accommodate temporarily disabled vehicles. Observations suggest that a heavy commercial vehicle in stopping traverses 100 to 200 m of shoulder and that each length of shoulder will be traversed by not more than 150 disabled commercial vehicles per year. In view of the light intensity of the commercial traffic and the absence of a need for a high-quality riding surface, early freeways in Britain were provided with granular shoulders, surfaced in some cases with a humus-rich stone-sand-soil layer intended to encourage the growth of grass. Such shoulders soon proved inadequate to support heavy vehicles, particularly in winter and in situations where jacking was necessary.

21.20. It has also been found necessary from time to time to divert slow-lane traffic onto the shoulder while carriageway maintenance and repairs are carried out. This has dictated a minimum width of shoulder of about 3.5 m and required a riding quality similar to that of the carriageway. Even with such use, the structural design of the shoulder will seldom need to accommodate more than 1 million standard axles during the life of the road. However, other considerations affect the design of the shoulder. The type 1 subbase provided under the carriageway is intended to act as a drainage as well as a structural layer, and to be effective it must be continued under the shoulder to connect with the side drainage system. Further, it is now normal practice to carry the carriageway surfacing across the shoulder for both concrete and flexible pavements, separation being defined by a continuous white line with red reflecting studs. These factors require a substantial thickness of subbase under the shoulder but permit a reduction in the thickness of the other layers. Where there is a probability that carriageway widening may be necessary there is a case for constructing the shoulder to carriageway standards.

Design of Housing Estate and Similar Roads

21.21. Traffic census data are generally not available for residential estate roads. However, the Transport and Road Research Laboratory has made limited surveys in several such estate complexes. The results, in terms of commercial traffic, are summarized in Table 21.5. Information on the damaging effect of the commercial vehicles is based on the types of vehicles using the different categories of road. The information given in Table 21.5 can be used in conjunction with

TABLE 21.5 Commercial Vehicle Flows Recommended for the Design of Estate Roads

Type of road	Estimated traffic flow of commercial vehicles per day in each direction	Estimated damaging effect of commercial vehicles (st. axles/veh.)
1. Culs-de-sac and minor residential roads	10	0.45
2. Through roads carrying up to 25 PSVs* per day in each direction	75	0.7
3. Through roads carrying 25 to 30 PSVs per day in each direction	175	1.0
4. Main shopping center of large development carrying delivery trucks and more than 50 PSVs per day in each direction	350	1.5

*Public service vehicles.

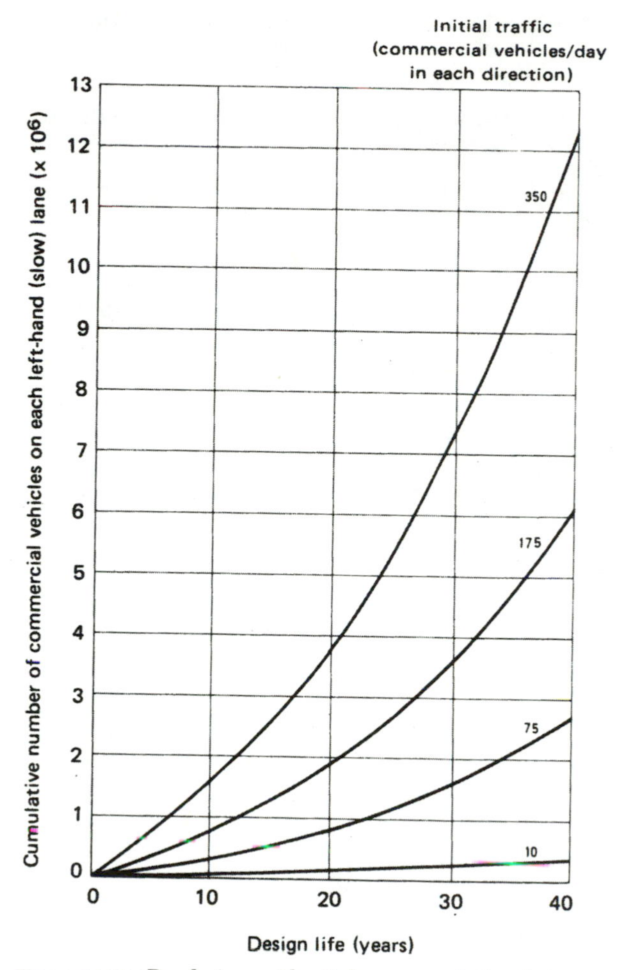

Figure 21.10 Roads in residential areas: relation between the cumulative number of commercial vehicles carried by each slow lane and design life.

Fig. 21.10 to derive the cumulative number of standard axles to be used in the design.

References

1. Transport and Road Research Laboratory: *A Guide to the Structural Design of Pavements for New Roads,* Department of the Environment, Road Note 29, 3d ed., HMSO, London, 1970.
2. Powell, W. D., J. F. Potter, H. C. Mayhew, and M. E. Nunn: The Structural Design of Bituminous Roads, *Transport and Road Research Laboratory Report* LR1132, TRRL, Crowthorne, 1984.

3. Mayhew, H. C., and H. M. Harding: Thickness Design of Concrete Roads, *Transport and Road Research Laboratory Report* RR87, TRRL, Crowthorne, 1987.
4. Department of Transport: Road Pavement Design, *Technical Memorandum* H6/78, HMSO, London, 1978.
5. Currer, E. W. H., and M. G. D. O'Connor: Commercial Traffic: Its Estimated Damaging Effect, 1945–2005, *Transport and Road Research Laboratory Report* 910, TRRL, Crowthorne, 1979.
6. Department of Transport: *The Motor Vehicle (Construction and Use) Regulations,* Statutory Instrument no. 24, HMSO, London.

Chapter

22

Current Design Procedures for Flexible and Concrete Pavements in the United States

Introduction

22.1. Unlike European countries, the United States of America includes a very wide range of climatic and geological conditions ranging from the permafrost regions of Alaska to the semitropical environment of Florida and southern California. Each state has its own highway and traffic engineering department responsible for developing design and maintenance procedures suited to its own environmental and traffic conditions. Many of the states, notably California, Michigan, and Illinois, have, however, made major contributions to the worldwide understanding of road pavement design, construction, and rehabilitation.

22.2. Founded in 1914, the AASHO (see Par. 2.20) has played a major part in coordinating research and field studies for the benefit of all the participating states. This information is disseminated mainly through regular meetings and through publications such as Standard Specifications for Transportation Materials and Methods of Sampling and Testing. This is revised annually and is issued in two parts, the first of which deals with specifications and the second with materials. The ASTM also covers road and paving materials in Volume 04.03, which is also revised annually.

22.3. Particularly relevant to this chapter is the AASHTO *Guide for Design of Pavement Structures* (1986),[1] which is discussed below. Also relevant to flexible pavement design and construction are the publications of the Asphalt Institute, particularly Manual MS-1, entitled *Thickness Design—Asphalt Pavements for Highways and Streets.*[2] The Portland Cement Association has produced a similar document, *Thickness Design for Concrete Highway and Street Pavements.*[3] These three publications are discussed in this chapter.

The AASHTO *Guide for Design of Pavement Structures*

22.4. The concepts used in this guide are those developed in connection with the AASHO road test, the results from which were published in 1962.[4] In Britain and most other European countries the failure condition of flexible pavements is defined in what can be broadly classified as engineering terms such as deformation and cracking. For concrete pavements the criterion is cracking and differential movement at cracks and joints. In the AASHO road test described in Chap. 17 failure was subjectively defined in terms of the quality of the ride experienced by the "average" road user, defined as the present serviceability rating (PSR), and interpreted by a panel of road users using a scale of 1 to 5 as defined in Fig. 17.3. It was the responsibility of the road engineer to interpret quantitatively the PSR in terms of the various factors such as deformation, cracking, patching, and surface regularity. The value so interpreted is defined as the present serviceability index (PSI), and its relationship with engineering terms has been defined in Eqs. (17.1) and (17.3), in relation, respectively, to flexible and concrete pavements.

Flexible Pavements

22.5. The basic design equation used in the *Guide* for flexible pavements is as follows:

$$\log W_{18} = Z_R S_0 + 9.36 \log(\text{SN}+1) - 0.20 + \frac{\log \dfrac{\Delta\text{PSI}}{4.2-1.5}}{0.40 + \dfrac{1094}{(\text{SN}+1)^{5.19}}} + 2.32 \log M_R - 8.07 \tag{22.1}$$

where W_{18} = predicted number of 18-kip equivalent single-axle load applications

Z_R = standard normal deviate

S_0 = combined standard error of the traffic prediction and performance prediction
ΔPSI = difference between the initial design serviceability index p_0 and the design terminal serviceability index p_t
M_R = resilient modulus, lb/in^2

SN is equal to the structural number indicative of the total pavement thickness required:

$$\text{SN} = a_1D_1 + a_2D_2m_2 + a_3D_3m_3$$

where a_i = ith layer coefficient
D_i = ith layer thickness, in
m_i = ith layer drainage coefficient

22.6. The values of the reliability factors Z_R and S_0 depend on the validity of the traffic and materials data used as input to the equation. Values of Z_R proposed for general use where a detailed analysis of such data is not made are given in Table 22.1. Where the variance of proposed future traffic is not to be considered in detail, the value of S_0 for flexible pavements is taken between 0.4 and 0.5.

TABLE 22.1 Standard Normal Deviate (Z_R) Values Corresponding to Selected Levels of Reliability

Reliability R, %	Standard normal deviate Z_R
50	−0.000
60	−0.253
70	−0.524
75	−0.674
80	−0.841
85	−1.037
90	−1.282
91	−1.340
92	−1.405
93	−1.476
94	−1.555
95	−1.645
96	−1.751
97	−1.881
98	−2.054
99	−2.327
99.9	−3.090
99.99	−3.750

22.7. The concept of the structural number SN as defined in Par. 22.5 was developed from the AASHO road test, in which various combinations of surfacing thickness D_1 (asphaltic concrete) were used in conjunction with base thickness D_2 (wet-mix stone, bitumen-coated aggregate, or cement-coated aggregate) and thickness D_3 of subbase (sandy gravel). It was concluded that if the coefficients a_1, a_2, and a_3 were correctly chosen, the structural number derived from any combination of surfacing, base, and subbase thicknesses would provide the same performance. A series of charts are provided, reproduced here as Figs. 22.1 to 22.5 from which the values of a_1, a_2, and a_3 can be obtained from applicable tests such as elastic modulus, Marshall stability, CBR, R value, and the results of triaxial and unconfined compression tests. There are in these charts useful implied equivalence relationships between the various test results. The values of m_2 and m_3 in the structural number equation depend on the drainage conditions which apply to the unbound materials, and the extent to which these conditions change seasonally. Where the surfacing is impermeable and side drains are provided to keep the water table below subbase level, the values of m_2 and m_3 in the base and subbase, respec-

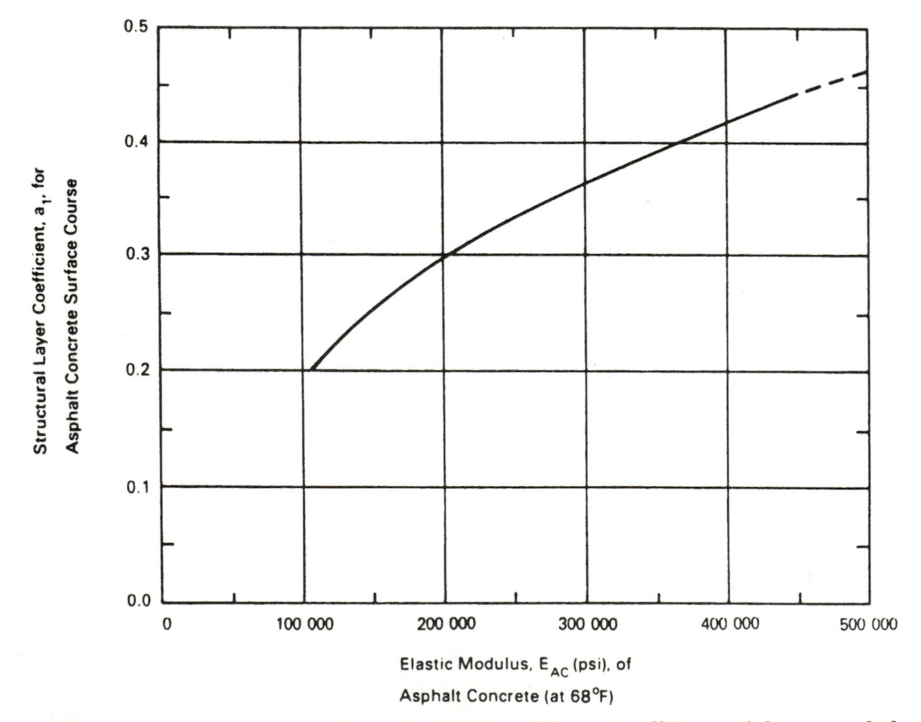

Figure 22.1 Chart for estimating the structural layer coefficient of dense-graded asphalt concrete based on the elastic (resilient) modulus.

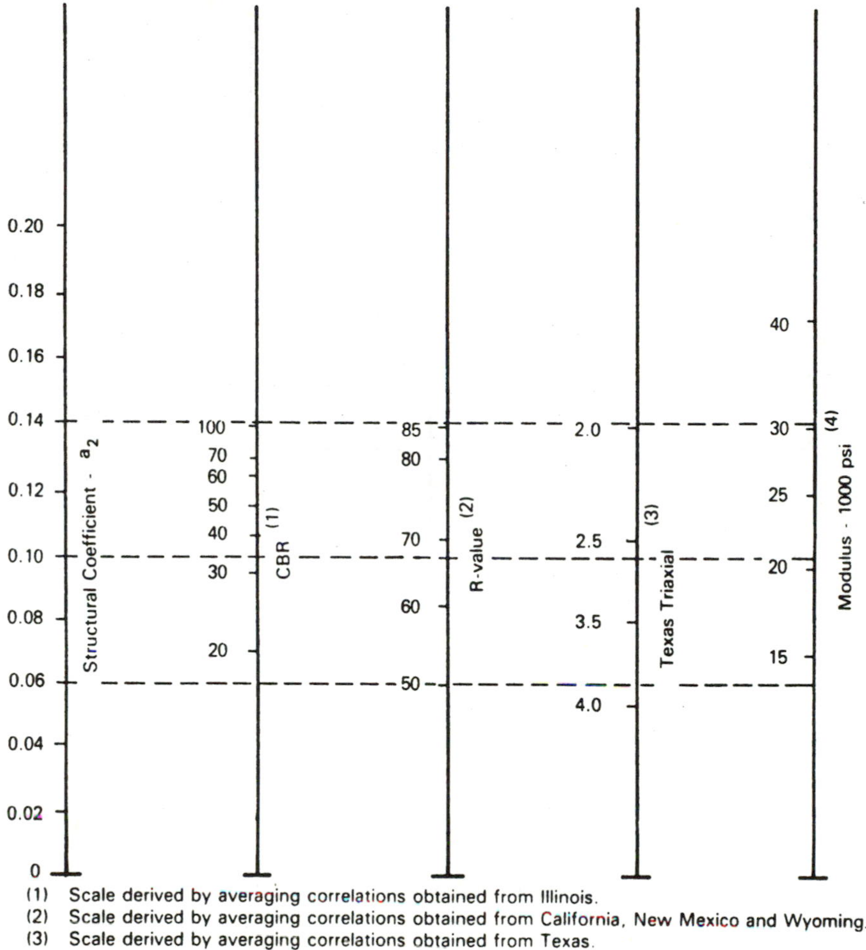

(1) Scale derived by averaging correlations obtained from Illinois.
(2) Scale derived by averaging correlations obtained from California, New Mexico and Wyoming.
(3) Scale derived by averaging correlations obtained from Texas.
(4) Scale derived on NCHRP project.

Figure 22.2 Variation in the granular base layer coefficient a_2 with various base strength parameters.

tively, will be between 1.0 and 1.4. However, where the water table during the annual weather cycle rises to a level where the unbound materials saturate, the value may fall to about 0.4.

22.8. In the AASHTO method of flexible pavement design the PSI value of a new pavement is assumed to be 4.2 and that of a failed pavement 1.5. The term $\Delta PSI/(4.2-1.5)$ therefore represents the point in the life span corresponding to W_{18} applications of a standard (18,000-lb or 8.2-t) axle.

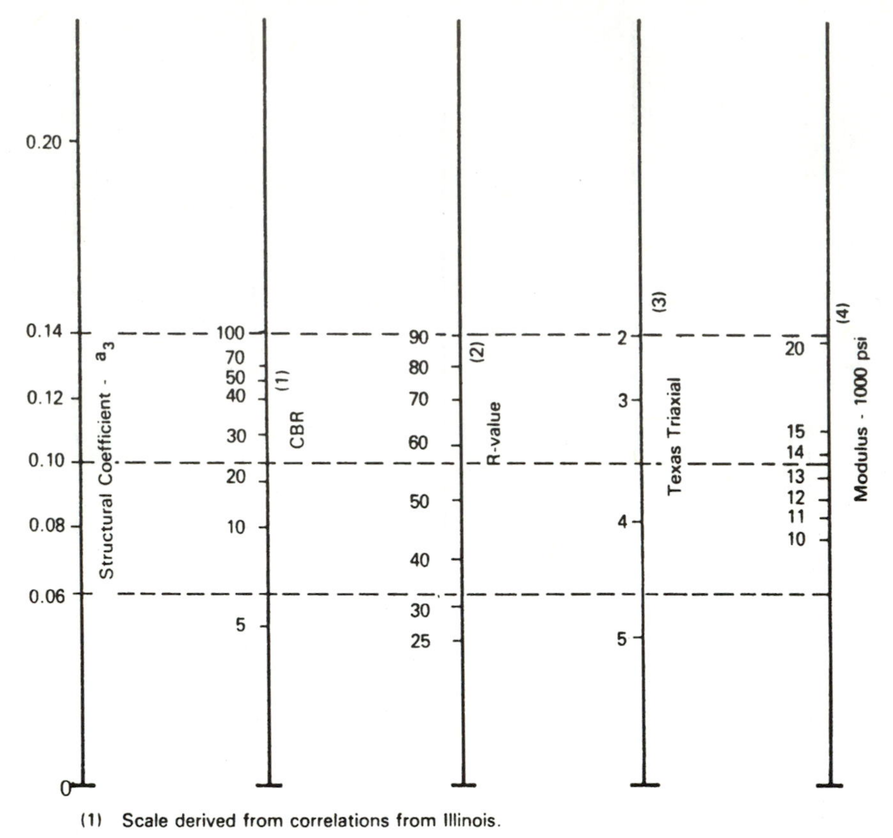

(1) Scale derived from correlations from Illinois.

(2) Scale derived from correlations obtained from The Asphalt Institute, California, New Mexico and Wyoming.

(3) Scale derived from correlations obtained from Texas.

(4) Scale derived on NCHRP project

Figure 22.3 Variation in the granular subbase layer coefficient a_3 with various subbase strength parameters.

22.9. The value used for the foundation modulus M_R needs careful consideration. In the United States frost penetration can be considerable, particularly in the northern states. As a consequence the foundation modulus is subject to a large increase above the average value during the winter, followed by a fall below the average value during the thaw. For this reason a procedure for obtaining an average value is described as part of the AASHTO design procedure. Where this effect is small, the average value of M_R can be obtained from the average CBR value of the subgrade in the manner discussed in Chap. 9.

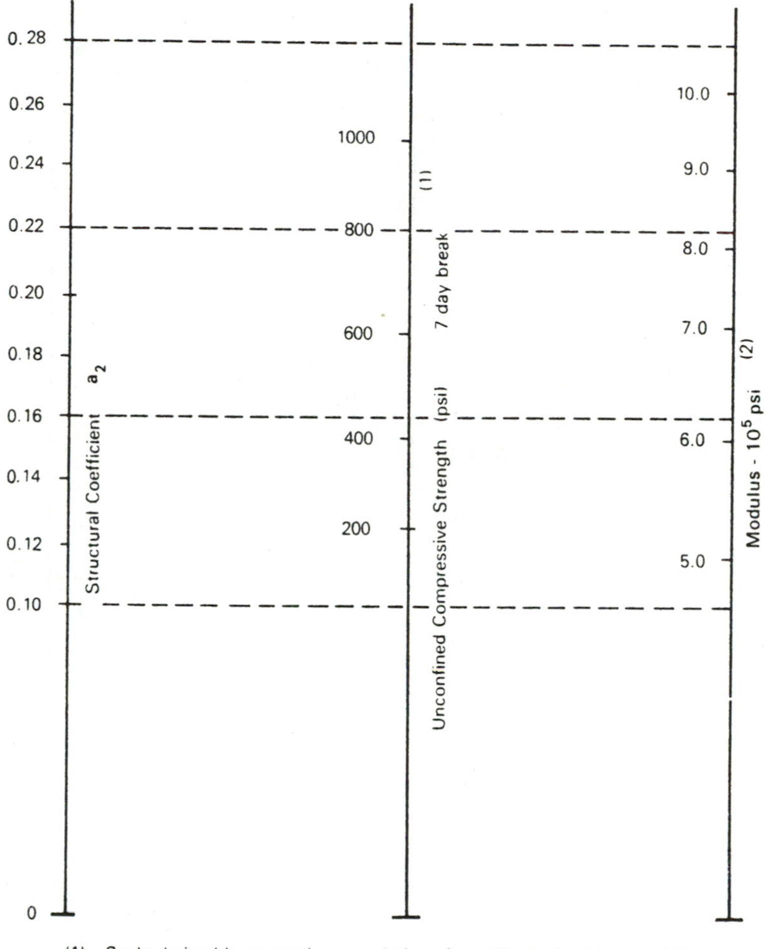

(1) Scale derived by averaging correlations from Illinois, Louisiana and Texas.
(2) Scale derived on NCHRP project

Figure 22.4 Variation in a for cement-treated bases with the base strength parameter.

Application of the AASHTO Design Procedure for Flexible Pavements to U.K. Experimental Roads

22.10. In Chap. 18 the performance of British full-scale pavement design experiments has been analyzed in terms of the number of standard axles carried. The predicted performance of two sections of the Alconbury Hill experiment using the AASHTO design procedure

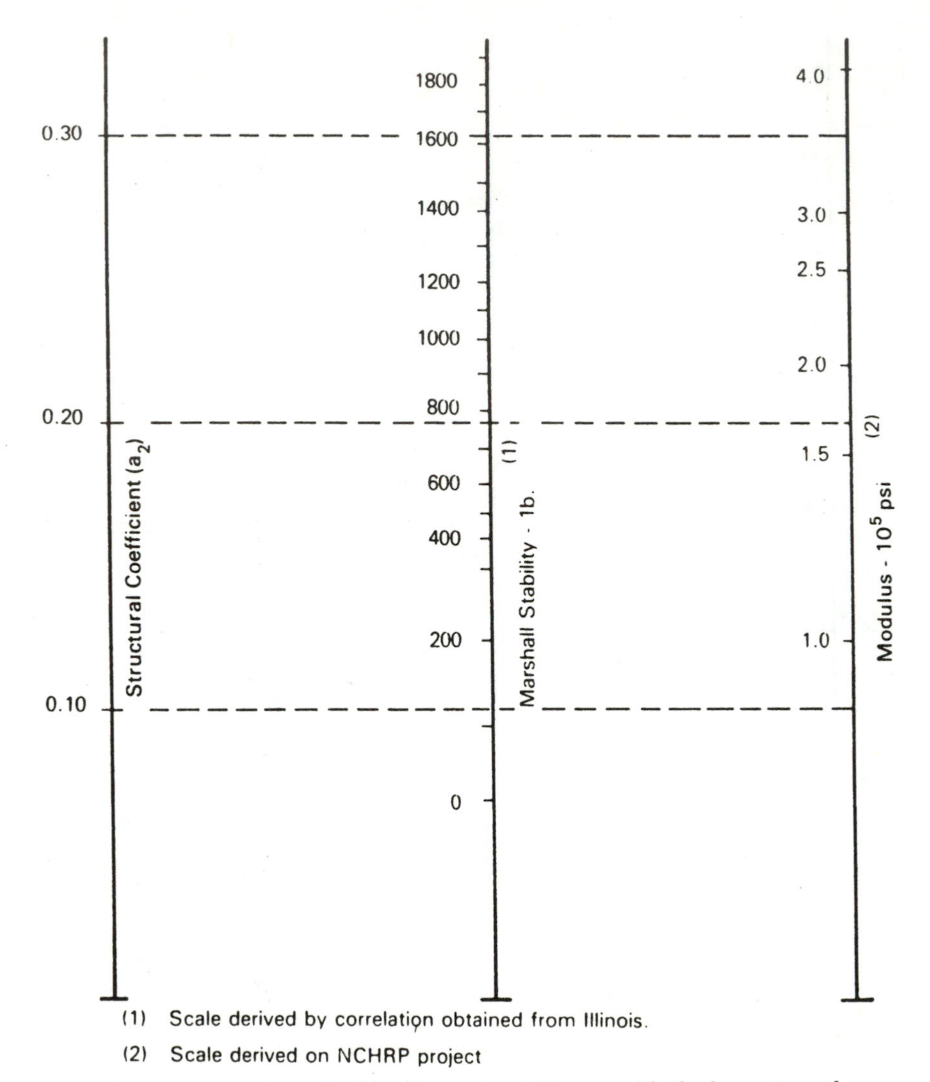

Figure 22.5 Variation in a_2 for bituminous-treated bases with the base strength parameter.

is compared in Fig. 22.6 with the actual performance determined in relation to rutting and cracking. The sections chosen for the analysis are 61 and 52 (see Fig. 18.3). The failure condition chosen is 20 mm of rutting accompanied by some cracking in the wheel paths. This equates to a PSI value of 2.0 using the data given in ref. 4.

22.11. The constitution of the pavements in terms of materials and layer thickness is shown in the upper part of Fig. 22.6. Below are the

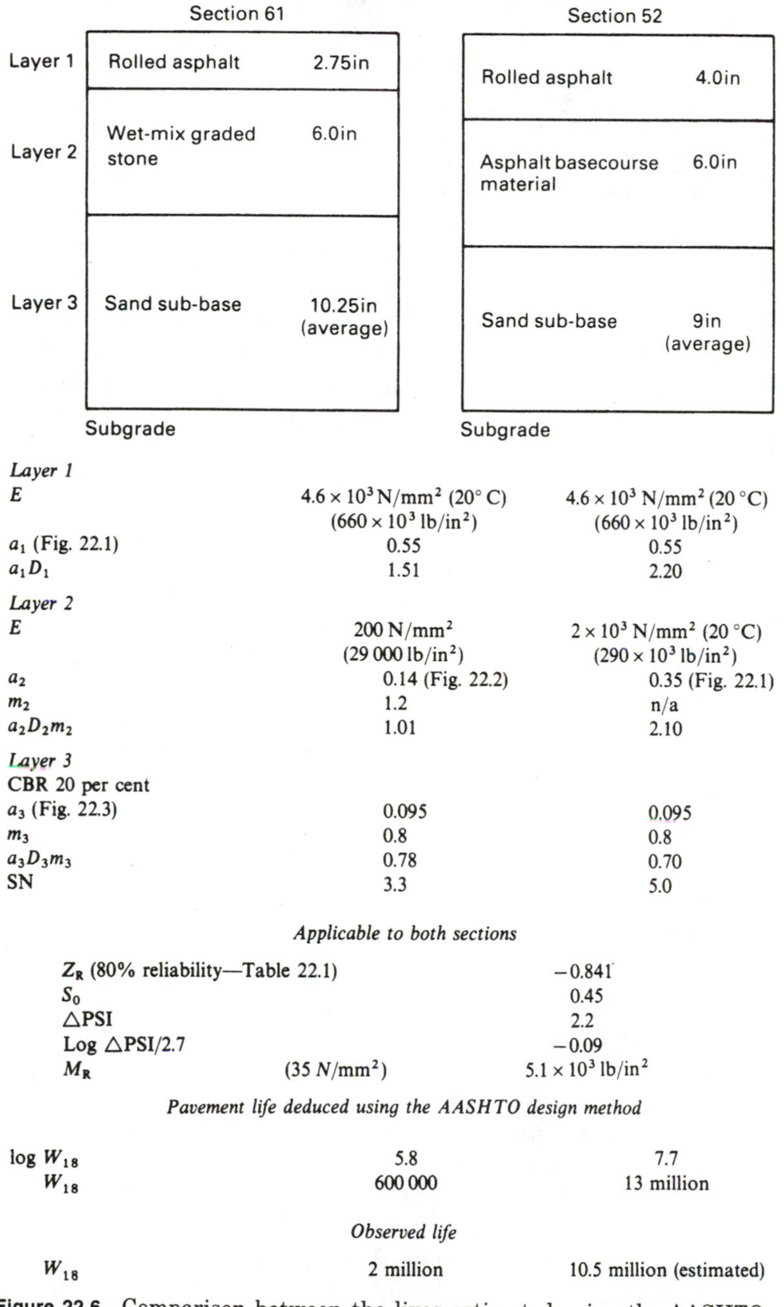

	Section 61	Section 52
Layer 1		
E	4.6×10^3 N/mm² (20° C) (660×10^3 lb/in²)	4.6×10^3 N/mm² (20 °C) (660×10^3 lb/in²)
a_1 (Fig. 22.1)	0.55	0.55
$a_1 D_1$	1.51	2.20
Layer 2		
E	200 N/mm² (29 000 lb/in²)	2×10^3 N/mm² (20 °C) (290×10^3 lb/in²)
a_2	0.14 (Fig. 22.2)	0.35 (Fig. 22.1)
m_2	1.2	n/a
$a_2 D_2 m_2$	1.01	2.10
Layer 3		
CBR 20 per cent		
a_3 (Fig. 22.3)	0.095	0.095
m_3	0.8	0.8
$a_3 D_3 m_3$	0.78	0.70
SN	3.3	5.0

Applicable to both sections

Z_R (80% reliability—Table 22.1)		−0.841
S_0		0.45
△PSI		2.2
Log △PSI/2.7		−0.09
M_R	(35 N/mm²)	5.1×10^3 lb/in²

Pavement life deduced using the AASHTO design method

	Section 61	Section 52
log W_{18}	5.8	7.7
W_{18}	600 000	13 million

Observed life

	Section 61	Section 52
W_{18}	2 million	10.5 million (estimated)

Figure 22.6 Comparison between the lives estimated using the AASHTO design procedure and the observed lives of flexible pavements.

inputs to the AASHTO equation [Eq. (22.1)] together with the resulting estimated lives in standard axles. The observed lives are shown for comparison. In the case of section 52, the pavement has not failed to date and an estimated life to PSI 2.0 is given based on the rate of deformation and cracking.

22.12. The agreement between the calculated and observed lives for section 52 is good, bearing in mind that the observed life had to be extrapolated. The agreement is poor in the case of section 61. Clearly, differences in the failure conditions cannot explain this since the middle term of Eq. (22.1) is very close to zero for terminal PSI values of 2.5 to 1.5. The explanation must lie in the coefficient a_2 used for the unbound base material. The material actually used for the base was crushed and graded blast furnace slag. This material has a mild cementing action. If as a result of this the value of a_2 rose from 0.14 to 0.2 then the predicted life would have risen close to the observed value of 2 million standard axles.

Concrete Pavements

22.13. The AASHTO equation for the design of concrete pavements is as follows:

$$\log W_{18} = Z_R S_0 + 7.35 \log (D + 1) - 0.06 + \frac{\log \dfrac{\Delta PSI}{4.5 \quad 1.5}}{1 + \dfrac{1.624 \times 10^7}{(D + 1)^{8.46}}}$$

$$+ (4.22 - 0.32p_t) \log \frac{S'_c C_d (D^{0.75} - 1.132)}{215.63J \left[D^{0.75} - \dfrac{18.42}{(E_c/k)^{0.25}} \right]} \qquad (22.2)$$

where W_{18} = predicted number of 18-kip equivalent single-axle load applications
Z_R = standard normal deviate
S_0 = combined standard error of the traffic prediction and performance prediction
D = thickness, in, of the pavement slab
ΔPSI = difference between the initial design serviceability index p_0 and the design terminal serviceability index p_t
S'_c = modulus of rupture, lbf/in^2 for portland cement concrete used on a specific project

J = load transfer coefficient used to adjust for load transfer characteristics of a specific design
C_d = drainage coefficient
E_c = modulus of elasticity, lbf/in^2, for portland cement concrete
k = modulus of subgrade reaction, lbf/in^3

22.14. Equation (22.2) is clearly of the same form as Eq. (22.1) for flexible pavements. Again, the middle term will be close to zero for pavements approaching the terminal condition. The terms J and C_d are deduced from Tables 22.2 and 22.3 and the value of k is obtained from Fig. 22.7 and the elastic moduli of the subbase and subgrade. The values of E_c and S_c must be obtained from tests carried out on the concrete being used.

Application of the AASHTO Design Procedure for Concrete Pavements to U.K. Experimental Roads

22.15. As for the flexible pavement design procedure, a comparison has been made between predicted and actual lives for concrete sections included in the Alconbury Hill experiment. The sections selected are both reinforced and of the same thickness of 175 mm (6.9 in). The difference is that one section (section 20) is made with concrete of strength 44 N/mm^2 at 28 days, and the other (section 5) has a 28-day compressive strength of 66 N/mm^2. The layout of the sections is shown in Fig. 19.4.

22.16. The values used for the material properties in Eq. (22.2) are shown in Fig. 22.8, expressed in imperial units. The figure also shows the pavement lives deduced using the AASHTO procedure compared with the observed lives (see Table 19.4). The observed lives are more

TABLE 22.2 Recommended Load Transfer Coefficient for Various Pavement Types and Design Conditions

Shoulder load transfer devices	Asphalt		Tied PCC	
	Yes	No	Yes	No
Pavement type				
Plain-jointed and jointed reinforced	3.2	3.8–4.4	2.5–3.4	3.6–4.2
CRCP	2.9–3.2	n.a.	2.3–2.9	n.a.

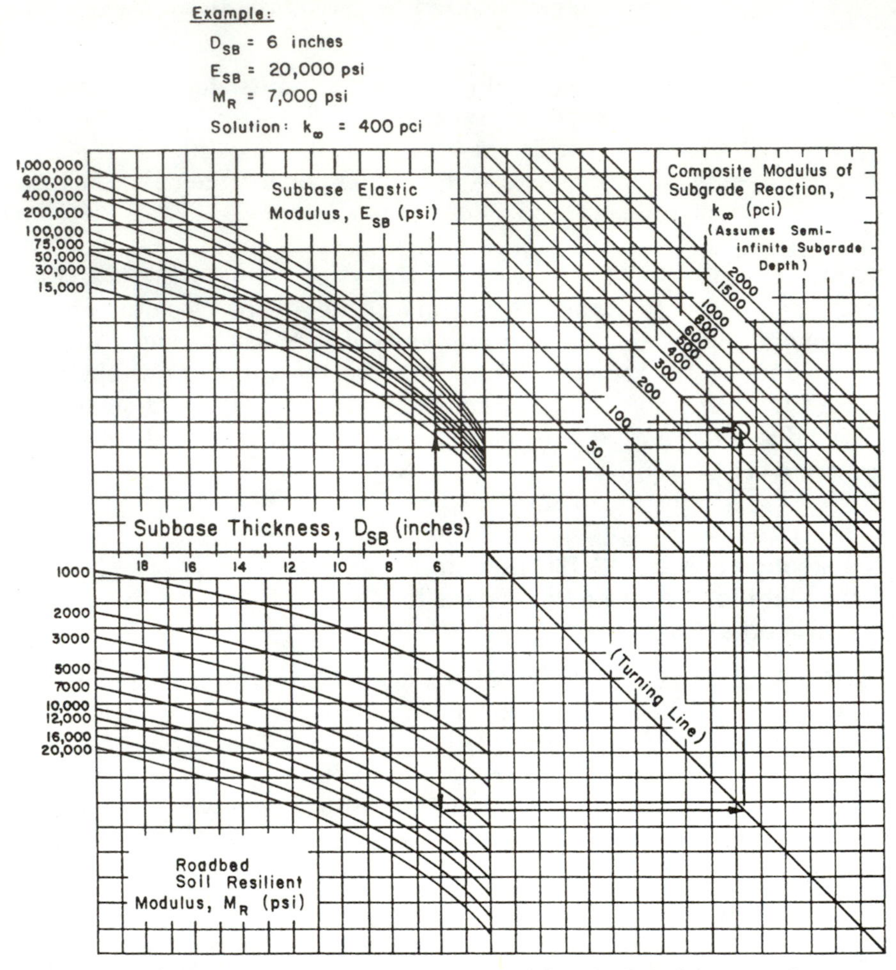

Figure 22.7 Chart for estimating the composite modulus of subgrade reaction k_∞ assuming a semi-infinite subgrade depth. (For practical purposes, a semi-infinite depth is considered to be greater than 10 ft below the surface of the subgrade.)

TABLE 22.3 Recommended Values of Drainage Coefficient C_d for Rigid Pavement Design

	Percentage of time pavement structure is exposed to moisture levels approaching saturation			
Quality of drainage	<1%	1–5%	6–25%	>25%
Excellent	1.25–1.20	1.20–1.15	1.15–1.10	1.10
Good	1.20–1.15	1.15–1.10	1.10–1.00	1.00
Fair	1.15–1.10	1.10–1.00	1.00–0.90	0.90
Poor	1.10–1.00	1.00–0.90	0.90–0.80	0.80
Very Poor	1.00–0.90	0.90–0.80	0.80–0.70	0.70

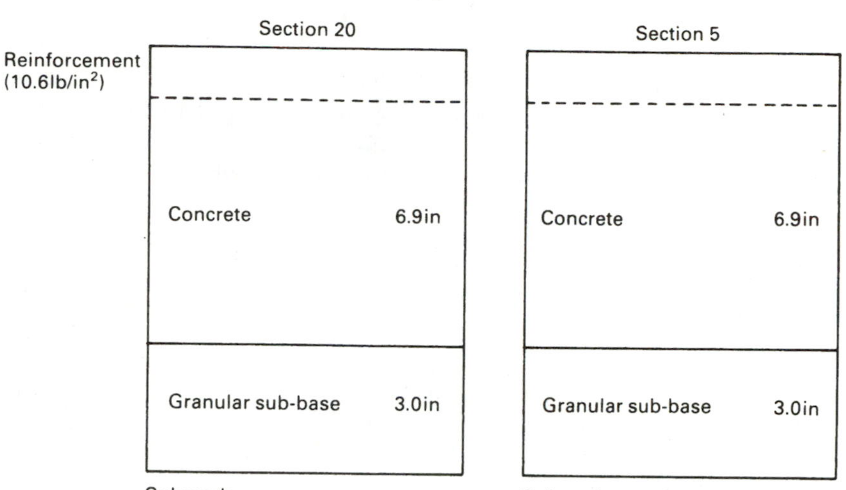

	Section 20	Section 5
Compressive strength of concrete (28 days)	6.3×10^3 lb/in^2	9.6×10^3 lb/in^2
Modulus of rupture of concrete (28 days) S_c	565 lb/in^2	981 lb/in^2
Modulus of elasticity of concrete – E_c	5.6×10^6 lb/in^2	6.5×10^6 lb/in^2

Applicable to both sections

E value of subgrade	5.1×10^3 lb/in^2
E value of sub-base	14.5×10^3 lb/in^2
k value of foundation (from Fig. 22.7)	250 lb/in^2/in
C_d (from Table 22.3)	0.95
J (from Table 22.2)	3.2
Z_R	-0.841
S_0	0.45
$\triangle$PSI	2.5
s	4.5

Pavement life deduced using the AASHTO design method

	Section 20	Section 5
$\log_{10} W_{18}$	6.322	6.65
W_{18}	2 million	5 million

Observed life

	Section 20	Section 5
W_{18}	10 million	27 million

Figure 22.8 Comparison between the lives estimated using the AASHTO design procedure and the observed lives of concrete pavements.

than five times those indicated by the AASHTO procedure, although the ratio of the lives for the two sections is much the same for the predicted and actual lives. A probable explanation for this marked difference in performance is that at the AASHTO test site (which provided the evidence for the AASHTO design procedure) heave due to frost and subgrade volume changes may have been a more important factor than cracking in determining the serviceability index.

General

22.17. The AASHTO guide is a very comprehensive document and it has been possible here to produce only a very brief summary of its contents. In particular it includes nomographs which permit designs to be prepared quickly. Engineers proposing to use the method are strongly advised to consult the full document.

The Asphalt Institute Design Procedure for Flexible Pavements

22.18. The Asphalt Institute Manual[2] relates the thickness and constitution of flexible pavements to the traffic to be carried during the design life, expressed as cumulative standard axles. The end of the design life is defined as the onset of the period when an overlay is required to seal the surface and restore the riding quality.

22.19. The traffic considered includes all commercial vehicles with the exception of all small vans and panel trucks. For use when the detailed constitution of the traffic is unknown, a table is provided giving the percentages of two-, three-, four-, and five-axle trucks likely to be found on different classes of highway in the United States. Information is also given concerning the distribution of commercial vehicles between the different lanes on single and dual carriageways. A growth rate table is provided to give cumulative totals of traffic for periods up to 35 years, for growth rates of 0 to 10 percent per annum. The procedure for calculating cumulative standard axles is therefore very similar to that adopted in the U.K. Road Note 29.[5] The main difference is that in Britain tandem axles are treated as two separate axles, each carrying half the total load on the tandem unit. The Asphalt Institute uses the equivalence values developed by Liddle from the AASHTO road test data,[6] which rates dual axles as being rather less damaging.

22.20. The thickness design charts included in the document are based on three- and four-layer elastic analyses. The three-layer solutions are for what is termed full-depth asphalt construction and the four-layer solution for the case where emulsified asphalt is used beneath the asphaltic concrete wearing course. Three types of emulsified asphalt mix are specified as follows:

Type I made with processed graded crushed aggregate

Type II made with clean crusher-run or pit-run materials

Type III made with as-dug sands or silty sands

22.21. A total of 10 design charts are provided, which relate subgrade resilient modulus M_r with the cumulative number of standard axles to be carried during the design period. These design charts are for full-depth asphaltic concrete, for the three types of emulsified asphalt, and for six thicknesses of untreated aggregate base-subbase layer. Where emulsified asphalt is used, the design includes a minimum thickness of asphaltic concrete wearing course of 50 to 130 mm depending on the design traffic.

22.22. There are several aspects of this design procedure likely to confuse British and European engineers. The term "full-depth asphaltic concrete" would generally be interpreted in Europe as a pavement of hot-mixed bituminous material laid directly on the subgrade. The design charts cater for soil resilient modulus values down to 20 MPa or CBR value 1.9 percent. Under U.K. climatic conditions bituminous laying plant could not operate on such a foundation and material would have to be spread by hand, with little chance of adequate compaction. The manual does state the following:

> Improved subgrade is normally not required in the design and construction of a full-depth asphalt pavement structure. It should be considered only when a subgrade that will not support construction equipment is encountered. In such cases it is used as a working platform for the construction of the pavement courses and does not affect the design thickness of the pavement structure.

This may be the case in the United States, but British experience is that a foundation of CBR not less than 20 percent is necessary to permit satisfactory performance of bituminous pavers and their supply trucks. For this reason a granular subbase of at least 200 mm thickness is used and this material is normally tipped and spread by grader.

22.23. Emulsified asphalt could presumably be laid by hand in a similar manner or processed in place, but this would be regarded as impractical in the wet climate of Britain and much of Europe, and engineers would prefer to lay any bituminous material on a properly designed subbase.

22.24. Table 22.4 compares the different forms of construction discussed in the manual when applied to a pavement designed to carry 10 msa. The CBR of the subgrade is assumed to be 5 percent. For comparison, the AASHTO and U.K. designs for the same traffic and soil conditions are included. The terminal PSI condition assumed for the AASHTO design is 2.0. The implication for the Asphalt Institute design is that beneath an asphaltic concrete wearing course the emulsified asphalt material performs, as a base, nearly as well as normal asphaltic concrete base course (or binder course) material. This is in line with the U.K. research finding that lean bituminous bases perform as well as rich ones under a rolled asphalt surfacing. The table also shows that the provision of a 100-mm crushed aggregate base-subbase layer allows the thickness of the type II emulsified asphalt upper base to be reduced by about 80 mm, but further increases in the thickness of the crushed aggregate permit much smaller reductions in the thickness of the type II material.

22.25. Comparison with the AASHTO and U.K. designs for the same traffic and foundation conditions indicates that the U.K. thickness requirement for the base thickness is about 10 percent less than the Asphalt Institute design and AASHTO thickness is about 17 percent greater.

Thickness Design for Concrete Highway and Street Pavements[3]

22.26. This document has been produced on behalf of the Portland Cement Association (PCA) in the United States. It follows the general approach adopted in Chap. 24 of considering the response of the pavement to the highest axle loads in the traffic spectrum. This is in contrast to the AASHTO procedure of using equivalent standard axle loads, which can be misleading when the traffic spectrum includes abnormally heavy axle loads. Therefore, it would be expected that the PCA would be more reliable for situations where heavy handling equipment, such as front and side-lift trucks, is to be used, e.g., dock terminal areas.

22.27. However, the document attempts to cover a very wide variety of design aspects such as erosion, load transfer, and shoulder design, by a series of nomographs and tables, the background to which is far from clearly stated. The reader has to take a great deal on trust, sometimes without adequate definition of the terms used. This makes it difficult to comment upon it in detail.

TABLE 22.4 Comparison of Flexible Pavement Designs Based on the Asphalt Institute, the AASHTO, and the U.K. Design Procedures, Assuming a Design Traffic 10 Million Standard Axles and a CBR of Subgrade 5 Percent

	Thickness, mm							
	Asphalt Institute designs							
	Full-depth asphaltic concrete	Emulsified asphalt base		Pavements with untreated aggregate base of thickness, mm				
	(Design Chart VI-1)	Type 1	Type II	100	200	300	AASHTO design	U.K. design
Surfacing	50	50	100	100	100	100	100	100*
Bituminous base and/or base course	308	310	316	235†	215†	205†	246‡	182
Untreated aggregate base/subbase	0	0	0	100	200	300	250	250

*Rolled asphalt.
†Emulsified asphalt type II.
‡Asphalt-treated base.

22.28. With regard to traffic, the document differentiates between single, tandem, and tridem axles. This is a complication unlikely to be of much importance in thickness design. It is probably an intake from the AASHO road test. In practice, dynamic weighbridges do not differentiate accurately between single and multiple axles, and this is not generally a factor covered by visual traffic surveys. From limited observations one can attempt to provide "average" contents of multiple axles in the commercial traffic flow, but the accuracy is likely to be low in the present situation where the use of multiple axles is changing in response to legislation on maximum vehicle weights. This is the current situation in Europe. In practice, little error is introduced into design thickness calculations if multiple axles are treated as two or three single axles carrying one-half or one-third of the total axle assembly load.

References

1. American Association of State Highway and Transportation Officials: *Guide for Design of Pavement Structures,* AASHO, Washington, D.C., 1986.
2. *Thickness Design—Asphalt Pavements for Highways and Streets,* Asphalt Institute Manual Series no. 1 (MS-1), Asphalt Institute, College Park, Md., 1984.
3. *Thickness Design for Concrete Highway and Street Pavements,* The Portland Cement Association, Skokie, Ill., 1984.
4. Highway Research Board: *The AASHO Road Test, Report 5: Pavement Research, Special Report* 61E, National Academy of Sciences, National Research Council, Publication 945, Washington, D.C., 1962.
5. Road Research Laboratory: *A Guide to the Structural Design of Pavements for New Roads,* Department of the Environment, Road Note 29, 3d ed., HMSO, London, 1970.
6. Liddle, W. J.: Application of AASHO Road Test Results to the Design of Flexible Pavement Structures, *Proc. Int. Conf. on the Structural Design of Asphalt Pavements, Ann Arbor, Michigan, 1962*, University of Michigan, Ann Arbor, pp. 42–51, 1962.

Part

5

Analytical Design Procedures Based on Elastic Theory

Chapter

23

The Structural Design of Flexible Pavements

Introduction

23.1. The methods of pavement design which have evolved worldwide since World War II have largely been derived empirically from the results of full-scale experiments and the detailed study of the performance of in-service roads. Since the forties there has also been a sustained interest in the development of more fundamental structural procedures which attempt to relate the stresses and strains caused by traffic loading in pavement materials with the performance of those materials under repetitive loading.

23.2. Despite concentrated efforts in such centers as the University of California, the TRRL, and the University of Nottingham and by many similar organizations, no fully satisfactory, comprehensive alternative to the empirical approach has yet been found for the design of flexible pavements. As is discussed in Chap. 24, the position is rather different for concrete pavements, largely because concrete, although a complex material chemically, has physical properties which are easier to study and quantify than is the case with materials bound with bitumen and tar.

23.3. The limitations of the structural approach do not mean that it ceases to be a valuable tool in assessing pavement life. Full-scale experiments on a sufficiently large scale and carried out at sites embracing a range of traffic and subgrade conditions are costly and slow. The theoretical method, appropriately calibrated by comparison with the lives of experimental pavements, allows a much better understanding of the manner in which flexible pavements perform, and the role played by the different layer thicknesses. The full-scale approach is of necessity limited by the axle loading of road vehicles, and only by rather risky extrapolation can it be extended to the design of industrial pavements intended for very heavy wheel loads, which are becoming important in relation to port terminals and to other heavy industrial complexes. The same limitations of empirical methods also apply to the design of pavements under abnormal temperature conditions. A structural analysis should form part of any pavement failure investigation.

23.4. The structural method of design has two main objectives:

1. To produce deflection spectra for the axle loading distribution and in particular for the 6350-kg axle loading used in deflection beam and deflectograph surveys
2. To compute stresses and strains in the pavement materials under the full range of axle loads and temperature conditions and hence, from fatigue and permanent deformation data, to determine the potential mode of failure and the potential life.

23.5. A modern highway pavement designed for a life of 20 years or more will, during its life, carry a cumulative total in excess of 50 million commercial axles. Using a maximum deformation of 25 mm to define failure (see Chap. 4), it follows that the average permanent deformation caused by each application of a loaded wheel is less than 10^{-6} mm and that stresses, strains, and deflections under a loaded wheel can be estimated with sufficient accuracy by elastic theory.

23.6. The material parameters required for predicting:

1. The surface deflection
2. The stresses and strains in each layer
3. The fatigue life and permanent deformation

are summarized in Chaps. 9 to 13 for the soil foundation, unbound granular materials, cement-bound subbases, and bases and bituminous bases and surfacings.

23.7. As bituminous materials are the most expensive element of a flexible pavement, much research effort has been concentrated on them. There is no doubt that as part of the failure process bituminous surfacings crack. The origin of this cracking is not clear. It is generally assumed that it originates at the underside of the bituminous material as a consequence of fatigue and fairly quickly spreads upward to the surface. In a brittle material like concrete this is clearly a likely mechanism, but with a viscoelastic material like a bituminous mixture compressive stress in the upper part of the material might be expected to delay or even heal cracks as they become apparent. There is evidence from many cores taken in bituminous materials by the TRRL that cracks in the surface material often extend only to a depth of 10 to 15 mm, suggesting that in some cases cracking originates from the surface and may be due to thermal changes. A recent reanalysis of deflection measurements, discussed in Chap. 20, has shown clearly that the stiffness of bituminous materials increases progressively over a period of at least 10 years. This information has been incorporated into the figures of Chap. 13 (Figs. 13.15 and 13.16) which present effective moduli for use in structural analyses, and as a result good predictions of early- and late-life deflections of flexible pavements can now be made. It would be expected that there would be similar effects on the fatigue relationships. Unfortunately, correspondingly detailed information on fatigue data is not currently available, and this may be a source of inaccuracy in the prediction of pavement life.

23.8. As part of the deflection studies referred to in Chap. 20 it was found that there was a significant phase lag in the response of bituminous materials to diurnal temperature changes. The effect of this lag was such that the modulus of the bituminous material was most satisfactorily modeled by taking the mean 24-hour temperature of the bituminous material. It is thus adequate in almost all cases to use the monthly average air temperature to determine the percentage of axle loads in each 10°C temperature interval and to consider the deformation and fatigue properties for each of these intervals.

23.9. Chapter 8 presents typical axle load spectra. In most cases, when a prediction is required of pavement life, stresses and strains must be predicted for each of the axle loading bands (bands are usually taken in 2-tonne intervals, as shown in Fig. 8.13). Chapter 8 also gives details of typical axle configurations and details of tire pressures and contact areas.

Principles of Structural Analysis of Pavements

23.10. It appears that Lord Kelvin in 1848 was the first to calculate the displacements which occur when a force is applied to an elastic half-space, i.e., to the surface of a homogeneous material infinite in area and depth.[1] In 1868, the French mathematician and engineer

Boussinesq gave his attention to the same subject and developed expressions for the stresses and displacements under a point load.[2] The stresses which arise when an area is loaded are readily obtained by elastic summation. In 1943, Burmister extended the analysis to two layers infinite in the horizontal direction, and as early as 1945, Burmister had extended his theoretical treatment to three layers[3] and 6 years later Acum and Fox[4] published the first axial solutions for this more complicated case. Westergaard in 1926 took a different approach, more applicable to concrete pavements, in which the pavement was modeled as a slab supported on springs representing the soil.[5] All these methods were useful in demonstrating to the engineer the method of behavior of the pavement and the relative importance of the various layers. They were not sufficiently detailed or flexible to be developed into an accurate design method. This was changed by the advent of the computer and the development of the finite-element method pioneered by Clough[6] and Zienkiewicz[7] in the middle sixties.

The Finite-Element Method

23.11. The versatility of the method allows the modeling of any number of layers which may have variations in structural properties with area and depth subjected to complex nonuniform loading patterns.

23.12. The principle of the method is that the region of interest, i.e., the pavement and subgrade, is discretized or divided into a number of elements. At the top center of the region of interest is a single loaded wheel. The elements extend horizontally and vertically from the wheel to include all the area within the influence of the wheel. The analysis of a pavement is almost invariably axisymmetric and the loading is assumed to be applied by a single wheel. The elements are annular in shape and need only be described in two dimensions on one side of the axis of symmetry. The interrelationship between multiple wheels and axles is readily considered by summation. The elements interconnect at nodes. The early element types were triangular in section with three nodes. These could model only a constant stress within the element and hence large numbers of elements were required to give accurate modeling. Increasingly more sophisticated element types have been introduced. A particularly stable and successful element is the eight-noded isoparametric element with reduced integration rule. The section through the pavement is divided into quadrilaterals (the majority of which are rectangles). Most pavement problems are satisfactorily modeled by 150 to 200 of these eight-noded elements. For a typical wheel loading it is adequate to extend the mesh to a radius of the order of 3.5 m and to a depth of 2.5 m. The boundaries of the mesh are vertical and horizontal and are assumed to be supported by rollers. The wheel loading is applied as a pressure acting on a circular area of the upper surface.

23.13. The output from the finite-element program consists of the following:

1. Displacements at each node
2. Strain tensor and principal strains and directions at a number of points within each element
3. Stress tensor and principal stresses and directions at the same points as the strains

23.14. From the geometry of a single element and its material properties a single matrix is readily developed which relates the load on each node of the element to the displacement of all the nodes of the element. This stiffness matrix for each element can then be assembled into a global stiffness matrix which relates the load at each node in the problem area to the displacement of all the nodes. The displacement of each node subject to boundary conditions is known and can be input to the equations. The global matrix can then be inverted using conventional matrix algebra, and the result is the calculated displacement of each node in the area of interest under the input loading. From the stiffness matrix of each separate element the strains and stresses within the element are directly calculated at any point.

23.15. The stiffness matrix of a single element is most easily generated by the multiplication of two separate matrices. The first, the strain matrix, relates the strains in the element to the displacements of the nodes and the second the elasticity matrix which relates the stresses in the element to the strains. In its simplest, most widely used form, all the terms in this elasticity matrix are functions of the modulus of elasticity and the Poisson ratio. The method is sufficiently flexible for it to be straightforward to include more complex relationships if the actual data are available (rarely the case). It is also easy to consider thermal effects, creep, etc. The finite-element method is routinely used for the analysis of a large number of geotechnical problems. In the latter it is often necessary to model the nonlinear behavior of soils at large strains, including shear strength cutoff. To model this correctly the program is usually run incrementally, the load being applied in stages and the stress and strain output from one stage being the input to the next stage. Although programs which include these features are readily available, the practicing pavement engineer will be adequately served by a robust linear elastic program using a stable element type.

23.16. Each element may be ascribed individual material properties. The different pavement layers and even variations of stiffness within a single layer are readily accommodated.

23.17. The finite-element approach enables pavement performance under traffic to be modeled comparatively easily. The validity of the computations depends critically on the elastic properties and particularly on the fatigue properties selected for the various layers of the pavement. These properties have been considered in Chaps. 9 to 14.

23.18. The transient deflection of pavements under the passage of wheel loads is easier to model reliably than long-term performance, and it provides a means of checking the validity of elastic properties which have been assumed. This aspect of structural analysis is considered first.

The Application of Structural Analysis to Pavement Deflection Measurements

23.19. Chapter 20 describes the two most common methods of measuring pavement deflection under a moving wheel load, i.e., the deflection beam and the deflectograph. Figure 20.1 shows the principal dimensions of the deflection beam used in the United Kingdom. The two-axle truck used for deflection studies has a rear-axle load to 6350 kg (14,000 lb) equally divided between the twin wheel assemblies at each end of the axle. The recommended tire size is 7.50 × 20 or 8.25 × 20 with an inflation pressure of 590 kPa.

23.20. In the United States the dimensions of the Benkelman beam are very similar to those used in the United Kingdom, but the axle load used is 8200 kg (18,000 lb), i.e., the standard axle defined in the AASHO road test. The tires used are 279 × 572 mm 12-ply inflated to 483 kPa [AASHTO Designation T256-77 (1986)]. An approximate correlation between U.S. and U.K. deflection measurements is achieved by increasing U.K. values in proportion to the axle loads used.

23.21. The tire-print areas for the deflection beam and for the deflectograph are shown in Fig. 20.5. In the case of the deflection beam the arrangement can be satisfactorily modeled by assuming a loaded radius of 92.5 mm for each wheel and a separation of 285 mm between the centers of the loaded wheels.

23.22. The finite-element program is used to compute the deflection of the pavement under consideration at a point 142 mm from the center of a single wheel with loading of 1588 kg. This deflection is then doubled to take into account the second wheel of the assembly. This represents the maximum deflection as the shoe of the beam passes between the twin wheels. When the shoe is in this position, which is 2.74 m from the twin feet of the beam assembly, the feet experience a small deflection *S*. When the wheels are in the initial position they are 1.30 m from the shoe and 1.44 m from the twin feet; the shoe then has an initial deflection *D*. Lister[8] calculated that the reading of the dial gauge would be smaller than the maximum absolute deflection by an amount d_2, where

$$d_2 = 1.4S - 0.5D$$

23.23. Similar calculations can be made for the deflectograph, but it is usual for the correlation between deflection beam and deflectograph measurements to be made experimentally, a mean relationship being established for a wide range of pavement constructions. Such a correlation is shown in Fig. 20.6.

The Application of the Finite-Element Method to Flexible Pavement Design

23.24. The use of the finite-element approach in the solution of pavement design problems is best illustrated by the analysis of some existing pavements which have been in service for some time and which have known construction and for which the constitution and intensity of the traffic is available. For this purpose three types of pavement construction which have been incorporated into a number of experimental sections from the TRRL's program of full-scale experiments have been considered. These include lean concrete, rolled asphalt, and wet-mix road bases. The sections considered have been taken from the following experiments:

Conington on trunk road A1

Alconbury bypass on trunk road A1

Wheatley bypass on trunk road A40

Nately Scures on trunk road A30

These full-scale experiments have been described in detail in Chap. 18.

23.25. The axle load distribution of the commercial vehicles using Alconbury bypass, Wheatley bypass, and Nately Scures (prior to

TABLE 23.1 Axle Load Spectra at Experiment Road Sites

Axle load, t	Conington, %	Alconbury bypass, Wheatley bypass, and Nately Scures
0–2	9	22
2–4	25	45
4–6	35	18
6–8	16	8
8–10	8	5
10–12	5	2
12–14	2	0.42
14–16	0.42	0.07
16–18	0.07	0.042

1972) is sufficiently similar to be represented by the axle load spectrum shown in Table 23.1. The damaging effect of this traffic over the past 20 years has corresponded to approximately 30 standard axles per 100 commercial axles. The axle load distribution at the Conington experiment is more damaging. The axle load spectrum is also included in Table 23.1. At this site the damaging effect per 100 commercial axles is approximately 60 standard axles.

23.26. The average numbers of commercial axles per year using the left-hand lanes at the four sites, based on a period of 10 to 15 years, are as follows:

Conington	2.14 million
Alconbury bypass	1.4 million
Wheatley bypass	1.2 million
Nately Scures	1.6 million (1964–1972)

23.27. Table 23.2 shows the percentage of the year for which the mean 24-hour temperature of asphalt surfacings falls between temperature bands covering the range 0 to 25°C. This is based on measurements made in southern Britain.

The Structural Design of Pavements with Bituminous Road Bases

23.28. The experiments referred to in Par. 23.24 include a number of sections with 100-mm asphalt surfacings on 150-mm dense bitumen macadam and rolled asphalt road bases. The sections are laid on 150-

TABLE 23.2 Asphalt Temperature Distribution

Temperature band, °C	Percentage of year
0–5	25
5–10	16
10–15	17
15–20	17
20–25	25

mm granular subbases, and the subgrade CBR is close to 5 percent. It is proposed here to follow through in detail the structural design process for a section with a rolled-asphalt road base 150 mm thick using the axle load distribution appropriate to Conington. The conclusions from similar analyses of pavements with lean concrete and wet-mix bases will also be based mainly on that axle load distribution.

23.29. The computer program is first used to analyze the maximum tensile strains induced in the asphalt for a wide range of temperature and axle load conditions. Because of the stiffening of the asphalt with time, it is necessary to differentiate between early-life and late-life strains. The elastic moduli used to cater for temperature and aging are shown in Table 23.3. These are derived from Figs. 13.14 and 13.15. In the following calculations it has been assumed that the early-life modulus applies to the first year and the late-life value is asymptotic as the tenth year's value.

TABLE 23.3 Asphalt Modulus Values Used in Analyses

Temperature band, °C	Vehicle speed	Asphalt age	Modulus, GPa
0–5	Normal traffic	Early life	10
		Late life	12
5–10	Normal traffic	Early life	6.5
		Late life	10
10–15	Normal traffic	Early life	5.5
		Late life	8
15–20	Normal traffic	Early life	2.2
		Late life	6.5
20–25	Normal traffic	Early life	1.8
		Late life	5
10	Deflection beam	Early life	4.5
		Late life	6
20	Deflection beam	Early life	1.8
		Late life	3.5
30	Deflection beam	Early life	0.45
		Late life	1.4

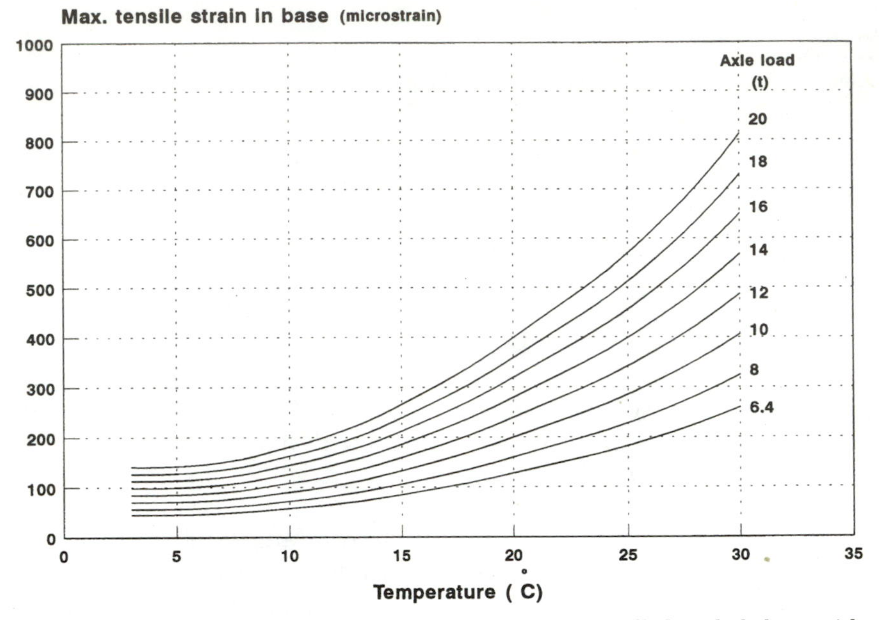

Figure 23.1 Variation of the maximum tensile strain in a rolled asphalt base with axle load and temperature: 250-mm rolled asphalt, early life.

23.30. Figures 23.1 and 23.2 show the maximum tensile strain–axle load–temperature relations for the 100-mm asphalt surfacing on 150-mm asphalt road base for the early- and late-life conditions. From the traffic and temperature data given in Pars. 23.25 to 23.27, Table 23.4 has been constructed to show the annual numbers of axles corresponding to six axle load and five temperature categories at the Conington site.

23.31. Using Figs. 23.1 and 23.2 and Table 23.4, Table 23.5 has been prepared to show the maximum tensile strain developed in the asphalt for each load band and each temperature band. Both the early- and late-life conditions are considered. Using the fatigue data for rolled asphalt given in Fig. 13.18 each strain value in Table 23.5 is related to the number of applications of that strain to cause fatigue failure. The proportion of the fatigue life absorbed during the year by that strain level is recorded. These proportions are summed for the early- and late-life years. The table shows that during the early-life period 0.15 of the fatigue life will be absorbed in 1 year. At the lower strain applicable to the late-life years, 0.016 of the fatigue life will be absorbed per year.

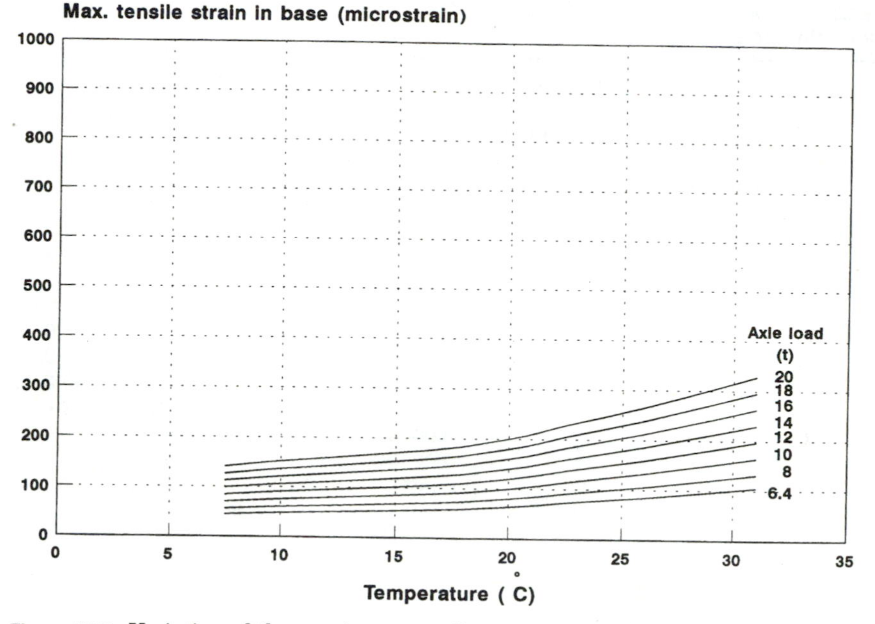

Figure 23.2 Variation of the maximum tensile strain in a rolled asphalt base with axle load and temperature: 250-mm rolled asphalt, late life.

23.32. The early life was assumed applicable to the first year and the late life condition after 10 years. On this assumption a curve was constructed relating the fraction of the fatigue life absorbed each year. Using this curve, Table 23.6 sets out the predicted fraction of life for each year. In the first 10 years 59 percent of the life is absorbed and the total life predicted is 35 years.

23.33. To date the relevant sections at the Conington experiment have had a life of 18 to 19 years and they are still in good condition.

TABLE 23.4 Numbers of Axle Loads per Year in Various Axle Load and Temperature Ranges—Conington

	Number of axle loads per year in temperature bands, °C				
Axle load category, t	0–5	5–10	10–15	15–20	20–25
6–8	86,000	55,000	58,000	58,000	86,000
8–10	43,000	27,000	29,000	29,000	43,000
10–12	27,000	17,000	18,000	18,000	27,000
12–14	11,000	6,700	7,200	7,200	11,000
14–16	2,200	1,400	1,500	1,500	2,200
16–18	380	240	255	255	380

TABLE 23.5 Strain Levels and Fatigue Life in Relation to Axle Loading and Temperature in the Conington Experiment

	Early life			Late life		
Temperature, °C	Strain, microstrain	Fatigue life, millions of applications	Fraction of fatigue life	Strain microstrain	Fatigue life, millions of applications	Fraction of fatigue life
6–8-t axle						
0–5	40	200	0.0004	40	200	0.0004
5–10	60	70	0.0008	50	140	0.0004
10–15	70	60	0.0009	60	100	0.0006
15–20	100	30	0.002	70	110	0.0005
20–25	180	4	0.02	80	110	0.0008
			Total 0.0241			Total 0.0027
						8–10-t axle
0–5	50	70	0.0006	50	70	0.0005
5–10	65	50	0.0006	65	50	0.0005
10–15	90	22	0.0013	75	56	0.0005
15–20	160	4.5	0.0065	80	70	0.0004
20–25	230	1.5	0.0285	100	50	0.0009
			Total 0.0375			Total 0.0028
						10–12-t axle
0–5	65	25	0.0011	65	25	0.001
5–10	75	27	0.0007	75	27	0.0006
10–15	130	4.5	0.0004	90	20	0.0009
15–20	180	2.7	0.0067	110	14	0.001
20–25	270	0.9	0.0297	130	20	0.001
			Total 0.0386			Total 0.0045
						12–14-t axle
0–5	90	8	0.0013	90	8	0.001
5–10	100	8	0.0009	100	8	0.0008
10–15	155	2.5	0.0029	110	8	0.0009
15–20	225	1.1	0.0066	120	10	0.0007
20–25	310	0.5	0.021	150	10	0.001
			Total 0.0327			Total 0.0044
						14–16-t axle
0–5	98	5	0.0004	98	5	0.0004
5–10	106	5	0.0003	106	6.5	0.0002
10–15	170	1.5	0.001	125	6	0.0002
15–20	260	0.6	0.0025	155	5.5	0.0003
20–25	360	0.3	0.0075	170	5	0.0004
			Total 0.0117			Total 0.0015
						16–18-t axle
0–5	115	3	0.0001	102	4.5	0.0000
5–10	134	2	0.0001	123	2.5	0.0001
10–15	190	2.2	0.0001	137	4	0.0000
15–20	276	0.3	0.0009	161	4	0.0001
20–25	415	0.15	0.0025	193	4	0.0003
			Total 0.0037			Total 0.0005
Totals for all axles/year			0.1483			0.0164

TABLE 23.6 Sections with 150-mm Rolled Asphalt Base—Fraction of Life Absorbed Each Year—Conington

Year	Fraction of life absorbed
1	0.15
2	0.10
3	0.075
4	0.06
5	0.05
6	0.04
7	0.035
8	0.03
9	0.025
10	0.020
11	0.016

The life of the four sections, 15, 18, 21, and 24, estimated from the present deformation, ranges from 25 to 200 msa with a mean of 80 msa (see Table 18.9). The mean number of equivalent standard axles per commercial vehicle for the site over the years 1971–1985 for this experiment is 1.80. Therefore, the estimated experiment life would be about 44 years, and there is reasonable agreement with the structural analysis.

23.34. Two important conclusions can be drawn from Table 23.5, as follows:

1. With a bituminous road base, the axles carrying less than 6 t contribute little to the structural damage of the pavement.
2. The amount and damaging effect of the traffic in the early life of such a pavement has a major influence on the life. If a pavement with a bituminous road base survives the first 5 years without serious damage then it is likely to have a long life.

23.35. The only other experimental section directly comparable with those at Conington, considered above, is section 44 at Alconbury bypass. This is still in excellent condition after carrying traffic for 22 years. However, the traffic at this site is rather less damaging than that at Conington, although both experiments are on trunk road A1. Rather thicker bases at Nately Scures had an estimated life of more than 20 years, prior to the opening of the M3 freeway (see Table 18.4).

23.36. The computer program was also used to predict the late-life deflection measured by the deflection beam for a pavement consisting

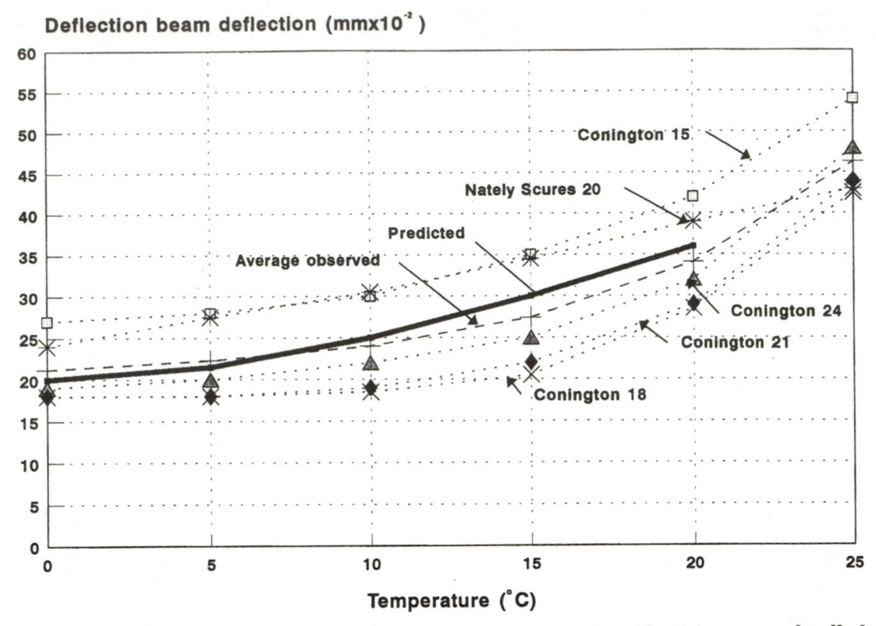

Figure 23.3 Predicted and observed deflections for sections with 250-mm total rolled asphalt on a 150-mm granular subbase: early life.

of 250 mm of rolled asphalt on a 150-mm granular subbase, using the modulus values used in the structural analysis referred to above. The predicted deflection–temperature relationship is shown in Figs. 23.3 and (23.4 for early- and late-life conditions) compared with the measured relationships on a number of relevant experimental pavements. There is some scatter due partly to the different conditions of the pavements and the different amounts of traffic they have carried. There were also differences in the composition of the asphalts. However, the mean curve is very similar to the computed relationship. A comparison of Fig. 23.3 with Fig. 23.4 shows the substantial differences between early- and late-life deflections.

A similar exercise was carried out for sections with a 150-mm dense bitumen macadam (DBM) base under 100-mm rolled asphalt surfacing. In this case, during the early life period 0.30 of the fatigue life will be absorbed in 1 year. At the lower rate applicable to the late-life years 0.033 of the fatigue life will be absorbed per year. The predicted life for these sections is 10 years. The relevant sections at Conington with this construction have an estimated life from the currently observed permanent deformation of between 11 and 100 msa with a mean of 47 msa. The implication is that the fatigue curves for DBM shown in Fig. 13.18 are probably overconservative.

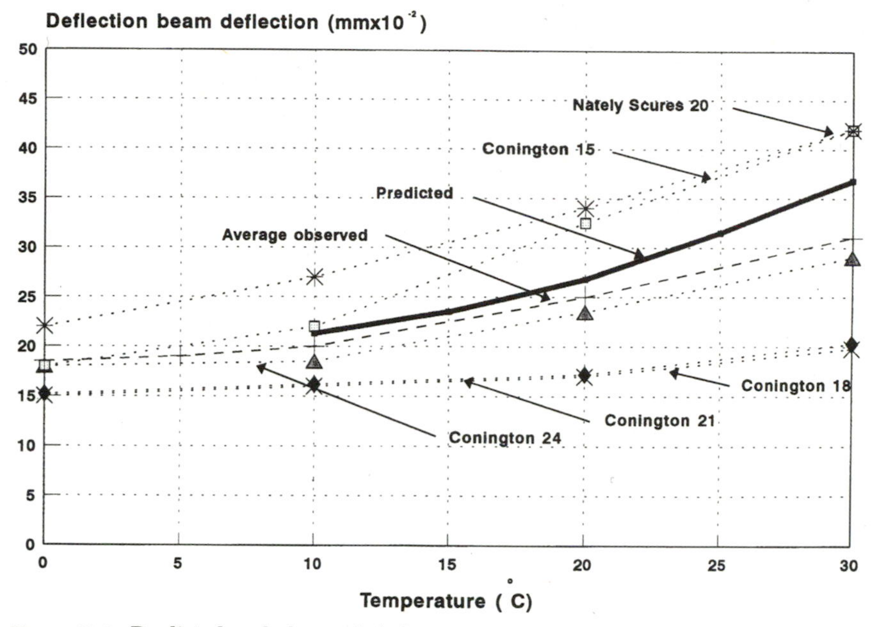

Figure 23.4 Predicted and observed deflections for sections with 250-mm total rolled asphalt on a 150-mm granular subbase: late life.

The Structural Design of Pavements with Wet-Mix Road Bases

23.37. An asphalt surfacing laid on a wet-mix or unbound stone road base is generally assumed to fail as a result of fatigue cracking in the surfacing. It should be possible, if this assumption is correct, to regard the construction as a thin asphalt pavement laid on an unbound material of properties superior to those of a normal subbase. Therefore, it should be feasible to analyze such a pavement in exactly the same way as is followed in the previous section of this chapter, dealing with flexible pavements with an asphalt road base.

23.38. As an example of the procedure the analyses below refer to a 100-mm asphalt surfacing laid on a wet-mix base 230 mm thick over a granular subbase 150 mm thick. Such analyses will model three sections included in the Conington experiment, sections 3, 6, and 9 (see Fig. 18.9). Because it is known that the subgrade under these sections was strengthened at the time of construction by rolling in granular material, two values of modulus have been assumed for the subgrade. These are 45 and 60 MPa. A value of modulus of 150 MPa was assumed for the wet-mix material, but as in the case of the sub-

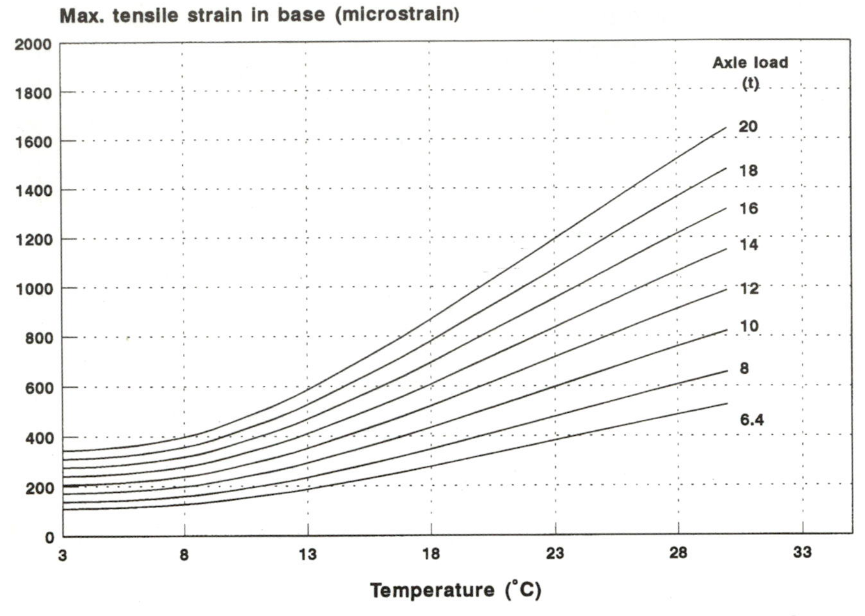

Figure 23.5 Maximum tensile strain in 100-mm rolled asphalt surfacing with 230-mm wet-mix base, 150-mm granular subbase, subgrade modulus 45 MPa: early life.

base under the thick asphalt pavement this value was adjusted by the computer program to ensure that unacceptable tensile strains were not developed in the unbound materials.

23.39. The early- and late-life strain-temperature-axle load relations are shown in Figs. 23.5 to 23.8. The conclusions from these analyses are shown in Tables 23.7 and 23.8. The implication is that over the weaker subgrade the traffic at Conington should have caused failure, using the early-life condition, in 2 to 3 months. For the stronger subgrade the life would have been approximately doubled. In fact, the performance was very different, as is shown in Table 23.9. Certain other sections included in the experiment using the same thickness of surfacing and base did fail earlier by cracking, but these used other than asphalt base courses. However, the life of no section was less than 10 years.

23.40. The conclusion must be that the structural approach cannot be applied in this relatively simple manner to pavements with unbound bases and thin surfacings. The reason may well be that a thin asphalt surfacing is capable of accommodating deformation by creep and that any tendency to crack at lower temperatures is controlled by the more viscous properties at higher temperatures.

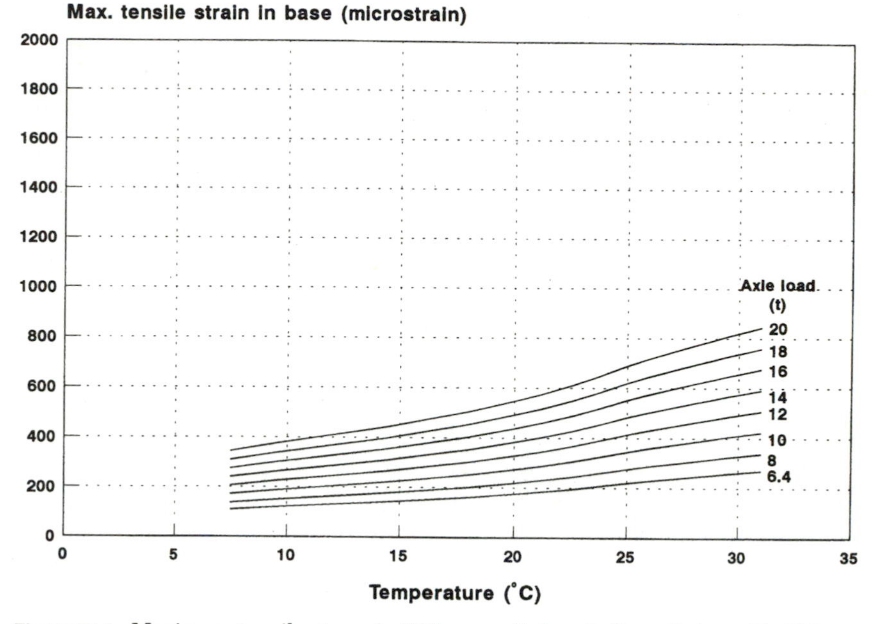

Figure 23.6 Maximum tensile stress in 100-mm rolled asphalt surfacing with 230-mm wet-mix base, 150-mm granular subbase, subgrade modulus 45 MPa: late life.

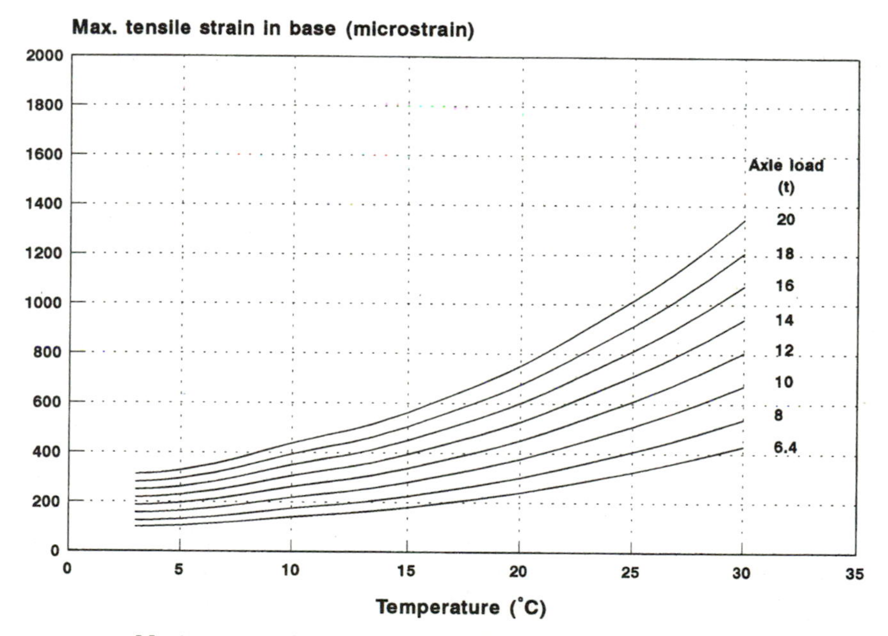

Figure 23.7 Maximum tensile strain in 100-mm rolled asphalt surfacing with 230-mm wet-mix base, 150-mm granular subbase, subgrade modulus 60 MPa: early life.

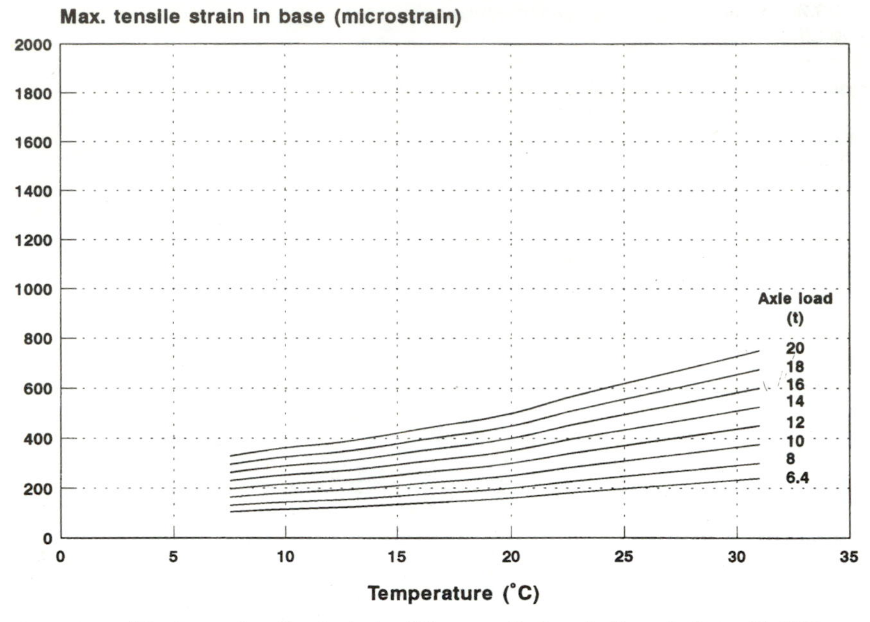

Figure 23.8 Maximum tensile strain in 100-mm rolled asphalt surfacing with 230-mm wet-mix base, 150-mm granular subbase, subgrade modulus 60 MPa: late life.

23.41. The computer program was again used to derive temperature-deflection relations for the two subbase modulus cases considered in the structural analysis of the sections with wet-mix bases. The predicted curves for the subgrades of modulus 45 and 60 MPa are compared with experimental curves for similar experimental sections in Fig. 23.9. The relationship for sections from the Conington experiment fall below the predicted and other experimental curves. This may reflect a rather stiffer foundation at the Conington site or it may be influenced by the wide variety of aggregates used in the base course of the surfacing at that site.

The Structural Design of Pavements with Lean Concrete Bases

23.42. In a pavement with an asphalt surfacing and a lean concrete base, the design criterion is the maximum tensile stress developed by traffic in the underside of the lean concrete. An asphalt surfacing reduces the stress in the lean concrete only marginally, and the influence of temperature fluctuations in the asphalt is small, and it is permissible therefore to carry out a structural analysis using a constant temperature of 20°C in the asphalt. The cases selected here for analy-

TABLE 23.7 Maximum Strain Levels in Relation to Axle Loading, Temperature, and Fatigue Life at Conington*

	Early life			Late life		
Temperature, °C	Calculated strain, microstrain	Fatigue life	Fraction of fatigue life	Calculated strain microstrain	Fatigue life	Fraction of fatigue life
6–8 tonne axle load						
0–5	110	2.5×10^6	0.03	100	4.6×10^6	0.02
5–10	160	9×10^5	0.06	110	6.5×10^6	0.008
10–15	220	5×10^5	0.12	150	3.3×10^6	0.02
15–20	320	2.5×10^5	0.12	180	2×10^6	0.03
20–25	410	1×10^5	0.86	220	1.6×10^6	0.05
8–10 tonne axle load						
0–5	140	1.3×10^6	0.03	130	1.8×10^6	0.02
5–10	200	5×10^5	0.05	150	1.5×10^6	0.02
10–15	290	2.2×10^5	0.13	170	1.3×10^6	0.02
15–20	390	1×10^5	0.29	230	1×10^6	0.03
20–25	510	8×10^4	0.54	270	8×10^5	0.05
10–12 tonne axle load						
0–5	150	8×10^5	0.03	160	8×10^5	0.03
5–10	240	2.5×10^5	0.07	190	6×10^5	0.03
10–15	350	1×10^5	0.18	240	5×10^5	0.04
15–20	460	6×10^4	0.30	270	5×10^5	0.04
20–25	650	3×10^4	0.90	350	4×10^5	0.07
12–14 tonne axle load						
0–5	200	3×10^5	0.04	190	2×10^5	0.06
5–10	270	2×10^5	0.03	240	4×10^5	0.02
10–15	400	6×10^4	0.12	270	2.6×10^5	0.03
15–20	560	3×10^4	0.24	330	2×10^5	0.04
20–25	750	1.4×10^4	0.78	380	2×10^5	0.06
14–16 tonne axle load						
0–5	230	1.4×10^5	0.02	200	2×10^5	0.01
5–10	330	8×10^4	0.02	250	2×10^5	0.007
10–15	450	2.5×10^4	0.06	330	1.8×10^5	0.008
15–20	650	1.6×10^4	0.09	370	1.6×10^5	0.009
20–25	870	1×10^4	0.22	460	8×10^4	0.03
Totals for all axle loads and temperatures			5.33			0.76
		Subgrade modulus = 45 MPa				

*Pavement structure: 100-mm asphalt surfacing, 230-mm wet-mix, 150-mm subbase.

sis are those of a 100-mm asphalt surfacing on a 150- and a 200-mm thickness of lean concrete. A granular subbase 150 mm thick is assumed. The structural properties adopted for the asphalt, the subbase, and the subgrade are the same as those used in the previous analysis of pavements with asphalt bases. Three 28-day target compressive strengths are assumed for the lean concrete. These are shown in Table 23.10 and the associated moduli of rupture in Table 23.11. The computed stresses in the lean concrete are shown in Table 23.12.

TABLE 23.8 Maximum Strain Levels in Relation to Axle Loading, Temperature, and Fatigue Life at Conington*

	Early life			Late life		
Temperature, °C	Calculated strain, microstrain	Fatigue life	Fraction of fatigue life	Calculated strain microstrain	Fatigue life	Fraction of fatigue life
6–8 tonne axle load						
0–5	100	4.5×10^6	0.02	100	4.5×10^6	0.02
5–10	140	1.8×10^6	0.03	120	2.6×10^6	0.02
10–15	190	1.2×10^6	0.05	130	6×10^6	0.01
15–20	260	5×10^5	0.1	160	4×10^6	0.01
20–25	340	3×10^5	0.3	200	2.4×10^6	0.04
8–10 tonne axle load						
0–5	110	2×10^6	0.02	130	1.8×10^6	0.02
5–10	170	1.2×10^6	0.02	150	1.7×10^6	0.02
10–15	240	3×10^5	0.01	180	1.2×10^6	0.02
15–20	340	2×10^5	0.15	200	1.7×10^6	0.02
20–25	420	1.4×10^5	0.31	260	1×10^6	0.04
10–12 tonne axle load						
0–5	160	4×10^5	0.07	150	7×10^5	0.04
5–10	210	2.5×10^5	0.07	180	8×10^5	0.02
10–15	300	2×10^5	0.09	210	5.5×10^5	0.03
15–20	400	1×10^5	0.20	260	4.8×10^5	0.04
20–25	520	5.5×10^4	0.50	300	8×10^5	0.03
12–14 tonne axle load						
0–5	200	3.2×10^5	0.03	180	4×10^5	0.03
5–10	250	2.2×10^5	0.03	220	3.6×10^5	0.02
10–15	330	1.4×10^5	0.05	260	3×10^5	0.02
15–20	460	6.5×10^4	0.17	300	3.5×10^5	0.02
20–25	620	3×10^4	0.40	360	3×10^5	0.04
14–16 tonne axle load						
0–5	240	1.4×10^5	0.02	200	3.2×10^5	0.006
5–10	310	8×10^4	0.02	250	2.1×10^5	0.007
10–15	410	5×10^4	0.03	300	2×10^5	0.008
15–20	550	4.5×10^4	0.03	350	6.5×10^5	0.002
20–25	700	1.8×10^4	0.12	430	6.5×10^4	0.03
Totals for all axle loads and temperatures			2.84			0.563
Subgrade modulus = 60 MPa						

*Pavement structure: 100-mm asphalt surfacing, 230-mm wet-mix base, 150-mm subbase.

TABLE 23.9 Performance of Sections 3, 6, and 9 in the Conington Experiment

Section no.	Condition in mid-1975	Condition in 1984	Life, years
3	10-mm deformation in wheel tracks. No cracking	18-mm deformation. No cracking	21
6	12-mm deformation. No cracking	20-mm deformation. No cracking	19
9	12-mm deformation. No cracking	20-mm deformation. No cracking	19

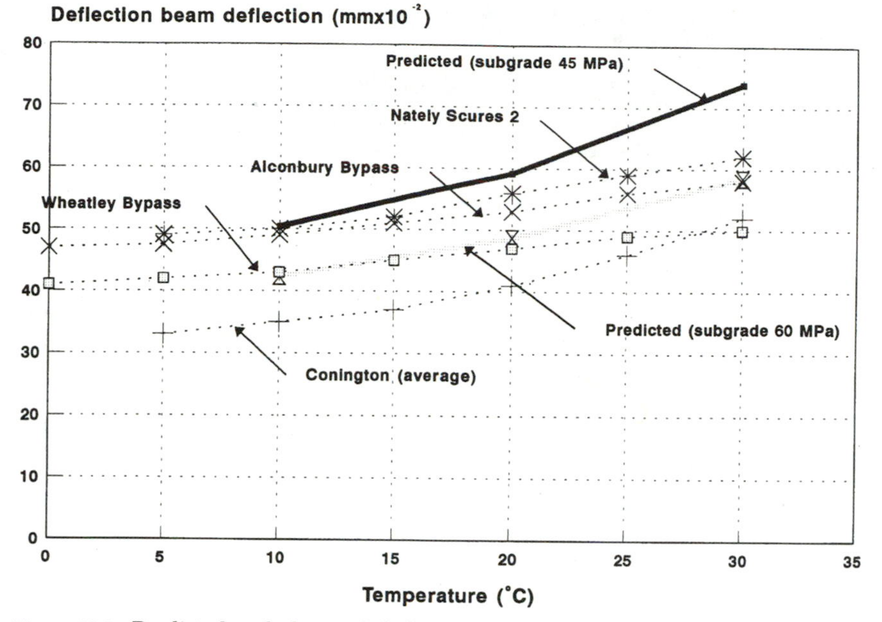

Figure 23.9 Predicted and observed deflection beam deflection for sections with 100 mm of rolled asphalt on 230 mm of wet-mix: late life.

23.43. For the three strengths of lean concrete considered in Tables 23.10 and 23.11, Figs. 23.10 to 23.12 show fatigue curves from which the probable lives of pavements with lean concrete bases can be evaluated.

23.44. The computed maximum tensile stresses induced in the lean concrete by axle loads between 6 and 18 tonnes are given in Table 23.12, together with the number of repetitions of each axle load which would be imposed by the axle load distribution and number of commercial axles at the Conington site. The numbers of axle loads are in periods of 3 months, 1 year, and 5 years. The table also gives the stress induced by a "standard" 8.16-tonne axle.

TABLE 23.10 Structural Properties Assumed for the Lean Concrete

28-day compressive strength, MPa		Modulus of rupture, MPa	Elastic modulus, GPa
Target	Field		
7.5	6.4	1.26	18
10	8.5	1.50	23
14	11.9	1.91	27

TABLE 23.11 Relationship between Field Compressive Strength of Lean Concrete and the Modulus of Rupture

Field compressive strength (28 days), MPa	Modulus of rupture, MPa			
	28 days	3 months	1 year	5 years
6.4	1.26	1.43	1.57	1.80
8.5	1.50	1.75	1.94	2.09
11.9	1.91	2.13	2.31	2.40

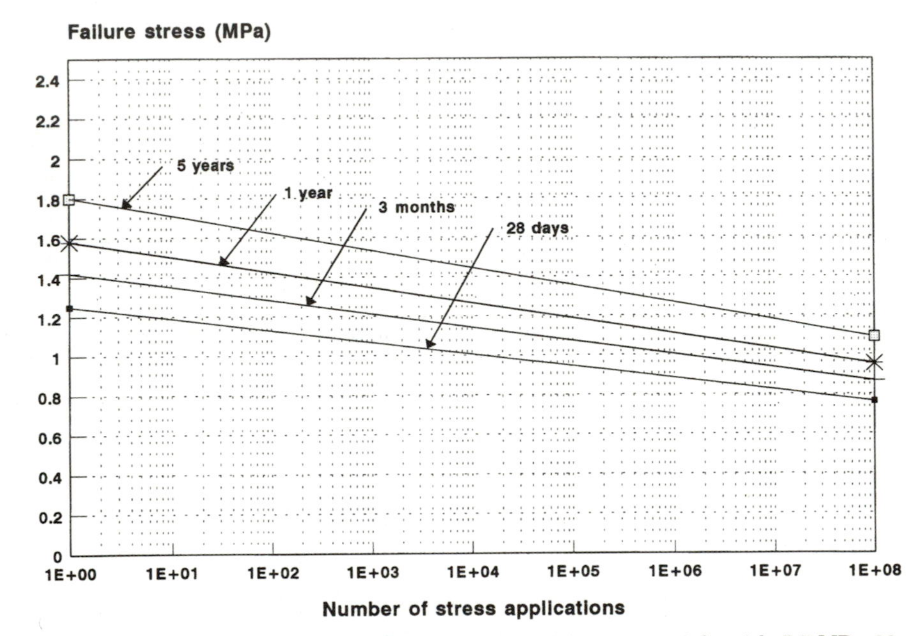

Figure 23.10 Fatigue relationship for cement-bound base materials with 7.5-MPa 28-day compressive strength.

23.45. Inspection of Table 23.12 in conjunction with Figs. 23.10 to 23.12 shows that the 150-mm thickness of lean concrete would fail as a result of most of the axle bands considered. The 200-mm thickness at the highest of the three strengths would accept the 14- to 18-tonne axles without failure and would accept easily the other axle loads. The life of the 200-mm pavement would be controlled by the 16- to 18-tonne axle to about 25 years. In designing a pavement for the traffic at Conington it would therefore be prudent to use the highest-strength lean concrete and to increase the thickness from 200 to about 230 mm. Fracturing of the lean concrete base under tensile stresses does not imply instantaneous failure. Heavily fractured lean

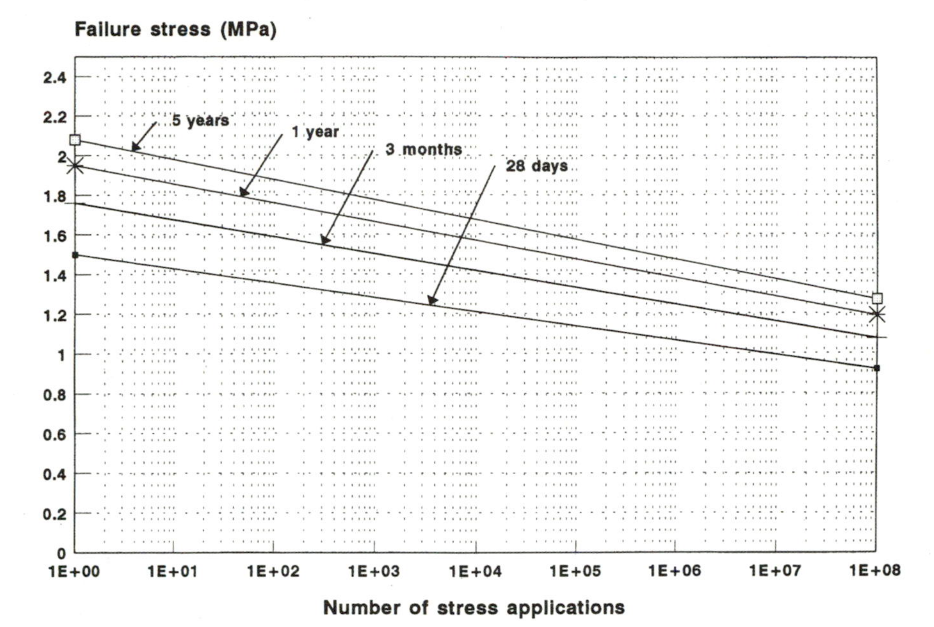

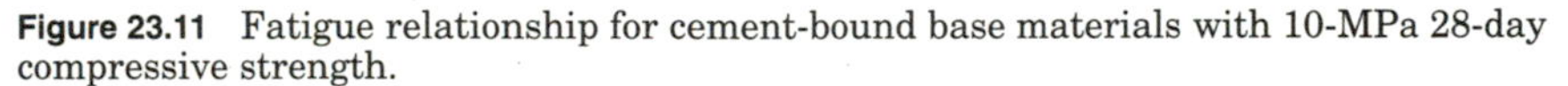

Figure 23.11 Fatigue relationship for cement-bound base materials with 10-MPa 28-day compressive strength.

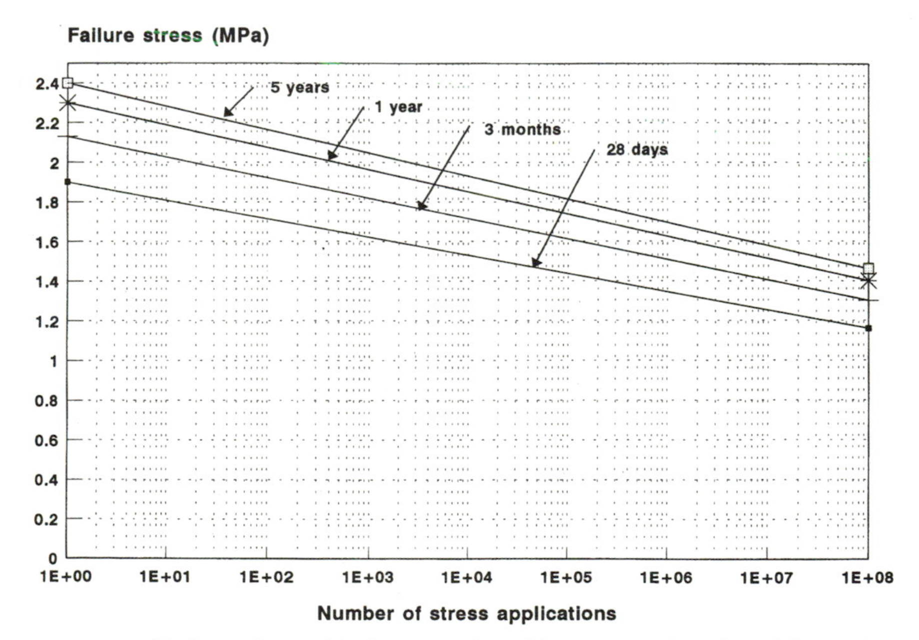

Figure 23.12 Fatigue relationship for cement-bound base materials with 14-MPa 28-day compressive strength.

TABLE 23.12 Computed Maximum Tensile Stress in Lean Concrete Road Bases of Thickness 150 and 200 mm under 100 mm of Rolled Asphalt

Axle load range, t	Field compressive strength (28-day), MPa	Computed max. tensile stress, MPa		No. of repetitions of axle load in		
		150 mm	200 mm	3 months	1 year	5 years
6–8	6.4	0.94	0.68	8.6×10^4	3.4×10^5	1.7×10^6
	8.5	0.95	0.73			
	11.9	0.99	0.77			
8–10	6.4	1.21	0.88	4.3×10^4	1.7×10^5	8.6×10^5
	8.5	1.27	0.93			
	11.9	1.32	0.99			
10–12	6.4	1.48	1.07	2.7×10^4	1.1×10^5	5.5×10^5
	8.5	1.55	1.25			
	11.9	1.62	1.21			
12–14	6.4	1.75	1.27	1.1×10^4	4.3×10^4	2.2×10^5
	8.5	1.83	1.35			
	11.9	1.91	1.43			
14–16	6.4	2.02	1.47	2.2×10^3	8.8×10^3	4.4×10^4
	8.5	2.11	1.56			
	11.9	2.21	1.65			
16–18	6.4	2.29	1.65	3.8×10^2	1.5×10^3	7.5×10^3
	8.5	2.39	1.77			
	11.9	2.50	1.87			
Standard 8.16-t axle	6.4	1.10	0.80			
	8.5	1.15	0.85			
	11.9	1.20	0.90			

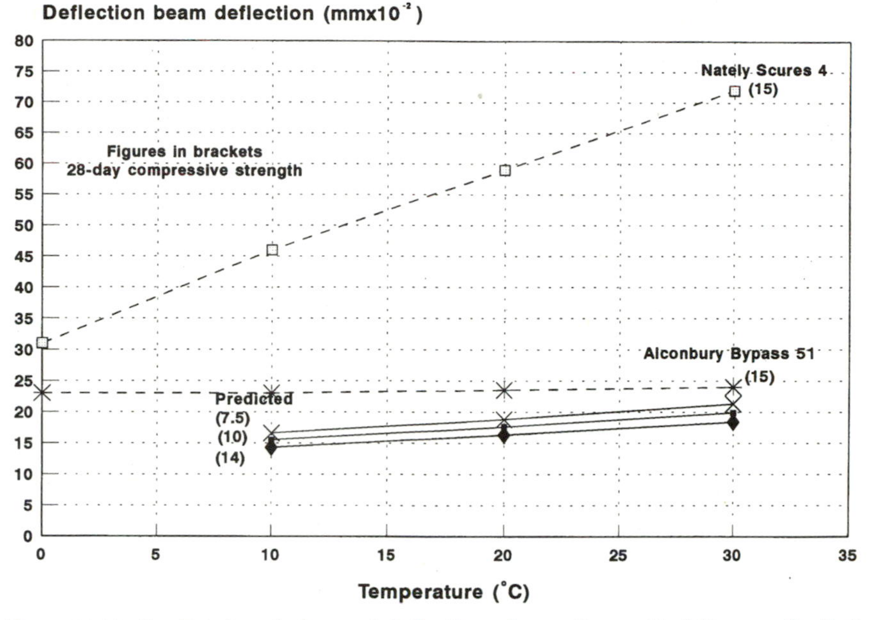

Figure 23.13 Predicted and observed deflections for sections with 100 mm of rolled asphalt on 150-mm lean concrete base.

concrete will behave as a rather inferior form of wet-mix, and hence the life expectancy may still be substantial.

23.46. Figures 23.13 and 23.14 show measured deflections for a number of experimental sections with 150- and 230-mm lean concrete bases of different strengths. As would be expected, the deflections are small but the computed values model the measured values quite well. A comparison of observed and predicted deflection is useful in determining the state of the lean concrete base. Section 1 at Conington, with 230-mm base thickness, has very close agreement between the predicted and observed deflection and there has been relatively little change of deflection with time. This would indicate that the base is still largely unfractured. Section 4, Nately Scures, with a 150-mm lean concrete base, had a deflection before it was overlaid more than double the predicted value, indicating that it was heavily fractured almost immediately after trafficking.

23.47. A 230-mm strong lean concrete base under a 100-mm asphalt surfacing laid at the Conington site in 1968 is still in excellent condition after 28 years and its present life expectancy is about 30 years under the present traffic. At Nately Scures a 150-mm lean concrete

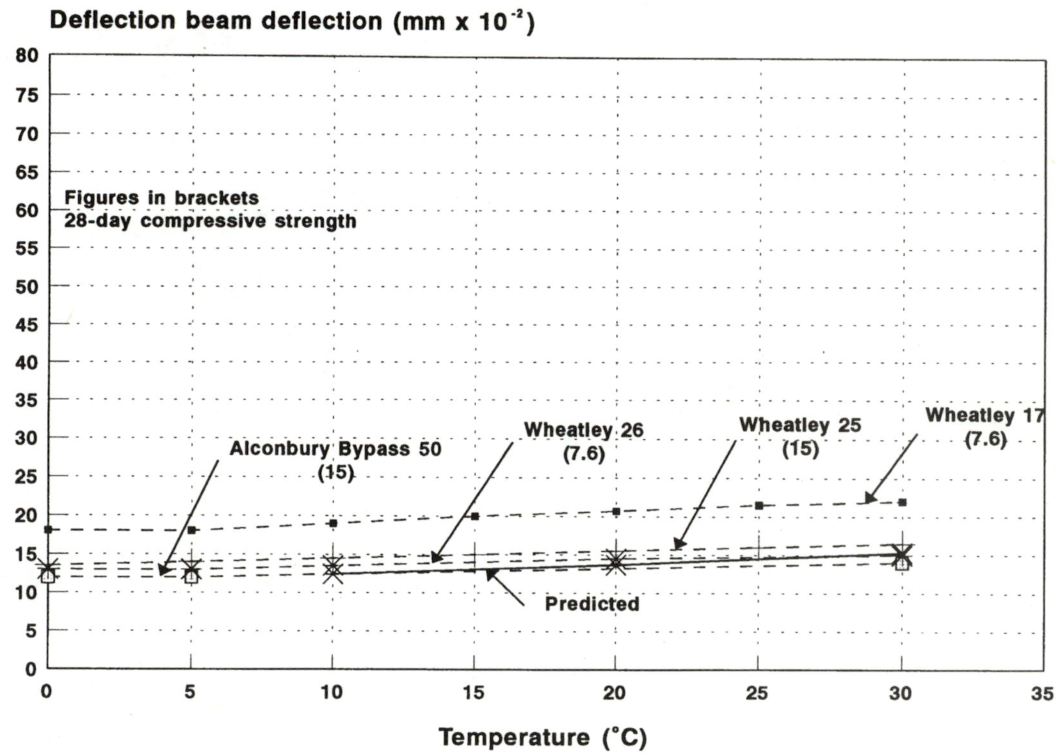

Figure 23.14 Predicted and observed deflections for sections with 100 mm of rolled asphalt on 200-mm lean-concrete base.

base laid under a 100-mm asphalt surfacing had, after 5 years' traffic (2.4 msa), extensive cracking and was overlaid. The section with a 200-mm base showed no sign of distress at the time of opening of the M3 freeway (which took most of the traffic off the experimental sections). This evidence gives considerable confidence in the use of the structural approach to the design of lean concrete bases.

References

1. Thomson, W.: Note on the Integration of the Equations of Equilibrium of an Elastic Solid, *Cambridge and Dublin Mathematical Journal,* 1848.
2. Boussinesq, V. J.: *Applications des potentiels a l'étude de l'équilibre et du mouvement des solides élastiques avec les notes étendues sur divers points de physique, mathématique et d'analyse,* Gauthier-Villars, Paris, 1885.
3. Burmister, D. M.: The Theory of Stresses and Displacement in Layered Systems and Applications to the Design of Airport Runways, *Proc. Highw. Res. Bd, Wash.,* vol. 23, pp. 126–144, 1943.
4. Acum, W. E. A., and L. Fox: Computation of Load Stresses in a Three-layer Elastic System, *Geotechnique, London,* vol. 2, no. 4, pp. 293–300, 1951.
5. Westergaard, H. M.: Stresses in Concrete Pavements Computed by Theoretical Analysis, *Pub. Rds., Washington,* vol. 7, no. 2, pp. 25–35, 1926.
6. Clough, R. W.: The Finite Element Method in Structural Mechanics, *Stress Analysis,* Wiley, London, 1965.
7. Zienkiewicz, O. C.: *The Finite Element Method in Structural and Continuum Mechanics,* McGraw-Hill, London, 1967.
8. Lister, N. W.: A Deflection Beam for Investigating the Behaviour of Pavements under Load, *Road Research Laboratory Research Note* RN/3842/NWL (unpublished), RRL, Crowthorne, 1960.

Chapter

24

The Analytical Design of Concrete Pavements

Introduction

24.1. The finite-element program used to compute the stresses in the various layers of a flexible pavement can also be applied to concrete pavements. The stress which primarily determines whether or not a concrete slab will crack is the maximum tensile stress which develops in the underside of the concrete. Because of the very considerable difference in the elastic properties of the concrete on the one hand and of the subbase and subgrade on the other, the latter two elements of the pavement have less influence than is the case with flexible pavements.

24.2. Since concrete is less able to deform in response to temperature changes than is bitumen, thermal stresses have to be taken into account as well as traffic stresses in analyzing the behavior of concrete pavements. This matter is considered first below.

Thermal Stresses in Concrete

24.3. Temperature stresses in concrete slabs can be conveniently divided into three categories as follows:

1. End-restraint compressive stress
2. Foundation-restraint compressive and tensile stress
3. Partially or completely restrained warping stresses

These are considered separately below.

24.4. End-restraint compressive stress. If a slab is located between rigid abutments without space for expansion, an increase of temperature above the temperature at the time of placing will induce an internal compressive stress. If the coefficient of linear expansion of the concrete is a and the rise of temperature is t, then the strain induced in the concrete will be at and the compressive stress induced in the concrete is given by

$$\sigma_{es} = Eat \tag{24.1}$$

where E is the elastic modulus of the concrete.

Shacklock[1] has given a useful summary of the coefficients of linear expansion of concretes made with different aggregates, and this is reproduced in Table 24.1.

24.5. The placing of pavement concrete is not permitted when the temperature is likely to fall below 3°C, and this can therefore be taken as the lowest temperature at the time of construction. Measurements made in the United Kingdom show that the maximum temperature at the middepth of a concrete slab is unlikely to exceed 30°C. The maximum temperature range which a concrete slab is likely to experience is + 3 to + 30°C. From Eq (24.1) for a concrete of coef-

TABLE 24.1 Influence of Aggregate Type on the Coefficient of Thermal Expansion of Concrete

Aggregate (geological group)	Coefficient of thermal expansion (per °C × 10^{-6})	
	Range	Mean
Chert	11.4–12.2	11.8
Quartzite	11.7–14.6	13.2
Quartz	9.0–13.2	11.1
Sandstone	9.2–13.3	11.3
Marble	4.1–7.4	5.8
Siliceous limestone	8.1–11.0	9.6
Granite	8.1–10.3	9.2
Basalt	7.9–10.4	9.2
Limestone	4.3–10.3	7.3
Gravel	9.0–13.7	11.4
		Mean 10.0

ficient of linear expansion of 10×10^{-6} and Young's modulus 25,000 N/mm^2,

$$\sigma_{es} = 35{,}000 \times 10^{-5} \times 27$$

$$= 9.45\ MN/m^2$$

In the southern states of the United States middepth slab temperatures may well be 5 to 10°C higher than in the United Kingdom, but the temperature of laying will also be correspondingly higher and the restrained compressive stress is unlikely to exceed 10 MN/m^2. This is small in comparison with the minimum compressive strength required of 40 to 50 MN/m^2 at 28 days. It follows that compression failures are very unlikely to occur in restrained slabs unless the concrete in the immediate vicinity of joints has been badly compacted or otherwise disturbed by the formation of the joints. Under such circumstances the restraint stress can give rise to spalling around the joints. However, end-restraint stresses can give rise to buckling or blowup-type failures in extreme temperature conditions, and for this reason it is usual practice in Britain to require expansion joints to be included in all concrete pavements constructed between Oct. 22 and April 20. For pavements constructed in the summer months, i.e., outside this period, the increase of temperature between the time of laying and the hottest part of the year is unlikely to exceed 10°C and the compressive stress generated will be less than 3.5 N/mm^2. Experience in Britain and America indicates that there is little risk of buckling with such a level of stress.

24.6. In reinforced concrete pavements the widely spaced joints can open up sufficiently to trap road grit, and thermal closure of the joint can then lead to stress concentration sufficient to cause spalling. This is less likely to occur with plain concrete pavements where the more frequent joints lead to correspondingly smaller joint movements. If joint grooves and arrises are well formed and properly compacted, the risk of damage from spalling is small.

24.7. Foundation-restraint stress. The opening and closing of joints in a concrete road entails sliding between the slabs and their foundation (subbase or subgrade), or in extreme cases shearing within the foundation itself. The friction between the slab and its foundation will thus generate, in a slab not subject to end restraint, a compressive or tensile stress depending on whether the slab is increasing or decreasing in temperature. The stress resulting from foundation restraint will be a maximum at the midpoint.

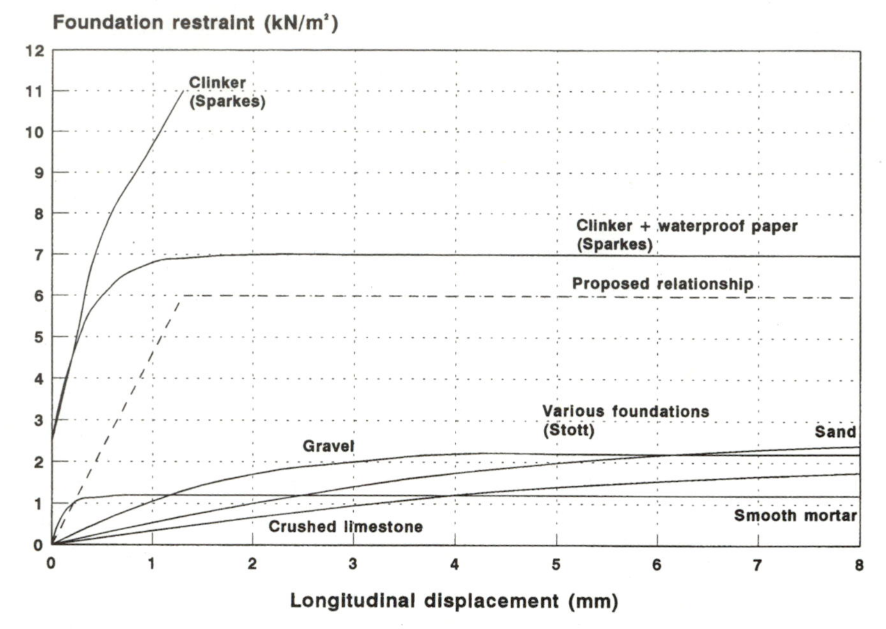

Figure 24.1 Relationships between foundation force and slab displacement (150-mm slabs).

24.8. The magnitude of the foundation restraint stress depends on the force necessary to cause sliding between the concrete and its foundation and the displacement necessary to initiate such sliding. This has been examined in some detail by Sparkes and Stott;[2,3] both studied the force necessary to move horizontal concrete slabs supported by different foundation materials, and the relationship between force and displacement. The results they obtained are summarized in Fig. 24.1.

24.9. There is a large difference between the values obtained by Sparkes using a clinker foundation and those of Stott using various other foundations. In each case the slab was cast on the foundation under test to simulate practical conditions and essentially the same procedure of moving the slabs by jacks was adopted. It is possible that the clinker used by Sparkes may have exhibited a slight pozzolanic effect which increased the bond between the waterproof paper and the foundation. On the other hand, Stott was consciously trying to reduce the restraint force in his tests and the surface finish which he obtained on the foundation may have been to a closer tolerance than that used by Sparkes. In the absence of other information it would seem reasonable to adopt a value of restraint which is the average of Sparkes's value and the mean of the values obtained by Stott. The relationship shown by the broken line on Fig. 24.1 is recommended. Clearly the type and smoothness of the foundation has some effect on the restraint stress, but it is impossible to quantify on this evidence. Both Sparkes and Stott used slabs 150 mm thick for their tests. The restraint stress will increase with increasing weight and hence thickness of slab. Stott made some tests on surcharged slabs and showed that the foundation restraint stress was approximately proportional to the thickness of slab and this enables restraint stresses to be calculated for thickness other than the 150 mm to which Fig. 24.1 applies.

24.10. The stress due to foundation restraint can be calculated as follows using the method originally developed in *Concrete Roads.*[4] Figure 24.2 shows in elevation one-half of a concrete slab of length L, C being the midpoint and A one extremity. On expansion or contraction, all points between C and A will tend to move away from or toward C, thereby generating the foundation restraint stress. There will be one point Y at which the movement will be just sufficient to mobilize the maximum restraint force Q, which will apply to the length between Y and A. Between C and Y the movement will be smaller and only part of the maximum restraint force will be mobilized, as indicated on Fig. 24.2. The actual proportion generated at any point can be deduced from the thermal displacement using Fig. 24.1.

24.11. The force acting on unit width of the slab to the right of point D (Fig. 24.2) is given by

$$\Sigma Q = Q\left(\frac{L}{2} - y\right) + Q\left(\frac{1 + x/y}{2}\right)(y - x) \qquad (24.2)$$

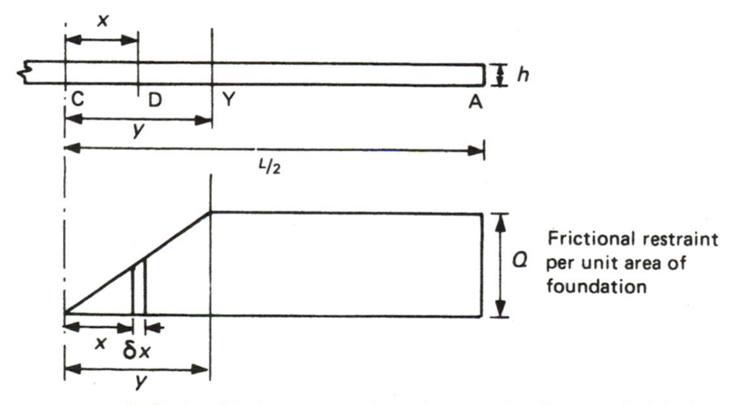

Figure 24.2 Relationship between subgrade restraint force and slab displacement (150-mm slabs).

If we consider the equilibrium of the length of slab between C and Y then the element of slab δx shown in Fig. 24.2 will be subject to a foundation restraint force $(x/y)/Q\ \delta x$. The elastic extension or contraction of length CD as a result of the restraint force acting on δx will be

$$\frac{Q}{Eh} \cdot \frac{x^2}{y}\ \delta x$$

and for the length CY the change in length due to foundation restraint will be

$$\frac{Q}{Ehy} \int_0^y x^2 dx$$

i.e.

$$\frac{Q}{Eh} \cdot \frac{y^2}{3}$$

The change in length of CY due to restraint acting over the length YA will be

$$\frac{Q}{Eh}\, y\left(\frac{L}{2} - y\right)$$

The total change of length of CY due to foundation restraint over the length CA will thus be

$$\frac{Q}{Eh}\, y\left(\frac{L}{2} - y\right) + \frac{Q}{Eh}\, \frac{y^2}{3} \qquad \text{or} \qquad \frac{Q}{Eh}\left(\frac{Ly}{2} - \tfrac{2}{3}y\right)$$

If the slab is subjected to a uniform temperature change θ, and the coefficient of linear expansion is a, then the unrestrained length change in CY would be $ya\theta$ and the restrained length change Δ over the length CY will be given by

$$\Delta = ya\theta - \frac{Q}{Eh}\, y\left(\frac{L}{2} - \tfrac{2}{3}y\right) \tag{24.3}$$

Equation (24.3) used in conjunction with Fig. 24.1 enables the value of y to be calculated. This is then used in Eq. (24.2) to deduce the restraint force acting over length DA, and hence the foundation restraint stress acting over the cross section of the slab at D. The maximum stress will be generated at the center of the slab, where $x = 0$, and where

$$\Sigma Q = Q\left(\frac{L-y}{2}\right)$$

If the foundation restraint stress σ_{fr} is assumed to act uniformly over the cross section of the slab,

$$\sigma_{fr} = \Sigma \frac{Q}{h} \tag{24.4}$$

Near the ends of the slab the eccentricity of loading will result in a marked nonuniformity of stress with depth, but Wallace has shown that the assumption of uniformity is valid except in the end meter of the slab.[5]

24.12. The values of σ_{fr}, deduced from Eqs. (24.2), (24.3), and (24.4), using Fig. 24.1 in conjunction with a value of slab thickness h of 150 mm will apply to all thicknesses of slab if it is assumed that the value of Q is proportional to h. Using values of E = 35,000 N/mm^2 and a = 10^{-6}/°C, which are representative of normal-strength pavement concrete, Fig. 24.3 shows relationships between foundation restraint stress, slab length, and temperature change, which, using the assumption above, will apply to slabs of any thickness.

Thermal Warping Stress

24.13. If at a given instant the measured temperature of a concrete slab decreases with depth then the greater thermal expansion at the surface will tend to produce warping such that the central area is

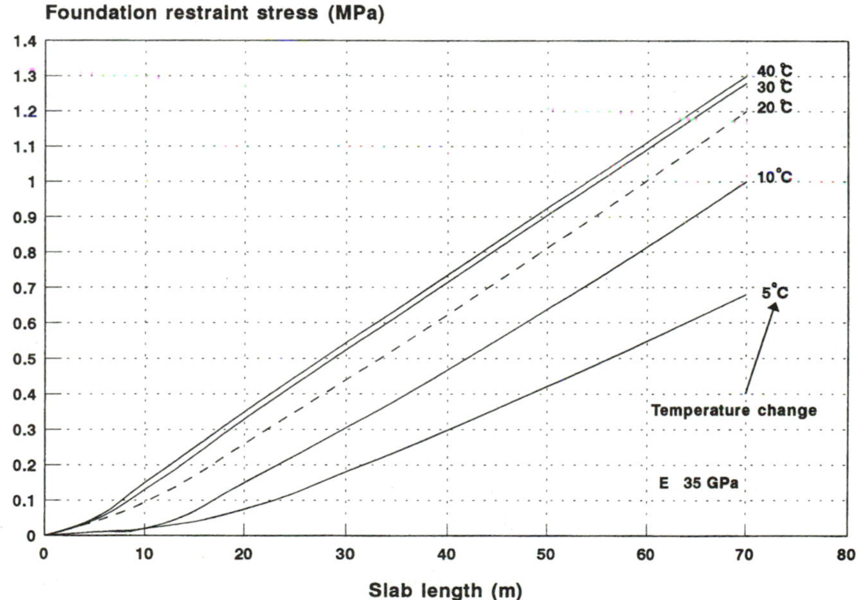

Figure 24.3 Relationship between the foundation restraint stress at the center of a slab and slab length.

higher than the edges. Similarly, if the temperature increases with depth then the slab will tend to warp upward at the edges. If the temperature change with the depth is linear then bending in an unrestrained slab will be symmetrical about the middepth of the concrete and no internal stress will be developed by the bending process. If vertical movement of the slab is restrained, as it will be in practice by the self-weight of the concrete, then restrained warping stresses will be created in the slab. If the temperature at a given instant decreases with depth then the horizontal stress in the surface will be compressive and the horizontal stress in the bottom of the slab will be tensile. Westergaard[6] showed that if the top face of a slab had a temperature $t/2°$ greater than that at the middepth, and the bottom face had a temperature $t/2°$ less than the middepth, then the restrained tensile warping stress in the undersurface of the slab would be given by

$$\sigma_{rw} = \frac{Eat}{2(1-v)} \tag{24.5}$$

where v is the Poisson ratio and the other symbols have the previous definitions.

It follows from Eq. (24.5) that the restrained warping stress defined in this way is independent of the dimensions of the slab.

24.14. Table 24.2 shows hourly temperature measurements made at the surface, the bottom, and the middepth at four equally spaced times of year to correspond to the four seasons. The measurements were made in the southern United Kingdom. At certain times of day the temperature gradient is approximately linear with depth. At 16.00 h on July 1, 1969, the temperature difference between the top and bottom faces of the 254-mm concrete slabs referred to in the table was 4.4°C. Assuming that E = 35,000 N/mm^2, a = 10 × 10^{-6}/°C, and v = 0.14, σ_{rw} deduced from Eq. (24.5) will be approximately 0.9 N/mm^2.

24.15. However, the variation of temperature with depth in a concrete slab is not normally linear, as is shown by Table 24.2. When the surface of the concrete has its maximum or minimum daily temperatures, the temperature difference between the surface and the middepth may be more than double the difference between the middepth and the underface. As a result, internal stresses are set up by the bending process and these tend to increase the restrained warping stress at the top of the slab and decrease it at the underface.

24.16. Thomlinson has made a detailed analysis of the warping stress in concrete slabs taking into account the effects of internal stress.[7] To generalize the treatment, he assumed that the surface of

TABLE 24.2 Temperature Measurements in a Concrete Road Pavement 254 mm Thick at Four Times a Year

Time from midnight, h	January 27, 1970				April 16, 1969				July 17, 1969				October 31, 1969			
	Temperature (°C) at depths shown (mm) and air temperature (°C)															
	0	127	254	Air	0	127	254	Air	0	127	254	Air*	0	127	254	Air
1	5.5	5.7	6.0	5.4	5.4	7.0	8.3	4.2	22.1	25.0	26.0		10.2	10.3	10.7	9.9
2	5.0	5.6	5.9	5.0	5.0	6.6	8.1	4.2	21.0	24.0	25.6		10.0	10.3	10.6	9.9
3	4.7	5.4	5.8	4.8	5.2	6.4	7.7	3.7	19.8	23.1	24.9		10.0	10.3	10.7	9.9
4	3.6	5.0	5.5	3.4	4.7	6.1	7.4	4.0	19.3	22.7	24.3		10.0	10.3	10.8	9.9
5	3.5	4.8	5.4	3.2	4.5	5.9	7.2	4.0	18.8	22.0	21.7		10.4	10.5	10.8	10.0
6	3.0	4.4	5.0	2.7	4.3	5.7	7.0	3.3	18.5	21.5	23.2		10.4	10.5	10.7	10.4
7	3.2	4.0	4.9	3.6	4.4	5.4	6.9	4.0	18.6	21.0	22.0		10.4	10.5	10.7	10.4
8	3.2	4.0	4.8	2.9	4.9	5.4	6.7	4.7	19.7	20.7	22.4		10.1	10.5	10.7	9.9
9	3.2	4.0	4.8	3.1	5.4	5.5	6.7	5.1	21.4	20.7	22.1		10.0	10.4	10.7	10.0
10	3.3	3.9	4.7	3.5	6.8	6.0	6.8	6.7	22.7	21.2	22.0		10.6	10.4	10.7	10.5
11	3.9	4.0	4.7	4.0	9.0	6.6	6.9	7.7	24.2	22.0	22.2		11.2	10.7	10.7	11.0
12	4.3	4.2	4.7	4.7	10.1	7.5	7.1	8.5	25.1	23.0	22.5		11.5	10.9	10.9	11.5
13	4.8	4.4	4.6	5.6	11.0	8.1	7.5	9.0	25.5	23.3	22.7		11.8	10.1	10.0	12.3
14	5.9	4.8	4.8	6.5	12.0	9.0	8.0	9.2	26.8	23.7	23.0		12.7	11.4	11.3	12.7
15	7.0	5.1	4.9	8.0	12.7	9.5	8.4	9.7	27.1	24.4	23.3		12.6	11.8	11.4	12.4
16	7.1	5.6	5.1	7.9	12.4	10.1	8.9	10.4	28.0	25.0	23.6		12.4	11.9	11.8	11.8
17	6.0	5.6	5.3	7.2	12.5	10.4	9.2	10.5	27.4	25.5	24.0		12.1	11.9	11.8	11.7
18	5.0	5.4	5.3	6.0	12.0	10.5	9.6	10.2	27.3	25.6	24.4		11.8	11.8	11.8	11.4
19	4.5	5.0	5.2	5.1	11.0	10.5	9.8	9.0	25.5	25.5	24.5		11.6	11.7	11.8	11.0
20	3.7	4.7	5.1	4.0	10.0	10.1	9.9	8.4	24.5	25.0	24.5		11.3	11.6	11.7	10.5
21	3.5	4.4	5.0	3.6	9.0	9.7	9.8	7.4	22.6	24.1	24.2		11.3	11.5	11.7	10.2
22	2.8	4.1	4.8	3.5	8.0	9.1	9.6	5.0	21.0	23.4	24.0		10.4	11.2	11.6	9.8
23	2.5	4.0	4.7	2.0	7.0	8.6	9.4	4.9	20.0	22.6	23.7		10.2	11.0	11.4	9.4
24	2.0	3.5	4.4	1.0	6.3	7.8	9.0	4.5	19.0	21.8	23.1		9.5	10.8	11.4	8.8

*Air temperatures for July 17, 1969, are not available.

the slab was subject to a sinusoidal variation of temperature such that the temperature at any instant was given by

$$\theta = \theta_0 \sin \frac{2\pi}{T} t \qquad (24.6)$$

where θ_0 = amplitude of the temperature variation (difference between maximum temperature and mean temperature)

θ = temperature at time t after the start of the temperature cycle (generally when the surface is at its mean temperature)

T = length of the time cycle (i.e., for diurnal variations, 24 hours)

Using the well-known thermal diffusion equations, Thomlinson calculated the temperature distribution with depth at any instant, assuming that the thermal diffusivities of the concrete and its foundation were equal. Normal elastic theory was then used to calculate the amplitude and phase of the internal stress variation in the top and bottom of the concrete for various slab thicknesses. The amplitude of the restrained warping stress was similarly calculated, and hence the amplitude and phase of the combined internal stress and restrained warping stress.

24.17. If a value of 0.009 cgs units is used for the thermal diffusivity of moist concrete then Thomlinson's method for calculating the thermal gradients in concrete slabs agrees quite closely with measured values such as those shown in Table 24.2, and therefore the method can be used with confidence to calculate daily and annual variations of warping stress. Thomlinson quoted the stress amplitudes in terms of coefficients to be multiplied by $Ea\theta_0/(1-v)$. Coefficients for slabs in the thickness range 200 to 300 mm are given in Table 24.3.

24.18. Figure 24.4 shows the variation of combined internal and restrained warping stress with time of day for a 250-mm slab subjected to a surface temperature amplitude of 7°C. This corresponds to the summer period when the temperature amplitude is likely to be greatest (see Table 24.2). The maximum tensile stress occurs in the top of the slab in the early hours of the morning. A smaller maximum tensile stress occurs in the bottom of the slab in the early afternoon. The values of E, a, and v used are shown in the figure and they are those previously used in Par. 24.14. In the early summer, days may occur when the temperature amplitude in the surface of the concrete is as great as 10°C. On such days the stress amplitude will be larger in proportion to the temperature amplitude. The stress amplitude in the top of the slab is slightly affected by slab thickness as indicated in

TABLE 24.3 Amplitude of Combined Internal and Restrained Warping Stresses for Concrete Slabs of Various Thicknesses

Slab thickness, mm	Coefficient of combined internal and restrained warping stress*	
	Top of slab	Bottom of slab
200	0.59	0.38
250	0.68	0.38
300	0.73	0.38

*To be multiplied by $\frac{Ea\theta_o}{1-v}$ to give actual amplitude of stress.

Fig. 24.4 but the stress amplitude at the bottom of the slab is not significantly affected by the thickness.

24.19. Thomlinson's method can also be used to calculate the warping stresses due to the annual temperature cycle and these can be compounded with the effects of the daily cycle. However, the temperature gradients associated with the annual cycle are small and the warping stresses they induce can generally be neglected.

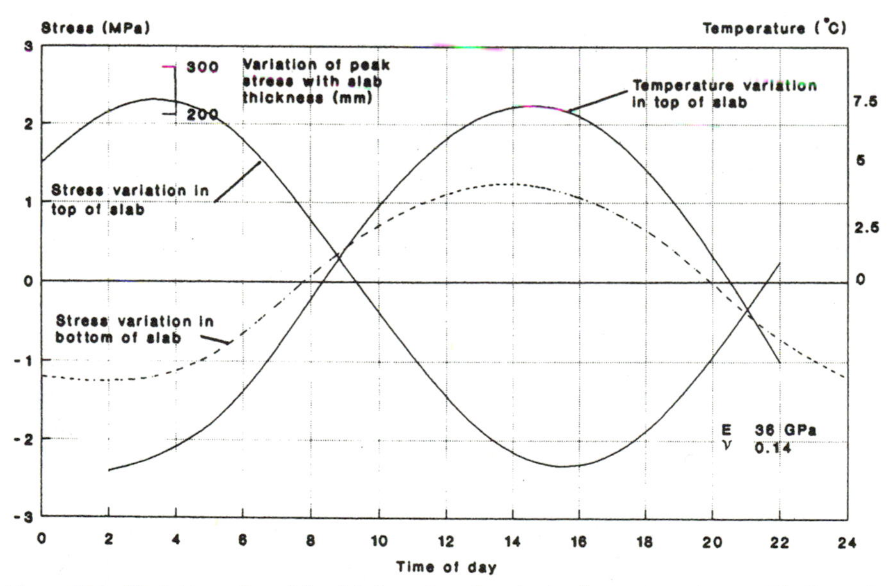

Figure 24.4 Variation of combined internal and restrained warping stress with time of day.

Combination of Warping, Foundation Restraint, and Traffic-Induced Stress

24.20. A reinforced concrete slab designed in the United Kingdom to Road Note 29[8] will normally have a maximum length between sliding joints of about 30 m. For such a slab constructed in the summer months the maximum temperature drop at the middepth in the following winter will, from Table 24.2, be about 20°C. From Fig. 24.3 the maximum tensile foundation restraint stress generated in the winter would be approximately 0.4 N/mm^2. In short unreinforced slabs of length about 5 m, Fig. 24.3 shows that the foundation restraint stress would be negligible.

24.21. Table 24.2 shows that in the summer in the United Kingdom the amplitude of the temperature fluctuation in the top of the slab is about 5°C, and from Fig. 24.4 it can be concluded that the associated maximum tensile warping stress will be about 0.8 N/mm^2. This will occur in the afternoon with an average value of about 0.4 N/mm^2 between 08.00 and 20.00.

24.22. It is not difficult, given the information in Chap. 8 relating to the traffic distribution on major roads during the 24-hour period, together with deductions from Figs. 24.3 and 24.4, to compute combined traffic and thermal stresses on a daily and seasonal basis during the life of a concrete road. However, a "safe" procedure in the United Kingdom would be to add 0.8 N/mm^2 to traffic stresses to cover the influence of thermal stresses for both reinforced and unreinforced pavements. Throughout the life of the road this would significantly overestimate the combined stress and provide a factor of safety. For more extreme climates this procedure would need to be reexamined based on Figs. 24.3 and 24.4.

Design Criterion for Concrete Pavements

24.23. If at any time during the life of a concrete pavement the tensile stress in the underside of the concrete, owing to the combined effects of traffic and temperature, exceeds the modulus of rupture at that time then cracking may be initiated. It has been seen in Chap. 14 that the modulus of rupture is related to the compressive strength, and as a consequence it increases with the age of concrete over a period of at least 5 years. The relation between modulus of rupture and compressive strength depends also to some extent on the aggregate used in the concrete. Values of modulus of rupture for concretes of compressive strength between 10 and 60 N/mm^2 made with gravel and crushed-rock aggregates have been shown in Table 14.4.

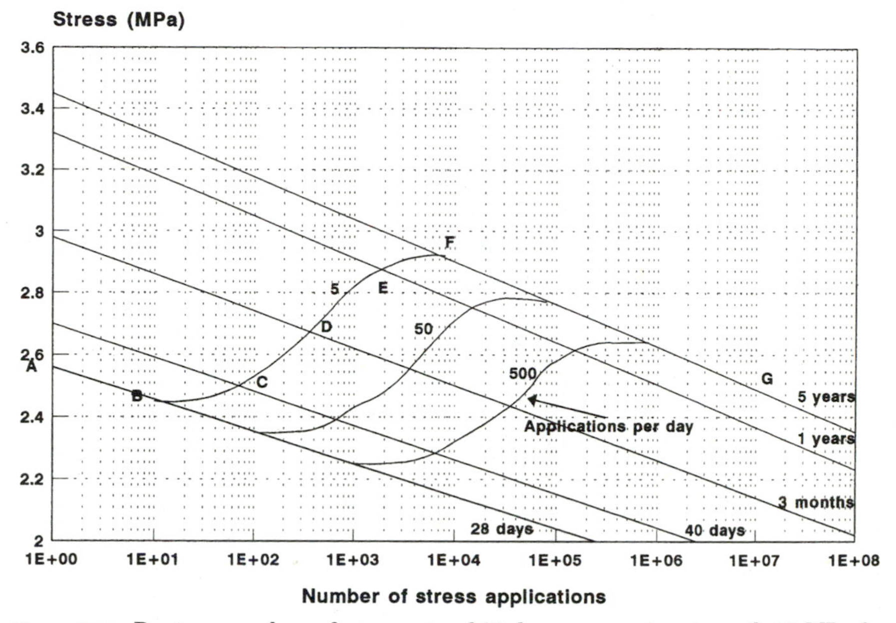

Figure 24.5 Rupture envelopes for concrete of 28-day compressive strength 20 MPa for three rates of loading.

24.24. Taking as an example a concrete with 28-day compressive strength 20 N/mm^2 made with gravel aggregate, Fig. 24.5 shows fatigue lines corresponding to ages of from 28 days to 5 years. In accordance with Par. 14.19 the fatigue strength at 10^5 applications is taken as 0.8 times the modulus of rupture at 28 days. Were stress applications applied so quickly that no significant gain of strength occurred then the behavior of the concrete would be represented by a single fatigue line. However, in practice a road pavement is designed for a long life and during the first 5 years a considerable change of strength will occur, as shown in Fig. 14.1. As a consequence of the increase in the modulus of rupture,the rupture envelope representing the behavior of the concrete will cross the fatigue lines representing different ages in the manner shown in Fig. 24.5. For a rate of five applications per day the rupture envelope is represented by the path *ABCDEFG* in the figure, and the material will then fatigue on approximately the 5-year fatigue line. The "safe" stress is represented by the point *B* and the life in terms of number of stress applications applied at that rate will correspond to the point *G*. As the rate of application of stress is increased the life in terms of number of applications is increased, although it is decreased in terms of time.

24.25. The procedure shown in Fig. 24.5 has been applied to all the concretes considered in Table 14.4, and the results are summarized in Table 24.3. The maximum "safe" stresses for rates of loading of 5, 50, and 500 applications per day are shown.

Computation of Traffic Stress

24.26. The finite-element computer program already discussed in connection with flexible pavement design was used to compute the maximum tensile stress generated by axle loads in the range 6 to 20 tonnes in concrete pavements of thickness 160 to 360 mm. The elastic properties assumed for the concrete are shown in Table 24.5. Two strengths of subgrade, CBR 2 and 5 percent, are considered. In accordance with the discussion in Chap. 14 a constant value of elastic modulus was assumed for the concrete, irrespective of its compressive strength. The computed traffic generated stresses are shown in Table 24.6. To these stresses a further tensile stress of 0.8 must be added in accordance with Par. 24.22 to obtain the maximum combined tensile stress due to traffic and temperature. This has been done in Fig. 24.6. This enables the "safe" stress to be read off for any thickness of slab and any axle load. As would be expected, the CBR of the subgrade has a relatively small influence on the tensile stress in the concrete.

24.27. The maximum safe levels of tensile stress given in Table 24.4 for each of the three rates of loading can be applied in Fig. 24.6 to obtain the relations between the thickness, axle load, and 28-day compressive strength. This has been done in Figs. 24.7, 24.8, and 24.9 for the three rates of loading of 5, 50, and 500 applications per day. These three figures apply only to concrete made with gravel aggregate. From the safe stresses given in Table 24.4 for concretes of the same compressive strength made with crushed-rock aggregate a companion set of curves can be developed.

Application of the Structural Design Approach to Concrete Road Pavements

24.28. The design procedure described in this chapter is based on the maximum axle load the pavement is likely to carry. A check should be carried out to verify that the design so obtained will be capable of carrying the appropriate number of other axle loads to be carried during the design life without the danger of fatigue failure. An example is used below to illustrate the procedure in detail.

TABLE 24.4 Maximum Acceptable Tensile Stress for Various Rates of Loading and Various Strengths of Concrete Made with Gravel and Crushed-Rock Aggregates, Together with Fatigue Lives

		Maximum acceptable tensile stress, N/mm^2		Life for maximum acceptable stress applications, number of applications	
28-day compressive strength of concrete, N/mm^2	Number of stress applications per day	Gravel aggregate	Crushed rock	Gravel aggregate	Crushed-rock aggregate
10	5	1.68	1.70	1.5×10^7	3.2×10^8
	50	1.61	1.65	8.5×10^7	2.5×10^9
	500	1.53	1.58	4.0×10^8	4.6×10^9
20	5	2.45	2.82	1.8×10^7	3.3×10^8
	50	2.35	2.68	1.0×10^8	8.2×10^8
	500	2.25	2.57	6.2×10^8	2.4×10^9
30	5	3.05	3.75	5.2×10^6	1.6×10^8
	50	2.94	3.58	2.5×10^7	5.0×10^8
	500	2.80	3.43	1.6×10^8	3.4×10^9
40	5	3.59	4.57	1.1×10^6	2.7×10^7
	50	3.43	4.38	8.0×10^6	1.3×10^8
	500	3.25	4.20	5.5×10^7	7.0×10^8
50	5	4.05	5.35	2.7×10^5	2.6×10^6
	50	3.89	5.12	1.9×10^6	1.6×10^7
	500	3.72	4.90	1.2×10^7	8.2×10^7
60	5	4.45	6.08	4.5×10^4	4.0×10^5
	50	4.25	5.75	4.4×10^5	2.2×10^6
	500	4.07	5.58	3.4×10^6	1.3×10^7

TABLE 24.5 Elastic Properties of the Pavement Materials

Pavement layer	Modulus of elasticity E, MN/m^2	Poisson's ratio
Concrete slab	40,000	0.15
Granular subbase	Maximum 300	0.40
Subgrade:		
CBR 2%	10	0.40
CBR 5%	30	0.40

24.29. The Alconbury Hill pavement design experiment on trunk road A1 is suitable for analysis since the performance over a period of 20 years has been documented.[9] The experiment is described in detail in Chap. 19. Concrete sections of various thicknesses were incorporated in the experiment and concretes made with gravel aggregate and crushed-rock aggregate were used, the strengths at 28 days being 44

TABLE 24.6 Computed Tensile Stress Caused by Various Axle Loads in Concrete Slabs of Various Thickness

Slab thickness, mm	CBR of subgrade, %	Maximum tensile stresses in concrete, MN/m^2, for axle loads shown, tonnes							
		6	8	10	12	14	16	18	20
140	2	1.53	1.93	2.49	2.98	3.56	4.04	4.65	5.30
	5	1.39	1.72	2.31	2.71	3.23	3.83	4.30	4.82
160	2	1.29	1.62	2.08	2.46	2.92	3.32	3.80	4.30
	5	1.16	1.47	1.93	2.25	2.67	3.11	3.52	3.94
180	2	1.06	1.41	1.77	2.11	2.46	2.82	3.16	3.54
	5	0.97	1.29	1.62	1.94	2.25	2.59	2.90	3.24
200	2	0.87	1.16	1.46	1.74	2.03	2.33	2.62	2.92
	5	0.82	1.09	1.37	1.63	1.90	2.18	2.48	2.74
220	2	0.74	0.99	1.24	1.48	1.72	1.98	2.22	2.48
	5	0.70	0.93	1.17	1.39	1.62	1.86	2.08	2.34
240	2	0.65	0.86	1.08	1.29	1.50	1.72	1.94	2.16
	5	0.61	0.81	1.01	1.21	1.40	1.62	1.82	2.02
260	2	0.57	0.75	0.94	1.13	1.31	1.51	1.70	1.88
	5	0.53	0.70	0.88	1.06	1.23	1.41	1.58	1.76
280	2	0.50	0.66	0.83	0.99	1.15	1.33	1.48	1.66
	5	0.45	0.60	0.76	0.90	1.05	1.21	1.40	1.54
300	2	0.44	0.59	0.74	0.88	1.02	1.18	1.32	1.48
	5	0.40	0.53	0.67	0.80	0.93	1.07	1.20	1.34
320	2	0.39	0.52	0.65	0.78	0.90	1.04	1.16	1.30
	5	0.35	0.47	0.59	0.70	0.82	0.94	1.06	1.18
340	2	0.35	0.47	0.59	0.70	0.82	0.94	1.06	1.18
	5	0.32	0.43	0.53	0.64	0.74	0.86	0.96	1.08
360	2	0.31	0.42	0.52	0.62	0.73	0.83	0.94	1.04
	5	0.29	0.39	0.49	0.58	0.68	0.78	0.88	0.98

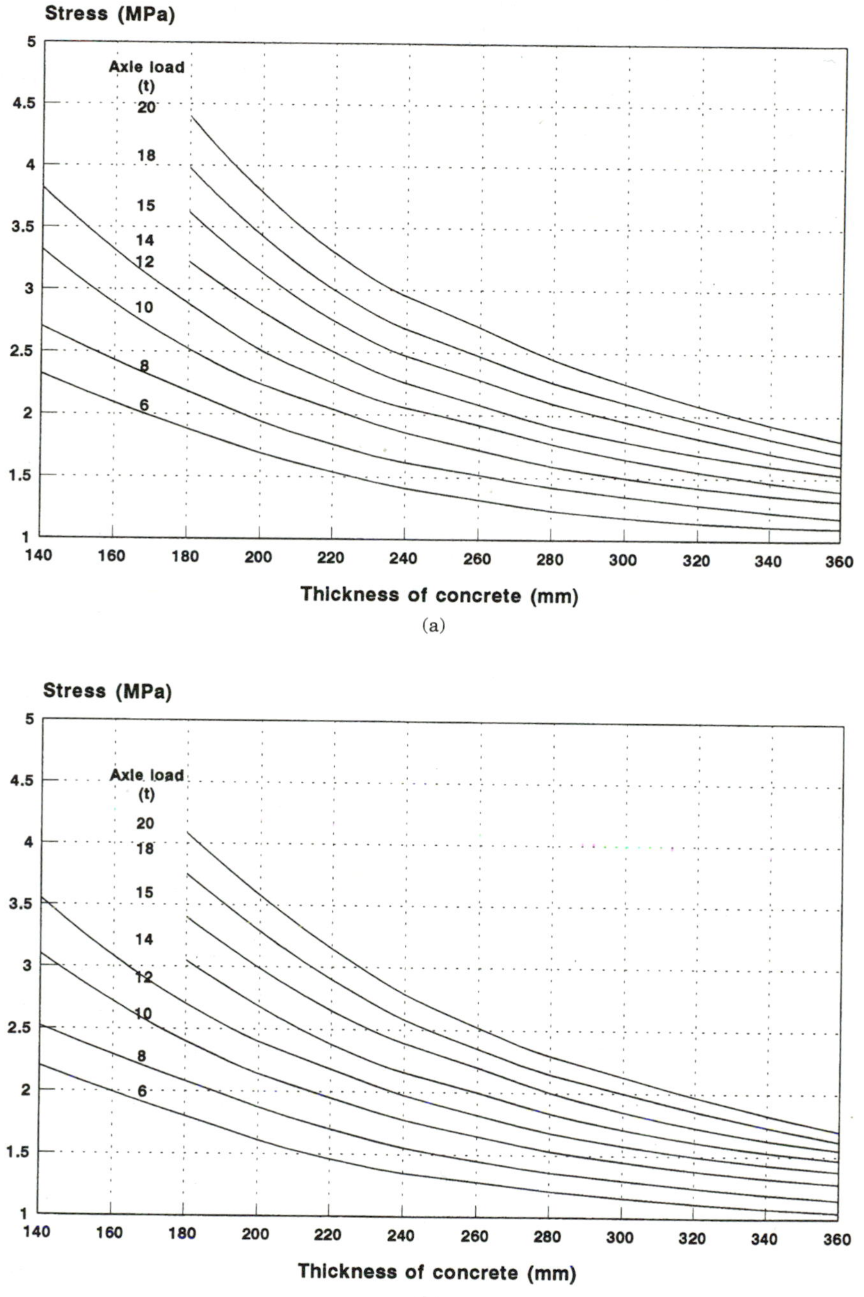

Figure 24.6 Relation between slab thickness and combined traffic/thermal stress for various axle loads: (*a*) foundation CBR 2 percent (*b*) foundation CBR 5 percent.

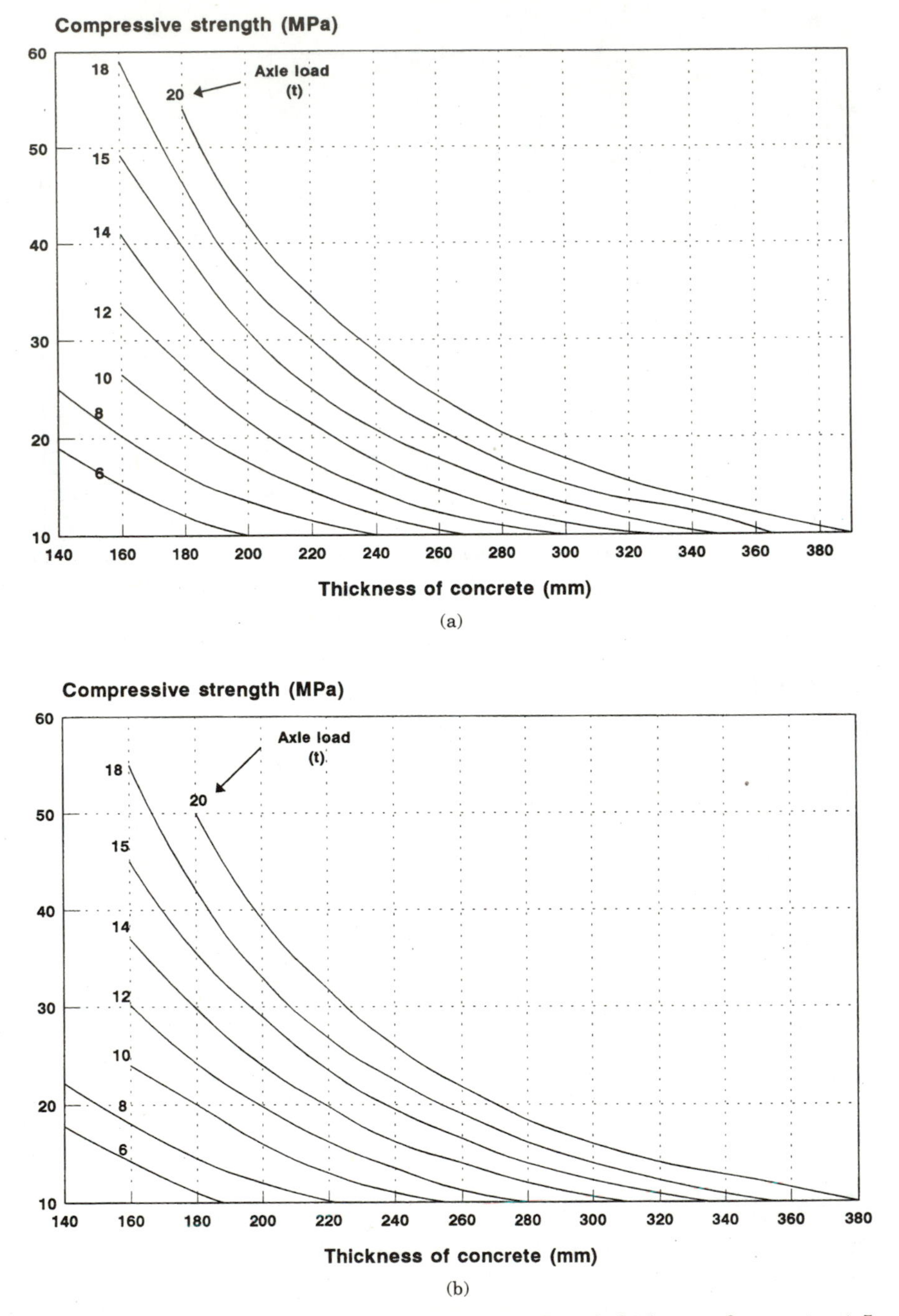

Figure 24.7 Relation between compressive strength and thickness of concrete at 5 applications/day: (*a*) foundation CBR 2 percent (*b*) foundation CBR 5 percent.

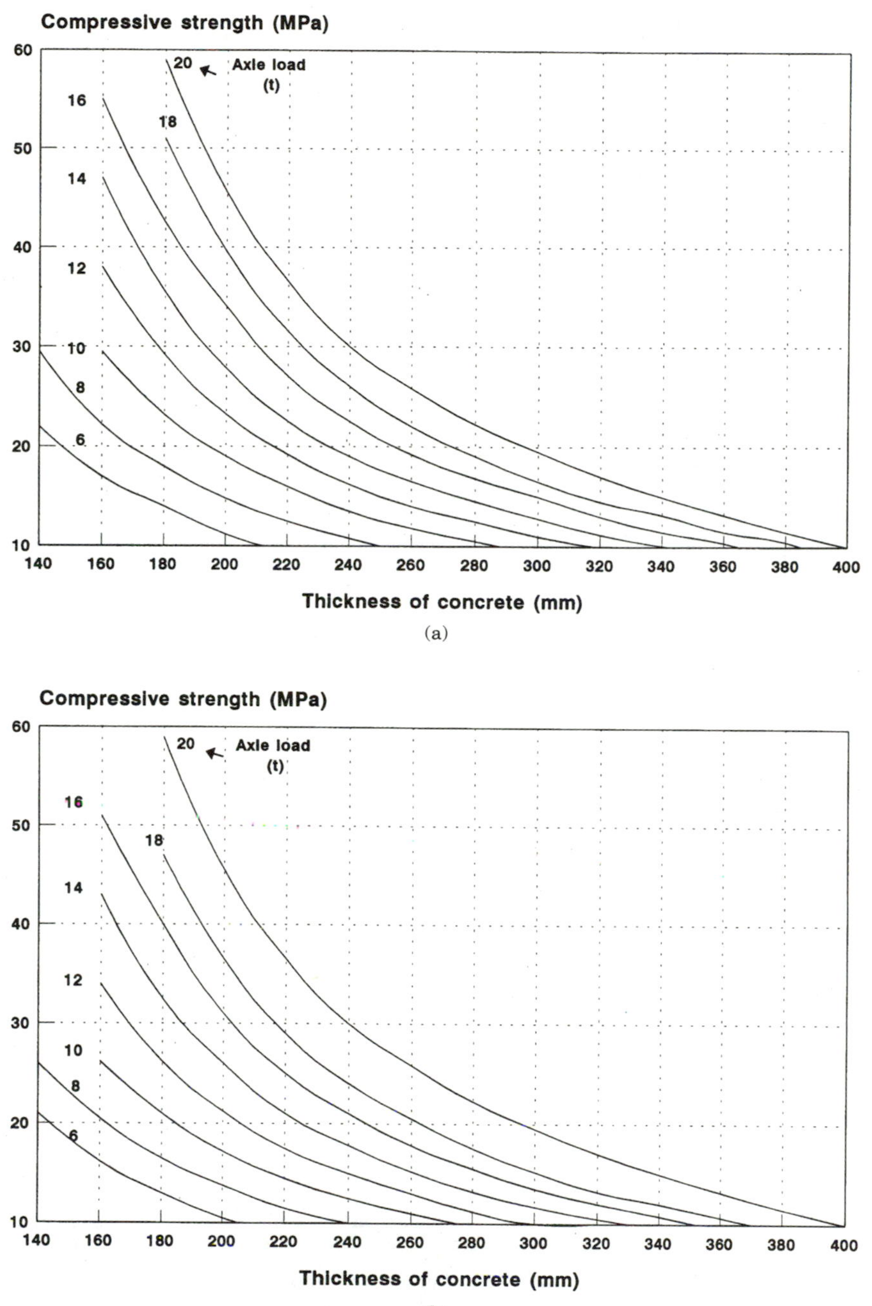

Figure 24.8 Relation between compressive strength and thickness of concrete at 50 applications/day: (*a*) foundation CBR 2 percent (*b*) foundation CBR 5 percent.

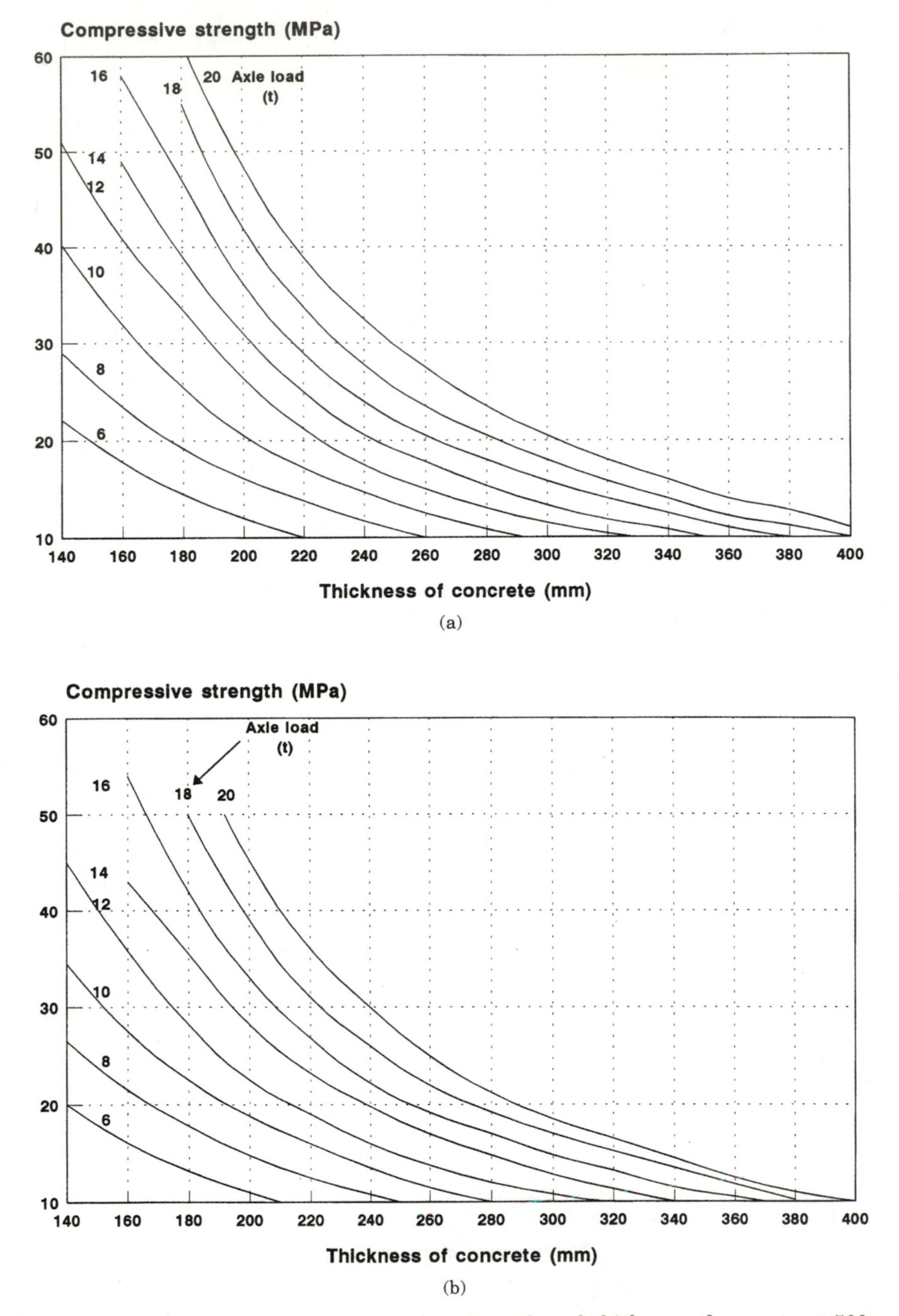

Figure 24.9 Relation between compressive strength and thickness of concrete at 500 applications/day: (*a*) foundation CBR 2 percent (*b*) foundation CBR 5 percent.

TABLE 24.7 Axle Loading at the Alconbury Hill Experimental Road, Trunk Road A1

(1) Axle load, tonnes	(2) Percentage	(3) Number of axles in 20 years	(4) Stress level, N/mm^2	(5) Fatigue life applications (40 N/mm^2 gravel aggregate)
1–2	34.75	11.5×10^7		
2–3	21.11	7.0×10^6		
3–4	16.08	5.3×10^6		
4–5	7.40	2.4×10^6		
5–6	6.17	2.0×10^6	1.6	$>10^{12}$
6–7	4.57	1.5×10^6		
7–8	3.21	1.06×10^6	1.8	$>10^{12}$
8–9	2.68	8.84×10^5		
9–10	1.42	4.68×10^5	2.2	$>10^{12}$
10–11	1.34	4.42×10^5		
11–12	0.54	1.78×10^5	2.45	$>10^{10}$
12–13	0.39	1.28×10^5		
13–14	0.13	4.29×10^4	2.7	$>10^9$
14–15	0.11	3.63×10^4		
15–16	0.11	3.63×10^4	2.95	$>10^9$
16–17	0.03	9.9×10^3		
17–18	0.007	2.3×10^3	3.25	10^8
18–19	0.016	5.2×10^3		
19–20	0.032	1.0×10^4	3.5	3×10^6

and 66 N/mm^2, respectively. The axle load distribution, determined from a recording weighbridge, is shown in Table 24.7, and the total number of commercial axles which passed over the sections in 20 years was 33 million.

24.30. The number of axles of loading in the range 1 to 20 tonnes which passed over the experimental section during the first 20 years is shown in column 3 of Table 24.7. There are a significant number of axles carrying 20 tonnes, some of which will be associated with heavy indivisible loads. The design must cater for these, although the number is only 1 to 2 per day. Reference to Figs. 24.7 and 24.8 shows that the thickness requirements for 5 and 50 stress applications per day are very similar and there is likely to be very little difference between the requirements for 2 and 5 applications per day. The design for this road would therefore be based on Fig. 24.7 for the sections of gravel aggregate. This approach gives the following design thicknesses for a 20-tonne maximum axle load on concrete pavements laid on a foundation of CBR 5 percent.

Aggregate	28-day compressive strength	Thickness, mm
Gravel	40 N/mm^2	200
Gravel	60 N/mm^2	172
Crushed rock	46 N/mm^2	165
Crushed rock	60 N/mm^2	135

24.31. Reference to Table 19.4 shows that the lower-strength concrete made with gravel aggregate of thickness 200 mm gave a life in excess of 20 years, while that with a thickness of 175 mm had a life of 16 to 18 years. The same table shows that 175 mm of the higher-strength concrete had a life in excess of 20 years while a thickness of 150 mm had a life of about 16 years. These results show close agreement between the theoretical and observed performance. With the high-strength crushed-rock concrete the actual life was a little less than the theoretical life, and this may well reflect the difficulty of compacting this material in situ.

24.32. In applying the structural approach it is advisable to check that the combined loading due to the axle loads less than the maximum used in the design will not lead to earlier fatigue failure. To provide this confirmation, Table 24.7 has been extended to show the maximum combined loading and thermal stress levels due to axle loads of between 6 and 18 tonnes. These stress levels are deduced from Fig. 24.7 for a 200-mm pavement. Figure 24.10 shows the 5-year fatigue relationships for concrete made with gravel aggregates. For such relationships the proportion of the fatigue life absorbed by each axle load can be estimated. The fatigue lives are shown in column 5 of Table 24.7. Inspection shows that less than 5 percent of the fatigue life would be absorbed by axle loads less than the maximum used in the design.

Structural Design and the Concept of Standard Axles

24.33. In recent years some research workers have used structural design procedures to design pavements on the basis of repetitions of standard axles. This is inadmissible in the case of concrete pavements. Reverting to the Alconbury Hill experiment referred to above, the concrete pavements at that site have carried 9.43 × 10^6 standard axles in 20 years, or about 1290 per day. Following the procedure discussed in this chapter, for a concrete made with a gravel aggregate of 28-day strength 40 N/mm^2, the maximum tensile stress which the concrete would accept without cracking is 3.2 N/mm^2 for this rate of

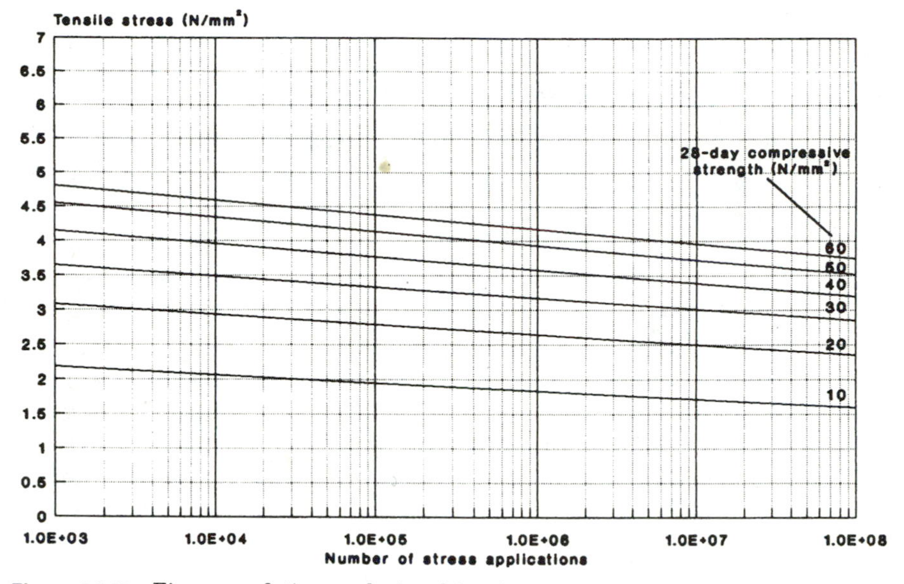

Figure 24.10 Five-year fatigue relationships for concretes of 28-day compressive strength 10 to 60 N/mm^2.

loading. Figure 24.6 shows that for this level of combined traffic and thermal stress, the thickness of concrete required would be about 105 mm. Figure 24.10 indicates that fatigue failure at this stress level would occur at about 10^8 applications, i.e., after a life much in excess of 20 years. However, the experimental evidence (Fig. 19.4) shows that a pavement of this thickness would have lasted less than 2 years. The conclusion is that concrete pavements must be designed for the maximum axle loads they are likely to carry.

24.34. This does not mean that standard axles should not be used, as in Road Note 29, to quantify traffic. The constituents of the traffic in terms of axle loading are inevitably taken into account when designs are based solely on experimental evidence from in-service roads.

24.35. The structural approach to the design of concrete roads highlights the critical influence of all the factors, namely, axle loading, thickness, and concrete strength, in formulating designs.

References

1. Shacklock, B. W.: *Concrete Constituents and Mix Proportions,* Cement and Concrete Association, London, 1974.

2. Sparkes, F. N.: Stresses in Concrete Road Slabs, *Struc. Engr.,* vol. 17, no. 2, pp. 98–116, 1939.
3. Stott, J. P.: Tests on Materials for Use in Sliding Layers under Concrete Road Slabs, *Civ. Engng., London,* vol. 56, no. 663, 1297, 1299, 1301; no. 664, 1466–1468; no. 665, 1603–1605, 1961.
4. Road Research Laboratory: *Concrete Roads—Design and Construction,* HMSO, 1955.
5. Wallace, K. B.: Subgrade Restraint Stresses in Concrete Roads, *Roads and Road Construction,* vol. 46, no. 505, pp. 303–304, 1968.
6. Westergaard, H. M.: Analysis of Stresses in Concrete Roads Caused by Variations of Temperature, *Publ. Rds. Wash.,* vol. 8, no. 3, pp. 54–60, 1927.
7. Thomlinson, J.: Temperature Variations and Consequent Stresses Produced by Daily and Seasonal Temperature Cycles in Concrete Slabs, *Conc. Constr. Engng,* vol. 35, no. 6, pp. 298–307; no. 7, pp. 352–360, 1940.
8. Transport and Road Research Laboratory: A Guide to the Structural Design of Pavements for New Roads, *Department of the Environment Road Note* 29, 3d ed., HMSO, London, 1970.
9. Nowak, J. R.: The Concrete Pavement Design Experiment on Trunk Road A1 at Alconbury Hill: Twenty Years' Performance, *Transport and Road Research Laboratory Report* LR887, TRRL, Crowthorne, 1979.

Chapter

25

The Design of Heavily Loaded Industrial Pavements

Introduction

25.1. The last 50 years have seen a revolution in the handling of heavy goods, particularly in port areas. The almost random packing of miscellaneous commodities in the holds of fleets of small ships has given way to the use of standard containers, which are lifted from large purpose-built container ships in a fraction of the time taken by dockers dealing with mixed cargoes. ISO containers are limited by convention to a maximum gross loaded weight of 20.3 tonnes for the shorter 6-m type and 30.5 t for the larger 12-m type. Weights of the empty boxes are approximately 2.5 and 4 t for the two sizes. Where the larger containers are to be transported by road the gross weight is of necessity limited by highway loading regulations.

25.2. A container ship is expected to be in port for only 2 to 3 days and effective handling dictates that the location and stacking both of imported and exported boxes must be controlled in the most efficient manner and computer control is necessary, so that the port manager knows the exact location of any box at any time.The transfer of containers between ship and quay is normally effected by a railed crane. However, a variety of plant may be used to move, stack, and retrieve containers which may be stacked up to five high in the transit areas.

25.3. The plant available for site movements includes the following:

1. Rubber-tired gantry cranes with clusters of four wheels on each of four legs

2. Straddle carriers with three or four wheels spaced on each side
3. Front-lift trucks with dual wheel assemblies on the front axle and single wheels on the rear axle

25.4. Rubber-tired gantry cranes. These have the ability to move containers sideways over distances of 15 to 20 m. They are therefore very robust with a tare weight of 90 to 100 t, and a maximum wheel load when carrying a loaded container of about 16 t.

Straddle carriers. These lift containers to the required height for stacking. Although they cannot move boxes sideways while stationary, they are highly mobile and can place containers accurately. They have a tare weight of about 45 t and a maximum wheel load when fully loaded of 12 to 14 t.

Front-lift trucks. These are made in various sizes. For container handling there are two main groups, with lifting capacities of 12 t for the smaller group and 30 to 35 t for the larger machines. The smaller type are used mainly for handling empty containers. The load on the front axle of the smaller type when fully loaded is likely to be about 30 t, i.e., 8 t per wheel. For the larger machines the front axle load can be 70 to 80 t or 20 t per wheel.

25.5. Irrespective of the method of handling, the layout of a container terminal incorporates "roadways" running at right angles to the quay. The arrangement of the stacking blocks depends on the handling plant. Straddle carriers require a spacing of about 1 m between the standing boxes to allow movement of the side frames. With gantry cranes the boxes can be packed more closely with rather larger gaps at intervals to allow passage for the crane legs. Front-lift trucks will pack the boxes close together, but to allow access the boxes need to be only two deep between access roadways. Each type of plant requires a different width of roadway for easy operation.

Design Approach for Heavily Loaded Industrial Pavements

25.6. An engineer commissioned to design pavements for a container depot or similar facility must become fully aware of the intended function of the installation and how it is to be operated, and on submitting the design should make clear all the assumptions which have been made. Stacking areas need to be marked out and numbered so that plant operators can locate a particular container with the minimum of delay. This tends to establish a layout in which the heavily trafficked roadways become permanent. In such a situation the design could use a variable thickness with the roadways thicker than the stacking areas. However, the client should be made fully aware of the limitations imposed by such an arrangement. In practice, clients generally prefer uniform construction.

25.7. Flexible pavements using bituminous or unbound bases are not recommended for pavements designed to accept very heavy wheel loads. To avoid unacceptable deformation and wheel tracking the bituminous material would need very careful design and the addition of expensive additives throughout the full depth. Concrete pavements or bituminous surfacings on a strong lean concrete base generally provide the best solution. In the case of the lean concrete base, the surfacing can be of precast concrete or ceramic blocks if preferred.

25.8. It is fairly common for haulers to reduce space charges at dockside terminals by constructing small depots outside the dock area. These are used as a storage area for empty containers or as a staging point for the distribution of loaded containers. In designing such a facility the engineer responsible must make it quite clear for which items of plant the pavement is designed.

Examples of Structural Design Applied to Industrial Pavements

25.9. The following examples refer to actual designs for heavily loaded pavements prepared during the past 10 to 15 years.

25.10. Lean concrete construction for straddle carrier use. The design relates to a site operated by straddle carriers of maximum static wheel load 10.5 t. These machines in turning tend to transfer load from the wheels on one side to those on the other. In braking the transfer is from the wheels at the back to those in the front. In addition, as with a normal road, unevenness of the surface tends to increase the maximum static wheel load. These effects produce a combined dynamic factor which may be in the range 1.2 to 1.6. However, it is unlikely in practice to exceed 1.3 when the vehicle is safely driven and the riding quality of the roadways is acceptable.

25.11. The pavement was designed to have a two-course rolled asphalt surfacing, with a high-stone-content wearing course for increased stability. The total thickness of the asphalt selected was 100 mm. The base was of lean concrete to the United Kingdom specification requiring a 28-day compressive strength of 10 N/mm^2. The deep sand foundation had an in situ CBR value of 15 percent. The finite-element computer program described in Chap. 23 was used to compute the maximum tensile stress induced in the bottom of the lean concrete base by the passage of wheel loads of 7.5 to 10.5 t with dynamic effects of 1.0 to 1.6. The thicknesses of lean concrete included in the analysis were in the range 200 to 350 mm. The results are summarized in Table 25.1, and beneath the table are shown the elastic properties ascribed to the various materials. A check was made to show that the stress generated by each wheel was not significantly augmented by that due to the adjacent wheels.

25.12. Figure 25.1 shows fatigue lines corresponding to ages of 28 days and 5 years for lean concrete of compressive strength 9 N/mm^2 at 28 days. These curves were derived using the equation for gravel aggregates shown in Par. 14.13. The value of 9 N/mm^2 for the crushing strength was chosen as more nearly representing the actual strength of the base as laid rather than the 10 N/mm^2 required for fully compacted cubes.

25.13. Table 25.1 in conjunction with Fig. 25.1 shows that for all the wheel loads considered the lean concrete base would fail by cracking if laid to a thickness of 200 mm. Assuming a dynamic factor of 1.3, a thickness of 250 mm would fail for all the wheel loads. Increasing the thickness to 300 mm would result in failure owing only to the 9.5- and 10.5-t wheel loads, in conjunction with a dynamic factor of 1.3. The 350-mm thickness would be safe for all the wheel loads and all the dynamic factors. For the reasons given in Chap. 24 fatigue failure would then occur on the 5-year fatigue line, and it is clear from Fig. 25.1 that the thickness of 350 mm would accept about 10^8 applications of the 10.5-t wheel load and would have a much greater fatigue life for the smaller wheel loads. Since rather less than 1 percent of the wheel loads were expected to be in the 10- to 11-t load bracket, it was clear that if the pavement was correctly constructed it should have a life well in excess of 100 years.

25.14. Concrete pavement suitable for the temporary storage of empty container boxes. This is an example of a yard designed to be operated by Hyster 250 or similar front-lift trucks loading empty container boxes onto and off articulated trucks. The design was for a 200-mm concrete slab laid on a 150-mm crushed-stone subbase placed on a 150-mm capping layer on a soil of CBR 6 percent. The concrete was to comply with the Department of Transport specification (minimum crushing strength 28 N/mm^2 at 28 days). Lifting a 12-m box, the load on the front axle was calculated to be 18 t with a rear axle load

TABLE 25.1 Maximum Tensile Stress at the Bottom of the Lean Concrete, N/mm², for Various Wheel Loads and Dynamic Factors

Static wheel load, t	7.5				8.5				9.5				10.5			
Dynamic factor	1.0	1.2	1.4	1.6	1.0	1.2	1.4	1.6	1.0	1.2	1.4	1.6	1.0	1.2	1.4	1.6
Thickness of lean concrete base, mm:																
200	1.58	1.90	2.21	2.53	1.80	2.16	2.52	2.88	2.18	2.62	3.05	3.49	2.53	3.04	3.54	4.05
250	1.24	1.49	1.74	1.98	1.40	1.68	1.96	2.24	1.56	1.87	2.18	2.50	1.73	2.08	2.42	2.77
300	0.91	1.09	1.27	1.46	1.03	1.24	1.44	1.68	1.15	1.38	1.61	1.84	1.27	1.52	1.78	2.03
350	0.68	0.82	0.95	1.09	0.77	0.92	1.08	1.23	0.86	1.03	1.20	1.40	0.95	1.14	1.33	1.52

Elastic properties of the materials:
Rolled asphalt $E = 2.5 \times 10^3$ N/mm² (20°C) Poisson's ratio 0.4
Lean concrete $E = 34 \times 10^3$ N/mm² Poisson's ratio 0.15
Sand foundation $E = 150$ N/mm²
Tire pressure 1 N/mm²

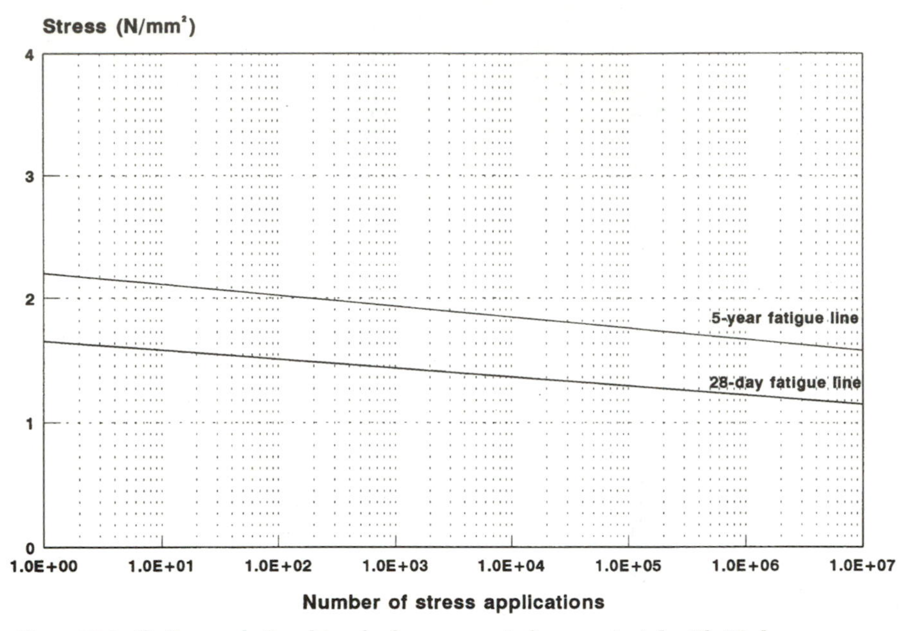

Figure 25.1 Fatigue relationships for lean concrete base material with 28-day compressive strength 9 N/mm^2 (gravel).

of 2.5 t. The maximum load on each front dual wheel assembly was thus 9 t.

25.15. The maximum tensile stress in the bottom of the slab due to the load was computed for a wheel load of 4.5 t. The spacing of the dual wheels was approximately 540 mm between the centers of the contact areas, and the stresses were compounded in the manner shown in Fig. 25.2. The spacing between the two dual-wheel assemblies was too large to give any stress interaction. In accordance with Pars. 24.21 and 24.22, a maximum tensile warping stress of 0.8 N/mm^2 was added to the computed load stress of 1.8 N/mm^2 taken from Fig. 25.2, to give a possible total tensile stress of 2.6 N/mm^2. The 28-day compressive strength of the concrete of 29 N/mm^2 corresponded to a modulus of rupture of 3.1 N/mm^2. The design therefore allowed for a substantial dynamic component of load. As the modulus of rupture increased with age the design would become safer, with no danger of fatigue failure.

25.16. Concrete pavement to accommodate large front-lift trucks operating with full containers. A similar calculation was made for one of the larger front-lift trucks carrying a loaded 12-m container. The calculated front axle load was 55 t, with 9 t on the rear axle. In this case the

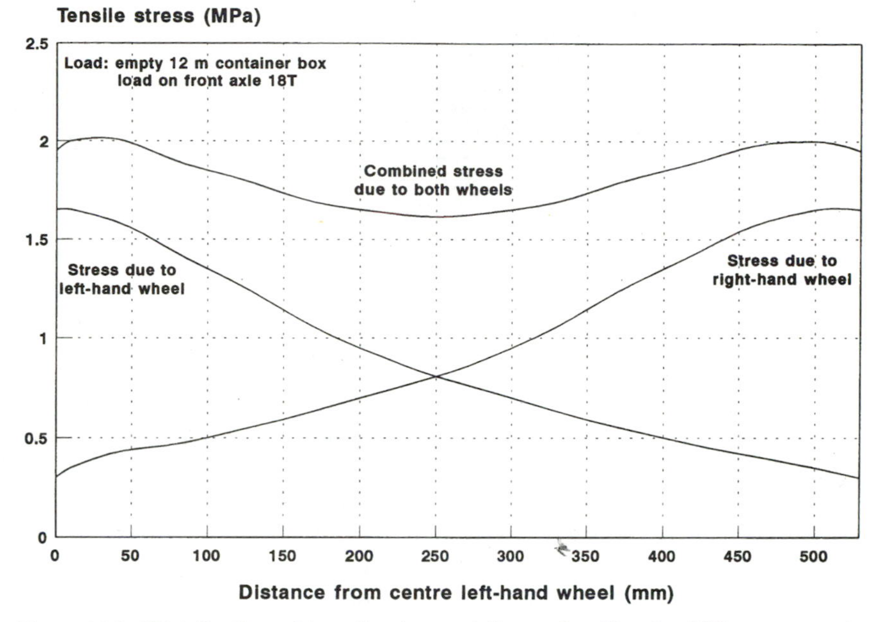

Figure 25.2 Distribution of tensile stress at the underside of a 200-mm concrete slab—dual-front-wheel Hyster Challenger.

27.5-t load on each dual-wheel assembly generated a maximum tensile loading stress midway between the two contact areas of the dual tires. This stress was calculated for three thicknesses of concrete as shown in Table 25.2. The stresses were also calculated for the unloaded machine.

For the loaded truck the total tensile stress including the warping stress would be 2.91 for the 350-mm slab. With the modulus of rupture of 3.1 N/mm^2 there would be only a small reserve in the design to allow for dynamic loading.

25.17. Dock pavement using a cement-stabilized base, mixed in situ, with a concrete block surfacing. In this case the pavement was to be used primarily by rubber tired gantry cranes with a cluster of four wheels on each of the four legs. The maximum load on each leg was to be 16 t (including a dynamic factor of 1.2). The configuration of the wheel clusters is shown inset in Fig. 25.4. In this example two variables were considered. These were the crushing strength of the stabilized material and the required thickness. A considerable depth of sandy gravel had been imported to provide the foundation, and the upper part of this material was to be stabilized in situ. In accordance with

TABLE 25.2 Tensile Stress Developed in the Underside of Concrete Slabs of Various Thicknesses by a Large Front-Lift Truck,0 N/mm²

Thickness of concrete, mm	Lancer Henley 68 truck	
	Fully loaded	Unloaded
200	5.10	2.70
300	2.76	1.45
350	2.11	1.14

the discussion in Chap. 15, an elastic modulus of 45 MPa was adopted for the concrete blocks and their sand foundation, of total thickness of 150 mm. The same modulus was assumed for the sandy-gravel foundation. The elastic modulus of the base is considered in Fig. 25.3. The dynamic modulus in relation to the compressive strength is taken from Table 12.3. In accordance with the recommendation in Par. 12.20, the mean value was adopted in the structural analysis.

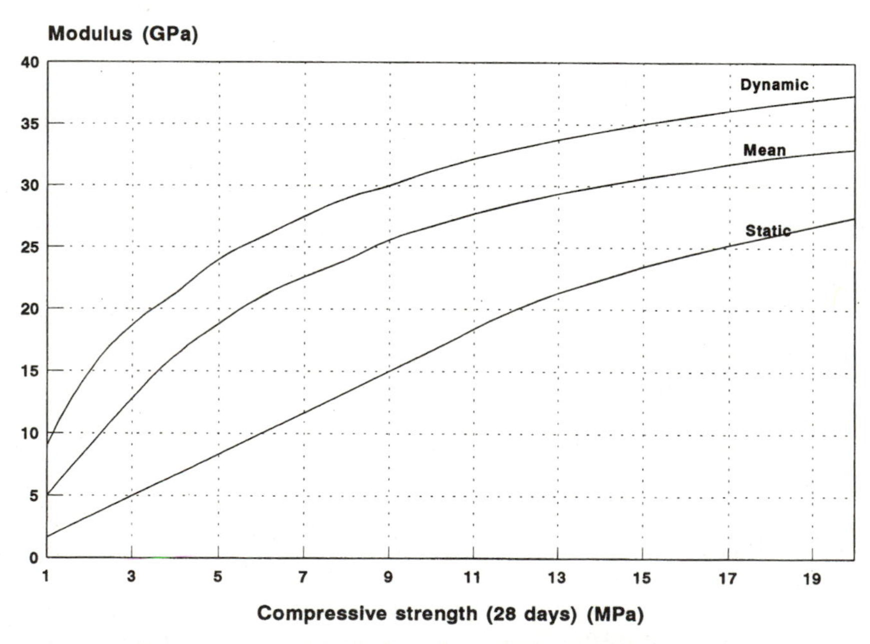

Figure 25.3 Port pavement with block surfacing. Relation between modulus and compressive strength of lean concrete.

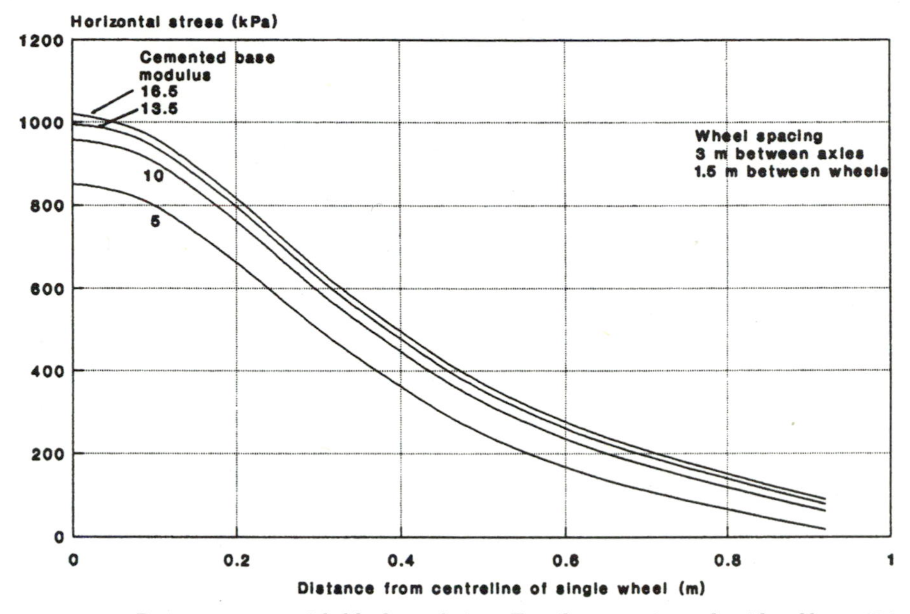

Figure 25.4 Port pavement with block surfacing. Tensile stress in underside of base 400 mm thick.

25.18. The radius of loading adopted was deduced from the tire size, the maximum wheel load, and the tire pressure applicable to the machine. The finite-element program was used to deduce the stress distribution beneath the center of the tire contact area. Figure 25.4 shows the maximum tensile stress both on and off the axis of loading, corresponding to four strengths of cemented material defined in terms of their modulus of elasticity. It was clear that there was no significant interaction of stress between the four wheels of each cluster. Figure 25.4 refers to a base thickness of 400 mm; similar families of curves were computed for base thicknesses of 300 and 350 mm. This permitted the maximum tensile stress developed to be related to the modulus of elasticity of the base as shown in Fig. 25.5.

25.19. By using Fig. 25.3 and Eq. (14.1) in Par. 14.12 the variation of modulus of rupture with base modulus of elasticity can be superimposed on Fig. 25.5, and the moduli of elasticity necessary to prevent rapid failure can be determined from the points of interception between the stress curves and the modulus of rupture. Compressive strengths can be added to the rupture curve, again by using Fig. 25.3. The fatigue situation can then be investigated, as in the first example

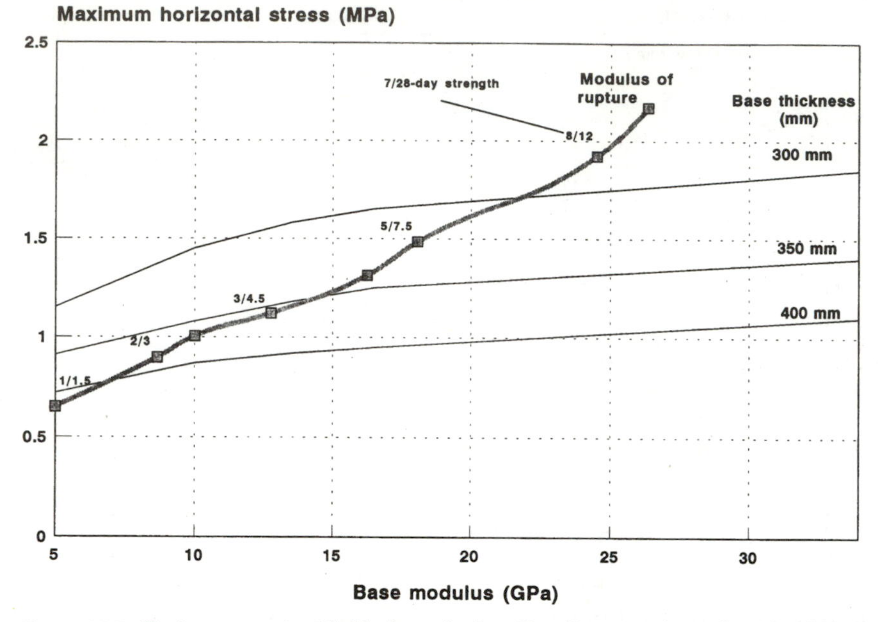

Figure 25.5 Port pavement with block surfacing. Tensile stress in underside of base related to base modulus and thickness.

considered above. In using this approach it must be remembered that with mix-in-place stabilization the actual strength obtained in the base may be considerably lower than the equivalent cube strength, and the preliminary trial should be carried out to investigate the efficiency of mixing throughout the full depth of the stabilized layer.

Part

6

Surface Characteristics of Pavements

Chapter

26

The Riding Quality of Pavements and Its Measurement

Introduction

26.1. In the twenties rural roads in the United Kingdom consisted of the original water-bound macadam to which liberal surface dressings of hot tar had been added. The riding quality by modern standards was extremely poor, and the situation was not helped by the tight suspension and indifferent shock absorbers of the vehicles of the period. It was at that time that attention was first given to the development of profilometers to quantify riding quality. The need for a rolling "fixed" datum against which surface undulations could be measured was recognized early, and by the thirties the multiwheeled profilometer shown in Fig. 26.1 had been developed. The multiplicity of hinged bogies provided the "fixed" datum, and profile measurements were made by a central recording wheel. This machine, which remained in service until the seventies, is discussed in Par. 26.8.

26.2. Riding quality, as interpreted by the road user, is highly subjective and therefore very difficult to quantity in engineering terms. Much the same problem also arose in the thirties when possible restrictions on the noise made by motor vehicles and motor horns came under consideration. Experienced observer panels were found to rate middle-order noises in a quite different order of nuisance on different days. With the increasing "average" smoothness of road surfacings the problem of subjective rating has increased and there is a tendency for tire and body noise to be confused with riding quality.

Figure 26.1 The multiwheeled profilometer.

26.3. In some ways it is to be regretted that in the United States the subjective approach was used to assess serviceability in the AASHO road test. This has made it difficult, without studying the original data sheets, to separate the engineering conclusions from the impressions of road users. An implication from the work of the statisticians involved in the road test is that the change of slope variance of the surfacing with time is the major factor in determining pavement serviceability, and that the panels of observers used were able to differentiate between the effects of slope variance and, for example, integrated up-and-down movements. As a consequence, since 1962 effort has been expended in developing equipment to measure slope variance with greater precision than was possible during the road test, and in refining the relations between cracking and deformation and PSI. In the late sixties the multiwheel profilometer referred to in Par. 26.1 was used on many of the full-scale road experiments described in Chap. 18 and the profiles were analyzed manually using the equivalent of a 230-mm (9-in) base line, as adopted in the Chloe profilometer used in the road test. In this way approximate correlations were obtained between PSI and the U.K. definitions of the "critical" and "failed" state of flexible and concrete pavements.

26.4. Modern roadlaying machines with sophisticated methods of level control, used in conjunction with a workable specification relating to the magnitude of acceptable surface undulations, should produce a satisfactory pavement, given an experienced contractor and proper supervision. However, the use of end-product specifications allows a certain amount of experimentation on the part of the contractor and this means that the riding quality may or may not be acceptable when the job is finished. At that stage it is difficult to make improvements.

Surface Finish Required for Pavements in the United Kingdom

26.5. To ensure that the various layers of a pavement are of adequate thickness, relative to the specification, and to facilitate the achievement of an adequate riding quality, tolerances are set for the surface levels of the various courses by the U.K. Department of Transport specification (clause 701). These tolerances are shown in Table 26.1, and they are controlled by level measurements made over a grid with specified spacings, the levels being checked at the same locations on all the pavement courses.

26.6. To enable these checks to be made quickly and accurately, a laser-controlled leveling system is available.[1] The equipment consists of a rotating laser source capable of generating a datum plane with an effective radius up to 300 m, and a leveling staff fitted with a movable optical receiver, sensitive to laser light. The receiver travels up and down the staff until it locks onto the laser plane. Longitudinal profiles obtained in this manner on the various layers of two flexible pavements are shown in Fig. 26.2. These examples show how irregularities in the lower courses are to some extent reflected in the running surface, and emphasize the need to get the best possible profiles at each level.

26.7. In addition to the tolerances of surface levels, the Department of Transport specification also limits the number of surface irregularities of magnitudes 4 and 7 mm which are permitted in a given length

TABLE 26.1 Tolerances in Surface Levels of Pavement Courses

Courses	
Road surfaces	± 6 mm
Base course	± 6 mm
Upper road base in pavements without base course	± 8 mm
Road base other than above	± 15 mm
Subbase under concrete pavement surface slabs laid full thickness in one operation by machines with surface compaction	± 10 mm
Subbases other than above	+ 10 mm − 30 mm

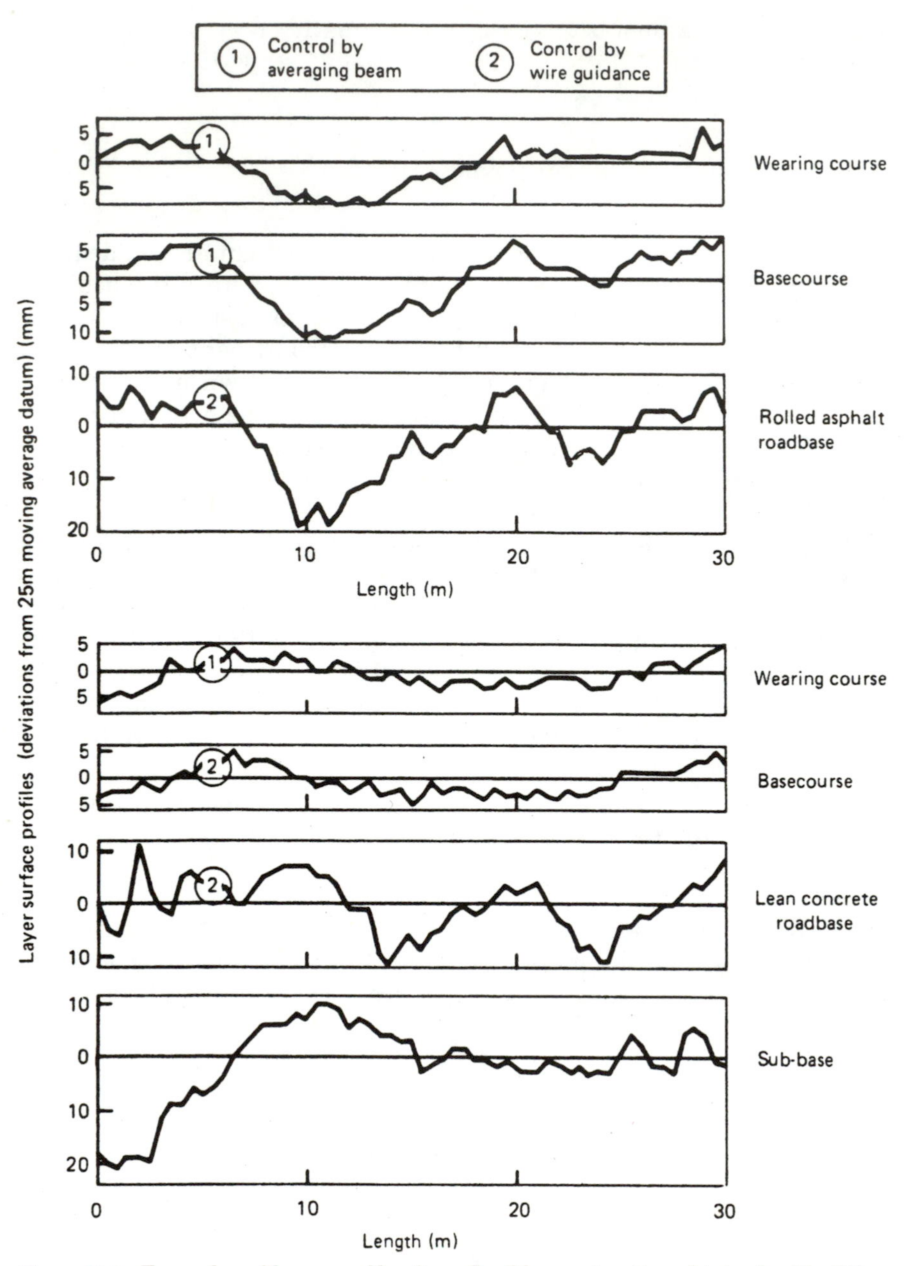

Figure 26.2 Examples of layer profiles from flexible construction obtained with different control methods (profile features longer than 25 m have been removed).

TABLE 26.2 Maximum Permitted Number of Surface Irregularities

	Surfaces of carriageways, hard strips and hard shoulders				Surfaces of lay-bys, service areas all bituminous base courses and upper road bases in pavements without base courses			
Irregularity	4 mm		7 mm		4 mm		7 mm	
Length, m	300	75	300	75	300	75	300	75
Category A roads	20	9	2	1	40	18	4	2
Category B roads	40	18	4	2	60	27	6	3

of road. These requirements are shown in Table 26.2, where category A roads are generally those which permit high speeds, such as trunk roads and freeways, and category B roads are more minor, where the speed is unlikely to exceed 50 mi/h. On newly constructed roads compliance with Table 26.2 is normally checked by a hand-propelled traveling straightedge of the type shown in Fig. 26.3.

Recent Developments in Profile Measurement

26.8. The multiwheel profilometer shown in Fig. 26.1 was essentially a research tool. It was propelled by hand and its speed had to be controlled to prevent bounce of the sensor wheel and overrunning of the mechanical recorder. Although it was invaluable during the changeover period between hand and machine laying of both bituminous and concrete roads, there was never any possibility that it could be used for routine acceptance testing.

26.9. Consideration was being given in Britain to the development of more robust, if less accurate, forms of towed profilometer based on the AASHTO Chloe design, when contactless sensor technology using the reflection of a pulsating laser beam became available. In the last 15 years highly sophisticated equipment has been developed at the TRRL using this principle to measure longitudinal profile, rut depth, and the macrotexture of road surfacings. The latter application is considered in Chap. 27.

26.10. The basic design of the contactless displacement transducer is shown in Fig. 26.4.[2] A parallel beam of pulsating laser light is project-

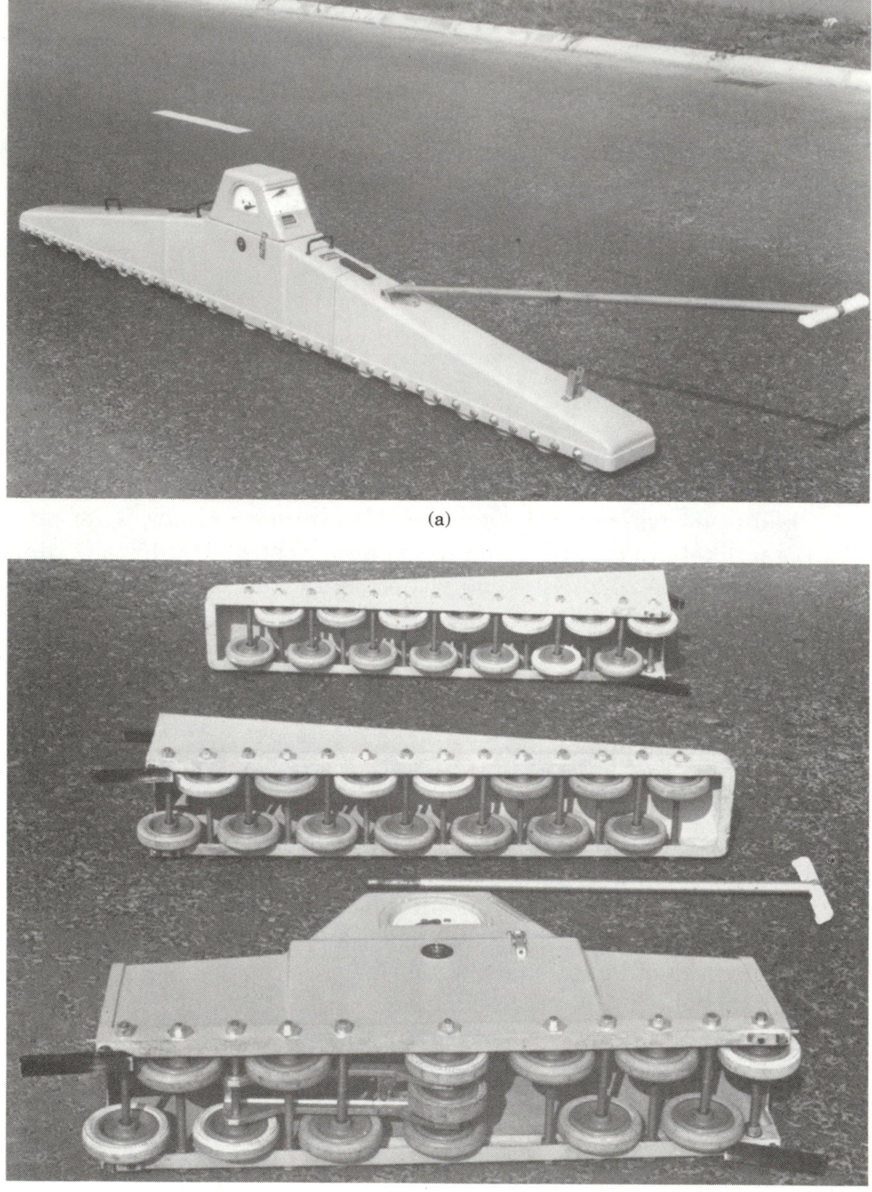

(a)

(b.)

Figure 26.3 The rolling straightedge: (*a*) assembled (*b*) component parts.

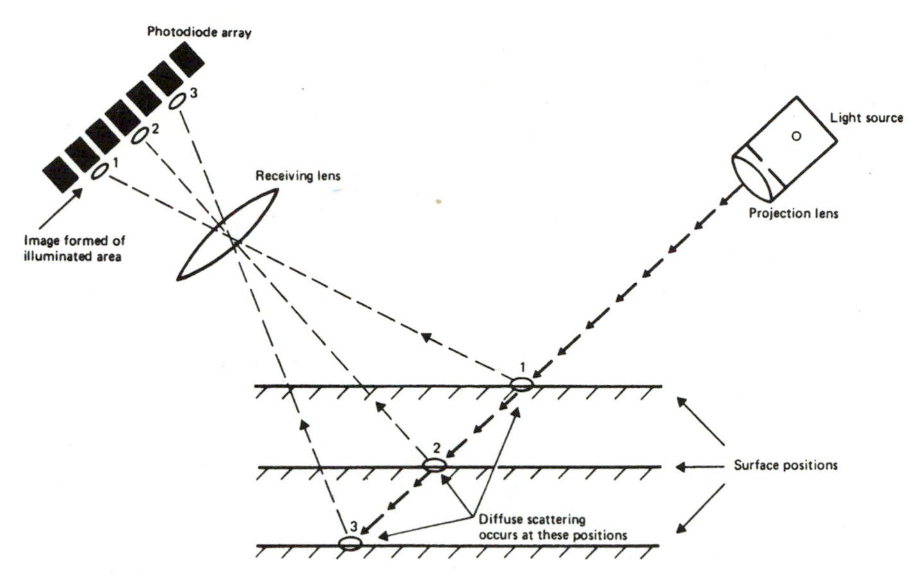

Figure 26.4 Basic contactless displacement transducer design.

ed at an angle of 45° onto the surface being examined. Diffuse scattering of the laser light occurs at the point of contact on the surface, and a proportion is focused by a receiving lens onto a linear array of photodiodes to actuate one or more of the diodes. If the reflecting surface drops to position 2 or position 3 in Fig. 26.4, then the image will move across the photodiodes to give a linear relationship between the level of the surface and the activated diode. The laser source emits pulses of light of duration 2×10^{-7} seconds. It is estimated that the projection lens collects and transmits 64 percent of the laser energy into an area on the road 0.25 mm wide by 3.8 mm long. The receiving lens focuses this area of light so that it falls on only one of the line of photodiodes over which the reflected beam sweeps. The individual photodiodes have a receiving area approximately 0.09 mm wide by 1 mm long, and this determines their spacing.

26.11. The original version of the TRRL high-speed profilometer using laser sensors had a 5-m beam fabricated from three alloy tubes stiffened with cross members as shown in Fig. 26.5. It had four sets of laser and receiver assemblies spaced along the length of the beam, which was mounted on a two-wheel unsprung axle. The end of the beam remote from the wheels was attached to the towing vehicle, which operated at speeds up to 80 km/h and carried the associated electronic units and the computer used for the analysis of the data.

Figure 26.5 TRRL High-Speed Road Monitor.

26.12. It will be appreciated that no profilometer of this type can give an accurate record of changes of road surface level in absolute terms, ranging from the texture of the surfacing to large-scale undulations associated with the ground contours. This would require accurate measurements from a horizontal plane above the highest point of the length of road under consideration. What such a profilometer can do is to give a reasonably accurate picture of the large and small undulations over a length of about 30 m. The machine does this by providing a running analysis of the surface irregularity over lengths, of, for example, 3 and 30 m. The following six paragraphs are reproduced from reference 3, with the paragraph, figure, and equation numbers modified to accommodate the system used in this chapter.

26.13. The operation of the multisensor profilometer is based upon the three-sensor device shown schematically in Fig. 26.6. Three sensors are mounted in a straight line on a rigid beam at P, Q, and R, and measure the distances to the road surface; the sensors are equidistantly spaced at a distance δ apart. Denoting these distances by v_1, v_2, and v_3, respectively, define the quantity u_k as

$$u_k = \tfrac{1}{2} v_3 - v_2 + \tfrac{1}{2} v_1 \tag{26.1}$$

when the front sensor P is at $x = k\delta$.

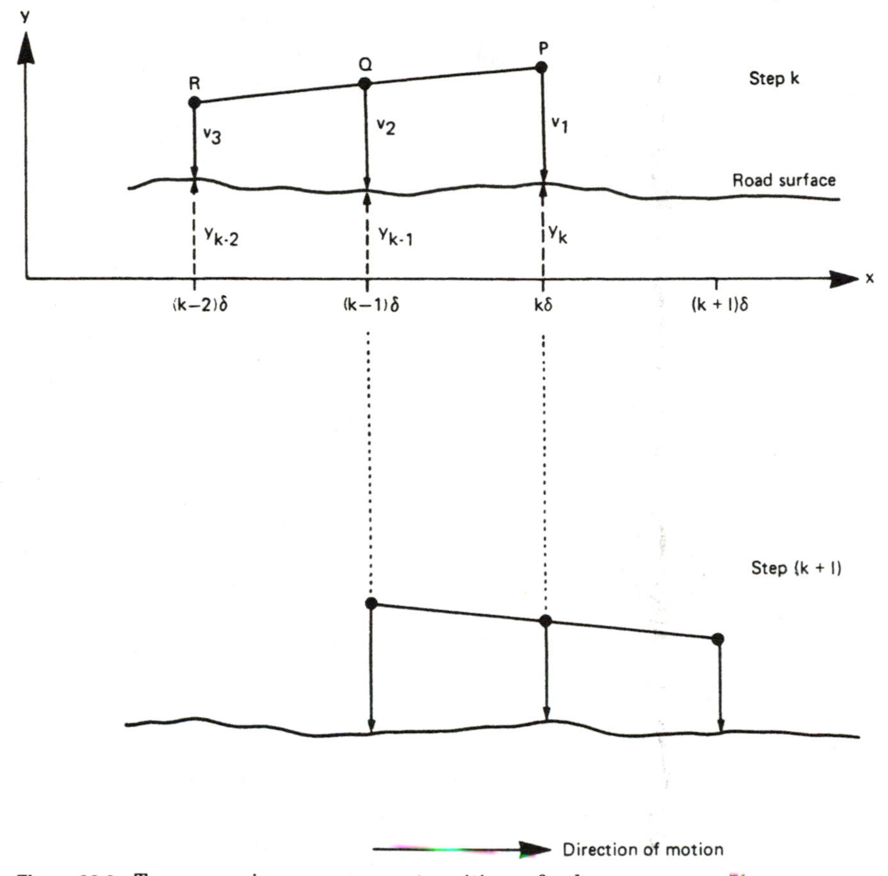

Figure 26.6 Two successive measurement positions of a three-sensor profilometer.

Since *PQR* is a straight line with *Q* as its midpoint, the height of *Q* above the x axis is midway between the heights of *P* and *R*; thus

$$v_2 + y_{k-1} = \tfrac{1}{2}\left[(v_3 + y_{k-2}) + (v_1 + y_k)\right] \qquad (26.2)$$

By combining Eqs (26.1) and (26.2), it may be shown that the profile height y_k at $x = k\delta$ is given by

$$y_k = -2u_k + 2y_{k-1} - y_{k-2} \qquad (26.3)$$

The device now moves forward in steps of δ, and successive values of u_k are found. These are, in fact, proportional to the second differences of the profile height y at intervals of δ.

26.14. At each step of the profilometer, the required value y_k may be obtained from u_k and the previous two values of y. Ideally, this recursive recovery process is perfect and the true height of the profile is found at the points δ apart where the sensors make their measurements. The recovery process is able to compensate for both the vertical motion and the pitching of the beam as it moves forward over the profile because the center and rear sensor measurements are made at the same points as those made, on the previous step, by the front and center sensors, respectively.

26.15. In practice, three types of error occur which result in deviations, increasing with the distance traveled, between the true profile and the measured one:

1. When the recursion process is started, two initial conditions are required, corresponding to the heights of the profile at the center and rear sensor positions. These values are usually taken as zero, but errors result unless the profilometer starts on a flat, horizontal surface. Incorrect initial conditions result in a linear deviation of the measured profile from the true one; however, for the prototype profilometer the erroneous slope is likely to be less than 1 in 50.
2. The beam carrying the sensors has to be rigid. Flexing will contribute curvature to the measured profile and a permanent deformation of the beam, unless exactly allowed for in the calculations, produces a parabolic error. For the prototype profilometer, the beam is rigid to within the resolution of the sensors.
3. As the profilometer moves forward, measurements are taken at distances nominally δ apart. Successive sensor measurements are not made at precisely the same points on the profile because of errors in step size, transverse movements of the beam, cornering, and the texture of the road surface. When cornering on a road with a crossfall, the measured vertical curvature of the road is greater than that which actually exists because the three sensors lie along a chord of the lateral curve rather than on the arc itself. To reduce the effect of the surface texture, which would be the dominant one, several sensor readings may be taken while the profilometer moves through the distance δ, and their average is then used to calculate an average profile height over this distance. Successive values are obtained at intervals of δ as before. The surface texture results in a random error in profile, the rms value of which increases with the distance traveled. For a profilometer of the type shown in Fig. 26.7*A*, having the dimension δ equal to 108 mm and

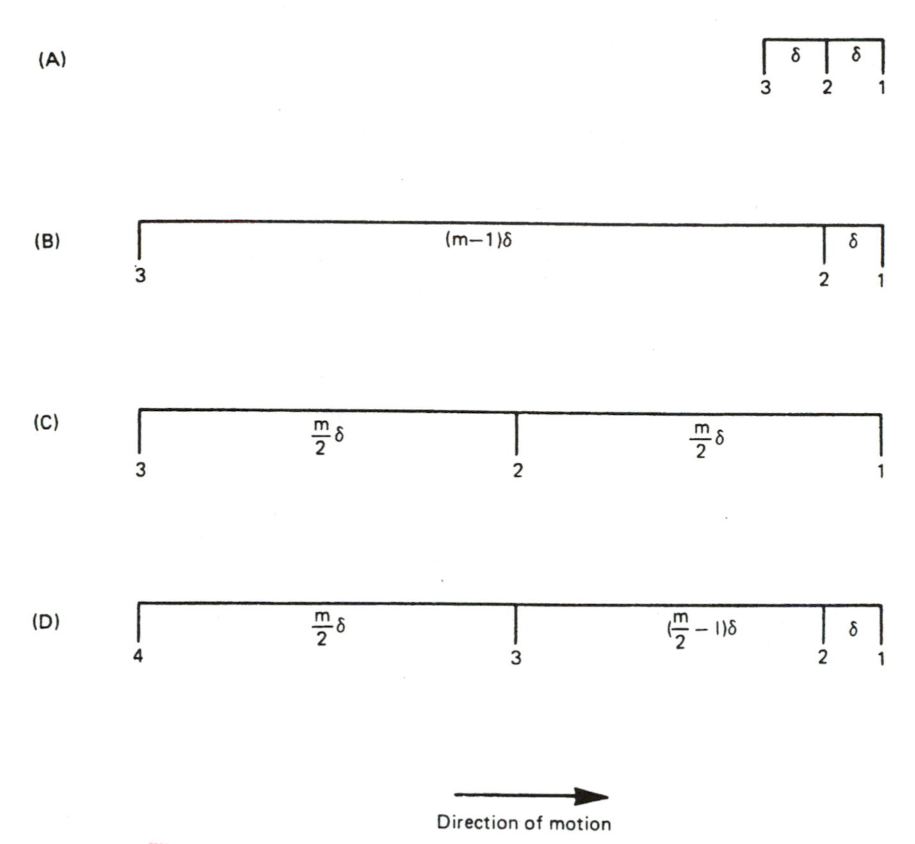

Figure 26.7 Three- and four-sensor profilometers.

moving at 50 km/h, with each sensor making 2000 measurements per second, the theoretical rms "texture error" as a function of distance is given by curve *A* of Fig. 26.8, where a road surface texture of 2 mm rms has been assumed. After traveling 10 m, the rms error would be 0.62 m.

26.16. This error may be reduced by increasing the spacing between the intermediate and rear sensors leading to an asymmetrical profilometer of overall length $m\delta$ as shown in Fig. 26.7*B*. The step size remains equal to the spacing of the front and intermediate sensors, but the quantity u_k is now defined as

$$u_k = \frac{1}{m}v_3 - v_2 + \left(1 - \frac{1}{m}\right)v_1 \qquad (26.4)$$

and the recursive recovery process is given by

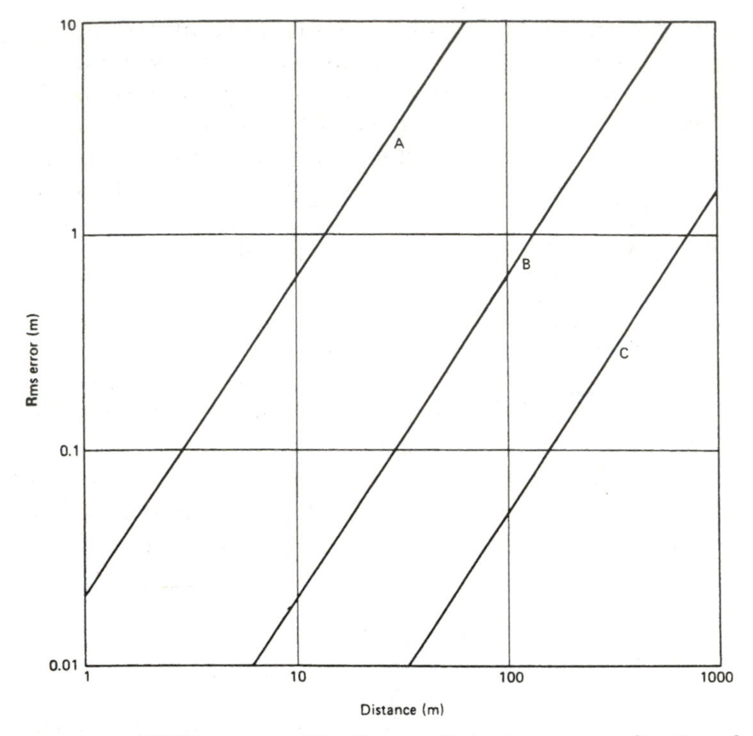

Figure 26.8 RMS error resulting from surfacing texture as a function of distance.

$$y_k = -\left(\frac{m}{m-1}\right)u_k + \left(\frac{m}{m-1}\right)y_{k-1} - \left(\frac{1}{m-1}\right)y_{k-m} \tag{26.5}$$

The texture errors occurring with a profilometer having m equal to 40 and operating under the same conditions as before are shown in Fig. 26.8, curve *B*.

26.17. Consideration of a symmetrical profilometer of overall length 40δ moving in steps of 20δ with averaging over this length (see Fig. 26.7*C*), leads to curve *C* of Fig. 26.8 where the error is only 0.25 percent of that for the first profilometer configuration. This improved performance is achieved at the expense of a wider spacing between the profile measurement points. A further result of the averaging process is that the effect of the finite resolution of the sensors is reduced.

26.18. It is desirable to retain the horizontal resolution of the profilometer shown in Fig. 26.7*B* while reducing the errors of those of that shown in Fig. 26.7*C*. This may be done by combining a symmetrical profilometer with an asymmetrical one, both of overall length 40δ, so that the arrangement of Fig. 26.7*D* is obtained involving the use of four sensors. It is then possible to use the asymmetrical profilometer (sensors 1, 2, and 4) to provide profile measurements at intervals of δ and to correct its drift at intervals of 20δ using the symmetrical profilometer (sensors 1, 3, and 4). The prototype profilometer shown in Fig. 26.5 was constructed to Fig. 26.7*D*. In use the framework is enveloped in polyurethane foam to damp mechanical vibrations.

26.19. As part of its proving trials, tests were carried out in Turkey to examine the performance of the high-speed profilometer under hot ambient temperature conditions.[4] Figure 26.9 shows details of one of many comparisons made between surface profiles determined by direct leveling and by the profilometer. To ensure that the same path was followed using each method, the profilometer was fitted, at the rear mid-center, with a water drip to define its path. It is clear that excellent agreement was found. The method discussed above, including all four sensors, was used in the case of the profilometer measurements.

Measurement of Rut Depth

26.20. On in-service roads, the high-speed profilometer can also be used to measure the depth of rutting in wheel tracks generated in the

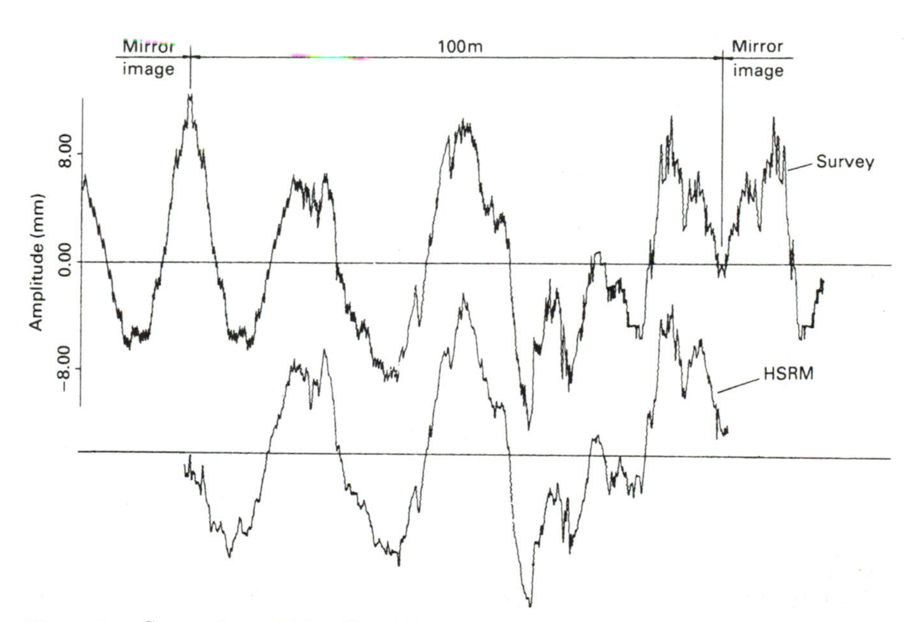

Figure 26.9 Comparison of High Speed Road Monitors and surveyed profiles for section 4.

slow traffic lane by heavy commercial vehicles. Observations made on trunk and principal roads have shown the centers of the two wheel paths for 70 percent of such vehicles are separated by 1.8 m. Later versions of the high-speed profilometer have been modified to have this track width and an additional central laser assembly has been fitted a little forward of the trailer axle, as shown in Fig. 26.10.[5]

26.21. It follows from Fig. 26.10 that if the distance between the rut-depth sensor and the road surface is h_0 when there is no rutting and is h_r when the profilometer wheels are in ruts of depth r_1 and r_2, then the mean rut depth is given by $(h_0 - h_r)$ divided by 2.

26.22. Accuracy and repeatability of the measurement of rut depth in this manner does depend on the care of the operator in following the alignment of the wheel tracks. Figure 26.11 shows that with a trained operator repeatability over a considerable length of road can be obtained.

The High-Speed Road Monitor

26.23. The three functions of (1) profile evaluation, (2) rut depth measurement, and (3) macrotexture determination (discussed in Chap. 27) have been integrated into a single machine called the TRRL High-Speed Road Monitor.[6] This machine allows all three functions to be carried out during a single run. To permit this, the four sensors necessary to measure surface profile and macrotexture have been moved to the left-hand side of the trailer, i.e., in the nearside wheel path when the vehicle is driven normally, and the additional sensor for rut depth measurement is between the wheel tracks where there will normally be little deformation. The layout of the sensors is shown in Fig. 26.12.

The High-Speed Road Monitor and Riding Quality

26.24. Work is currently being carried out at the TRRL to correlate rut depth and surface profile with subjective riding quality. Figure 26.13 shows histograms of rut depth and profile variance (square of standard deviation) determination; a moving average length of 3 m was used in the analysis. On the basis of this work an approximate relationship between profile variance and riding quality has been pro-

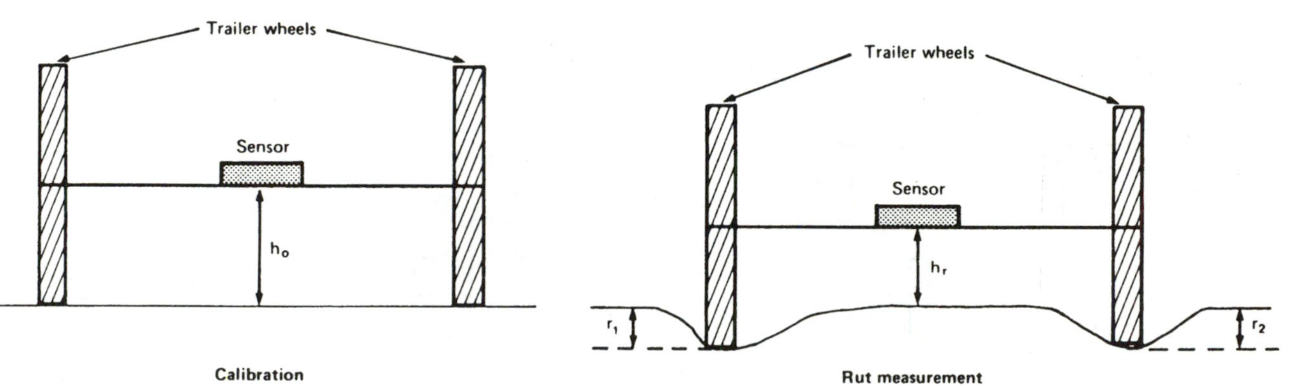

$$h_o - h_r = \frac{r_1 + r_2}{2} = \text{rut depth averaged over both wheel paths}$$

Figure 26.10 Rut-depth measurement.

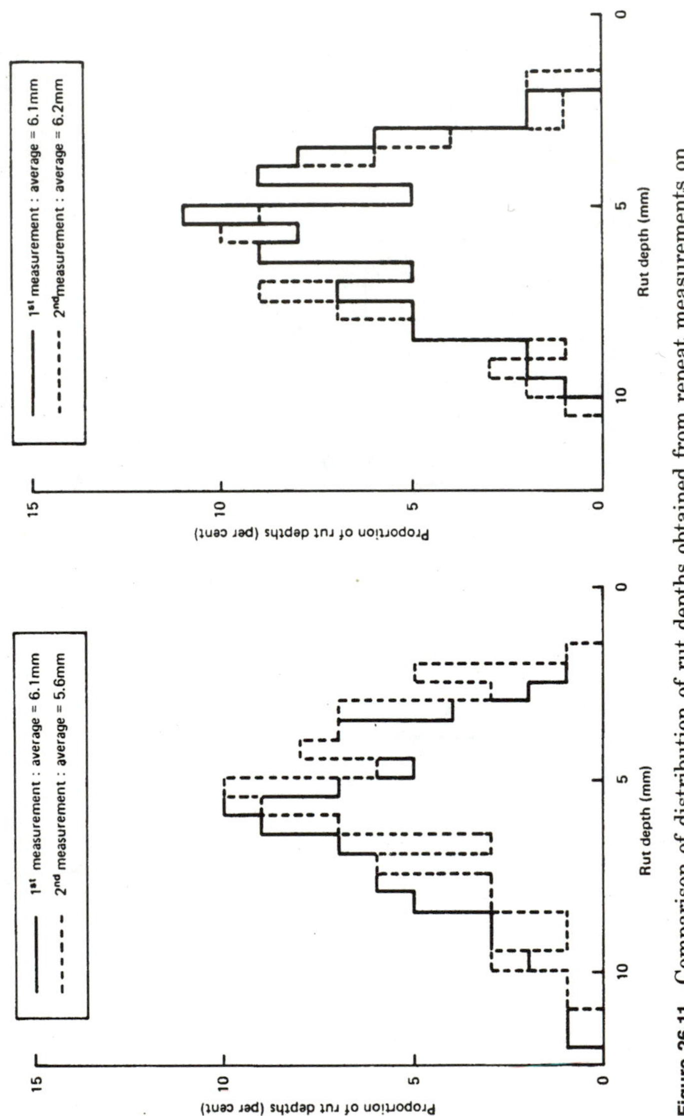

Figure 26.11 Comparison of distribution of rut depths obtained from repeat measurements on each of two 20-km lengths of freeway.

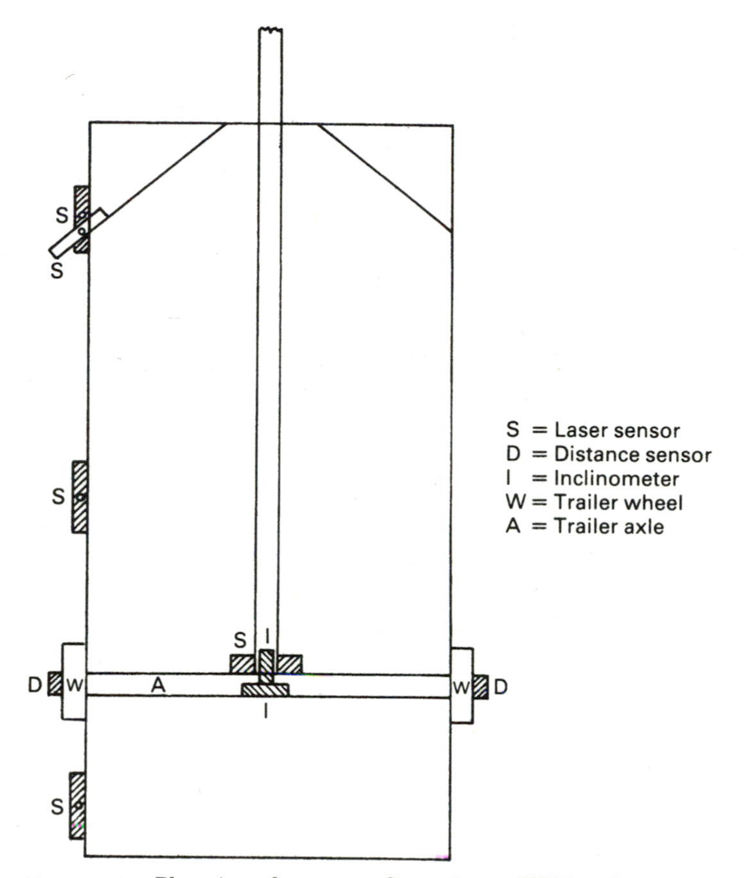

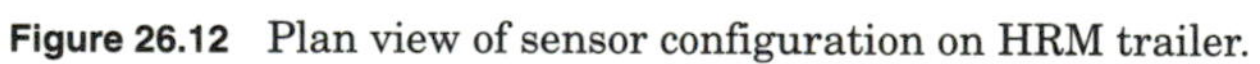

Figure 26.12 Plan view of sensor configuration on HRM trailer.

posed[6] (Table 26.3). This will no doubt be further defined as more evidence becomes available.

TABLE 26.3 Evenness/Ride Criteria

Moving-average length, m	Variance levels, mm², to give a ride that is: Good/acceptable	Poor/very poor
3	≤ 1.0	≥ 3.0
10	≤ 4.0	> 16.0
30	≤55.00	≥150.0

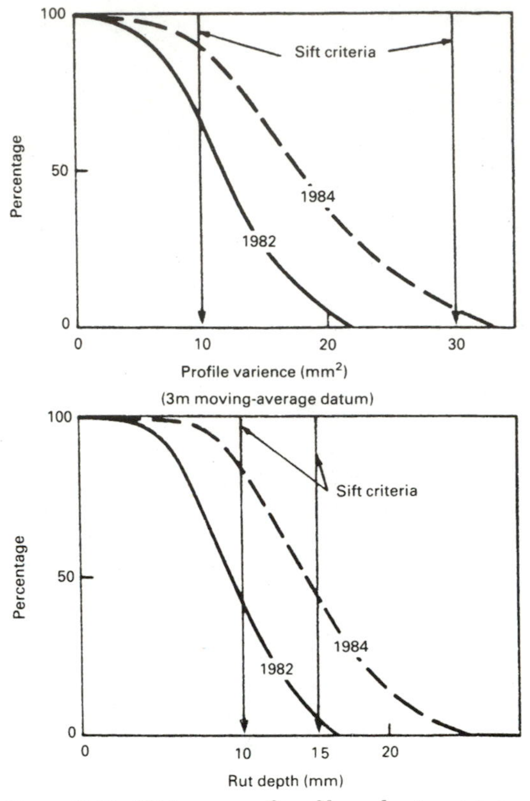

Figure 26.13 Histograms of profile and rut surveys.

The Bump Integrator

26.25. Another relatively rapid method for assessing the irregularity of road surfaces is the Bump Integrator (BI) shown in Fig. 26.14. Originally designed in the United States, the machine has been in use in Europe for about 40 years.

26.26. The trailer, a heavy rectangular chassis, is supported on a central wheel supported by two single-leaf springs positioned one on each side of the wheel. Two dashpot assemblies positioned between the chassis and the wheel axle provide viscous damping. In operation, the downward movement of the wheel relative to the chassis is summed by a mechanical integrator unit fitted to the chassis. The unevenness index r is given by the integrated vertical movements divided by the distance traveled. The integrator unit measures only in inches per mile (1 in/mi = 1.58 cm/km).

Figure 26.14 The Bump Integrator.

26.27. The standard speed of operation is 20 mi/h (32 km/h). However, recent research has shown[7] that higher speeds can be used if the following corrections are made:

For uneven surfaces and operating speeds 20 to 65 km/h and for even surfaces and operating speeds 20 to 32 km/h,

$$r_{32} = \sqrt{\frac{V}{32}}\,(r_v - 30) + 30 \text{ in/mi} \tag{26.6}$$

for even surfaces and operating speeds 32 to 65 km/h

$$r_{32} = \frac{V}{32}\,(r_v - 30) + 30 \text{ in/mi} \tag{26.7}$$

where r_{32} is the standard value at 32 km/h and r_v and V refer to the higher speed of operation.

In practice, Eq. (26.7) would be applicable in the great majority of cases where speed corrections would be needed. For operating speeds that deviate from the standard speed by less than 10 km/h, the accuracy of estimation of the index r_{32} using Eq. (26.6) or (26.7) is within 10 percent of the true value. The accuracy reduces to within 20 percent for a deviation of 30 km/h; the correction procedures are applicable only within the speed range 20 to 65 km/h. In the practical operation of the BI, maintaining a constant speed is often difficult and the

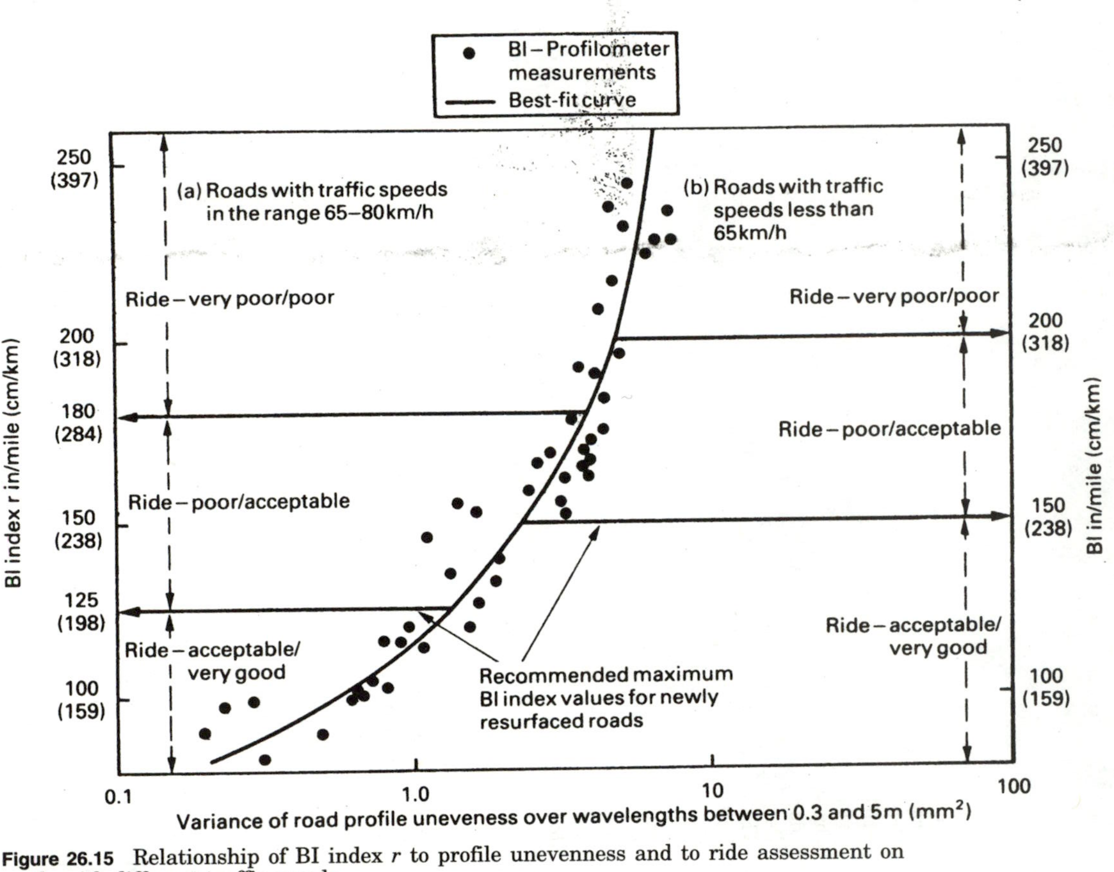

Figure 26.15 Relationship of BI index r to profile unevenness and to ride assessment on roads with different traffic speeds.

correction procedures given above then improve the flexibility of use of the machine.

26.28. A very useful correlation has been made between the variance of the road profile as determined by the High-Speed Road Monitor and the Bump Integrator r value. This is shown in Fig. 26.15.

References

1. McLellan, J. C.: Pavement Thickness, Surface Evenness and Construction Practice, *Transport and Road Research Laboratory Supplementary Report* 706, TRRL, Crowthorne, 1982.
2. Still, P. B., and M. A. Winnett: Development of a Contactless Displacement Transducer, *Transport and Road Research Laboratory Report* 690, TRRL, Crowthorne, 1975.
3. Dickerson, R. S., and D. G. W. Mace: A High-speed Road Profilometer—Preliminary Description, *Transport and Road Research Laboratory Supplementary Report* 182, TRRL, Crowthorne, 1976.
4. Petts, R. C.: The TRRL High-speed Road Monitor. Evaluation of Performance in Hot Ambient Conditions in Turkey, *Transport and Road Research Laboratory Contractor's Report* 3, TRRL, Crowthorne, 1984. (Prepared by Howard Humphreys and Partners.)
5. Jordan, P. G., and P. B. Still: Measurement of Rut-depth in Road Surfaces by the TRRL Highspeed Profilometer, *Transport and Road Research Laboratory Supplementary Report* 1037, TRRL, Crowthorne, 1982.
6. Jordan, P. G., B. W. Ferne, and D. R. C. Cooper: An Integrated System for the Evaluation of Road Pavements. *Proc. 6th Int. Conf. on the Structural Design of Asphalt Pavements, Ann Arbor, Michigan, 1987,* University of Michigan, Ann Arbor, 1987.
7. Jordan, P. G., and J. C. Young: Developments in the Calibration and Use of the Bump-integrator for Ride Assessment, *Transport and Road Research Laboratory Supplementary Report* 604, TRRL, Crowthorne, 1980.

Chapter

27

The Skid Resistance of Pavements and Its Measurement

Introduction

27.1. Prior to the introduction of mechanically propelled transport the slipperiness of road surfaces was important only in relation to the ability of horses to retain an adequate foothold. The higher speeds of motor vehicles and the comparatively low friction between rubber tires and the types of urban road surfaces then in use soon produced an acute skidding problem indicated by a steady increase in the number and seriousness of skidding accidents. As early as 1906 a Parliamentary Select Committee was set up in Britain to investigate the causes and control of skidding.

27.2. Since then research into the skidding problem has been two-pronged. First, it has been necessary to establish from accident statistics limits of slipperiness which can be tolerated on various types of road, bearing in mind, on the one hand, the cost of accidents, and, on the other, the increased cost of providing high-quality nonskid road surfaces. Second, the attributes necessary in the road surface both to obtain and to retain adequate resistance to skidding have needed close study. Both these facets of the skidding problem require reliable methods of measuring the skid resistance of road surfaces.

Measurement of the Slipperiness of Road Surfaces

27.3. The Special Advisory Committee of the Road Board (see Par. 2.18) organized long-term studies into the skidding problem nearly 70 years ago. The work was initially carried out at the National Physical Laboratory. It was early recognized that static measurements of the coefficient of friction could do no more than place road materials in an approximate order of slipperiness. The dependence of skid resistance on vehicle speed and weather conditions necessitated standardized tests by which these factors could be investigated systematically. By the late twenties the concepts of sideway force coefficient and braking force coefficient had been introduced and machines for their measurement had been constructed.

27.4. The Sideway Force Coefficient (SFC) was originally measured using a motorcycle combination in which the sidecar wheel could be fixed during measurements at an angle of 20° to the direction of travel (see Fig. 27.1). The force at right angles to the plane of the inclined wheel expressed as a fraction of the vertical force acting on the wheel is defined as the Sideway Force Coefficient (SFC). Simultaneous measurement of the two forces enables a continuous record of SFC to be obtained. To standardize the process, a smooth tire of standard hardness was specified and for routine testing a standard speed of 30 mi/h (48 km/h) has always been used. After World War II the skidding motorcycle was replaced by a series of test cars working on the same

Figure 27.1 Motorcycle and sidecar for measuring sideway force coefficient.

Figure 27.2 Sideway force test car.

principle, in which the inclined wheel was located within the vehicle chassis (see Fig. 27.2). With rapid advances in electronics the recording methods used have been subject to continuous improvement.

27.5. Since skidding is largely a wet-road problem, routine skid-resistance tests normally require prior wetting of the road surface by a water browser. The SCRIM (Sideway Force Coefficient Routine Investigation Machine) equipment more recently developed at TRRL (Fig. 27.3) carries its own water supply, the road surface being wetted in advance of the test wheel. With this equipment tests can be carried out with the machine operating as part of the normal road traffic.

27.6. The Braking Force Coefficient (BFC) was originally measured by locking the wheels of a vehicle in motion and measuring the braking torque when skidding occurred. Subsequently, various types of towed-wheel braking force machines were developed, one of which is shown in Fig. 27.4. From the measured torque, the force between the tire and the road surface is deduced and this is expressed as a fraction of the vertical load on the wheel, to give the BFC. In use, the towed wheel is locked for periods of about 2 s and then released to give a series of isolated readings rather than a continuous record. This type of machine has generally been used for high-speed testing. Because of differences in the test procedure, the SFC and BFC of

Figure 27.3 The Sideway Force Coefficient Routine Investigation Machine (SCRIM).

identical surfaces measured under the same conditions are not numerically equal. The BFC is approximately 0.8 times the SFC.

27.7. Another machine widely used to assess the slipperiness of road surfaces is the Portable Skid-Resistance Tester shown in Fig. 27.5.[1] This is often referred to as the Pendulum Tester. A pad of tire-tread rubber mounted at the end of the pendulum arm slides over the road

Figure 27.4 Braking force trailer.

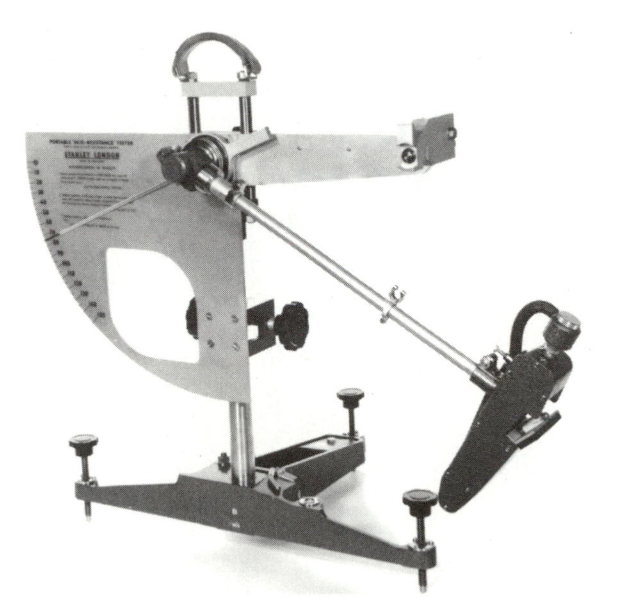

Figure 27.5 Portable Skid-Resistance Tester.

surface on which the machine is placed. The difference in height of the center of gravity of the slider head between the horizontal release position and the highest point of the swing after the slider has passed over the road is used to calculate the loss of energy arising from friction. The test conditions, which must be closely observed and controlled, have been chosen so that the values read off the calibrated scale of the instrument correspond to the Skid-Resistance Value (SRV) of a patterned tire skidding at 30 mi/h (48 km/h). The test is carried out with the pavement surface wetted in a standard manner, and a number of tests spaced at 5- to 10-m intervals are required to give an average value.

27.8. The Pendulum Tester assesses an area of the road surface which may not be large in relation to the coarse texture. For this and other reasons the SRV does not necessarily correlate closely with SFC or BFC measurements. However, an approximate correlation with SFC for rough-looking and medium-textured surfaces is shown in Fig. 27.6.

27.9. The principal attributes of the machine are its portability and simplicity. It is particularly useful in laboratory research aimed at the development of skid-resistant surfaces and it is valuable to the road engineer for investigating potential accident sites.

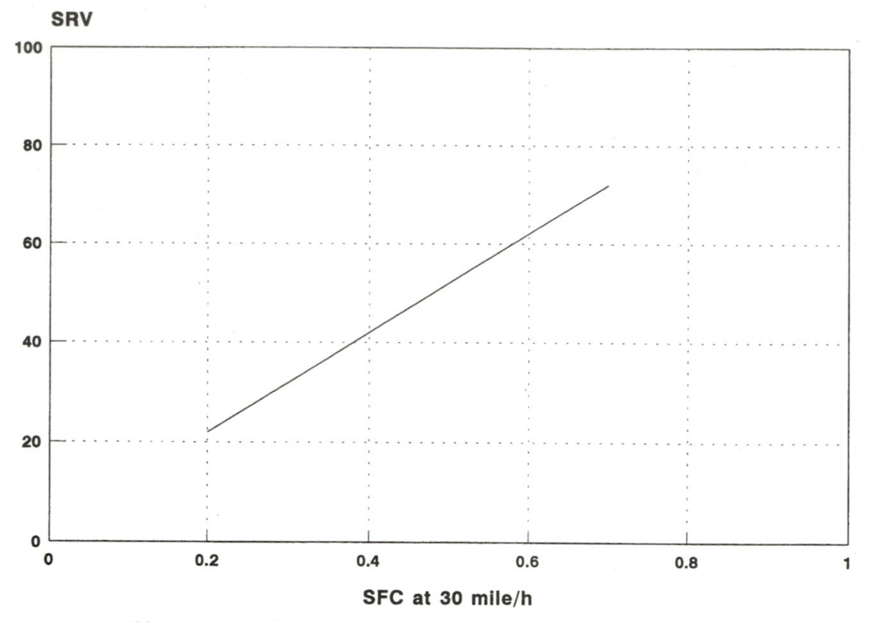

Figure 27.6 Observed correlation between SRV and SFC in 63 tests on rough-looking and medium-textured surfacing.

The Development of Skid-Resistance Criteria for Different Types of Road

27.10. In the fifties Giles developed skid-resistance criteria for different types of road based on the detailed testing of sites from which the police had reported wet-weather skidding accidents.[2] He found fairly well established thresholds of sideway force coefficient for the different types of road, below which the possibility of skidding accidents increased markedly. These criteria were modified and to some extent simplified by Sabey in 1968.[3] Her recommendations were accepted by the Marshall Committee in 1970 as target values for skidding resistance to be used as a guide in the preparation of maintenance schedules.[4] These values and the types of road to which they apply are reproduced in Table 27.1.

27.11. The road engineer must provide pavements which will meet the requirements of Table 27.1 and which will continue to satisfy those requirements with a minimum of surface maintenance. To do this requires an understanding of the pavement factors which influence skidding.

TABLE 27.1 Categories of Sites and Suggested Target Values for the Sideway Force Coefficient Proposed by the Marshall Committee

		Sideway force coefficient	
Category of site	Type of site	Test speed, km/h (mi/h)	SFC
A	Most difficult sites: 1. Roundabouts 2. Bends with radius less than 150 m (500 ft) on unrestricted roads 3. Gradients of 5% (1 in 20) or steeper or longer than 100 m (330 ft) 4. Approaches to traffic signals on unrestricted roads	50 (30)	0.55
B	Average sites: 1. Freeways and other high-speed roads, i.e, speeds in excess of 95 km/h (60 mi/h)	50 (30) 80 (50)	0.50 0.45
	2. Trunk and principal roads, and other roads with more than 200 vehicles per day in urban areas (sum in both directions)	50 (30)	0.50
C	Other sites: Straight roads with easy gradients and curves without junctions and free from any feature such as mixed traffic especially liable to create conditions of emergency	50 (30)	0.40

Factors Which Affect the Skid Resistance of Road Pavements

27.12. Although skidding can occur on dry roads, it is a major cause of accidents only when the pavement is wet. Research into skidding has for this reason been confined largely to wet surfaces. The mechanism has been succinctly discussed by Sabey, and the following three paragraphs are reproduced from one of her publications.[3]

27.13. The friction coefficient between two sliding surfaces can generally be expressed as the sum of two terms. The first arises from the adhesion at the points where the surfaces are in contact, and in the case of a wet road this implies that at such points of contact the lubricating layer of water on the road must have been broken through and areas of dry contact established. While drainage channels, provided by the large-scale texture of the road or by a pattern on the tire, assist in getting rid of the main bulk of the water, the ultimate penetration of the water film can be achieved only by the presence of the fine-scale sharp edges in the road, on which high pressures are built up.

27.14. The second component of friction arises if the irregularities in one surface produce appreciable deformation of the other, and at least some of the energy of deformation is irrecoverable. This deformation can occur in the presence of a lubricant, even if no actual contact between the surfaces is established.

27.15. When vehicles are traveling at speeds up to 50 km/h (30 mi/h), the fine-scale texture is the dominant factor determining the skidding resistance; the adhesion component predominates. At higher speeds, it becomes increasingly difficult to penetrate the water film in the time available. The resistance to skidding then depends largely on the deformation component of friction, and projections in the road surface must be sufficiently large and angular to deform the surface of the tire tread, even though water may still be present on the surface. At slow speeds, therefore, the microtexture of the road surface or its constituents (mainly the stone) is the major factor determining the level of skidding resistance: at high speeds, its macrotexture, the size and shape of the visible asperities, is equally important.

27.16. Research into the various factors which determine the slipperiness of road surfaces has been in progress for more than 60 years. During that time the speed capability of motor vehicles has progressively increased and this capability has been increasingly utilized with the introduction of grade-separated dual-carriageway roads and freeways. Research into skidding was formerly directed mainly at the slower-speed situation relevant to urban areas, in which the bulk of skidding accidents occur. Only in the last 30 years has research been specifically directed toward improving the skid resistance of pavements at the very high speeds at which skidding accidents, although less frequent, tend to be serious in their consequences.

Research Studies into the Factors Which Influence the Skid Resistance of Pavements

27.17. In the early thirties Bird and Scott reported systematic measurements of skid resistance made on selected roads over a period of years.[5] The motorcycle and sidecar combination shown in Fig. 27.1 was used to make the measurements of SFC. Typical results for two experimental sections on the Kingston bypass are shown in Fig. 27.7. Section 1, with an unchipped rolled asphalt wearing course, shows an increase of SFC from 0.3 to 0.7 over the 5-year period, while section 2,

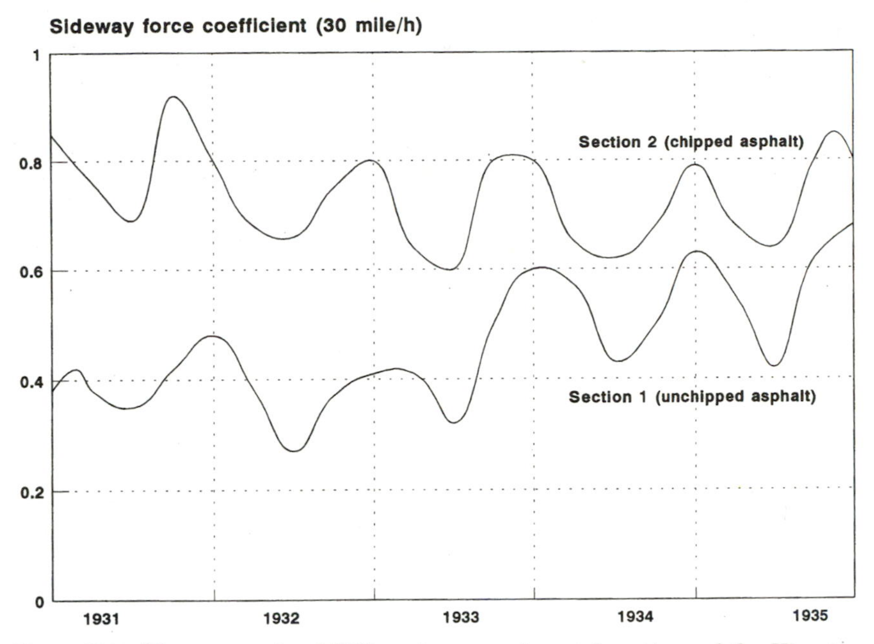

Figure 27.7 Measurements of SFC on two experimental sections of the Kingston bypass (1931–1933).

in which uncoated chippings were rolled into the surface, shows a corresponding decline from 0.9 to 0.8. Although at that time the processes involved in producing these changes of SFC were not fully understood, it is now clear that in the case of the unchipped asphalt it took several years (under the comparatively modest volume of heavy traffic then being carried) for the thick bituminous film to wear away to expose the microtexture of the slag aggregate. The uncoated chippings (also of slag) used in the second section had an exposed microtexture from the start of the tests, and the small decline in SFC can be attributed to a modification of this microtexture by the process now known as polishing.

27.18. In standardizing the test procedure to be used in making SFC measurements, Bird and Scott repeated measurements on selected areas at various times after the commencement of rain and in this way produced idealized curves of the type shown in Fig. 27.8. They found that the magnitude of the reduction component d defined in the diagram depended on the weather conditions immediately prior to the particular period of rain being studied. They also observed that artificial watering of the road by sprinklers consistently produced lower readings of SFC than were obtained on the same roads after pro-

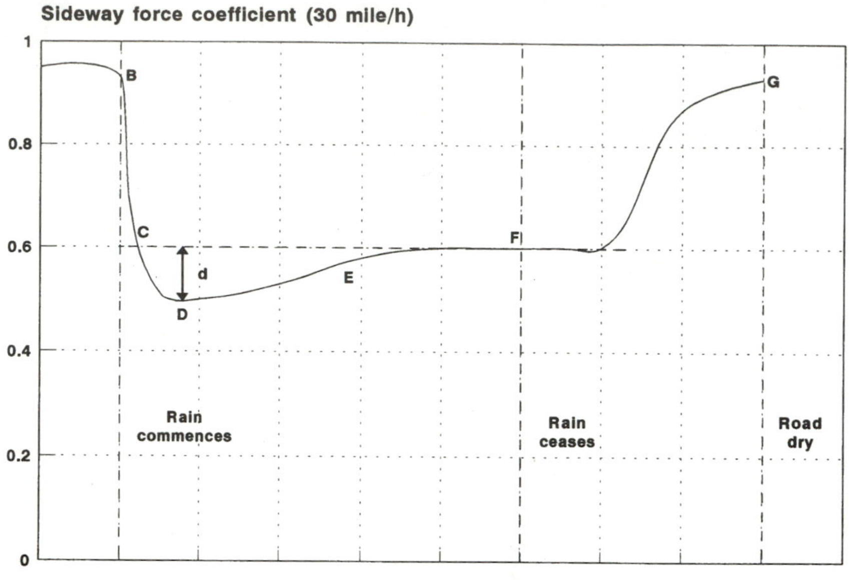

Figure 27.8 Variation of SFC from commencement of rain to road dry condition at constant speed.

longed rain. Because measurements after watering were generally carried out immediately after wetting, they concluded that these measurements correspond to point *D* in Fig. 27.8 while observations on naturally wet roads were more likely to correspond to the condition represented by *EF*. Since watering appeared to reproduce the most dangerous condition relative to skidding they decided to use the method of artificial wetting for producing skidding standards.

27.19. Practical observations of the extreme slipperiness of some early road surfaces immediately after the commencement of rain resulted in the adoption of the term "greasy" to describe this condition. This led to the erroneous conclusion that slightly wet roads were necessarily more slippery than pavements wetted to a condition at which runoff was occurring. The explanation is that the runoff of rainwater tends to remove lubricating agents left adhering to the road surface during the previous dry period. These agents probably consist of a mixture of rubber and oil deposited mainly by heavy vehicles. It is this cleansing process which accounts for the increase in SFC between the points *D* and *F* in Fig. 27.8. There is evidence that the processes of contamination, polishing, and scouring are going on continuously. In the spring and early summer when the rainfall in Britain tends to be low and evaporation rates are rising, runoff is at its lowest and deposition tends to increase. Autumnal conditions of high rainfall and decreasing evaporation favor scouring. This is probably the main reason for the apparently seasonal change in SFC in Fig. 27.7. Recent research has shown that there may be other seasonal factors affecting the microtexture of road aggregates.[6] Because of the seasonal effect it is usual to make measurements of SFC during the months May–September, and quoted figures will normally relate to this period.

27.20. The measured SFC of a wet road surface tends to decrease with increasing speed because of the time factor involved in the exclusion of water between the tire and the pavement. This is intimately associated with both the micro- and macrotextures of the road surface and it is therefore difficult in any discussion to divorce the speed factor from textural considerations. (In practice vehicle tires are not smooth, as in the SFC determination, and tire tread also plays a part in determining the actual resistance to skidding between vehicle wheels and the pavement.) As would be expected, the reduction of SFC with increasing speed is greater on smooth-textured surfacings than is the case with surfaces of rougher texture. Figure 27.9 shows results reported by Giles which illustrate this effect.[2] The curves refer

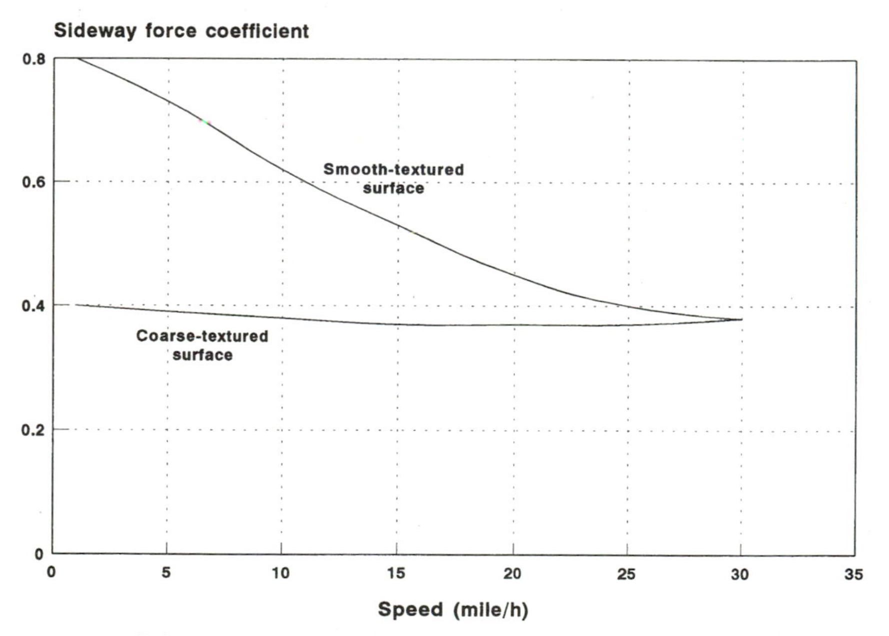

Figure 27.9 Relationships between SFC and speed for two surfaces with different textures.

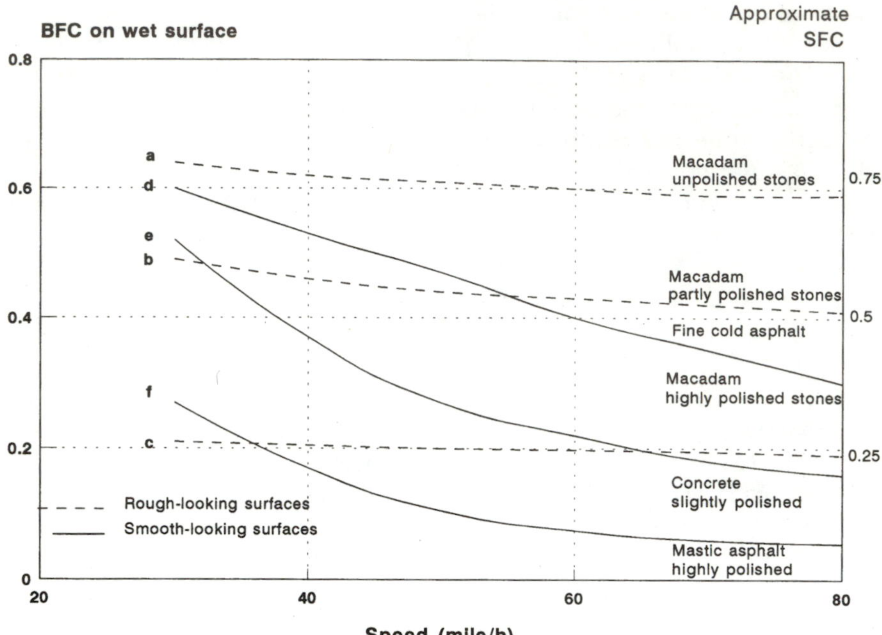

Figure 27.10 Change of BFC with speed on different wet surfaces.

to two particular surfacings which gave substantially the same level of SFC at 30 mi/h (48 km/h).

27.21. It must not be inferred from Fig. 27.9 that all coarse-textured surfacings would give a lower value of SFC than smoother surfacings at speeds lower than 30 mi/h (48 km/h). Figure 27.10 shows later data reported by Sabey,[3] which relates BFC with speeds up to 80 mi/h (129 km/h). The curves illustrate the wide range of BFC values encountered in both "rough-looking" and "smooth-looking" surfaces. Each of the pairs of curves *a* and *d*, *e* and *b*, *f* and *c* has approximately the same value of BFC at 30 mi/h. The divergence of the curves at higher speeds follows the trend at lower speeds indicated in Fig. 27.9. An approximate SFC scale has been added to Fig. 27.10, to permit ready comparison of the two diagrams. The textural appearances of the six surfaces referred to in Fig. 27.10 are shown in Fig. 27.11.

The Surface Characteristics Influencing Resistance to Skidding

27.22. It follows from the above discussion that if a road surface is to provide an acceptable resistance to skidding then the following conditions must exist:

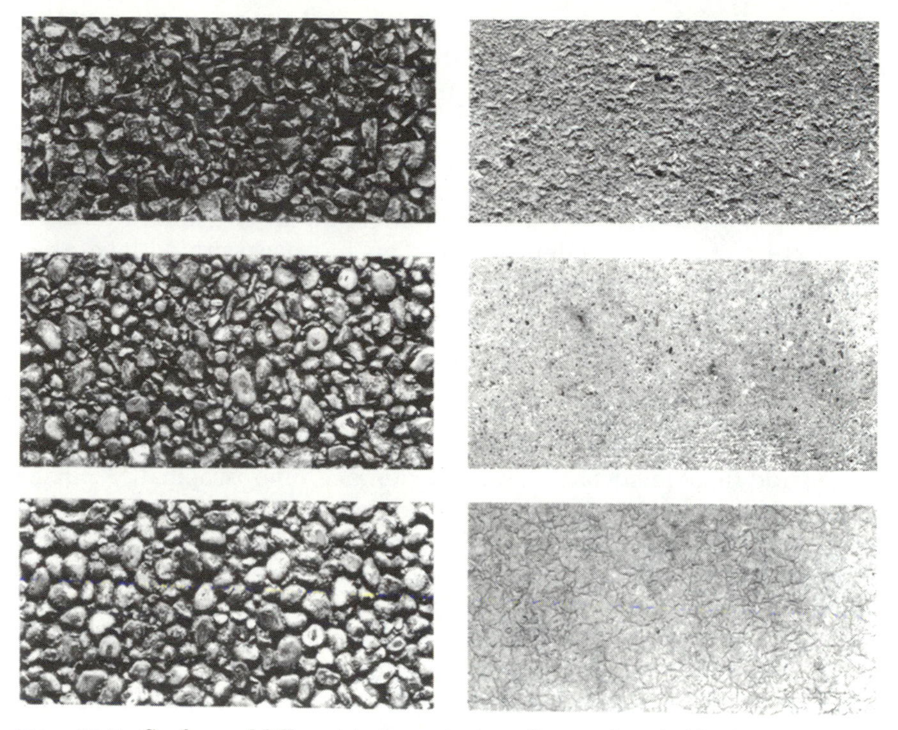

Figure 27.11 Surfaces of different textures; test results are given in Fig. 27.10.

1. Those areas of the surface in contact with the tire must have a sufficiently pronounced microtexture to provide an adequate coefficient of friction.
2. The microtexture must not be readily removed by the action of traffic.
3. Between the tire and the road there must be areas which are not in contact to facilitate the displacement of water. These areas are provided by the coarse texture on the road and by the tire tread.
4. The coarse texture must not be readily removed by the action of traffic.

Requirements 1 and 2 are essential to all pavements; requirements 3 and 4 are particularly important on high-speed roads.

27.23. In flexible pavements the microtexture is provided mainly by the surface of the exposed aggregate or by the surface of the chippings in rolled asphalt surfacings and surface dressings. Coarse texture is provided by the spaces between the exposed aggregate or chippings.

In concrete pavements the microtexture is principally associated with the sand/mortar fraction (rather than the coarse aggregate) and the coarse texture is provided by the surface brush marking or wet grooving.

27.24. To be able to meet (or attempt to meet) the SFC values shown in Table 27.1 the engineer must be able to quantify them in terms of the texture of the pavement surface. This has necessitated laboratory research into methods of measuring texture and full-scale research to relate those measurements to observed levels of SFC on the road.

Fine Texture or Microtexture

27.25. Fine texture of the exposed aggregate, important in the design of nonskid flexible surfaces, can be examined microscopically,[6] but for routine purposes it is generally assessed by measurements of the coefficient of friction determined by a rubber slider moving over a prepared sample of the aggregate set in a cement mortar bed.[7] The test sample is cast in a flat or cylindrical mold to ensure that the exposed surface of the aggregate conforms to a specified flat or curved profile. The determination of coefficient of friction is made with the surface of the aggregate wet, using a machine very similar to the Portable Skid Resistance Tester referred to in Pars. 27.7 to 27.9. The sample preparation and test procedure are described in detail in British Standard 812:1975.[8]

27.26. Under the action of traffic certain aggregates tend to lose their fine texture quickly and are said to "polish." To determine the liability of aggregates to polish, an accelerated-polishing machine has been developed, and this is also described in British Standard 812:1975. The machine, shown in Fig. 27.12*a*, comprises a 16-in (406-mm) diameter wheel, having a periphery 2.5 in (64 mm) wide, around which curved specimens of the aggregate under test (Fig. 27.12*b*) are mounted to form the "road" surface, which is in contact with an 8-in (203-mm) diameter pneumatic-tired wheel, with an inflation pressure of 45 lbf/in^2 (310 kN/m^2). By means of a lever arm the tire is pressed onto the surface of the aggregate with a normal load of 88 lb (40 kg). The specimen wheel is driven at 320 rpm, which gives a peripheral speed of about 15 mi/h (24 km/h). The loading wheel is free to rotate on its axis. Grit or other polishing agents can be fed from a hopper onto the aggregate surface just before it passes under the loaded tire, and water is supplied through the same chute. (It is to accommodate samples from this machine that the friction measuring device referred to in Par. 27.25 was designed to operate on curved as well as

(a)

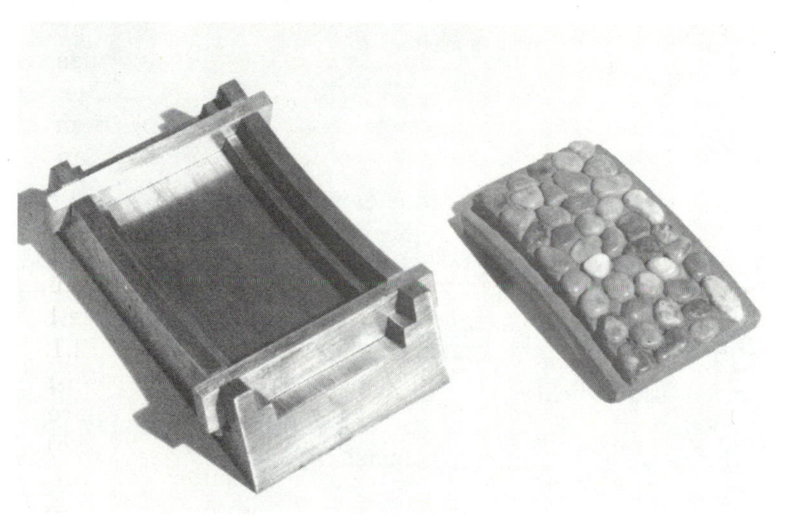

(b)

Figure 27.12 (*a*) Apparatus for accelerated polishing test. (*b*) Mould and specimen for accelerated polishing test.

flat specimens.) The test period used is 6 hours; during the first 3 hours the machine is fed with No. 36 corn emery and water both at the rate of 20 to 35 g/min, and during the second 3 hours with air-floated emery powder at a rate of 2 to 4 g/min and water at a rate of 4 to 8 g/min. This method of testing was originally selected as repre-

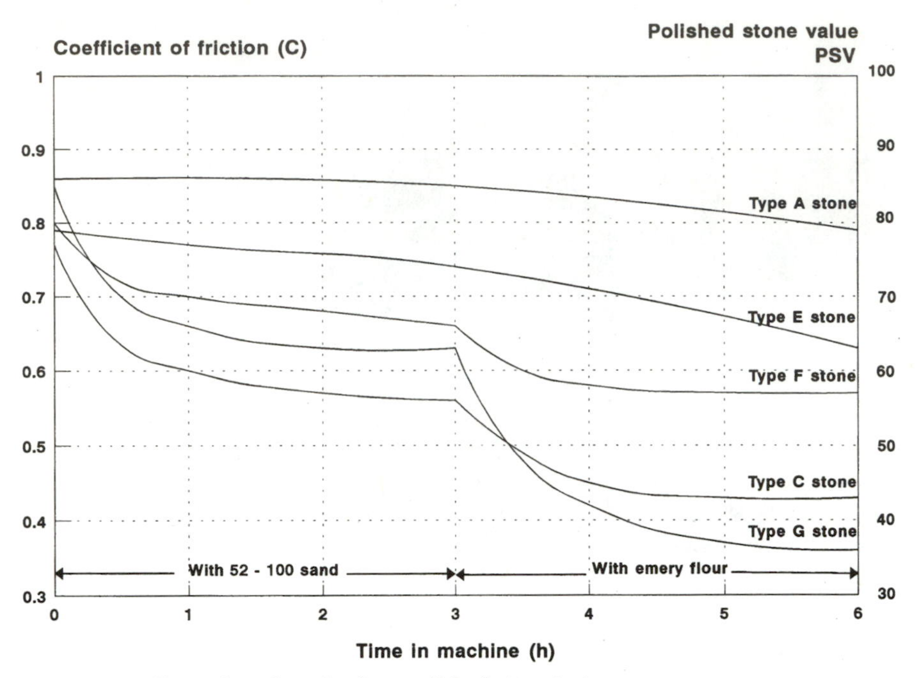

Figure 27.13 Examples of results from polished stone tests.

senting the polishing effect likely to be experienced on a heavily trafficked road, but it is preferable to regard the test as a purely comparative one to be related by long-term experience to what happens on the road. The measured coefficient of friction at the end of the 6-hour period was originally defined as the Polished Stone Value (PSV). Figure 27.13 shows typical variations of coefficient of friction during the test period and indicates that different types of stone all having an initial coefficient of friction of the order 0.75 to 0.85 may show markedly different polishing characteristics, the PSV after the test varying between 36 and 80.

27.27. Figure 27.14 shows measured distributions of PSVs for different groups of stone tested at the TRRL. It is clear that some groups polish more readily than others, but within all the groups there is considerable variation. In addition to having a high PSV, the exposed stone must be durable. Current recommendations for category A and B sites, as defined in Table 27.1, are given in Table 27.2.

27.28. Recent research reported by Szatkowski and Hosking shows that the mean summer values of SFC for a new road tend to fall with-

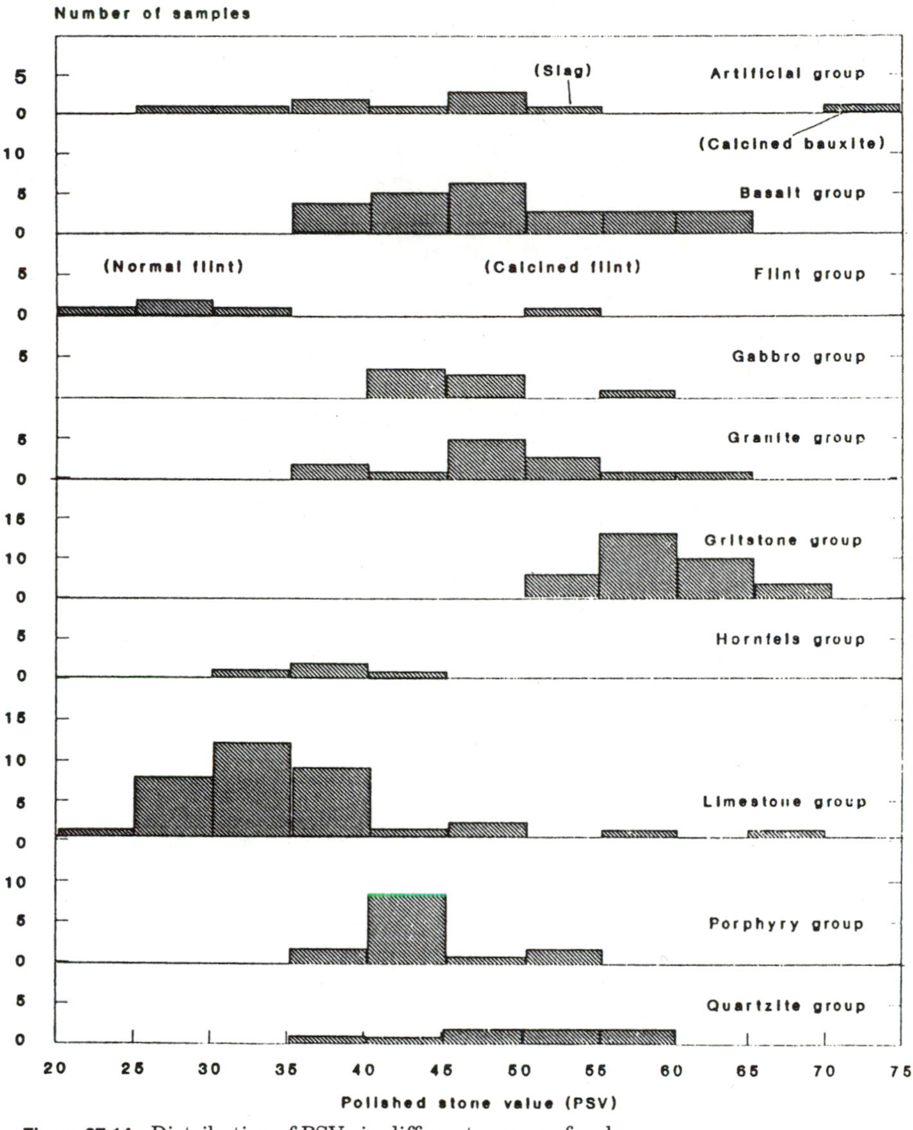

Figure 27.14 Distribution of PSVs in different groups of rock.

in 1 year to a level which thereafter changes very little.[9] This level appears to depend on the PSV of the stone and the intensity of the commercial traffic using the lane on which the measurements are made. Typical results for rolled asphalt surfacings with precoated chippings of PSV 58 to 60 are shown in Fig. 27.15. It appears that the processes of contamination, polishing, and scouring referred to in Par.

TABLE 27.2 Aggregate Requirements for Category A and B Sites

Category of site	PSV	Aggregate abrasion value (BS 812)
A	62 min	10 max
B	59 min	12 max

NOTE: There are currently no similar recommendations for category C sites, but it is usual to specify a PSV value of not less than 50.

27.19 may be responsible for the establishment of this "equilibrium" SFC condition associated with the level of commercial traffic. In confirmation of this hypothesis these authors have reported a case where the daily commercial traffic decreased from 2750 to 730 vehicles per day (owing to the construction of a bypass) and the SFC increased from 0.43 to 0.58.

27.29. Following tests of this type on a variety of flexible pavements with stones of different PSV used in asphalt surfacings and surface dressings, Szatkowski and Hosking have proposed the relationship shown in Fig. 27.16 between SFC, PSV, and intensity of commercial traffic. This shows that for category A and B sites carrying very heavy

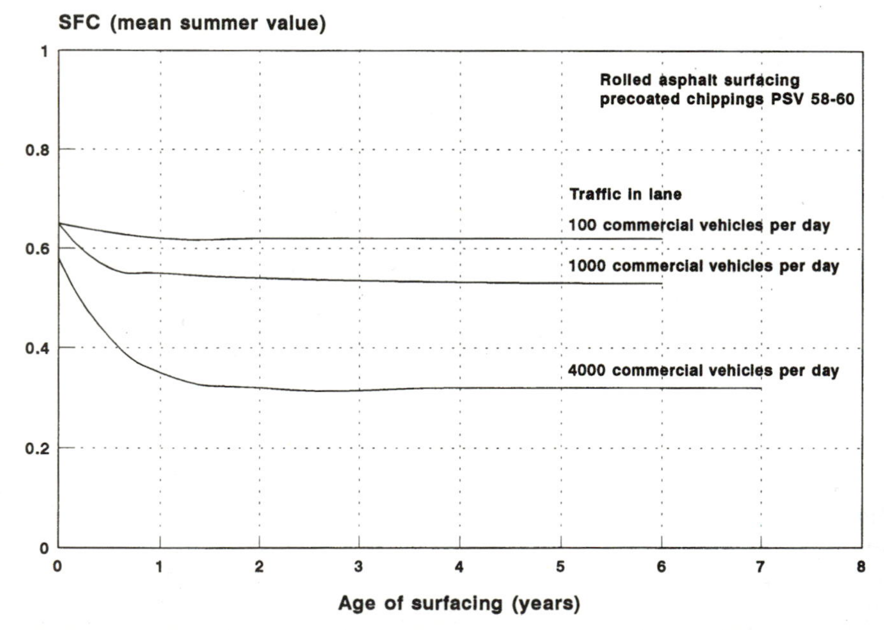

Figure 27.15 Effect of traffic on skidding resistance of freeway-type surfacing.

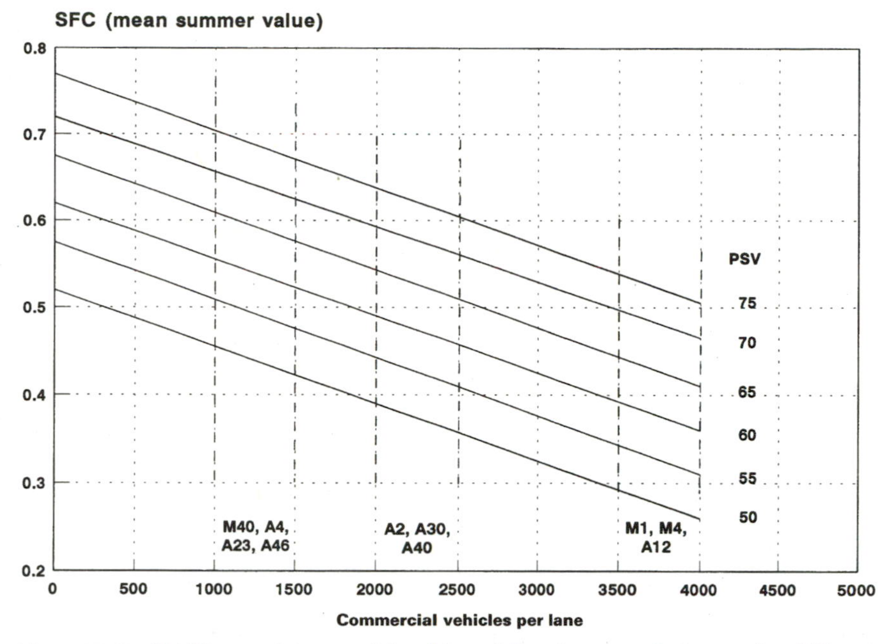

Figure 27.16 Skidding resistance achievable on bituminous surfacings under different traffic conditions.

traffic the PSVs shown in Table 27.1 will be inadequate to ensure the required level of SFC and that PSVs as high as 70 to 75 will be necessary. This has led to considerable research into the properties of synthetic aggregates with high resistance to polishing.[10,11] Such aggregates incorporated in surface dressings have had a dramatic effect on urban accidents at sites such as the approaches to pedestrian crossings, traffic signals, and roundabouts,[12] where the relatively high cost is justified.

27.30. Although approximately half of the weight of materials constituting a concrete mix is accounted for by the coarse aggregate, its degree of exposure in the road surface is normally quite small. As a consequence, the coarse aggregate is of little importance in determining the fine texture and the resistance to skidding. Franklin and Calder have found that the PSV and aggregate abrasion value of the fine aggregate largely determine the SFC of concrete pavements at moderate speeds.[13] They studied, over a period of 4 to 5 years, the change in skid-resistance value of concrete cores inserted in the left-hand lane wheel tracks of a freeway and a major trunk road. The concrete mixes used included coarse aggregates of PSV in the range 28 to 71 and fine aggregate of PSV 28 to 73. (Since the PSV test could not be directly applied to the small particles constituting the fine aggre-

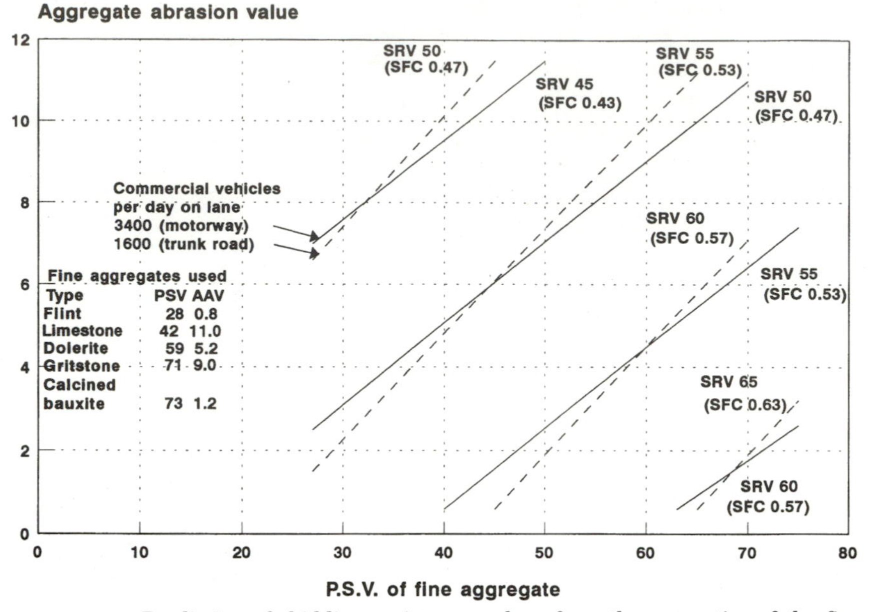

Figure 27.17 Prediction of skidding resistance values from the properties of the fine aggregate–concrete pavements.

gate, it was assumed that the PSV of that fraction was equal to that of the rock from which it was derived.) The results of this research are summarized in Fig. 27.17. Details of the PSV and AAV of the fine aggregates used in the study are given in the figure. The values of SRV quoted are terminal values after 5 years, but the annual measurements indicated no significant reduction of SRV with time. As with flexible pavements, however, the terminal value was higher for the trunk road traffic of 1600 commercial vehicles per day on the lane studied than on the freeway where the corresponding traffic was 3400 commercial vehicles per day.

27.31. The data given in Fig. 27.16 have been used to express SRV in terms of SFC in Fig. 27.17. It can be inferred from the figure that with certain types of fine aggregate it would be difficult to achieve and maintain the targets of SFC quoted in Table 27.1. This has been confirmed by routine measurements of SFC on modern concrete roads.

Coarse Texture or Macrotexture

27.32. Coarse texture is necessary to prevent a loss of resistance to skidding at high speeds. It is generally quantified in terms of the average texture depth measured by the sand patch method.[14] In this test a metal cylindrical container of diameter 19 mm and height 84 mm is filled in a controlled manner with a natural rounded sand of particle size between 150 and 300 μm. This sand is poured onto the pavement surface in a conical heap which is then spread by the circular motion of a flat wooden disk of 64 mm in diameter provided with a rubber facing. With the plane of the disk parallel with the road surface the sand is spread over an approximately circular area to fill all the depressions within that area. The mean radius of the sand-filled patch is determined with dividers, and the average texture depth is obtained from the volume of sand used, divided by the area of the sand patch.

27.33. The sand patch method of assessing texture depth is not suited to the routine assessment of large lengths of highway, and this role has now been taken over by the high-speed texture meter based on the contactless depth-measuring procedure discussed in detail in Chap. 26. One of the three functions of the High-Speed Road Monitor referred to in Par. 26.23 is to measure the texture depth of both flexible and concrete pavements. However, there is a separate version known as the High-Speed Texture Meter (HSTM) which is much smaller and more convenient when only texture depth is under investigation.

27.34. The HSTM has only one contactless sensor, which concentrates on the depressions incorporated in the surfacing to increase the high-speed skid resistance. How this is done is shown in Fig. 27.18.[15]

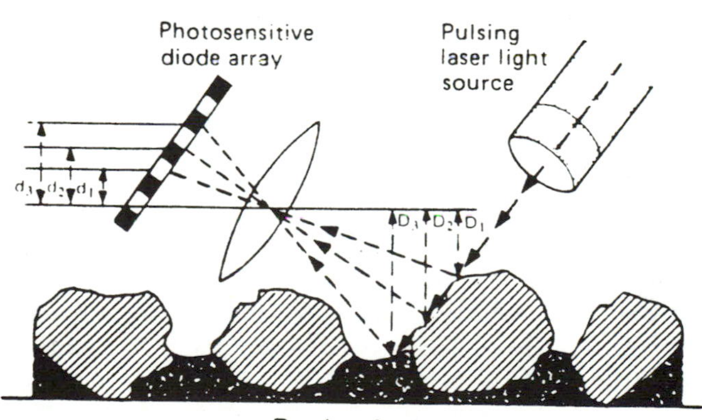

Figure 27.18 Principle of the laser-based contactless sensor for measuring texture depth.

A rapidly pulsing semiconductor laser, producing infrared light (wavelength 906 nm), is projected onto the road surface. Light reflected from the spot so formed is focused by a receiving lens onto a linear array of 256 photodiodes. The position of the diode receiving most light gives a measure of the distance to the road surface at that instant, and the depth of the texture is computed from a series of such measurements.

27.35. The sensor system moves over the road surface in a plane parallel to the road surface as in Fig. 27.18, and rays of laser light reflected from different points in the texture D_1, D_2, and D_3 are detected by appropriate diodes d_1, d_2, and d_3 in the receiving array, thus giving a measure of the depth of the points. The laser pulses at approximately 3.5 kHz and the number of the illuminated diode is transmitted to the computer on board the towing vehicle. The texture depth is computed and expressed as a root-mean-square value with averages recorded for every 10 m traveled.

27.36. The sensor is mounted in a trailer rather than directly on the vehicle in order to isolate the sensor from any vibration from the engine and transmission. The towing vehicle carries all the equipment needed to control the texture-measuring process. The center of the system is an Interdata 16-bit minicomputer which is connected to the various elements of the system. These are: the sensor in the trailer, a distance meter which generates pulses from a unit attached to the gearbox final drive, a visual display unit (VDU) with keyboard, for immediate display of results, a paper-tape reader to allow rapid programming of the computer, and a control box with pushbuttons for starting and stopping texture measurements and inserting codes for location referencing.

27.37. Measurements are generally made in the left-hand wheel track of the carriageway, and therefore the results are likely to represent the minimum texture depth for the road studied. As the wheel tracks are the parts of the road likely to be involved when vehicles are skidding, it is there that the surface texture is most relevant. Normally, test speeds are in the range 45 to 65 km/h, but higher speeds (up to 80 km/h) are used on freeways; however, sometimes traffic or road conditions necessitate lower speeds being used. Earlier studies with the machine showed that texture results were sensibly constant over a speed range of 15 to 100 km/h.

27.38. Although the trailer has a rigid suspension, slight bounce on the tires is sometimes experienced when passing over certain types of surfaces; this is reflected in the sensor outputs being disposed along an apparently curved profile. To eliminate the effect of this motion on the readings, the sensor outputs are grouped into blocks of data. A quadratic least-squares regression technique is used to draw a curve through the data points of any one data block. The variance and standard deviation of the measured points from this curve are then calculated. The standard deviation of sensor measurements is defined as the Sensor-Measured Texture Depth (SMTD); although it is dimensionally the same as the measurement made by the sand patch method, the latter defines a mean depth of texture below a plane through the high points on the road surface. The two methods would not therefore be expected to give the same numerical results. Figure 27.19 shows the relationship between contactless sensor and sand patch measurements made as far as practicable at the same points of the road surface.[16] From this relationship it can be concluded that the ratio between the two methods of measurement is 1.7. For the time being the sand patch method is regarded as the standard.

27.39. Typical values of texture depth for new asphalt surfacings with precoated chippings lie between 1 and 2 mm. Newly laid surface dressings generally have a texture depth between 3 and 4 mm. For

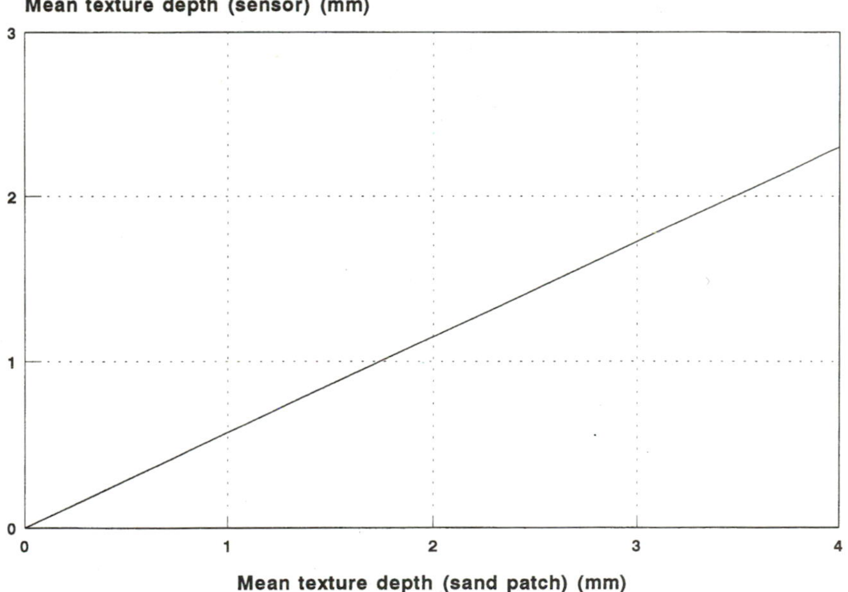

Figure 27.19 Texture depth comparison of determinations made by contactless sensor and sand patch method.

unchipped asphalt and asphaltic concrete pavements the texture depth may be as low as 0.2 mm. The Department of Transport specification for highway works does not give texture depth requirements for new flexible roads. Other requirements of the specification are considered to ensure an adequate initial texture depth. For concrete roads the specification calls for an initial texture depth of not less than 0.75 mm.

27.40. From studies of the SFC of flexible and concrete pavements having a range of texture depths, Salt and Szatkowski have published the information given in Table 27.3.[17] A given change of skidding resistance with speed is associated with a smaller texture depth in a concrete pavement than in a flexible pavement. This probably reflects the continuous nature of the channels obtained by the brushing or grooving processes used to texture concrete.

27.41. It can be inferred from Table 27.3 that to satisfy the maximum reduction of 10 percent in SFC between 50 and 80 km/h for freeways, which is contained in Table 27.1, a texture depth of about 0.6 mm would need to be maintained on concrete surfaces and one of 1.3 mm on flexible surfaces.

27.42. A decrease in texture depth under the action of traffic may be effected in flexible pavements by compaction during service or by the removal or embedment of chippings by traffic. On concrete surfaces wear of the brushmarks or surface grooving is responsible. Experience shows that close adherence to the specification and particularly to those parts relating to laying and rolling temperatures is essential to the production and maintenance of a satisfactory texture depth in bituminous surfaces. Comprehensive research by Weller and Maynard on existing concrete pavements, and using a wear wheel rather similar to the machine used in the PSV test, has shown that if

TABLE 27.3 The Effect of Coarse Texture on the Change of Skidding Resistance with Speed

Drop in skidding resistance with speed change from 50 to 130 km/h, %	Texture depth, mm	
	Flexible	Concrete*
0	2.0	0.8
10	1.5	0.7
20	1.0	0.5
30	0.5	0.4

*When textured predominantly transversely.

a texture depth of 0.7 mm is to be maintained for 4 to 5 years on a heavily trafficked concrete road the initial texture depth will in general need to be of the order of 1.5 mm.[18] Such a texture depth is difficult to achieve by brushing, and this has led to the development of alternative wet-grooving techniques. It must be appreciated that an unnecessarily heavy texture on both flexible and concrete pavements may result in unacceptable noise both inside and outside motor vehicles, and it will tend to increase tire wear.

27.43. To obtain more information relating to the texture depth of in-service roads than could have previously been obtained using the sand patch method, the Transport and Road Research Laboratory has made a 3-year study in three counties in the United Kingdom and for four classes of road, as indicated in Table 27.4.[15] The texture depths shown are all sensor-measured, and the value must therefore be multiplied by 1.7 to make a comparison with the target values given in Pars. 27.39 and 27.40. The results show that the texture depths are generally satisfactory when compared with the typical values given in

TABLE 27.4 Average Texture Depth (SMTD) for Various Categories of Road[15]

		County A			County B			County C		
Road category	Year of test	Length tested, km	SMTD, mm Mean	SMTD, mm s.d.	Length tested, km	SMTD, mm Mean	SMTD, mm s.d.	Length tested, km	SMTD, mm Mean	SMTD, mm s.d.
Freeways	1982	22.1	0.71	0.25	80.4	0.67	0.17			
	1983	32.3	0.80	0.24	83.4	0.75	0.19	78.0*	0.58	0.22
	1984	30.2	0.91	0.30	82.9	0.76	0.19	80.2*	0.58	0.22
		31.3	0.87	0.26						
Class A	1982	556.7	0.77	0.30	275.8	0.67	0.26			
	1983	546.0	0.79	0.30	277.0	0.70	0.27	637.6	0.67	0.25
	1984	560.0	0.87	0.32	272.2	0.71	0.29	630.5	0.68	0.28
		567.8	0.81	0.32						
Class B	1982	75.7	0.69	0.29	80.7	0.64	0.29			
	1983	75.2	0.70	0.28	77.7	0.65	0.28	101.8	0.63	0.23
	1984	76.3	0.74	0.28	70.1	0.66	0.35	103.8	0.63	0.27
		70.6	0.67	0.25						
Unclassified	1982	26.5	0.77	0.29	4.3	0.59	0.19			
	1983	26.7	0.68	0.25	4.2	0.47	0.14	12.4	0.66	0.22
	1984	26.4	0.71	0.24	4.3	0.44	0.12	14.8	0.66	0.22
		26.5	0.78	0.24						
Concrete	1982	4.4	0.48	0.07						
	1983	4.3	0.47	0.09				48.7	0.48	0.14
	1984	4.3	0.46	0.07				46.7	0.48	0.13
		4.3	0.46	0.06						

*Much of the freeway tested length in county C was concrete.

Par. 27.39. The results for the three counties are surprisingly similar for the various classes of roads.

27.44. The cheapest method of restoring the coarse texture of structurally sound flexible pavements is surface dressing. Early experience with the surface dressing of smooth concrete pavements was generally disappointing. Surface grooving by closely spaced diamond saws has been increasingly used during the last 10 years to retexture concrete. Random grooving is desirable to minimize unpleasant noise effects. The cost of such grooving was originally high (several times that of surface dressing), but with improvements in the technique the process has become more competitive. Recent full-scale experiments have led to much-improved specifications for the surface dressing of concrete; some examples have performed excellently under several years of intense freeway traffic.

References

1. Giles, C. G., B. E. Sabey, and K. H. F. Cardes: Development and Performance of the Portable Skid-resistance Tester, *Road Research Technical Paper* 66, HMSO, London, 1964.
2. Giles, C. G.: The Skidding Resistance of Roads and the Requirements of Modern Traffic, *Proc. Instn. Civ. Engrs.*, vol. 6, pp. 216–242, 1957; *Crushed Stone J.*, vol. 32, no. 2, pp. 8–10, 15, 1957.
3. Sabey, B. E.: The Road Surface and Safety of Vehicles, *Symposium on Vehicle and Road Design for Safety, Cranfield 3–4 July 1968,* Institution of Mechanical Engineers, London, 1968.
4. Ministry of Transport: *Report of the Committee on Highway Maintenance,* HMSO, London, 1970.
5. Bird, G., and W. J. O. Scott: Road Surface Resistance to Skidding, *Road Research Technical Paper* 1, HMSO, London, 1936.
6. Neville, G.: A Study of the Mechanism of Polishing of Roadstones by Traffic, *Transport and Road Research Laboratory Report* LR621, TRRL, Crowthorne, 1974.
7. Maclean, D. J., and F. A. Shergold: The Polishing of Roadstone in Relation to Skidding of Bituminous Road Surfacings, *Road Research Technical Paper* 43, HMSO, London, 1958.
8. British Standards Institution: Method for Sampling and Testing of Mineral Aggregates, Sands and Fillers, British Standard 812:1975, BSI, London, 1975.
9. Szatkowski, W. S., and J. R. Hosking: The Effect of Traffic and Aggregate on the Skidding Resistance of Bituminous Surfacings, *Transport and Road Research Laboratory Report* LR504, TRRL, Crowthorne, 1972.
10. James, J. G.: Calcined Bauxite and Other Artificial, Polish-resistant, Roadstones, *Road Research Laboratory Report* LR84, RRL, Crowthorne, 1968.
11. Hosking, J. R.: Synthetic Aggregates of High Resistance to Polishing, Part 1—Gritty Aggregates, *Road Research Laboratory Report* LR350, RRL, Crowthorne, 1970.
12. Hatherly, L. W., J. H. Mahffy, and A. Tweddle: The Skid-resistance of City Streets and Road Safety, *J. Instn. Highw. Engrs.,* vol. 16, no. 4, pp. 3–12, 1969.
13. Franklin, R. E., and A. J. J. Calder: The Skidding Resistance of Concrete: The Effect of Materials under Site Conditions, *Transport and Road Research Laboratory Report* LR640, TRRL, Crowthorne, 1974.
14. Weller, D. E., and D. P. Maynard: Methods of Texturing New Concrete Road

Surfaces to Provide Adequate Skidding Resistance, *Road Research Laboratory Report* 290, RRL, Crowthorne, 1970.
15. Roe, P. G., L. W. Tubey, and G. West: Surface Texture Depth Measurements on Some British Roads, *Transport and Road Research Laboratory Report* RR143, TRRL, Crowthorne, 1988.
16. Cooper, D. R. C.: Measurement of Road Surface Texture by a Contactless Sensor, *Transport and Road Research Laboratory Report* LR639, TRRL, Crowthorne, 1974.
17. Salt, G. F., and W. S. Szatkowski: A Guide to Levels of Skidding Resistance for Roads, *Transport and Road Research Laboratory Report* LR510, TRRL, Crowthorne, 1973.
18. Weller, D. E., and D. P. Maynard: The Use of an Accelerated Wear Machine to Examine the Skidding Resistance of Concrete Surfaces, *Transport and Road Research Laboratory Report* LR333, TRRL, Crowthorne, 1976.

Chapter

28

Antisplash Surfacings

28.1. On multi-lane dual-carriageway freeways in Britain, heavy trucks are permitted by law to operate at speeds up to 60 mi/h (97 km/h) in the slow traffic lanes. Nominally the differential between the maximum speed for cars and trucks is 10 mi/h (16 km/h), but in fact owing to lack of enforcement trucks are normally driven at speeds around 70 mi/h (113 km/h). Passing two trucks moving side by side is a somewhat difficult operation in a small car because of the "piston effect" tending to draw the car into the center lane. However, it becomes highly dangerous in heavy rain when the spray thrown up by the trucks obscures the visibility of other road users. The use of extended mudflaps and wheel valances does little to improve the situation because a great deal of water is thrown sideways around the contact area between the truck tire and the road.

28.2. When the problem first became apparent some 30 years ago, it was felt that increasing the crossfall of the pavement might offer a solution. A full-scale section of pavement was constructed which could be tilted to have slopes in the range 1:400 to 1:24. Under simulated heavy-rainfall conditions the average depth of water in the "fast lane" remained virtually constant for all slopes within the range referred to above. It was concluded, therefore, that changing the slope within the practical range of 1:50 to 1:35 would not provide a solution.

28.3. The problem of splash on concrete roads with a well-developed transverse texture is less marked than is the case with flexible pavements with asphalt wearing courses. This has led to a series of full-scale experiments using very open-textured bituminous surfacings.

The Use of Open-Textured Wearing-Course Material

28.4. The pavement research carried out in Britain during the past 50 years has led to the conclusion that bituminous surfacings must be both dense and impervious. It was clear, therefore, that roads with antisplash surfacings would need to conform to these basic requirements. It was decided, therefore, to lay experimental open-textured surfacings on top of normal asphalt surfacings. The concept was that the open-textured material would have a large storage capacity for rainwater and allow some sideways drainage toward the low side of the road.

28.5. The first trial section was laid on the M40 freeway in 1967.[1] The grading of the aggregate used is shown in Fig. 28.1. The binder was rubberized bitumen of combined penetration 100 and the thickness was 40 mm laid on a rolled asphalt wearing course (unchipped). The air voids in the open-textured material amounted to about 20 percent. If the voids were interconnected this would permit the storage of about 8 mm of rainwater, neglecting possible drainage.

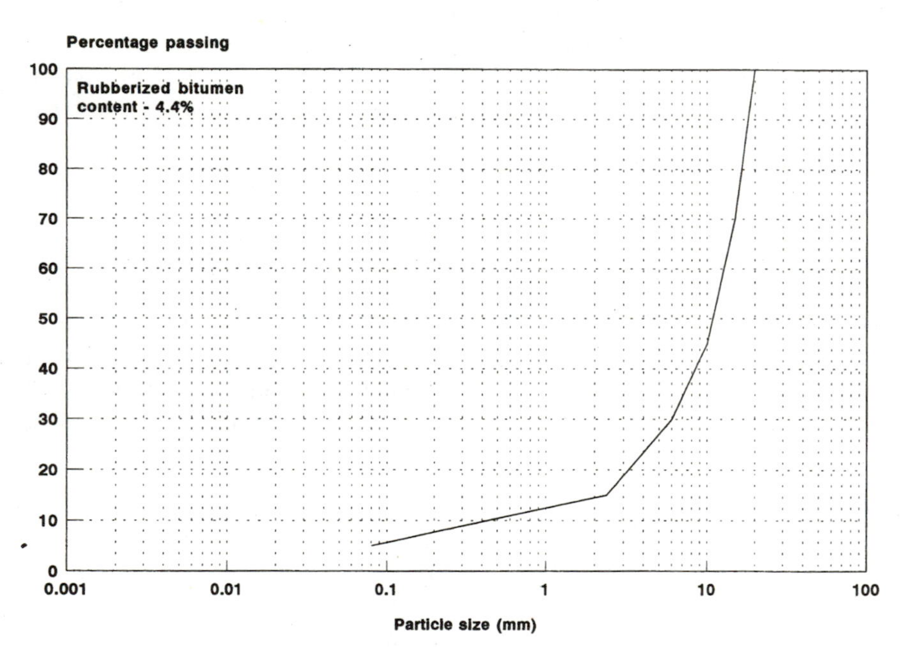

Figure 28.1 Grading and binder content of antisplash surfacing laid on the M40 freeway, 1967.

28.6. The experiment was under observation for several years and showed pronounced antisplash properties, except during abnormal heavy rainstorms.

28.7. A further experiment was constructed in 1970 on the heavily trafficked A45 trunk road in the Midlands of England.[2] The objective was to study the performance of a closely graded aggregate of maximum size 10 mm compared with a similar coarser material as laid at the M40 freeway. The grading of the finer material is shown in Fig. 28.2.

28.8. In addition to the two maximum sizes of aggregate, various binders were used, as indicated in Table 28.1. The thicknesses of material laid were 40 mm for the 19-mm nominal size material and 10 mm for the 10-mm material. Both materials were laid on an existing pavement with a sound impermeable asphalt surfacing. At the low side of the road a shallow drainage channel was left so that sideways drainage of the pervious materials was not impeded.

28.9. Since this experiment was not on a freeway it was possible to gain more frequent access to take samples and carry out tests.

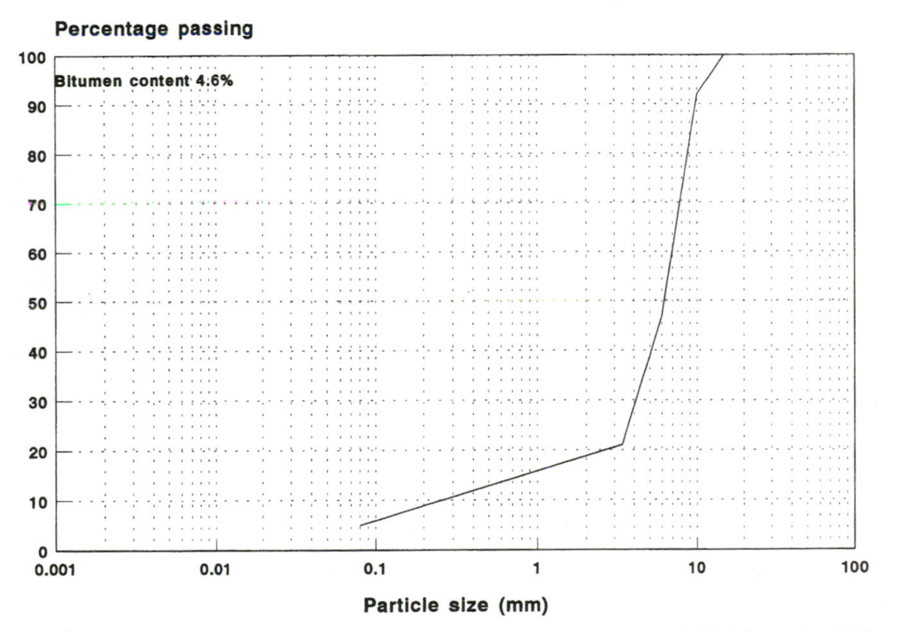

Figure 28.2 Grading and binder content of finer aggregate material laid on the A45 trunk road in 1970.

TABLE 28.1 Binders Used in Open-Textured Antisplash Surfacings

Section No.	Type of binder	Nominal size of gritstone aggregate, mm	Nominal thickness of surfacing, mm
1	100-pen straight-run Venezuelan bitumen	19	40
2	200-pen straight-run Middle East bitumen	19	40
3	100-pen straight-run Middle East bitumen	19	40
4	100-pen rubberized Middle East bitumen	19	40
5	100-pen rubberized Middle East bitumen	10	20
6	100-pen straight-run Middle East bitumen	10	20

Changes in the permeability percentage voids, texture depth, and sideway force coefficient were made, and results are shown in Fig. 28.3.

28.10. The four following paragraphs giving the conclusions from this experiment are reproduced from Reference 2.

28.11. The permeability values generally fell to between about one-half and one-fifth of their initial values after 22 months of trafficking, while the percentage of air voids in general decreased to about two-thirds of their original value. While this means that a 40-mm-thick surfacing with originally 25 percent of voids and capable of accepting 10 mm of rainfall after 22 months accepted only 6 mm of rain because of the reduced void content. This is, however, adequate under most rainfall conditions to prevent spray. The rate at which rain can permeate through the surfacing has been reduced so that the material now takes between two and five times longer to drain. Unless two very heavy showers occur within a few hours of each other or the rainfall is heavy and prolonged the reduced flow rate will not seriously affect the spray reducing properties of the surfacing. Changes of a decade or more in the permeability are required to seriously affect the flow rate since, in the United Kingdom, there is usually sufficient time between heavy showers for most of the water to drain away.

28.12. Because the pervious surfacing depends upon its voids to provide drainage, its performance will be impaired by loss of these voids through compaction under traffic or filling by detritus. In the experimental sections described here an attempt has been made to overcome the first of these by using harder binders than normal and by adding rubber to the binder. But the loss in efficiency due to silting up of the pores and oil droppings from vehicles may prove to be a more serious problem than compaction under traffic.

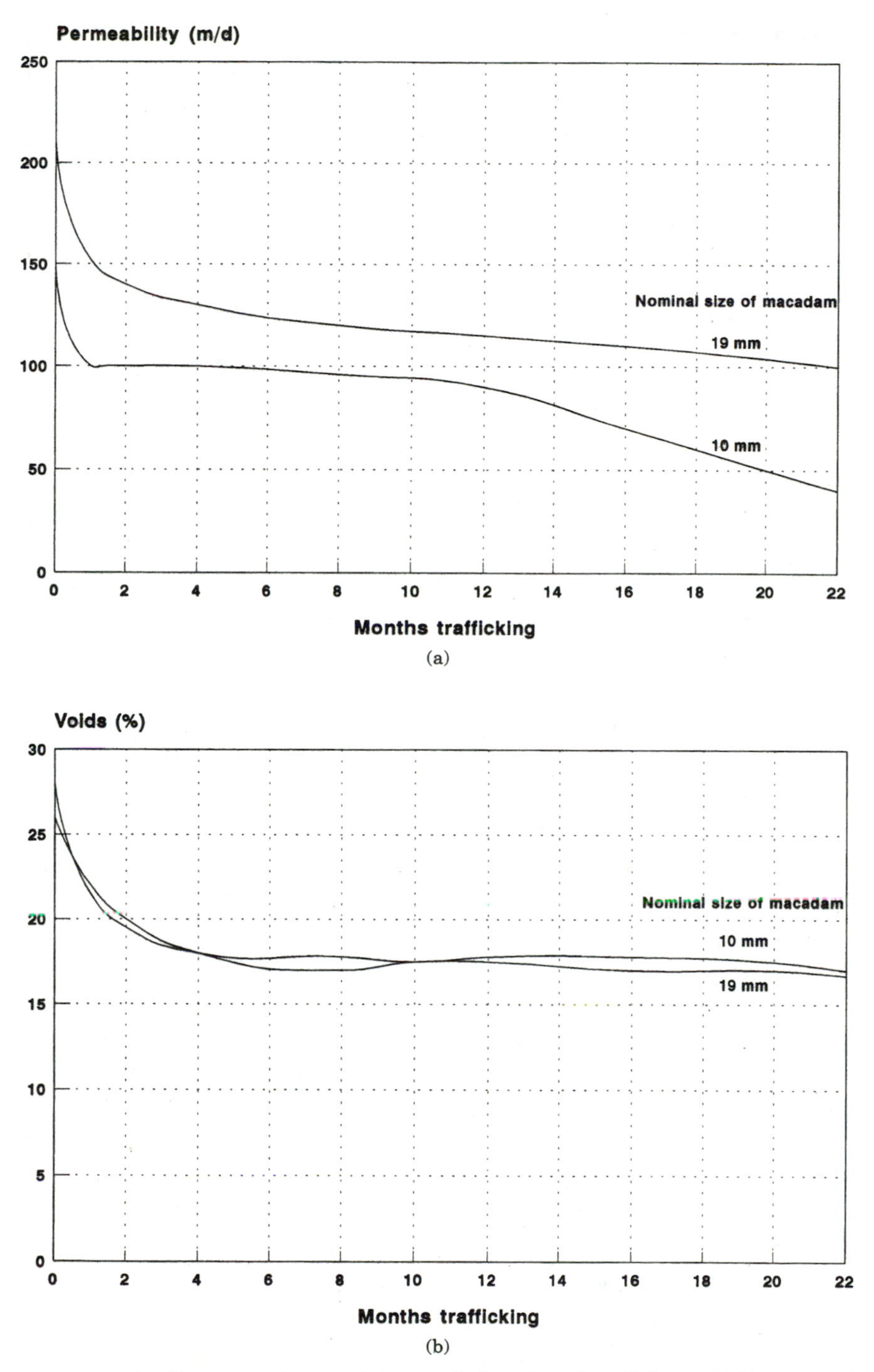

Figure 28.3 (*a*) Variation of permeability with duration of trafficking. (*b*) Variation of percentage voids with duration of trafficking.

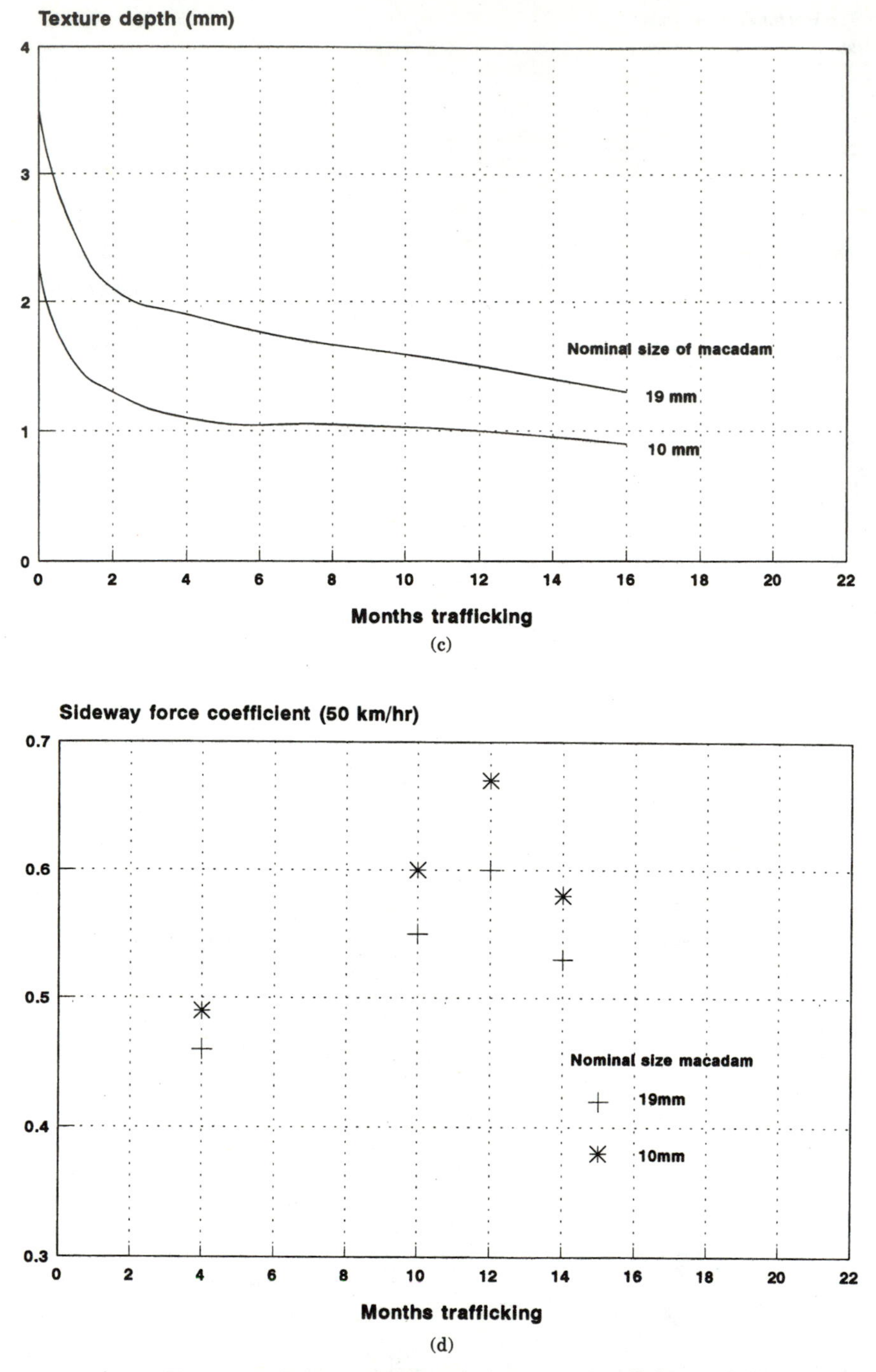

Figure 28.3 (*c*) Variation of texture depth with duration of trafficking. (*d*) Variation of SFC at 50 km/h with duration of trafficking.

28.13. Generally the permeability in the wheel tracks has, as expected, gradually reduced during the 2 years. That reductions in the permeability of a similar order have also occurred between the wheel tracks is somewhat surprising and may well be associated with the total waste lubrication system now becoming common on commercial vehicles.

28.14. The 19-mm nominal-size material manufactured using 100-penetration ME bitumen shows the least reduction in voids and permeability and that containing 100-penetration VEN bitumen the greatest reduction. The addition of rubber to the bitumen has not prevented compaction during the early life of the material.

28.15. Recently two further experiments have been constructed on the M1 freeway and on the A38 trunk road. Figures 28.4 and 28.5 compare spray from a surfacing of impermeable asphalt with that from an adjacent section with surfacing conforming to Table 28.2.

Influence of the Antisplash Layer on Pavement Strength

28.16. The provision of an antisplash surfacing 40 or 60 mm thick is clearly a significant addition to the cost of major roads and freeways. In the case of new flexible roads to current design standards, the

Figure 28.4 Spray from normal impervious asphalt surfacing.

Figure 28.5 Reduced spray from pervious macadam surfacing.

presence of this additional thickness over the required impermeable asphalt surfacing will remove the need for precoated chippings in the asphalt but will provide some increased strength above that required by the traffic type and intensity. It is reasonable therefore to assume that some reduction could be made in the thickness of asphalt or coated macadam incorporated in the design.

28.17. This matter has been investigated by the TRRL using a structural design procedure.[3] Beams 400 mm long, 100 mm wide, and 50 mm deep were sawn from 450-mm-diameter cores taken from an

TABLE 28.2 Tentative Specifications for Open-Textured Antisplash Surfacings

Aggregate grading—percent by weight passing	19-mm nominal-size bitumen macadam, %	10-mm nominal-size bitumen macadam, %
28-mm BS sieve	100	
20-mm BS sieve	90–100	
14-mm BS sieve	50–80	100
10-mm BS sieve		90–100
6.3-mm BS sieve	25–35	40–55
3.35-mm BS sieve	10–20	22–28
75-μm 200 BS sieve	3–6	3–5
Binder content, %	4.0–4.4	4.4–4.8

experimental area of antisplash surfacing laid on a rolled-asphalt surfacing. The grading of the material conformed closely to that shown in Fig. 28.1 for the nominal 19-mm maximum size material. The binder used was 100-penetration bitumen and the binder content 3.8 percent (i.e., a little lower than the values used in the full-scale road experiments).

28.18. The modulus of elasticity of the material was determined at temperatures between 9 and 33°C at loading frequencies of 0.1 to 80 Hz. The results are shown in Fig. 28.6. The modulus curve for the antisplash material is compared with that for a rolled asphalt wearing-course material. Also shown in the same figure are values for rolled asphalt and dense-coated macadam taken from another TRRL publication.[4] The rolled asphalt values are very similar over the range investigated in reference 3 but the modulus of the pervious macadam is significantly lower than that for dense bitumen macadam. This would be expected.

28.19. The result of the structural analysis shows that for a normal design of 80 to 220 mm of asphalt (wearing course and base

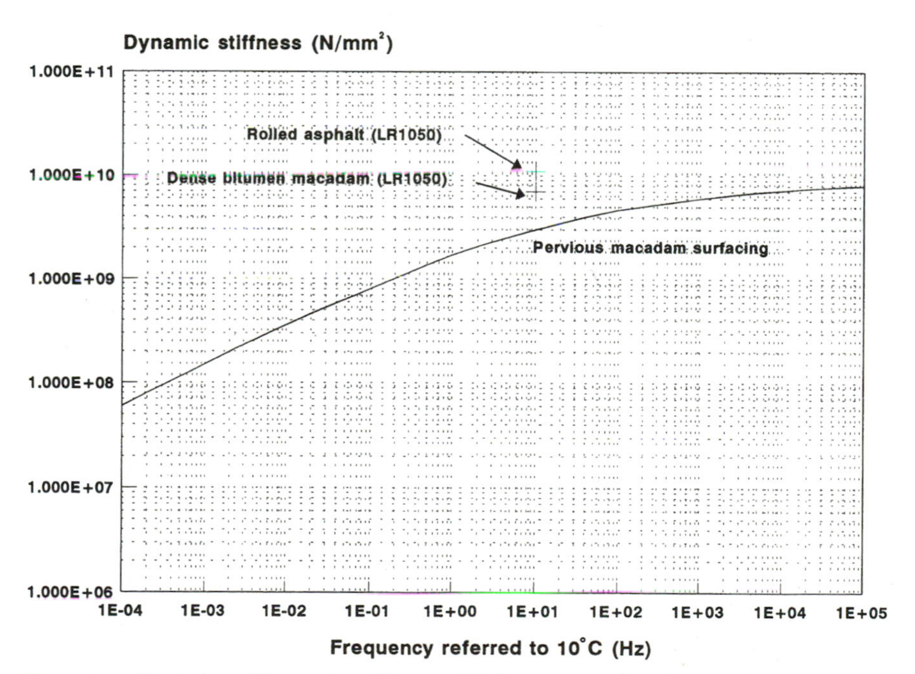

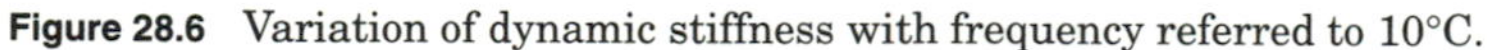

Figure 28.6 Variation of dynamic stiffness with frequency referred to 10°C.

course/base) on a 150-mm granular subbase over a soil of CBR 5.5 percent, the provision of a 40-mm antisplash surfacing on the asphalt wearing course would permit a reduction of 16 to 20 mm in the bituminous base course/base layer. This would probably result in only a small increase of total cost when the antisplash surfacing is used.

Influence of Antisplash and Other Surfacings on Road Noise

28.20. In recent months various claims have been made that antisplash surfacings reduce road noise and as a consequence are more environmentally friendly than other materials. This matter needs further discussion.

28.21. In the 5 years prior to the outbreak of World War II a great deal of research effort at the National Physical Laboratory in the United Kingdom was devoted to noise problems caused by road transport. Legislation had in fact been planned to limit such noise to 90 dB, when measured at a distance of 10 m from the nearside verge of the slow traffic lane. The noise measured included that from the engine and that related to the interaction of the tires and the road surface.

28.22. Since that time, noise-measuring equipment has been much improved, although correlation with subjective observations is often poor. All noisemeters consist essentially of a microphone connected via a multistage amplifier to a meter calibrated in decibels (dB). The noisemeters used in the early work were calibrated against a 1000-Hz frequency for which the sound particle pressure at the threshold of hearing is approximately 0.0002 dyne/cm^2. A sound generating r times this pressure can be converted to the dB scale using the equation

$$\text{dB} = 10 \log r \tag{28.1}$$

Using this equation, a sound involving a pressure of 20,000 dynes/cm^2 would be equivalent to 80 dB.

28.23. Because the human ear is approximately logarithmic in its response to loudness, it can accommodate a very large range of sound pressures without damage. However, ears vary considerably in their response to various frequencies and a noisemeter can at best only represent the average impression of a large number of observers. In particular, human ears are not very sensitive to low and high frequencies. In noisemeters this behavior is simulated by the use of filter circuits incorporated in the various stages of amplification. In particular the use of the dB(A) scale in noise measurements indicates that the low-frequency response has been suppressed. This scale is normally used in vehicle noise measurements.

28.24. Figure 28.7 shows noisemeter measurements made at a distance of 7.5 m from the slow traffic lane of a freeway and a normal urban road.[5] For the freeway, the total noise from a heavy commercial vehicle traveling at 80 km/h was about 86 dB(A). For the same vehicle the noise measured at the same speed but with the engine switched off was 77.5 dB(A). It would be quite incorrect to assume that road noise is the predominant factor. Because of the logarithmic nature of the dB scale, values cannot be added or subtracted arithmetically. It is necessary to convert the total noise and the road noise to pressure ratios using Eq. (28.1). The road noise ratio is then subtracted from the total noise ratio to give the ratio which applies to engine noise only. When this is done, the engine noise becomes 85 dB(A), i.e., only 1 dB(A) less than the total noise. It follows that any change in the surfacing used for a road (including an antisplash surfacing) would have no noticeable effect on noise levels observed close to freeways.

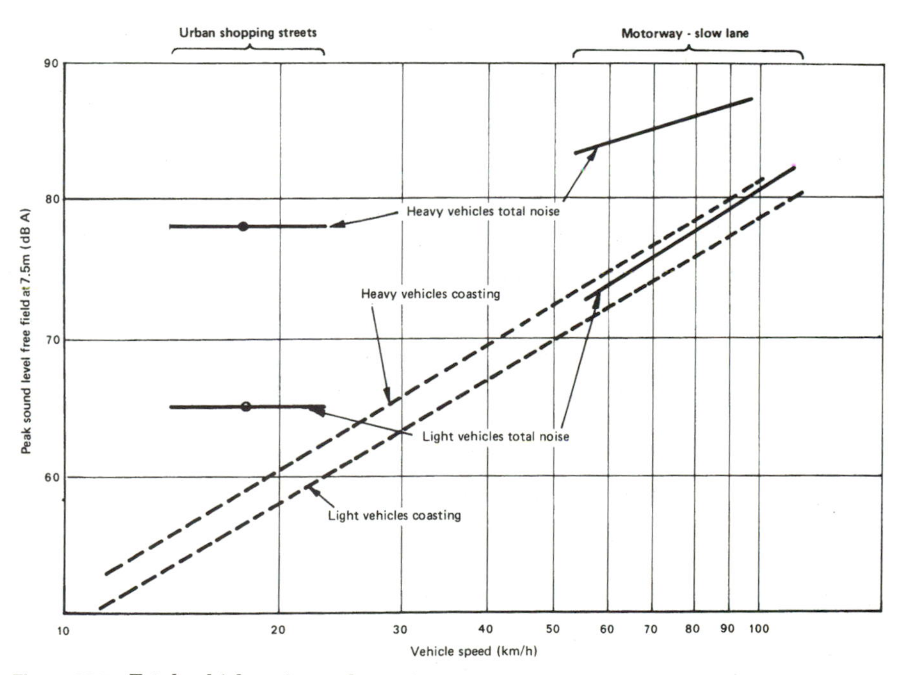

Figure 28.7 Total vehicle noise and coasting noise.

28.25. The situation observed by the driver and passengers in a road vehicle is another matter. Modern cars in which the axles are connected to the body shell to obviate the use of a chassis often magnify road noise. This is particularly the case on concrete roads with deep grooving. There is some evidence that antisplash surfacings produce less in-car noise than is the case with rolled asphalt surfacings with upstanding chippings.

References

1. Please, A. B., J. O'Connell, and B. F. Buglass: A Bituminous Surface-texturing Experiment, High Wycombe By-Pass (M40), *Transport and Road Research Laboratory Report* LR307, TRRL Crowthorne, 1970.
2. Brown, J. R.: Pervious Bitumen Macadam Surfacings Laid to Reduce Splash and Spray at Stonebridge, Warwickshire, *Transport and Road Research Laboratory Report* LR562, TRRL, Crowthorne, 1973.
3. Potter, J. E., and A. R. Halliday: The Contribution of Pervious Macadam Surfacing to the Structural Performance of Roads, *Transport and Road Research Laboratory Report* LR1022, TRRL, Crowthorne, 1981.
4. Goddard, R. T. N.: Fatigue Resistance of a Bituminous Road Pavement Designed for Very Heavy Traffic, *Transport and Road Research Laboratory Report* LR1050, TRRL, Crowthorne, 1982.
5. Harland, D. G.: Rolling Noise and Vehicle Noise, *Transport and Road Research Laboratory Report* LR652, TRRL, Crowthorne, 1974.

Index

Note: Numbers refer to boldface paragraph numbers in text, *not* to page numbers.

ABOUT THE AUTHORS

DAVID CRONEY is Director of Croney Associates, a geotechnical and highway engineering consulting firm based in the United Kingdom. He was previously head of the pavement design division of the Transport and Road Research Laboratory of England.

PAUL CRONEY is also with the engineering consulting firm of Croney Associates, where his primary responsibility is managing overseas operations for the company.